# AUTOMOTIVE FUEL AND EMISSIONS CONTROL SYSTEMS

## FOURTH EDITION

### James D. Halderman

**PEARSON**

Boston   Columbus   Indianapolis   New York   San Francisco   Hoboken
Amsterdam   Cape Town   Dubai   London   Madrid   Milan   Munich   Paris   Montreal
Toronto   Delhi   Mexico City   São Paulo   Sydney   Hong Kong   Seoul   Singapore   Taipei   Tokyo

**Editorial Director:** Andrew Gilfillan
**Editorial Assistant:** Nancy Kesterson
**Director of Marketing:** David Gesell
**Marketing Manager:** Darcy Betts
**Team Lead of Program Management:** Laura Weaver
**Program Manager:** Holly Shufeldt
**Team Lead of Project Management:** JoEllen Gohr
**Project Manager:** Rex Davidson
**Procurement Specialist:** Deidra Skahill
**Senior Art Director:** Diane Ernsberger
**Cover Designer:** Studio Montage
**Digital Studio Project Manager:** Noelle Chun
**Full-Service Project Management and
   Composition:** Integra Software Services, Inc.
**Printer:** Courier Kendallville
**Cover Printer:** Courier Kendallville

**Library of Congress Cataloging-in-Publication Data**

Halderman, James D.
Automotive fuel and emissions control systems / James D. Halderman.
   pages cm
Includes index.
ISBN 978-0-13-379949-1 (alk. paper)—ISBN 0-13-379949-2 (alk. paper)   1. Automobiles—
Fuel systems.   2. Automobiles—Pollution control devices.   I. Title.
TL214.F8H35 2016
629.25'3—dc23

2014043770

10  9  8  7  6  5  4  3  2  1

ISBN 10:      0-13-379949-2
ISBN 13: 978-0-13-379949-1

# PREFACE

**PROFESSIONAL TECHNICIAN SERIES** Part of Pearson Automotive Professional Technician Series, the fourth edition of **Automotive Fuel and Emissions Control Systems** represents the future of automotive textbooks. The series is a full-color, media-integrated solution for today's students and instructors. The series includes textbooks that cover all eight areas of ASE certification, plus additional titles covering common courses. The series is peer reviewed for technical accuracy.

**UPDATES TO THE FOURTH EDITION** The following changes and updates have been made to this new fourth edition based on requests from instructors and readers across North America.

- Updated throughout and correlated to the latest NATEF and ASE tasks.
- Many new full-color line drawings and photos have been added to help bring the subject to life.
- New OSHA hazardous chemical labeling requirements added to Chapter 2.
- Atkinson Cycle engine design and kilowatt engine rating system content added to Chapter 3.
- Expanded diesel engine diagnosis information added to Chapter 4.
- Additional content on top tier gasoline added to Chapter 5.
- Variable and dual intake manifold information added to Chapter 8.
- Permanent (Mode $0A) diagnostic trouble codes added to Chapter 11.
- GM low-speed GMLAN information added to Chapter 12.
- Additional fuel pump diagnosis information added to Chapter 19.
- Additional content on gasoline direct injection (GDI) added to Chapter 21.
- More detailed fuel injection diagnosis information added to Chapter 23.

- Updated and expanded OBD II diagnosis included in Chapter 30.
- New content on electric vehicle (EV) and plug-in hybrid electric vehicle (PHEV) charging information added to Chapter 32.
- New Appendix 1, which has a sample A8 Engine Performance ASE-type certification test with answers.
- New Appendix 2, which is a NATEF correlation chart that shows all MLR, AST, and MAST tasks for engine performance (A8) all in one chart.

**ASE AND NATEF CORRELATED** NATEF certified programs need to demonstrate that they use course material that covers NATEF and ASE tasks. All Professional Technician textbooks have been correlated to the appropriate ASE and NATEF task lists. These correlations can be found in the appendix.

**A COMPLETE INSTRUCTOR AND STUDENT SUPPLEMENTS PACKAGE** All Professional Technician textbooks are accompanied by a full set of instructor and student supplements. Please see page vi for a detailed list of supplements.

**A FOCUS ON DIAGNOSIS AND PROBLEM SOLVING** The Professional Technician Series has been developed to satisfy the need for a greater emphasis on problem diagnosis. Automotive instructors and service managers agree that students and beginning technicians need more training in diagnostic procedures and skill development. To meet this need and demonstrate how real-world problems are solved, "Real World Fix" features are included throughout and highlight how real-life problems are diagnosed and repaired.

The following pages highlight the unique core features that set the Professional Technician Series book apart from other automotive textbooks.

# IN-TEXT FEATURES

## chapter 1 — SERVICE INFORMATION, TOOLS, AND SAFETY

**LEARNING OBJECTIVES:** After studying this chapter, the reader should be able to: • Understand how vehicles are identified and how vehicle service information is retrieved. • Explain how threaded fasteners are sized. • Discuss how to safely use hand tools. • Understand the usage of electrical hand tools, and air and electrically operated tools. • List the personal protective equipment (PPE) that all service technicians should wear and the safety precautions to be taken in a workshop. • Describe the procedure and the equipment to hoist a vehicle safely. • Explain the purpose of using fire extinguishers, fire blankets, and eye wash stations. • List the precautions to be taken when working on hybrid electric vehicles.

**KEY TERMS:** Bench grinder 21 • Bolts 4 • Breaker bar 9 • Bump cap 21 • Calibration codes 2 • Campaign 4 • Casting number 2 • Cheater bar 11 • Chisels 16 • Drive sizes 9 • Extensions 9 • Eye wash station 29 • Files 15 • Fire blanket 28 • Fire extinguisher classes 27 • GAWR 2 • Grade 5 • GVWR 2 • Hacksaws 16 • Hammers 12 • HEV 29 • LED 19 • Metric bolts 4 • Nuts 6 • PPE 21 • Pinch weld seam 24 • Pitch 4 • Pliers 13 • Punches 15 • Ratchet 9 • Recall 4 • Screwdrivers 11 • Snips 15 • Socket 9 • Socket adapter 11 • Spontaneous combustion 23 • SST 19 • Stud 4 • Tensile strength 6 • Trouble light 19 • TSB 3 • UNC 4 • UNF 4 • Universal joint 9 • VECI 2 • VIN 1 • Washers 7 • Wrenches 7

### VEHICLE IDENTIFICATION

**MAKE, MODEL, AND YEAR** All service work requires that the vehicle and its components be properly identified. The most common identification is the make, model, and year of manufacture of the vehicle.

    **Make:** e.g., Chevrolet
    **Model:** e.g., Impala
    **Year:** e.g., 2008

**VEHICLE IDENTIFICATION NUMBER** The year of the vehicle is often difficult to determine exactly. A model may be introduced as the next year's model as soon as January of the previous year. Typically, a new model year starts in September or October of the year prior to the actual new year, but not always. This is why the **vehicle identification number**, usually abbreviated VIN, is so important. ● SEE FIGURE 1–1.

    Since 1981, all vehicle manufacturers have used a VIN that is 17 characters long. Although every vehicle manufacturer assigns various letters or numbers within these 17 characters, there are some constants, including:

- The first number or letter designates the country of origin. ● SEE CHART 1–1.
- The fourth or fifth character is the car line/series.
- The sixth character is the body style.
- The seventh character is the restraint system.
- The eighth character is often the engine code. (Some engines cannot be determined by the VIN number.)
- The tenth character represents the year on all vehicles. ● SEE CHART 1–2.

FIGURE 1–1 Typical vehicle identification number (VIN) as viewed through the windshield.

SERVICE INFORMATION, TOOLS, AND SAFETY   **1**

**OBJECTIVES AND KEY TERMS** appear at the beginning of each chapter to help students and instructors focus on the most important material in each chapter. The chapter objectives are based on specific ASE and NATEF tasks.

---

 **TECH TIP**

### Right to Tighten

It is sometimes confusing which way to rotate a wrench or screwdriver, especially when the head of the fastener is pointing away from you. To help visualize while looking at the fastener, say "righty tighty, lefty loosey."

**TECH TIPS** feature real-world advice and "tricks of the trade" from ASE-certified master technicians.

---

 **SAFETY TIP**

### Shop Cloth Disposal

Always dispose of oily shop cloths in an enclosed container to prevent a fire. ⊠ SEE FIGURE 1–69. Whenever oily cloths are thrown together on the floor or workbench, a chemical reaction can occur, which can ignite the cloth even without an open flame. This process of ignition without an open flame is called **spontaneous combustion.**

**SAFETY TIPS** alert students to possible hazards on the job and how to avoid them.

---

 **REAL WORLD FIX**

### The Rich-Running Toyota

A Toyota failed an enhanced emission test for excessive carbon monoxide, which is caused by a rich (too much fuel) air–fuel ratio problem. After checking all of the basics and not finding any fault in the fuel system, the technician checked the archives of the International Automotive Technicians Network (www.iatn.net) and discovered that a broken spring inside the airflow sensor was a possible cause. The sensor was checked, and a broken vane return spring was discovered. Replacing the airflow sensor restored the engine to proper operating conditions, and it passed the emission test.

**REAL WORLD FIXES** present students with actual automotive scenarios and shows how these common (and sometimes uncommon) problems were diagnosed and repaired.

---

 **FREQUENTLY ASKED QUESTION**

### How Many Types of Screw Heads Are Used in Automotive Applications?

There are many, including Torx, hex (also called Allen), plus many others used in custom vans and motor homes. ● SEE FIGURE 1–9.

**FREQUENTLY ASKED QUESTIONS** are based on the author's own experience and provide answers to many of the most common questions asked by students and beginning service technicians.

**NOTE:** Most of these "locking nuts" are grouped together and are commonly referred to as *prevailing torque nuts.* This means that the nut will hold its tightness or torque and not loosen with movement or vibration.

**NOTES** provide students with additional technical information to give them a greater understanding of a specific task or procedure.

**CAUTION:** *Never* use hardware store (nongraded) bolts, studs, or nuts on any vehicle steering, suspension, or brake component. Always use the exact size and grade of hardware that is specified and used by the vehicle manufacturer.

**CAUTIONS** alert students about potential damage to the vehicle that can occur during a specific task or service procedure.

☠ **WARNING**

Do not use incandescent trouble lights around gasoline or other flammable liquids. The liquids can cause the bulb to break and the hot filament can ignite the flammable liquid which can cause personal injury or even death.

**WARNINGS** alert students to potential dangers to themselves during a specific task or service procedure.

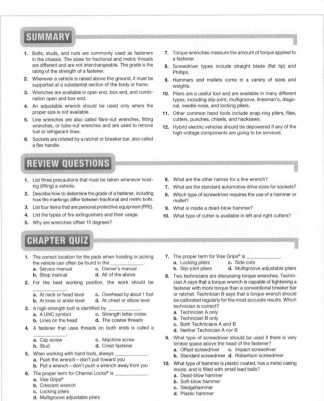

THE SUMMARY, REVIEW QUESTIONS, AND CHAPTER QUIZ at the end of each chapter help students review the material presented in the chapter and test themselves to see how much they've learned.

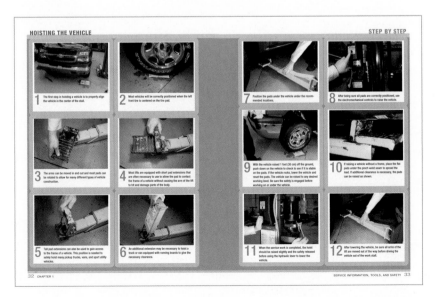

STEP BY STEP photo sequences show in detail the steps involved in performing a specific task or service procedure.

## RESOURCES IN PRINT AND ONLINE
### *Automotive Fuel and Emissions Control Systems*

| NAME OF SUPPLEMENT | PRINT | ONLINE | AUDIENCE | DESCRIPTION |
|---|---|---|---|---|
| **Instructor Resource Manual** 0133799530 | | ✔ | Instructors | **NEW!** The Ultimate teaching aid: Chapter summaries, key terms, chapter learning objectives, and lecture resources. |
| **TestGen** 0133799794 | | ✔ | Instructors | Test generation software and test bank for the text. |
| **PowerPoint Presentation** 0133799859 | | ✔ | Instructors | Slides include a lecture outline of the text to help instructors with in class instruction. |
| **Image Bank** 0133799824 | | ✔ | Instructors | All of the images from the textbook to create customized lecture slides. |
| **NATEF Correlated Task Sheets – For instructors** 0133799808 | | ✔ | Instructors | Downloadable NATEF task sheets for easy customization and development of unique task sheets. |
| **NATEF Correlated Task Sheets – For Students** 0133799816 | ✔ | | Students | Study activity manual that correlates NATEF Automobile Standards to chapters and page numbers in the text. Available to students at a discounted price when packaged with the text. |
| **CourseSmart eText** 0133799506 | | ✔ | Students | An alternative to purchasing the print textbook, students can subscribe to the same content online and save up to 50% off the suggested list price of the print text. Visit **www.coursesmart.com** |

All online resources can be downloaded from the Instructor's Resource Center: **www.pearsonhighered.com/irc**

# ACKNOWLEDGMENTS

A large number of people and organizations have cooperated in providing the reference material and technical information used in this text. The author wishes to express sincere thanks to the following individuals for their special contributions:

Bill Fulton, Ohio Automotive Technology

Doug Garrett, Linder Technical Services

Dan Marinucci, Communique'

Dave Scaler, Mechanic's Education Association

Dr. Norman Nall

Jim Linder- Linder Technical Services

John Thornton, Autotrain

Mark Warren

Randy Dillman

Rick Escalambre, Skyline College

Jim Morton, Automotive Training center (ATC)

Scot Manna

## TECHNICAL AND CONTENT REVIEWERS
The following people reviewed the manuscript before production and checked it for technical accuracy and clarity of presentation. Their suggestions and recommendations were included in the final draft of the manuscript. Their input helped make this textbook clear and technically accurate while maintaining the easy-to-read style that has made other books from the same author so popular.

**Jim Anderson**
Greenville High School

**Victor Bridges**
Umpqua Community College

**Darrell Deeter**
Saddleback College

**Matt Dixon**
Southern Illinois University

**Dr. Roger Donovan**
Illinois Central College

**A. C. Durdin**
Moraine Park Technical College

**Herbert Ellinger**
Western Michigan University

**Al Engledahl**
College of Dupage

**Larry Hagelberger**
Upper Valley Joint Vocational School

**Oldrick Hajzler**
Red River College

**Betsy Hoffman**
Vermont Technical College

**Richard Krieger**
Michigan Institute of Technology

**Steven T. Lee**
Lincoln Technical Institute

**Carlton H. Mabe, Sr.**
Virginia Western Community College

**Roy Marks**
Owens Community College

**Tony Martin**
University of Alaska Southeast

**Kerry Meier**
San Juan College

**Fritz Peacock**
Indiana Vocational Technical College

**Dennis Peter**
NAIT (Canada)

**Kenneth Redick**
Hudson Valley Community College

**Jeff Rehkopf**
Florida State College

**Omar Trinidad**
Southern Illinois University

**Mitchell Walker**
St. Louis Community College at Forest Park

**Jennifer Wise**
Sinclair Community College

## SPECIAL THANKS
The author wishes to thank Mike Garblik, and Chuck Taylor of Sinclair Community College in Dayton, Ohio, and James (Mike) Watson who helped with many of the photos. A special thanks to Dick Krieger for his detailed and thorough review of the manuscript before publication. Most of all, I wish to thank Michelle Halderman for her assistance in all phases of manuscript preparation.

—James D. Halderman

# ABOUT THE AUTHOR

**JIM HALDERMAN** brings a world of experience, knowledge, and talent to his work. His automotive service experience includes working as a flat-rate technician, a business owner, and a professor of automotive technology at a leading U.S. community college for more than 20 years. He has a Bachelor of Science degree from Ohio Northern University and a Master's degree in Education from Miami University in Oxford, Ohio. Jim also holds a U.S. Patent for an electronic transmission control device. He is an ASE-certified Master Automotive Technician, and is also Advanced Engine Performance (L1) ASE certified. Jim is the author of many automotive textbooks, all published by Pearson. He has presented numerous technical seminars to national audiences including the California Automotive Teachers (CAT) and the Illinois College Automotive Instructor Association (ICAIA). He is also a member and presenter at the North American Council of Automotive Teachers (NACAT). Jim was named Regional Teacher of the Year by General Motors Corporation and is an outstanding alumnus of Ohio Northern University. Jim and his wife, Michelle, live in Dayton, Ohio. They have two children. You can reach Jim at

jim@jameshalderman.com

http://jameshalderman.com

# BRIEF CONTENTS

# CONTENTS

# SERVICE INFORMATION, TOOLS, AND SAFETY

**LEARNING OBJECTIVES:** **After studying this chapter, the reader should be able to:** • Understand how vehicles are identified and how vehicle service information is retrieved. • Explain how threaded fasteners are sized. • Discuss how to safely use hand tools. • Understand the usage of electrical hand tools, and air and electrically operated tools. • List the personal protective equipment (PPE) that all service technicians should wear and the safety precautions to be taken in a workshop. • Describe the procedure and the equipment to hoist a vehicle safely. • Explain the purpose of using fire extinguishers, fire blankets, and eye wash stations. • List the precautions to be taken when working on hybrid electric vehicles.

**KEY TERMS:** Bench grinder 21 • Bolts 4 • Breaker bar 9 • Bump cap 21 • Calibration codes 2 • Campaign 4 • Casting number 2 • Cheater bar 11 • Chisels 16 • Drive sizes 9 • Extensions 9 • Eye wash station 29 • Files 15 • Fire blanket 28 • Fire extinguisher classes 27 • GAWR 2 • Grade 5 • GVWR 2 • Hacksaws 16 • Hammers 12 • HEV 29 • LED 19 • Metric bolts 4 • Nuts 6 • PPE 21 • Pinch weld seam 24 • Pitch 4 • Pliers 13 • Punches 15 • Ratchet 9 • Recall 4 • Screwdrivers 11 • Snips 15 • Socket 9 • Socket adapter 11 • Spontaneous combustion 23 • SST 19 • Stud 4 • Tensile strength 6 • Trouble light 19 • TSB 3 • UNC 4 • UNF 4 • Universal joint 9 • VECI 2 • VIN 1 • Washers 7 • Wrenches 7

## VEHICLE IDENTIFICATION

**MAKE, MODEL, AND YEAR** All service work requires that the vehicle and its components be properly identified. The most common identification is the make, model, and year of manufacture of the vehicle.

**Make:** e.g., Chevrolet

**Model:** e.g., Impala

**Year:** e.g., 2008

**VEHICLE IDENTIFICATION NUMBER** The year of the vehicle is often difficult to determine exactly. A model may be introduced as the next year's model as soon as January of the previous year. Typically, a new model year starts in September or October of the year prior to the actual new year, but not always. This is why the **vehicle identification number,** usually abbreviated VIN, is so important. ● **SEE FIGURE 1–1.**

Since 1981, all vehicle manufacturers have used a VIN that is 17 characters long. Although every vehicle manufacturer assigns various letters or numbers within these 17 characters, there are some constants, including:

■ The first number or letter designates the country of origin. ● **SEE CHART 1–1.**

■ The fourth or fifth character is the car line/series.

■ The sixth character is the body style.

■ The seventh character is the restraint system.

■ The eighth character is often the engine code. (Some engines cannot be determined by the VIN number.)

■ The tenth character represents the year on all vehicles. ● **SEE CHART 1–2.**

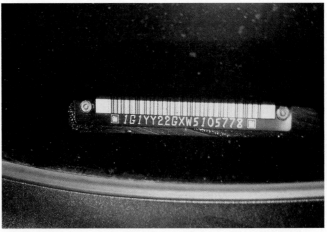

**FIGURE 1–1** Typical vehicle identification number (VIN) as viewed through the windshield.

| | | |
|---|---|---|
| 1 = United States | J = Japan | U = Romania |
| 2 = Canada | K = Korea | V = France |
| 3 = Mexico | L = China | W = Germany |
| 4 = United States | M = India | X = Russia |
| 5 = United States | P = Philippines | Y = Sweden |
| 6 = Australia | S = England | Z = Italy |
| 8 = Argentina | R = Taiwan | |
| 9 = Brazil | T = Czechoslovakia | |

**CHART 1–1**

The first number or letter in the VIN identifies the country where the vehicle was made.

| | | |
|---|---|---|
| A = 1980/2010 | L = 1990/2020 | Y = 2000/2030 |
| B = 1981/2011 | M = 1991/2021 | 1 = 2001/2031 |
| C = 1982/2012 | N = 1992/2022 | 2 = 2002/2032 |
| D = 1983/2013 | P = 1993/2023 | 3 = 2003/2033 |
| E = 1984/2014 | R = 1994/2024 | 4 = 2004/2034 |
| F = 1985/2015 | S = 1995/2025 | 5 = 2005/2035 |
| G = 1986/2016 | T = 1996/2026 | 6 = 2006/2036 |
| H = 1987/2017 | V = 1997/2027 | 7 = 2007/2037 |
| J = 1988/2018 | W = 1998/2028 | 8 = 2008/2038 |
| K = 1989/2019 | X = 1999/2029 | 9 = 2009/2039 |

**CHART 1–2**

The pattern repeats every 30 years for the year of manufacture.

**VEHICLE SAFETY CERTIFICATION LABEL**    A vehicle safety certification label is attached to the left side pillar post on the rearward-facing section of the left front door. This label indicates the month and year of manufacture as well as the **gross vehicle weight rating (GVWR),** the **gross axle weight rating (GAWR),** and the vehicle identification number (VIN).

**VECI LABEL**    The **vehicle emissions control information (VECI)** label under the hood of the vehicle shows informative settings and emission hose routing information. ● **SEE FIGURE 1–2.**

The VECI label (sticker) can be located on the bottom side of the hood, the radiator fan shroud, the radiator core support, or on the strut towers. The VECI label usually includes the following information:

- Engine identification
- Emissions standard that the vehicle meets
- Vacuum hose routing diagram
- Base ignition timing (if adjustable)
- Spark plug type and gap

**FIGURE 1–2** The vehicle emissions control information (VECI) sticker is placed under the hood.

**FIGURE 1–3** A typical calibration code sticker on the case of a controller. The information on the sticker is often needed when ordering parts or a replacement controller.

- Valve lash
- Emission calibration code

**CALIBRATION CODES**    **Calibration codes** are usually located on powertrain control modules (PCMs) or other controllers. Whenever diagnosing an engine operating fault, it is often necessary to use the calibration code to be sure that the vehicle is the subject of a technical service bulletin or other service procedure. ● **SEE FIGURE 1–3.**

**CASTING NUMBERS**    When an engine part such as a block is cast, a number is put into the mold to identify the casting. ● **SEE FIGURE 1–4.** These **casting numbers** can be used to identify the part and check dimensions such as the cubic inch displacement and other information, such as the year of manufacture. Sometimes changes are made to the mold, yet

**FIGURE 1-4** Casting numbers on major components can be either cast or stamped.

the casting number is not changed. Most often the casting number is the best piece of identifying information that the service technician can use for identifying an engine.

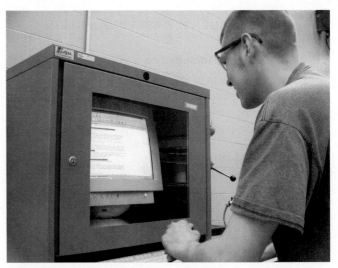
**FIGURE 1-5** Electronic service information is available from aftermarket sources such as All-Data and Mitchell-on-Demand, as well as on websites hosted by vehicle manufacturers.

## SERVICE INFORMATION

### SERVICE MANUALS
Service information is used by the service technician to determine specifications and service procedures and any needed special tools.

Factory and aftermarket service manuals contain specifications and service procedures. While factory service manuals cover just one year and one or more models of the same vehicle, most aftermarket service manufacturers cover multiple years and/or models in one manual.

Included in most service manuals are the following:

- Capacities and recommended specifications for all fluids
- Specifications including engine and routine maintenance items
- Testing procedures
- Service procedures including the use of special tools when needed

### ELECTRONIC SERVICE INFORMATION
Electronic service information is available mostly by subscription and provides access to an Internet site where service manual–type information is available. ● **SEE FIGURE 1–5.** Most vehicle manufacturers also offer electronic service information to their dealers and to most schools and colleges that offer corporate training programs.

### TECHNICAL SERVICE BULLETINS
**Technical service bulletins,** often abbreviated **TSB,** sometimes called *technical service information bulletins (TSIB),* are issued by the vehicle manufacturer to notify service technicians of a problem and include the necessary corrective action. Technical service

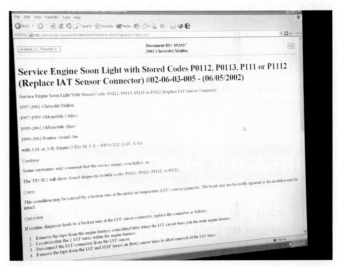
**FIGURE 1-6** Technical service bulletins (TSB) are issued by vehicle manufacturers when a fault occurs that affects many vehicles with the same problem. The TSB then provides the fix for the problem including any parts needed and detailed instructions.

bulletins are designed for dealership technicians but are republished by aftermarket companies and made available along with other service information to shops and vehicle repair facilities. ● **SEE FIGURE 1–6.**

### INTERNET
The Internet has opened the field for information exchange and access to technical advice. One of the most useful websites is the International Automotive Technician's Network at **www.iatn.net**. This is a free site, but service technicians must register to join. If a small monthly sponsor fee is paid, the shop or service technician can gain access to the archives, which include thousands of successful repairs in the searchable database.

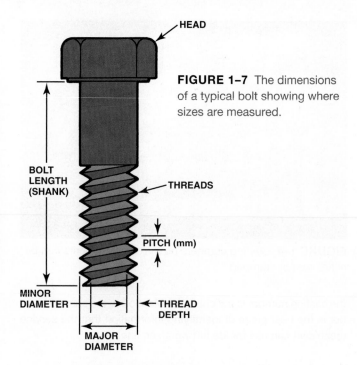

**What Should Be Included on a Work Order?**

A work order is a legal document that should include the following information:

1. Customer information
2. Identification of the vehicle including the VIN
3. Related service history information
4. The "three Cs":
   - Customer concern (complaint)
   - Cause of the concern
   - Correction or repairs that were required to return the vehicle to proper operation.

**FIGURE 1–7** The dimensions of a typical bolt showing where sizes are measured.

**RECALLS AND CAMPAIGNS** A **recall** or **campaign** is issued by a vehicle manufacturer and a notice is sent to all owners in the event of a safety-related fault or concern. While these faults may be repaired by shops, it is generally handled by a local dealer. Items that have created recalls in the past have included potential fuel system leakage problems, exhaust leakage, or electrical malfunctions that could cause a possible fire or the engine to stall. Unlike technical service bulletins whose cost is covered only when the vehicle is within the warranty period, a recall or campaign is always done at no cost to the vehicle owner.

# THREADED FASTENERS

**BOLTS AND THREADS** Most of the threaded fasteners used on vehicles are **bolts.** Bolts are called *cap* screws when they are threaded into a casting. Automotive service technicians usually refer to these fasteners as *bolts,* regardless of how they are used. In this chapter, they are called bolts. Sometimes, studs are used for threaded fasteners. A **stud** is a short rod with threads on both ends. Often, a stud will have coarse threads on one end and fine threads on the other end. The end of the stud with coarse threads is screwed into the casting. A nut is used on the opposite end to hold the parts together.

The fastener threads *must* match the threads in the casting or nut. The threads may be measured either in fractions of an inch (called fractional) or in metric units. The size is measured across the outside of the threads, called the *crest* of the thread. ●**SEE FIGURE 1–7.**

**FRACTIONAL BOLTS** Fractional threads are either coarse or fine. The coarse threads are called **unified national coarse (UNC),** and the fine threads are called **unified national fine (UNF).** Standard combinations of sizes and number of threads per inch (called **pitch**) are used. Pitch can be measured with a thread pitch gauge as shown in ●**FIGURE 1–8.** Bolts are identified by their diameter and length as measured from below the

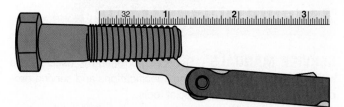

**FIGURE 1–8** Thread pitch gauge used to measure the pitch of the thread. This bolt has 13 threads to the inch.

**How Many Types of Screw Heads Are Used in Automotive Applications?**

There are many, including Torx, hex (also called Allen), plus many others used in custom vans and motor homes. ●**SEE FIGURE 1–9.**

head and not by the size of the head or the size of the wrench used to remove or install the bolt.

Fractional thread sizes are specified by the diameter in fractions of an inch and the number of threads per inch. Typical UNC thread sizes would be 5/16-18 and 1/2-13. Similar UNF thread sizes would be 5/16-24 and 1/2-20. ●**SEE CHART 1–3.**

**METRIC BOLTS** The size of a **metric bolt** is specified by the letter *M* followed by the diameter in millimeters (mm) across the outside (crest) of the threads. Typical metric sizes would be M8 and M12. Fine metric threads are specified by the thread

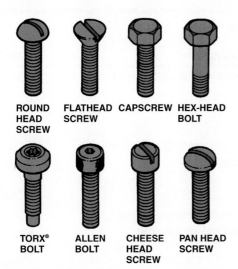

**ROUND HEAD SCREW** | **FLATHEAD SCREW** | **CAPSCREW** | **HEX-HEAD BOLT**

**TORX® BOLT** | **ALLEN BOLT** | **CHEESE HEAD SCREW** | **PAN HEAD SCREW**

**FIGURE 1–9** Bolts and screws have many different heads which determine what tool is needed.

## METRIC HEXAGON HEAD CAP SCREWS
### ALL MEASUREMENTS IN MILLIMETERS

M = NOMINAL THREAD DIAMETER
P = PITCH
D = HEAD SIZE ACROSS FLATS

| M | P | D | M | P | D | M | P | D |
|---|---|---|---|---|---|---|---|---|
| 1.6 | 0.35 | 3.2 | 10 | 1.00 | 17 | 20 | 1.50 | 30 |
| 1.7 | 0.35 | 3.5 | 10 | 1.25 | 17 | 20 | 2.50 | 30 |
| 2 | 0.40 | 4 | 10 | 1.50 | 17 | 22 | 1.50 | 32 |
| 2.3 | 0.40 | 4.5 | 12 | 1.25 | 19 | 22 | 2.50 | 32 |
| 2.5 | 0.45 | 5 | 12 | 1.50 | 19 | 24 | 2.00 | 36 |
| 3 | 0.50 | 5.5 | 12 | 1.75 | 19 | 24 | 3.00 | 36 |
| 3.5 | 0.60 | 6 | 14 | 1.50 | 22 | 27 | 3.00 | 41 |
| 4 | 0.70 | 7 | 14 | 2.00 | 22 | 30 | 3.50 | 46 |
| 5 | 0.80 | 8 | 16 | 1.50 | 24 | 33 | 3.50 | 50 |
| 6 | 1.00 | 10 | 16 | 2.00 | 24 | 36 | 4.00 | 55 |
| 7 | 1.00 | 11 | 18 | 1.50 | 27 | 39 | 4.00 | 60 |
| 8 | 1.00 | 13 | 18 | 2.50 | 27 | 42 | 4.50 | 65 |
| 8 | 1.25 | 13 | | | | 45 | 4.50 | 70 |

**FIGURE 1–10** The metric system specifies fasteners by diameter, length, and pitch.

diameter followed by X and the distance between the threads measured in millimeters (M8 X 1.5). ● **SEE FIGURE 1–10.**

## GRADES OF BOLTS
Bolts are made from many different types of steel, and for this reason some are stronger than others. The strength or classification of a bolt is called the **grade.** The bolt heads are marked to indicate their grade strength.

The actual grade of bolts is two more than the number of lines on the bolt head. Metric bolts have a decimal number to indicate the grade. More lines or a higher grade number indicate a stronger bolt. In some cases, nuts and machine screws have similar grade markings. Higher grade bolts usually have threads that are rolled rather than cut, which also makes them stronger. ● **SEE FIGURE 1–11.**

| SIZE | THREADS PER INCH NC UNC | THREADS PER INCH NF UNF | OUTSIDE DIAMETER INCHES |
|---|---|---|---|
| 0 | .. | 80 | 0.0600 |
| 1 | 64 | .. | 0.0730 |
| 1 | .. | 72 | 0.0730 |
| 2 | 56 | .. | 0.0860 |
| 2 | .. | 64 | 0.0860 |
| 3 | 48 | .. | 0.0990 |
| 3 | .. | 56 | 0.0990 |
| 4 | 40 | .. | 0.1120 |
| 4 | .. | 48 | 0.1120 |
| 5 | 40 | .. | 0.1250 |
| 5 | .. | 44 | 0.1250 |
| 6 | 32 | .. | 0.1380 |
| 6 | .. | 40 | 0.1380 |
| 8 | 32 | .. | 0.1640 |
| 8 | .. | 36 | 0.1640 |
| 10 | 24 | .. | 0.1900 |
| 10 | .. | 32 | 0.1900 |
| 12 | 24 | .. | 0.2160 |
| 12 | .. | 28 | 0.2160 |
| 1/4 | 20 | .. | 0.2500 |
| 1/4 | .. | 28 | 0.2500 |
| 5/16 | 18 | .. | 0.3125 |
| 5/16 | .. | 24 | 0.3125 |
| 3/8 | 16 | .. | 0.3750 |
| 3/8 | .. | 24 | 0.3750 |
| 7/16 | 14 | .. | 0.4375 |
| 7/16 | .. | 20 | 0.4375 |
| 1/2 | 13 | .. | 0.5000 |
| 1/2 | .. | 20 | 0.5000 |
| 9/16 | 12 | .. | 0.5625 |
| 9/16 | .. | 18 | 0.5625 |
| 5/8 | 11 | .. | 0.6250 |
| 5/8 | .. | 18 | 0.6250 |
| 3/4 | 10 | .. | 0.7500 |
| 3/4 | .. | 16 | 0.7500 |
| 7/8 | 9 | .. | 0.8750 |
| 7/8 | .. | 14 | 0.8750 |
| 1 | 8 | .. | 1.0000 |
| 1 | .. | 12 | 1.0000 |
| 1 1/8 | 7 | .. | 1.1250 |
| 1 1/8 | .. | 12 | 1.1250 |
| 1 1/4 | 7 | .. | 1.2500 |
| 1 1/4 | .. | 12 | 1.2500 |
| 1 3/8 | 6 | .. | 1.3750 |
| 1 3/8 | .. | 12 | 1.3750 |
| 1 1/2 | 6 | .. | 1.5000 |
| 1 1/2 | .. | 12 | 1.5000 |
| 1 3/4 | 5 | .. | 1.7500 |
| 2 | 4 1/2 | .. | 2.0000 |
| 2 1/4 | 4 1/2 | .. | 2.2500 |
| 2 1/2 | 4 | .. | 2.5000 |
| 2 3/4 | 4 | .. | 2.7500 |
| 3 | 4 | .. | 3.0000 |
| 3 1/4 | 4 | .. | 3.2500 |
| 3 1/2 | 4 | .. | 3.5000 |
| 3 3/4 | 4 | .. | 3.7500 |
| 4 | 4 | .. | 4.0000 |

**CHART 1–3**

American standard is one method of sizing fasteners.

| SAE BOLT DESIGNATIONS | | | | |
|---|---|---|---|---|
| SAE GRADE NO. | SIZE RANGE | TENSILE STRENGTH, PSI | MATERIAL | HEAD MARKING |
| 1 | 1/4 through 1 1/2 | 60,000 | Low or medium carbon steel | 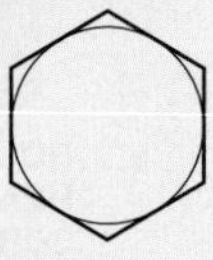 |
| 2 | 1/4 through 3/4 | 74,000 | | |
| | 7/8 through 1 1/2 | 60,000 | | |
| 5 | 1/4 through 1 | 120,000 | Medium carbon steel, quenched and tempered | 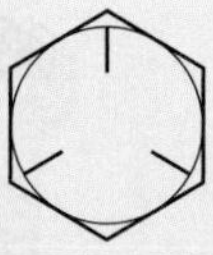 |
| | 1-1/8 through 1 1/2 | 105,000 | | |
| 5.2 | 1/4 through 1 | 120,000 | Low carbon martensite steel,* quenched and tempered | 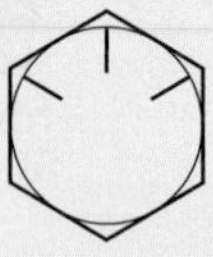 |
| 7 | 1/4 through 1 1/2 | 133,000 | Medium carbon alloy steel, quenched and tempered | 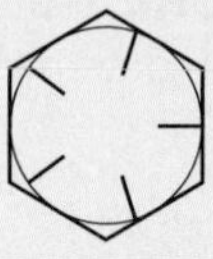 |
| 8 | 1/4 through 1 1/2 | 150,000 | Medium carbon alloy steel, quenched and tempered | 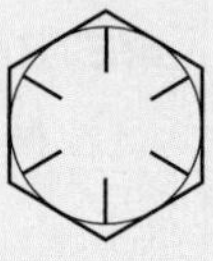 |
| 8.2 | 1/4 through 1 | 150,000 | Low carbon martensite steel,* quenched and tempered | 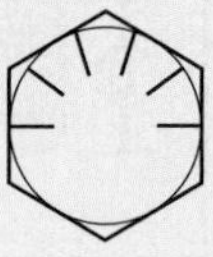 |

**CHART 1–4**

The tensile strength rating system as specified by the Society of Automotive Engineers (SAE).
*Martensite steel is steel that has been cooled rapidly, thereby increasing its hardness. It is named after a German metallurgist, Adolf Martens.

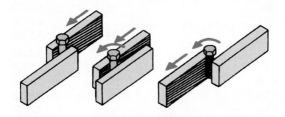

**ROLLING THREADS**

**FIGURE 1–11** Stronger threads are created by cold-rolling a heat-treated bolt blank instead of cutting the threads, using a die.

**CAUTION:** *Never* **use hardware store (nongraded) bolts, studs, or nuts on any vehicle steering, suspension, or brake component. Always use the exact size and grade of hardware that is specified and used by the vehicle manufacturer.**

**TENSILE STRENGTH OF FASTENERS** Graded fasteners have a higher tensile strength than nongraded fasteners. **Tensile strength** is the maximum stress used under tension (lengthwise force) without causing failure of the fastener. Tensile strength is specified in pounds per square inch (psi).

The strength and type of steel used in a bolt is supposed to be indicated by a raised mark on the head of the bolt. The type of mark depends on the standard to which the bolt was manufactured. Most often, bolts used in machinery are made to SAE Standard J429. ● **SEE CHART 1–4,** which shows the grade and specified tensile strength.

Metric bolt tensile strength property class is shown on the head of the bolt as a number, such as 4.6, 8.8, 9.8, and 10.9; the higher the number, the stronger the bolt. ● **SEE FIGURE 1–12.**

**NUTS** **Nuts** are the female part of a threaded fastener. Most nuts used on cap screws have the same hex size as the cap screw head. Some inexpensive nuts use a hex size larger than the cap screw head. Metric nuts are often marked with dimples to show their strength. More dimples indicate stronger nuts. Some nuts and cap screws use interference fit threads to keep them from accidentally loosening. This means that the shape of the nut is slightly distorted or that a section of the threads is deformed. Nuts can also be kept from loosening with a nylon washer fastened in the nut or with a nylon patch or strip on the threads. ● **SEE FIGURE 1–13.**

| METRIC CLASS | 4.6 | 8.8 | 9.8 | 10.9 |
|---|---|---|---|---|
| APPROXIMATE MAXIMUM POUND FORCE PER SQUARE INCH | 60,000 | 120,000 | 130,000 | 150,000 |

**FIGURE 1–12** Metric bolt (cap screw) grade markings and approximate tensile strength.

| HEX NUT | JAM NUT | NYLON LOCK NUT | CASTLE NUT | ACORN NUT |

**FIGURE 1–13** Nuts come in a variety of styles, including locking (prevailing torque) types, such as the distorted thread and nylon insert type.

| FLAT WASHER | LOCK WASHER | STAR WASHER | STAR WASHER |

**FIGURE 1–14** Washers come in a variety of styles, including flat and serrated used to help prevent a fastener from loosening.

**FIGURE 1–15** A forged wrench after it has been forged but before the flashing, which is the extra material around the wrench, has been removed.

**NOTE: Most of these "locking nuts" are grouped together and are commonly referred to as *prevailing torque nuts*. This means that the nut will hold its tightness or torque and not loosen with movement or vibration. Most prevailing torque nuts should be replaced whenever removed to ensure that the nut will not loosen during service. Always follow the manufacturer's recommendations. Anaerobic sealers, such as Loctite, are used on the threads where the nut or cap screw must be both locked and sealed.**

**WASHERS**  **Washers** are often used under cap screw heads and under nuts. ● **SEE FIGURE 1–14.** Plain flat washers are used to provide an even clamping load around the fastener. Lock washers are added to prevent accidental loosening. In some accessories, the washers are locked onto the nut to provide easy assembly.

## HAND TOOLS

**WRENCHES**  Wrenches are the most used hand tool by service technicians. **Wrenches** are used to grasp and rotate threaded fasteners. Most wrenches are constructed of forged alloy steel, usually chrome-vanadium steel. ● **SEE FIGURE 1–15.**

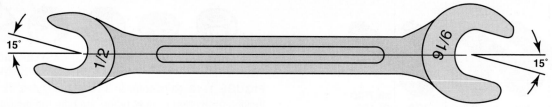

**FIGURE 1–16** A typical open-end wrench. The size is different on each end, and notice that the head is angled 15 degrees at the end.

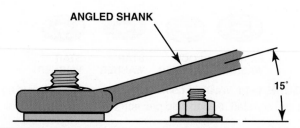

**FIGURE 1–17** The end of a box-end wrench is angled 15 degrees to allow clearance for nearby objects or other fasteners.

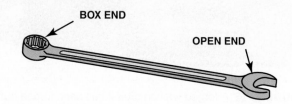

**FIGURE 1–18** A combination wrench has an open end at one end and a box end at the other end.

After the wrench is formed, the wrench is hardened, then tempered to reduce brittleness, and then chrome plated. There are several types of wrenches.

**OPEN-END WRENCH.** An *open-end wrench* is usually used to loosen or tighten bolts or nuts that do not require a lot of torque. Because of the open end, this type of wrench can be easily placed on a bolt or nut with an angle of 15 degrees, which allows the wrench to be flipped over and used again to continue to rotate the fastener. The major disadvantage of an open-end wrench is the lack of torque that can be applied due to the fact that the open jaws of the wrench contact only two flat surfaces of the fastener. An open-end wrench has two different sizes; one at each end. ● **SEE FIGURE 1–16.**

**BOX-END WRENCH.** A *box-end wrench,* also called a *closed-end wrench,* is placed over the top of the fastener and grips the points of the fastener. A box-end wrench is angled 15 degrees to allow it to clear nearby objects.

Therefore, a box-end wrench should be used to loosen or to tighten fasteners because it grasps around the entire head of the fastener. A box-end wrench has two different sizes; one at each end. ● **SEE FIGURE 1–17.**

Most service technicians purchase *combination wrenches,* which have the open end at one end and the same size box end on the other end. ● **SEE FIGURE 1–18.**

A combination wrench allows the technician to loosen or tighten a fastener using the box end of the wrench, turn it

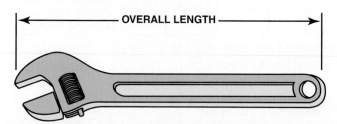

**FIGURE 1–19** An adjustable wrench. Adjustable wrenches are sized by the overall length of the wrench and not by how far the jaws open. Common sizes of adjustable wrenches include 8, 10, and 12 inch.

**FIGURE 1–20** The end of a typical line wrench, which shows that it is capable of grasping most of the head of the fitting.

around, and use the open end to increase the speed of rotating the fastener.

**ADJUSTABLE WRENCH.** An *adjustable wrench* is often used where the exact size wrench is not available or when a large nut, such as a wheel spindle nut, needs to be rotated but not tightened. An adjustable wrench should not be used to loosen or tighten fasteners because the torque applied to the wrench can cause the movable jaws to loosen their grip on the fastener, causing it to become rounded. ● **SEE FIGURE 1–19.**

**LINE WRENCHES.** *Line wrenches* are also called *flare-nut wrenches, fitting wrenches,* or *tube-nut wrenches* and are designed to grip almost all the way around a nut used to retain a fuel or refrigerant line and yet be able to be installed over the line. ● **SEE FIGURE 1–20.**

**SAFE USE OF WRENCHES** Wrenches should be inspected before use to be sure they are not cracked, bent, or damaged. All wrenches should be cleaned after use before being returned to the tool box. Always use the correct size of wrench for the fastener being loosened or tightened to help prevent the rounding of the flats of the fastener. When attempting to loosen a fastener, pull a wrench—do not push a wrench. If a wrench is pushed, your knuckles can be hurt when forced into another

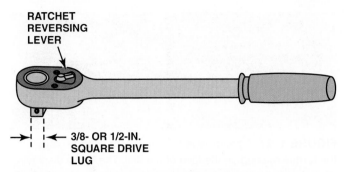

FIGURE 1–21 A typical ratchet used to rotate a socket. A ratchet makes a ratcheting noise when it is being rotated in the opposite direction from loosening or tightening. A knob or lever on the ratchet allows the user to switch directions.

FIGURE 1–22 A typical flex handle used to rotate a socket, also called a breaker bar because it usually has a longer handle than a ratchet and therefore can be used to apply more torque to a fastener than a ratchet.

 TECH TIP

### Hide Those from the Boss

An apprentice technician started working for a shop and put his top tool box on a workbench. Another technician observed that, along with a complete set of good-quality tools, the box contained several adjustable wrenches. The more experienced technician said, "Hide those from the boss." The boss does not want any service technician to use adjustable wrenches. If any adjustable wrench is used on a bolt or nut, the movable jaw often moves or loosens and starts to round the head of the fastener. If the head of the bolt or nut becomes rounded, it becomes that much more difficult to remove.

object if the fastener breaks loose or if the wrench slips. Always keep wrenches and all hand tools clean to help prevent rust and to allow for a better, firmer grip. Never expose any tool to excessive heat. High temperatures can reduce the strength ("draw the temper") of metal tools.

Never use a hammer on any wrench unless you are using a special "staking face" wrench designed to be used with a hammer. Replace any tools that are damaged or worn.

## RATCHETS, SOCKETS, AND EXTENSIONS A **socket** fits over the fastener and grips the points and/or flats of the bolt or nut. The socket is rotated (driven) using either a long bar called a **breaker bar** (flex handle) or a ratchet. ● **SEE FIGURES 1–21 AND 1–22.**

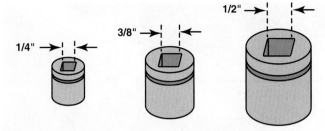

FIGURE 1–23 The most commonly used socket drive sizes include 1/4-, 3/8-, and 1/2-inch drive.

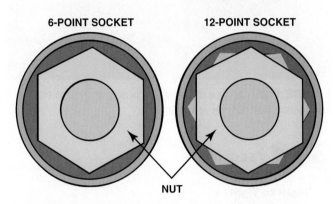

FIGURE 1–24 A 6-point socket fits the head of a bolt or nut on all sides. A 12-point socket can round off the head of a bolt or nut if a lot of force is applied.

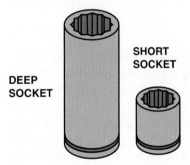

FIGURE 1–25 A socket allows access to the nut that has a stud. Sockets are also used in other locations needing great depth, such as spark plugs.

A **ratchet** is a tool that turns the socket in only one direction and allows the rotating of the ratchet handle back and forth in a narrow space. Socket **extensions** and **universal joints** are also used with sockets to allow access to fasteners in restricted locations.

**DRIVE SIZE.** Sockets are available in various **drive sizes,** including 1/4-, 3/8-, and 1/2-inch. sizes for most automotive use. ● **SEE FIGURES 1–23 AND 1–24.**

Many heavy-duty truck and/or industrial applications use 3/4- and 1-inch sizes. The drive size is the distance of each side of the square drive. Sockets and ratchets of the same size are designed to work together.

**REGULAR AND DEEP WELL.** Sockets are available in regular length for use in most applications or in a deep well design that allows for access to a fastener that uses a long stud or other similar conditions. ● **SEE FIGURE 1–25.**

FIGURE 1–26 Using a clicker-type torque wrench to tighten connecting rod nuts on an engine.

 TECH TIP

**Right to Tighten**

It is sometimes confusing which way to rotate a wrench or screwdriver, especially when the head of the fastener is pointing away from you. To help visualize while looking at the fastener, say "righty tighty, lefty loosey."

**TORQUE WRENCHES** Torque wrenches are socket turning handles that are designed to apply a known amount of force to the fastener. The two basic types of torque wrenches are the following:

1. **Clicker type.** This type of torque wrench is first set to the specified torque, and then it "clicks" when the set torque value has been reached. When force is removed from the torque wrench handle, another click is heard. The setting on a clicker-type torque wrench should be set back to zero after use and checked for proper calibration regularly. ● **SEE FIGURE 1–26.**

2. **Beam-type.** This type of torque wrench is used to measure torque, but instead of presenting the value, the actual torque is displayed on the dial of the wrench as the fastener is being tightened. Beam-type torque wrenches are available in 1/4-, 3/8-, and 1/2-inch drives and both English and metric units. ● **SEE FIGURE 1–27.**

**SAFE USE OF SOCKETS AND RATCHETS** Always use the proper size socket that correctly fits the bolt or nut. All sockets and ratchets should be cleaned after use before being placed back into the toolbox. Sockets are available in short and deep well designs. Never expose any tool to excessive heat. High temperatures can reduce the strength ("draw the temper") of metal tools.

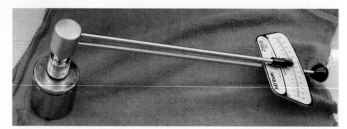

FIGURE 1–27 A beam-type torque wrench that displays the torque reading on the face of the dial. The beam display is read as the beam deflects, which is in proportion to the amount of torque applied to the fastener.

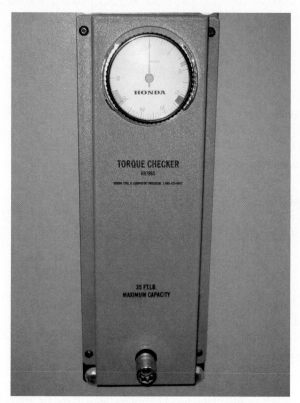

FIGURE 1–28 Torque wrench calibration checker.

 TECH TIP

**Check Torque Wrench Calibration Regularly**

Torque wrenches should be checked regularly. For example, Honda has a torque wrench calibration setup at each of their training centers. It is expected that a torque wrench be checked for accuracy before every use. Most experts recommend that torque wrenches be checked and adjusted as needed at least every year and more often if possible. ● **SEE FIGURE 1–28.**

Never use a hammer on a socket handle unless you are using a special "staking face" wrench designed to be used with a hammer. Replace any tools that are damaged or worn.

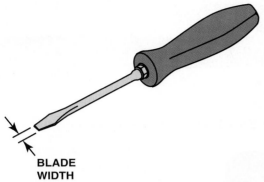

**FIGURE 1–29** A flat-tip (straight-blade) screwdriver. The width of the blade should match the width of the slot in the fastener being loosened or tightened.

**FIGURE 1–30** Two stubby screwdrivers that are used to access screws that have limited space above. A straight blade is on top and a #2 Phillips screwdriver is on the bottom.

 TECH TIP

**Use Socket Adapters with Caution**

A **socket adapter** allows the use of one size of socket and another drive size ratchet or breaker bar. Socket adapters are available and can be used for different drive size sockets on a ratchet. Combinations include the following:

- 1/4 in. drive—3/8 in. sockets
- 3/8 in. drive—1/4 in. sockets
- 3/8 in. drive—1/2 in. sockets
- 1/2 in. drive—3/8 in. sockets

Using a larger drive ratchet or breaker bar on a smaller size socket can cause the application of too much force to the socket, which could crack or shatter. Using a smaller size drive tool on a larger socket will usually not cause any harm but would greatly reduce the amount of torque that can be applied to the bolt or nut.

 TECH TIP

**Avoid Using "Cheater Bars"**

Whenever a fastener is difficult to remove, some technicians will insert the handle of a ratchet or a breaker bar into a length of steel pipe sometimes called a **cheater bar.** The extra length of the pipe allows the technician to exert more torque than can be applied using the drive handle alone. However, the extra torque can easily overload the socket and ratchet, causing them to break or shatter, which could cause personal injury.

**CAUTION: Do not use a screwdriver as a pry tool or as a chisel. Screwdrivers are hardened steel only at the tip and are not designed to be pounded on or used for prying because they could bend easily. Always use the proper tool for each application.**

**PHILLIPS SCREWDRIVER.** Another type of commonly used screwdriver is called a Phillips screwdriver, named for Henry F. Phillips, who invented the crosshead screw in 1934. Because the shape of the crosshead screw and screwdriver, a Phillips screw can be driven with more torque than can be achieved with a slotted screw.

A Phillips head screwdriver is specified by the length of the handle and the size of the point at the tip. A #1 tip has a sharp point, a #2 tip is the most commonly used, and a #3 tip is blunt and is used only for larger sizes of Phillips head fasteners. For example, a #2 3-inch Phillips screwdriver would typically measure 6 inches from the tip of the blade to the end of the handle (3-in. long handle and 3-in. long blade) with a #2 tip.

Both straight-blade and Phillips screwdrivers are available with a short blade and handle for access to fasteners with limited room. ● **SEE FIGURE 1–30.**

**OFFSET SCREWDRIVERS.** Offset screwdrivers are used in places where a conventional screwdriver cannot fit. An offset screwdriver is bent at the ends and is used similar to a wrench.

Also select the appropriate drive size. For example, for small work, such as on the dash, select a 1/4-inch drive. For most general service work, use a 3/8-inch drive, and for suspension and steering and other large fasteners, select a 1/2-inch drive. When loosening a fastener, always pull the ratchet toward you rather than push it outward.

## SCREWDRIVERS

**STRAIGHT-BLADE SCREWDRIVER.** Many smaller fasteners are removed and installed by using a **screwdriver.** Screwdrivers are available in many sizes and tip shapes. The most commonly used screwdriver is called a *straight blade* or *flat tip.*

Flat-tip screwdrivers are sized by the width of the blade, and this width should match the width of the slot in the screw. ● **SEE FIGURE 1–29.**

FIGURE 1–31 An offset screwdriver is used to install or remove fasteners that do not have enough space above to use a conventional screwdriver.

FIGURE 1–32 An impact screwdriver used to remove slotted or Phillips head fasteners that cannot be broken loose using a standard screwdriver.

Most offset screwdrivers have a straight blade at one end and a Phillips end at the opposite end. ● SEE FIGURE 1–31.

**IMPACT SCREWDRIVER.** An *impact screwdriver* is used to break loose or tighten a screw. A hammer is used to strike the end after the screwdriver holder is placed in the head of the screw and rotated in the desired direction. The force from the hammer blow does two things: It applies a force downward holding the tip of the screwdriver in the slot and then applies a twisting force to loosen (or tighten) the screw. ● SEE FIGURE 1–32.

## SAFE USE OF SCREWDRIVERS
Always use the proper type and size screwdriver that matches the fastener. Try to avoid pressing down on a screwdriver because if it slips, the screwdriver tip could go into your hand, causing serious personal injury. All screwdrivers should be cleaned after use. Do not use a screwdriver as a pry bar; always use the correct tool for the job.

## HAMMERS AND MALLETS
Hammers and mallets are used to force objects together or apart. The shape of the back part of the hammer head (called the *peen*) usually determines

FIGURE 1–33 A typical ball-peen hammer.

FIGURE 1–34 A rubber mallet used to deliver a force to an object without harming the surface.

 **FREQUENTLY ASKED QUESTION**

### What Is a Torx?

A Torx is a six-pointed star-shaped tip that was developed by Camcar (formerly Textron) to offer higher loosening and tightening torque than is possible with a straight blade (flat tip) or Phillips. Torx is very commonly used in the automotive field for many components. Commonly used Torx sizes from small to large include: T15, T20, T25, and T30.

Some Torx fasteners include a round projection in the center requiring that a special version of a Torx bit be used. These are called security Torx bits, which have a hole in the center to be used on these fasteners. External Torx fasteners are also used as engine fasteners and are labeled E instead of T, plus the size, such as E45.

the name. For example, a ball-peen hammer has a rounded end like a ball, and it is used to straighten oil pans and valve covers, using the hammer head, and for shaping metal, using the ball peen. ● SEE FIGURE 1–33.

**NOTE: A claw hammer has a claw used to remove nails and is not used for automotive service.**

A hammer is usually sized by the weight of the head of the hammer and the length of the handle. For example, a commonly used ball-peen hammer has an 8-ounce head with an 11-inch handle.

**MALLETS.** *Mallets* are a type of hammer with a large striking surface, which allows the technician to exert force over a larger area than a hammer, so as not to harm the part or component. Mallets are made from a variety of materials, including rubber, plastic, or wood. ● SEE FIGURE 1–34.

FIGURE 1–35 A dead-blow hammer that was left outside in freezing weather. The plastic covering was damaged, which destroyed this hammer. The lead shot is encased in the metal housing and then covered.

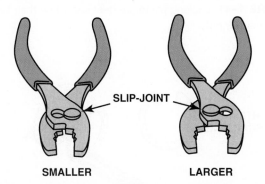

SLIP-JOINT

SMALLER          LARGER

FIGURE 1–36 Typical slip-joint pliers are a common household pliers. The slip joint allows the jaws to be opened to two different settings.

**DEAD-BLOW HAMMER.** A shot-filled plastic hammer is called a *dead-blow hammer.* The small lead balls (shot) inside a plastic head prevent the hammer from bouncing off of the object when struck. ● SEE FIGURE 1–35.

## SAFE USE OF HAMMERS AND MALLETS
All mallets and hammers should be cleaned after use and not exposed to extreme temperatures. Never use a hammer or mallet that is damaged in any way and always use caution to avoid doing damage to the components and the surrounding area. Always follow the hammer manufacturer's recommended procedures and practices.

## PLIERS

**SLIP-JOINT PLIERS.** **Pliers** are capable of holding, twisting, bending, and cutting objects and are an extremely useful classification of tools. The common household type of pliers are called *slip-joint pliers.* There are two different positions where the junction of the handles meets to achieve a wide range of sizes of objects that can be gripped. ● SEE FIGURE 1–36.

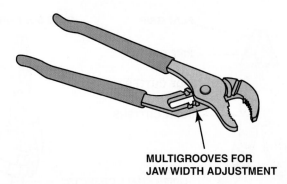

MULTIGROOVES FOR
JAW WIDTH ADJUSTMENT

FIGURE 1–37 Multigroove adjustable pliers are known by many names, including the trade name "Channel Locks®."

---

🔧 **TECH TIP**

### Pound with Something Softer

If you must pound on something, be sure to use a tool that is softer than what you are about to pound on to avoid damage. Examples are given in the following table.

| The Material Being Pounded | What to Pound With |
|---|---|
| Steel or cast iron | Brass or aluminum hammer or punch |
| Aluminum | Plastic or rawhide mallet or plastic-covered dead-blow hammer |
| Plastic | Rawhide mallet or plastic dead-blow hammer |

---

**MULTIGROOVE ADJUSTABLE PLIERS.** For gripping larger objects, a set of *multigroove adjustable pliers* is a commonly used tool of choice of many service technicians. Originally designed to remove the various size nuts holding rope seals used in water pumps, the name *water pump pliers* is also used. These types of pliers are commonly called by their trade name *Channel Locks®.* ● SEE FIGURE 1–37.

**LINESMAN'S PLIERS.** *Linesman's pliers* are a hand tool specifically designed for cutting, bending, and twisting wire. While commonly used by construction workers and electricians, linesman's pliers are a very useful tool for the service technician who deals with wiring. The center parts of the jaws are designed to grasp round objects such as pipe or tubing without slipping. ● SEE FIGURE 1–38.

**DIAGONAL PLIERS.** *Diagonal pliers* are designed to cut only. The cutting jaws are set at an angle to make it easier to cut wires. These pliers are also called *side cuts* or *dikes.* These pliers are made of hardened steel, and they are used mostly for cutting wire. ● SEE FIGURE 1–39.

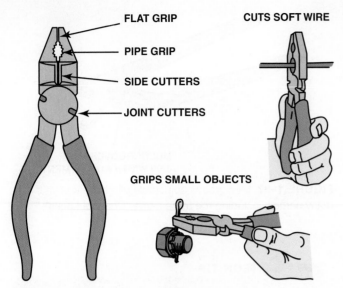

FLAT GRIP

PIPE GRIP

SIDE CUTTERS

JOINT CUTTERS

CUTS SOFT WIRE

GRIPS SMALL OBJECTS

**FIGURE 1–38** Linesman's pliers are very useful because they can help perform many automotive service jobs.

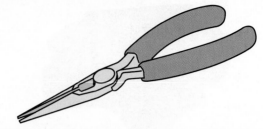

**FIGURE 1–40** Needle-nose pliers are used where there is limited access to a wire or pin that needs to be installed or removed.

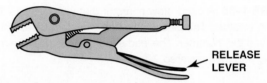

RELEASE LEVER

**FIGURE 1–41** Locking pliers are best known by their trade name Vise Grips®.

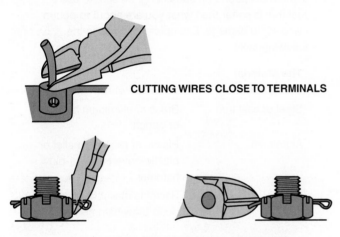

CUTTING WIRES CLOSE TO TERMINALS

PULLING OUT AND SPREADING COTTER PIN

**FIGURE 1–39** Diagonal-cut pliers are another common tool that have many names.

**NEEDLE-NOSE PLIERS.** *Needle-nose pliers* are designed to grip small objects or objects in tight locations. These pliers have long, pointed jaws, which allow the tips to reach into narrow openings or groups of small objects. ● SEE FIGURE 1–40.

Most needle-nose pliers have a wire cutter located at the base of the jaws near the pivot. There are several variations of needle nose pliers, including right-angle jaws or slightly angled to allow access to certain cramped areas.

**LOCKING PLIERS.** *Locking pliers* are adjustable pliers that can be locked to hold objects from moving. Most locking pliers also have wire cutters built into the jaws near the pivot point. Locking pliers come in a variety of styles and sizes and are commonly referred to by the trade name *Vise Grips®*. The size is the length of the pliers, not how far the jaws open. ● SEE FIGURE 1–41.

**SNAP-RING PLIERS.** *Snap-ring pliers* are used to remove and install snap rings. Many snap-ring pliers are designed to be able to remove and install both inward and outward expanding snap

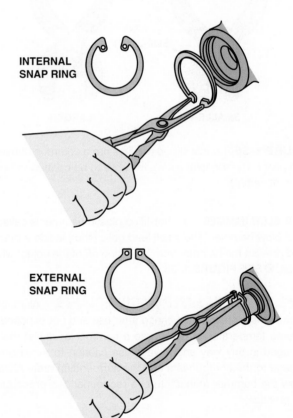

INTERNAL SNAP RING

EXTERNAL SNAP RING

**FIGURE 1–42** Snap-ring pliers are also called lock-ring pliers, and most are designed to remove internal and external snap rings (lock rings).

rings. Some snap-ring pliers can be equipped with serrated-tipped jaws for grasping the opening in the snap ring, while others are equipped with points that are inserted into the holes in the snap ring. ● SEE FIGURE 1–42.

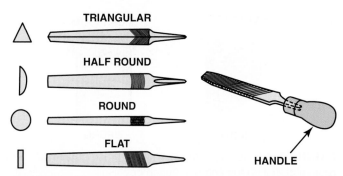

FIGURE 1–43 Files come in many different shapes and sizes. Never use a file without a handle.

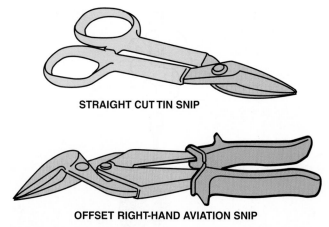

FIGURE 1–44 Tin snips are used to cut thin sheets of metal or carpet.

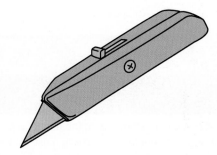

FIGURE 1–45 A utility knife uses replaceable blades and is used to cut carpet and other materials.

🔧 TECH TIP

**Brand Name versus Proper Term**

Technicians often use slang or brand names of tools rather than the proper term. This results in some confusion for new technicians. Some examples are given in the following table.

| Brand Name | Proper Term | Slang Name |
|---|---|---|
| Crescent wrench® | Adjustable wrench | Monkey wrench |
| Vise Grips® | Locking pliers | |
| Channel Locks® | Water pump pliers or multigroove adjustable pliers | Pump pliers |
| | Diagonal cutting pliers | Dikes or side cuts |

**SAFE USE OF PLIERS**    Pliers should not be used to remove any bolt or other fastener. Pliers should be used only when specified for use by the vehicle manufacturer.

**FILES**    **Files** are used to smooth metal and are constructed of hardened steel with diagonal rows of teeth. Files are available with a single row of teeth called a *single cut file,* as well as two rows of teeth cut at an opposite angle called a *double cut file.* Files are available in a variety of shapes and sizes, such as small flat files, half-round files, and triangular files. ● SEE FIGURE 1–43.

**SAFE USE OF FILES**    Always use a file with a handle. Because files only cut when moved forward, a handle must be attached to prevent possible personal injury. After making a forward strike, lift the file and return the file to the starting position; avoid dragging the file backward.

**SNIPS**    Service technicians are often asked to fabricate sheet metal brackets or heat shields and need to use one or more types of cutters available called **snips.** *Tin snips* are the simplest and are designed to make straight cuts in a variety of materials, such as sheet steel, aluminum, or even fabric. A variation of the tin snips is called *aviation tin snips.* There are three designs of aviation snips, including one designed to cut straight (called a *straight cut aviation snip*), one designed to cut left (called an *offset left aviation snip*), and one designed to cut right (called an *offset right aviation snip*). The handles are color coded for easy identification. These include yellow for straight, red for left, and green for right. ● SEE FIGURE 1–44.

**UTILITY KNIFE**    A *utility knife* uses a replaceable blade and is used to cut a variety of materials such as carpet, plastic, wood, and paper products, such as cardboard. ● SEE FIGURE 1–45.

**SAFE USE OF CUTTERS**    Whenever using cutters, always wear eye protection or a face shield to guard against the possibility of metal pieces being ejected during the cut. Always follow recommended procedures.

**PUNCHES**    A **punch** is a small-diameter steel rod that has a smaller-diameter ground at one end. A punch is used to drive a pin out that is used to retain two components. Punches come in a variety of sizes, which are measured across the diameter of

FIGURE 1-46 A punch used to drive pins from assembled components. This type of punch is also called a pin punch.

FIGURE 1-47 Warning stamped on the side of a punch warning that goggles should be worn when using this tool. Always follow safety warnings.

the machined end. Sizes include 1/16, 1/8, 3/16, and 1/4 inch. ●SEE FIGURE 1-46.

CHISELS    A **chisel** has a straight, sharp cutting end that is used for cutting off rivets or to separate two pieces of an assembly. The most common design of chisel used for automotive service work is called a *cold chisel.*

SAFE USE OF PUNCHES AND CHISELS    Always wear eye protection when using a punch or a chisel because the hardened steel is brittle and parts of the punch could fly off and cause serious personal injury. See the warning stamped on the side of this automotive punch in ●FIGURE 1-47.

The tops of punches and chisels can become rounded off from use, which is called "mushroomed." This material must be ground off to help avoid the possibility of the overhanging material being loosened and becoming airborne during use. ●SEE FIGURE 1-48.

HACKSAWS    A **hacksaw** is used to cut metals, such as steel, aluminum, brass, or copper. The cutting blade of a hacksaw is replaceable, and the sharpness and number of teeth can be varied to meet the needs of the job. Use 14 or 18 teeth per inch (TPI) for cutting plaster or soft metals, such as aluminum and copper. Use 24 or 32 teeth per inch for steel or pipe. Hacksaw blades should be installed with the teeth pointing away from the handle. This means that a hacksaw cuts

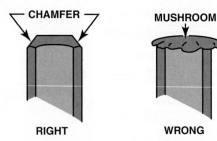

FIGURE 1-48 Use a grinder or a file to remove the mushroom material on the end of a punch or chisel.

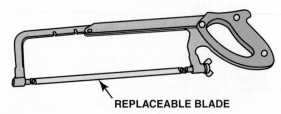

FIGURE 1-49 A typical hacksaw that is used to cut metal. If cutting sheet metal or thin objects, a blade with more teeth should be used.

only while the blade is pushed in the forward direction. ●SEE FIGURE 1-49.

SAFE USE OF HACKSAWS    Check that the hacksaw is equipped with the correct blade for the job and that the teeth are pointed away from the handle. When using a hacksaw, move the hacksaw slowly away from you, then lift slightly and return for another cut.

## BASIC HAND TOOL LIST

The following is a typical list of hand tools every automotive technician should possess. Specialty tools are not included.

Safety glasses

Tool chest

1/4 in. drive socket set (1/4 in. to 9/16 in. standard and deep sockets; 6-mm to 15-mm standard and deep sockets)

1/4 in. drive ratchet

1/4 in. drive 2 in. extension

1/4 in. drive 6 in. extension

1/4 in. drive handle

3/8 in. drive socket set (3/8 in. to 7/8 in. standard and deep sockets; 10-mm to 19-mm standard and deep sockets)

3/8 in. drive Torx set (T40, T45, T50, and T55)

3/8 in. drive 13/16 in. plug socket

3/8 in. drive 5/8 in. plug socket

3/8 in. drive ratchet

3/8 in. drive 1 1/2 in. extension

3/8 in. drive 3 in. extension

3/8 in. drive 6 in. extension

3/8 in. drive 18 in. extension

3/8 in. drive universal

1/2 in. drive socket set (1/2 in. to 1 in. standard and deep sockets)

1/2 in. drive ratchet

1/2 in. drive breaker bar

1/2 in. drive 5 in. extension

1/2 in. drive 10 in. extension

3/8 in. to 1/4 in. adapter

1/2 in. to 3/8 in. adapter

3/8 in. to 1/2 in. adapter

Crowfoot set (fractional in.)

Crowfoot set (metric)

3/8 in. through 1 in. combination wrench set

10-mm through 19-mm combination wrench set

1/16 in. through 1/4 in. hex wrench set

2-mm through 12-mm hex wrench set

3/8 in. hex socket

13-mm to 14-mm flare-nut wrench

15-mm to 17-mm flare-nut wrench

5/16 in. to 3/8 in. flare-nut wrench

7/16 in. to 1/2 in. flare-nut wrench

1/2 in. to 9/16 in. flare-nut wrench

Diagonal pliers

Needle pliers

Adjustable-jaw pliers

Locking pliers

Snap-ring pliers

Stripping or crimping pliers

Ball-peen hammer

Rubber hammer

Dead-blow hammer

Five-piece standard screwdriver set

Four-piece Phillips screwdriver set

#15 Torx screwdriver

#20 Torx screwdriver

Center punch

Pin punches (assorted sizes)

Chisel

Utility knife

Valve core tool

Filter wrench (large filters)

Filter wrench (smaller filters)

Test light

Feeler gauge

Scraper

Pinch bar

Magnet

## TOOL SETS AND ACCESSORIES

A beginning service technician may wish to start with a small set of tools before purchasing an expensive tool set. ● SEE FIGURES 1–50 AND 1–51.

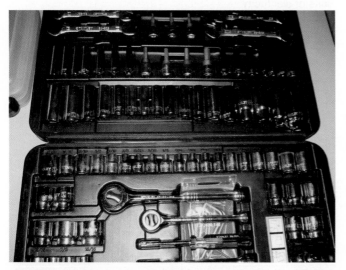

**FIGURE 1–50** A typical beginning technician tool set that includes the basic tools to get started.

**FIGURE 1–51** A typical large tool box, showing just one of many drawers.

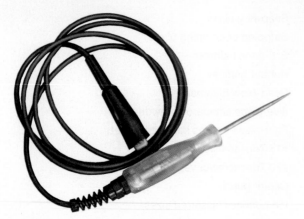

**FIGURE 1–52** A typical 12-volt test light.

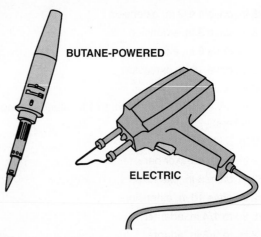

BUTANE-POWERED

ELECTRIC

**FIGURE 1–53** An electric and butane-powered soldering guns used to make electrical repairs. Soldering guns are sold by the wattage rating. The higher the wattage, the greater the amount of heat created. Most solder guns used for automotive electrical work usually fall within the 60- to 160-watt range.

**TECH TIP**

**Need to Borrow a Tool More Than Twice? Buy It!**

Most service technicians agree that it is okay for a beginning technician to borrow a tool occasionally. However, if a tool has to be borrowed more than twice, then be sure to purchase it as soon as possible. Also, whenever a tool is borrowed, be sure that you clean the tool and let the technician you borrowed the tool from know that you are returning the tool. These actions will help in any future dealings with other technicians.

# ELECTRICAL HAND TOOLS

**TEST LIGHT** A test light is used to test for electricity. A typical automotive test light consists of a clear plastic screwdriver-like handle that contains a lightbulb. A wire is attached to one terminal of the bulb, which the technician connects to a clean metal part of the vehicle. The other end of the bulb is attached to a point that can be used to test for electricity at a connector or wire. When there is power at the point and a good connection at the other end, the lightbulb lights. ● **SEE FIGURE 1–52.**

## SOLDERING GUNS

**ELECTRIC SOLDERING GUN.** This type of soldering gun is usually powered by 110-volt AC and often has two power settings expressed in watts. A typical electric soldering gun will produce from 85 to 300 watts of heat at the tip, which is more than adequate for soldering.

**ELECTRIC SOLDERING PENCIL.** This type of soldering iron is less expensive and creates less heat than an electric soldering gun. A typical electric soldering pencil (iron) creates 30 to 60 watts of heat and is suitable for soldering smaller wires and connections.

**BUTANE-POWERED SOLDERING IRON.** A butane-powered soldering iron is portable and very useful for automotive service work because an electrical cord is not needed. Most butane-powered

soldering irons produce about 60 watts of heat, which is enough for most automotive soldering. ● **SEE FIGURE 1–53.**

**ELECTRICAL WORK HAND TOOLS** In addition to a soldering iron, most service technicians who do electrical-related work should have the following:

- Wire cutters
- Wire strippers
- Wire crimpers
- Heat gun for heat shrink tubing

**DIGITAL METER** A digital meter is a necessary tool for any electrical diagnosis and troubleshooting. A digital multimeter, abbreviated DMM, is usually capable of measuring the following units of electricity:

- DC volts
- AC volts
- Ohms
- Amperes

# HAND TOOL MAINTENANCE

Most hand tools are constructed of rust-resistant metals, but they can still rust or corrode if not properly maintained. For best results and long tool life, the following steps should be taken:

- Clean each tool before placing it back into the toolbox.
- Keep tools separated. Moisture on metal tools will start to rust more readily if the tools are in contact with another metal tool.

**What Is an "SST"?**

Vehicle manufacturers often specify a **special service tool (SST)** to properly disassemble and assemble components, such as transmissions and other components. These tools are also called special tools and are available from the vehicle manufacturer or their tool supplier, such as Kent-Moore and Miller tools. Many service technicians do not have access to special service tools, so they use generic versions that are available from aftermarket sources.

- Line the drawers of the toolbox with a material that will prevent the tools from moving as the drawers are opened and closed. This helps to quickly locate the proper tool and size.
- Release the tension on all "clicker-type" torque wrenches.
- Keep the toolbox secure.

## TROUBLE LIGHTS

**INCANDESCENT**  *Incandescent lights* use a filament that produces light when electric current flows through the bulb. This was the standard **trouble light,** also called a *work light* for many years until safety issues caused most shops to switch to safer fluorescent or LED lights. If incandescent lightbulbs are used, try to locate bulbs that are rated "rough service," which are designed to withstand shock and vibration more than conventional lightbulbs.

**FIGURE 1–54** A fluorescent trouble light operates cooler and is safer to use in the shop because it is protected against accidental breakage where gasoline or other flammable liquids would happen to come in contact with the light.

Do not use incandescent trouble lights around gasoline or other flammable liquids. The liquids can cause the bulb to break, and the hot filament can ignite the flammable liquid, which can cause personal injury or even death.

**FLUORESCENT**  A trouble light is an essential piece of shop equipment and, for safety, should be fluorescent rather than incandescent. Incandescent lightbulbs can scatter or break if gasoline were to be splashed onto the bulb, creating a serious fire hazard. Fluorescent light tubes are not as likely to be broken and are usually protected by a clear plastic enclosure. Trouble lights are usually attached to a retractor, which can hold 20 to 50 feet of electrical cord. ● SEE FIGURE 1–54.

**LED TROUBLE LIGHT**  **Light-emitting diode (LED)** trouble lights are excellent to use because they are shock resistant and long lasting and do not represent a fire hazard. Some trouble lights are battery powered and therefore can be used in places where an attached electrical cord could present problems.

## AIR AND ELECTRICALLY OPERATED TOOLS

**IMPACT WRENCH**  An impact wrench, either air or electrically powered, is a tool that is used to remove and install fasteners. The air-operated 1/2 in. drive impact wrench is the most commonly used unit. ● SEE FIGURE 1–55.

**FIGURE 1–55** A typical 1/2 in. drive air impact wrench. The direction of rotation can be changed to loosen or tighten a fastener.

Electrically powered impact wrenches commonly include the following:

- Battery-powered units. ● **SEE FIGURE 1–56.**
- 110-volt AC-powered units. This type of impact is very useful, especially if compressed air is not readily available.

 **WARNING**

Always use impact sockets with impact wrenches and always wear eye protection in case the socket or fastener shatters. Impact sockets are thicker walled and constructed with premium alloy steel. They are hardened with a black oxide finish to help prevent corrosion and distinguish them from regular sockets. ● **SEE FIGURE 1–57.**

**AIR RATCHET** An air ratchet is used to remove and install fasteners that would normally be removed or installed using a ratchet and a socket. ● **SEE FIGURE 1–58.**

**DIE GRINDER** A die grinder is a commonly used air-powered tool that can also be used to sand or remove gaskets and rust. ● **SEE FIGURE 1–59.**

**BENCH- OR PEDESTAL-MOUNTED GRINDER** These high-powered grinders can be equipped with a wire brush wheel and/or a stone wheel:

- **Wire brush wheel.** This type is used to clean threads of bolts as well as to remove gaskets from sheet metal engine parts.
- **Stone wheel.** This type is used to grind metal or to remove the mushroom from the top of punches or chisels. ● **SEE FIGURE 1–60.**

**FIGURE 1–56** A typical battery-powered 3/8 in. drive impact wrench.

**FIGURE 1–58** An air ratchet is a very useful tool that allows fast removal and installation of fasteners, especially in areas that are difficult to reach or do not have room enough to move a hand ratchet or wrench.

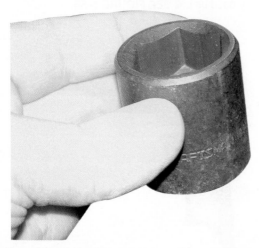

**FIGURE 1–57** A black impact socket. Always use an impact-type socket whenever using an impact wrench to avoid the possibility of shattering the socket, which could cause personal injury.

**FIGURE 1–59** This typical die grinder surface preparation kit includes the air-operated die grinder as well as a variety of sanding disks for smoothing surfaces or removing rust.

FIGURE 1–60 A typical pedestal grinder with a wire wheel on the left side and a stone wheel on the right side. Even though this machine is equipped with guards, safety glasses or a face shield should always be worn whenever using a grinder or wire wheel.

☠ WARNING

Always wear a face shield when using a wire wheel or a grinder.

Most **bench grinders** are equipped with a grinder wheel (stone) on one end and a wire brush wheel on the other end. A bench grinder is a very useful piece of shop equipment, and the wire wheel end can be used for the following:

- Cleaning threads of bolts
- Cleaning gaskets from sheet metal parts, such as steel valve covers

CAUTION: Use a steel wire brush only on steel or iron components. If a steel wire brush is used on aluminum or copper-based metal parts, it can remove metal from the part.

The grinding stone end of the bench grinder can be used for the following:

- Sharpening blades and drill bits
- Grinding off the heads of rivets or parts
- Sharpening sheet metal parts for custom fitting

## PERSONAL PROTECTIVE EQUIPMENT

Service technicians should wear **personal protective equipment (PPE)** to prevent personal injury. The personal protection devices include the following:

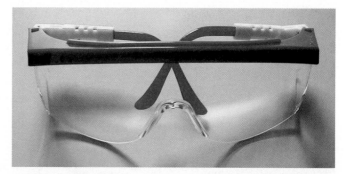

FIGURE 1–61 Safety glasses should be worn at all times when working on or around any vehicle or servicing any components.

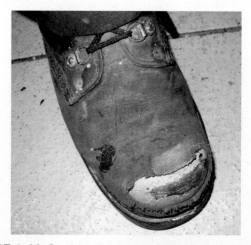

FIGURE 1–62 Steel-toed shoes are a worthwhile investment to help prevent foot injury due to falling objects. Even these well-worn shoes can protect the feet of this service technician.

**SAFETY GLASSES**   Wear safety glasses at all times while servicing any vehicle and be sure that they meet standard ANSI Z87.1. ● SEE FIGURE 1–61.

**STEEL-TOED SAFETY SHOES**   ● SEE FIGURE 1–62. If steel-toed safety shoes are not available, then leather-topped shoes offer more protection than canvas or cloth.

**BUMP CAP**   Service technicians working under a vehicle should wear a **bump cap** to protect the head against under-vehicle objects and the pads of the lift. ● SEE FIGURE 1–63.

**HEARING PROTECTION**   Hearing protection should be worn if the sound around you requires that you raise your voice (sound level higher than 90 dB). For example, a typical lawnmower produces noise at a level of about 110 dB. This means that everyone who uses a lawnmower or other lawn or garden equipment should wear ear protection.

**GLOVES**   Many technicians wear gloves not only to help keep their hands clean but also to help protect their skin from the effects of dirty engine oil and other possibly hazardous materials.

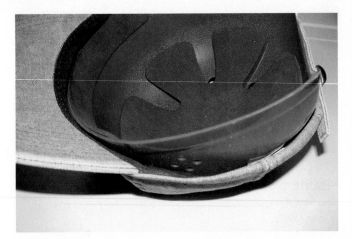

**FIGURE 1–63** One version of a bump cap is a molded plastic insert that is worn inside a regular cloth cap.

**FIGURE 1–64** Protective gloves are available in several sizes and materials.

Several types of gloves and their characteristics include the following:

- **Latex surgical gloves.** These gloves are relatively inexpensive but tend to stretch, swell, and weaken when exposed to gas, oil, or solvents.

- **Vinyl gloves.** These gloves are also inexpensive and are not affected by gas, oil, or solvents.

- **Polyurethane gloves.** These gloves are more expensive yet very strong. Even though these gloves are also not affected by gas, oil, or solvents, they do tend to be slippery.

- **Nitrile gloves.** These gloves are exactly like latex gloves but are not affected by gas, oil, or solvents, yet they tend to be expensive.

- **Mechanic's gloves.** These gloves are usually made of synthetic leather and spandex and provide thermo protection as well as protection from dirt and grime. ● **SEE FIGURE 1–64.**

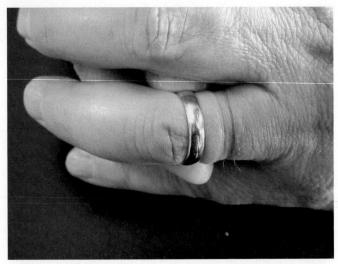

**FIGURE 1–65** Remove all jewelry before performing service work on any vehicle.

## SAFETY PRECAUTIONS

Besides wearing personal safety equipment, there are also many actions that should be performed to keep safe in the shop. These actions include the following:

- Remove jewelry that may get caught on something or act as a conductor to an exposed electrical circuit. ● **SEE FIGURE 1–65.**

- Take care of your hands. Keep your hands clean by washing with soap and hot water that is at least 110°F (43°C).

- Avoid loose or dangling clothing.

- When lifting any object, get a secure grip with solid footing. Keep the load close to your body to minimize the strain. Lift with your legs and arms, not your back.

- Do not twist your body when carrying a load. Instead, pivot your feet to help prevent strain on the spine.

- Ask for help when moving or lifting heavy objects.

- Push a heavy object rather than pull it. (This is opposite to the way you should work with tools—never push a wrench! If you do and a bolt or nut loosens, your entire weight is used to propel your hand[s] forward. This usually results in cuts, bruises, or other painful injury.)

- Always connect an exhaust hose to the tailpipe of any running vehicle to help prevent the buildup of carbon monoxide inside a closed garage space. ● **SEE FIGURE 1–66.**

- When standing, keep objects, parts, and tools with which you are working between chest height and waist height. If seated, work at tasks that are at elbow height.

- Always be sure the hood is securely held open.

FIGURE 1–66 Always connect an exhaust hose to the tail-pipe of a vehicle to be run inside a building.

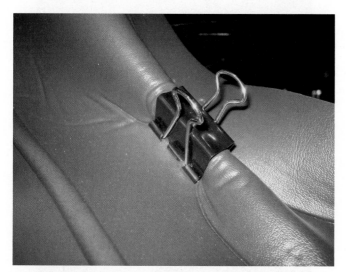

FIGURE 1–67 A binder clip being used to keep a fender cover from falling off.

FIGURE 1–68 Covering the interior as soon as the vehicle comes in for service helps improve customer satisfaction.

 **SAFETY TIP**

**Shop Cloth Disposal**

Always dispose of oily shop cloths in an enclosed container to prevent a fire. ● **SEE FIGURE 1–69.** Whenever oily cloths are thrown together on the floor or workbench, a chemical reaction can occur, that can ignite the cloth even without an open flame. This process of ignition without an open flame is called **spontaneous combustion.**

## VEHICLE PROTECTION

**FENDER COVERS** Whenever working under the hood of any vehicle, be sure to use fender covers. They not only help protect the vehicle from possible damage but also provide a clean surface to place parts and tools. The major problem with using fender covers is that they tend to move and often fall off the vehicle. To help prevent the fender covers from falling off, secure them to a lip of the fender using a *binder clip* available at most office supply stores. ● **SEE FIGURE 1–67.**

**INTERIOR PROTECTION** Always protect the interior of the vehicle from accidental damage or dirt and grease by covering the seat, steering wheel, and floor with a protective covering. ● **SEE FIGURE 1–68.**

## SAFETY LIFTING (HOISTING) A VEHICLE

Many chassis and underbody service procedures require that the vehicle be hoisted or lifted off the ground. The simplest methods involve the use of drive-on ramps or a floor jack and safety (jack) stands, whereas in-ground or surface-mounted lifts provide greater access.

*Setting the pads is a critical part of this hoisting procedure.* All vehicle service information, including service, shop, and owner's manuals, include recommended locations to be used when hoisting (lifting) a vehicle. Newer vehicles have a triangle decal on the driver's door indicating the recommended lift points. The recommended standards for the lift points and lifting procedures are found in SAE Standard JRP-2184. ● **SEE FIGURE 1–70.**

These recommendations typically include the following points:

1. The vehicle should be centered on the lift or hoist so as not to overload one side or put too much force either forward or rearward. ● **SEE FIGURE 1–71.**

2. The pads of the lift should be spread as far apart as possible to provide a stable platform.

3. Each pad should be placed under a portion of the vehicle that is strong and capable of supporting the weight of the vehicle.

   a. Pinch welds at the bottom edge of the body are generally considered to be strong.

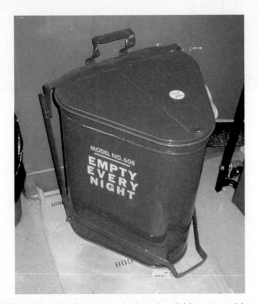

**FIGURE 1–69** All oily shop cloths should be stored in a metal container equipped with a lid to help prevent spontaneous combustion.

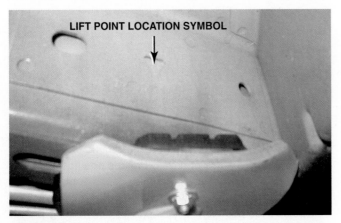

**FIGURE 1–70** Most newer vehicles have a triangle symbol indicating the recommended hoisting lift location.

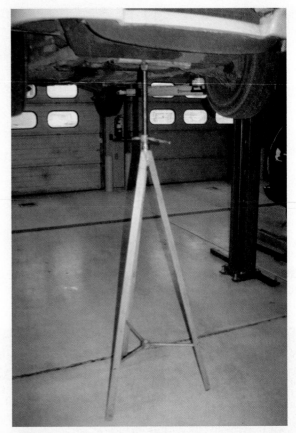

(a)

(b)

**FIGURE 1–71** (a) Tall safety stands can be used to provide additional support for the vehicle while on the hoist. (b) A block of wood should be used to avoid the possibility of doing damage to components supported by the stand.

**CAUTION: Even though pinch weld seams are the recommended location for hoisting many vehicles with unitized bodies (unit-body), care should be taken not to place the pad(s) too far forward or rearward. Incorrect placement of the vehicle on the lift could cause the vehicle to be imbalanced, and the vehicle could fall. This is exactly what happened to the vehicle in ● FIGURE 1–72.**

   b. Boxed areas of the body are the best places to position the pads on a vehicle without a frame. Be careful

**FIGURE 1–72** This training vehicle fell from the hoist because the pads were not set correctly. No one was hurt, but the vehicle was damaged.

to note whether the arms of the lift might come into contact with other parts of the vehicle before the pad touches the intended location. Commonly damaged areas include the following:

(1) Rocker panel moldings

(2) Exhaust system (including catalytic converter)

(3) Tires or body panels (● **SEE FIGURES 1–73 AND 1–74.**)

4. The vehicle should be raised about a foot (30 centimeters [cm]) off the floor, then stopped and shaken to check for stability. If the vehicle seems to be stable when checked at a short distance from the floor, continue raising the vehicle and continue to view the vehicle until it has reached the desired height. The hoist should be lowered onto the mechanical locks and then raised off of the locks before lowering.

**CAUTION: Do not look away from the vehicle while it is being raised (or lowered) on a hoist. Often one side or one end of the hoist can stop or fail, resulting in the vehicle being slanted enough to slip or fall, creating physical damage not only to the vehicle and/or hoist but also to the technician or others who may be nearby.**

**HINT: Most hoists can be safely placed at any desired height. For ease while working, the area in which you are working should be at chest level. When working on brakes or suspension components, it is not necessary to work on them down near the floor or over your head. Raise the hoist so that the components are at chest level.**

5. Before lowering the hoist, the safety latch(es) must be released and the direction of the controls reversed. The speed downward is often adjusted to be as slow as possible for additional safety.

(a)

(b)

**FIGURE 1–73** (a) An assortment of hoist pad adapters that are often needed to safely hoist many pickup trucks, vans, and sport utility vehicles (SUVs). (b) A view from underneath a Chevrolet pickup truck showing how the pad extensions are used to attach the hoist lifting pad to contact the frame.

## JACKS AND SAFETY STANDS

Floor jacks properly rated for the weight of the vehicle being raised are a common vehicle lifting tool. Floor jacks are portable and relatively inexpensive and must be used with safety (jack) stands. The floor jack is used to raise the vehicle off the ground, and safety stands should be placed under the frame on the body of the vehicle. The weight of the vehicle should never

(a)

(b)

**FIGURE 1–74** (a) The pad arm is just contacting the rocker panel of the vehicle. (b) The pad arm has dented the rocker panel on this vehicle because the pad was set too far inward underneath the vehicle.

be kept on the hydraulic floor jack because a failure of the jack could cause the vehicle to fall. ● **SEE FIGURE 1–75.** The jack is then slowly released to allow the vehicle weight to be supported on the safety stands. If the front or rear of the vehicle is being raised, the opposite end of the vehicle must be blocked.

**CAUTION: Safety stands should be rated higher than the weight they support.**

## DRIVE-ON RAMPS

Ramps are an inexpensive way to raise the front or rear of a vehicle. ● **SEE FIGURE 1–76.** Ramps are easy to store, but they can be dangerous because they can "kick out" when driving the vehicle onto the ramps.

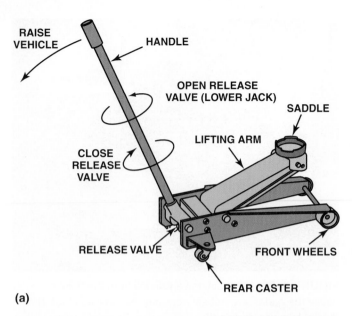

RAISE VEHICLE

HANDLE

OPEN RELEASE VALVE (LOWER JACK)

SADDLE

LIFTING ARM

CLOSE RELEASE VALVE

RELEASE VALVE

FRONT WHEELS

REAR CASTER

(a)

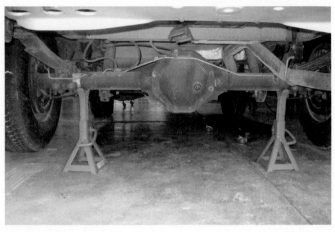
(b)

**FIGURE 1–75** (a) A typical 3-ton (6,000-pound) capacity hydraulic jack. (b) Whenever a vehicle is raised off the ground, a safety stand should be placed under the frame, axle, or body to support the weight of the vehicle.

**CAUTION: Professional repair shops do not use ramps because they are dangerous to use. Use only with extreme care.**

## ELECTRICAL CORD SAFETY

Use correctly grounded three-prong sockets and extension cords to operate power tools. Some tools use only two-prong plugs. Make sure these are double insulated and repair or replace any electrical cords that are cut or damaged to prevent the possibility of an electrical shock. When not in use, keep electrical cords off the floor to prevent tripping over them. Tape the cords down if they are placed in high-foot-traffic areas.

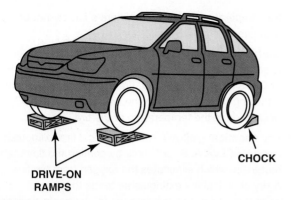

**FIGURE 1–76** Drive-on-type ramps are dangerous to use. The wheels on the ground level must be chocked (blocked) to prevent accidental movement down the ramp.

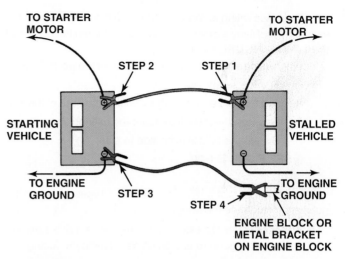

**FIGURE 1–77** Jumper cable usage guide. Follow the same connections if using a portable jump box.

# JUMP-STARTING AND BATTERY SAFETY

To jump-start another vehicle with a dead battery, connect good-quality copper jumper cables as indicated in ● **FIGURE 1–77** or a jump box. The last connection made should always be on the engine block or an engine bracket as far from the battery as possible. It is normal for a spark to be created when the jumper cables finally complete the jumping circuit, and this spark could cause an explosion of the gases around the battery. Many newer vehicles have special ground connections built away from the battery just for the purpose of jump-starting. Check the owner's manual or service information for the exact location.

Batteries contain acid and should be handled with care to avoid tipping them greater than a 45-degree angle. Always remove jewelry when working around a battery to avoid the possibility of electrical shock or burns, which can occur when the metal comes in contact with a 12-volt circuit and ground, such as the body of the vehicle.

**FIGURE 1–78** The air pressure going to the nozzle should be reduced to 30 PSI or less to help prevent personal injury.

---

**+ SAFETY TIP**

**Air Hose Safety**

Improper use of an air nozzle can cause blindness or deafness. Compressed air must be reduced to less than 30 PSI (206 kPa). ● **SEE FIGURE 1–78.** If an air nozzle is used to dry and clean parts, make sure the airstream is directed away from anyone else in the immediate area. Coil and store air hoses when they are not in use.

# FIRE EXTINGUISHERS

There are four **fire extinguisher classes.** Each class should be used on specific fires only:

- **Class A** is designed for use on general combustibles, such as cloth, paper, and wood.
- **Class B** is designed for use on flammable liquids and greases, including gasoline, oil, thinners, and solvents.
- **Class C** is used only on electrical fires.
- **Class D** is effective only on combustible metals, such as powdered aluminum, sodium, or magnesium.

The class rating is clearly marked on the side of every fire extinguisher. Many extinguishers are good for multiple types of fires. ● **SEE FIGURE 1–79.**

When using a fire extinguisher, remember the word "PASS":

P = Pull the safety pin.

A = Aim the nozzle of the extinguisher at the base of the fire.

S = Squeeze the lever to actuate the extinguisher.

S = Sweep the nozzle from side to side.

● **SEE FIGUR E 1–80.**

## TYPES OF FIRE EXTINGUISHERS
Types of fire extinguishers include the following:

- **Water.** A water fire extinguisher, usually in a pressurized container, is good to use on Class A fires by reducing

the temperature to the point where a fire cannot be sustained.

- **Carbon dioxide (CO$_2$).** A carbon dioxide fire extinguisher is good for almost any type of fire, especially Class B or Class C materials. A CO$_2$ fire extinguisher works by removing the oxygen from the fire, and the cold CO$_2$ also helps reduce the temperature of the fire.

- **Dry chemical (yellow).** A dry chemical fire extinguisher is good for Class A, B, or C fires by coating the flammable materials, which eliminates the oxygen from the fire. A dry chemical fire extinguisher tends to be very corrosive and will cause damage to electronic devices.

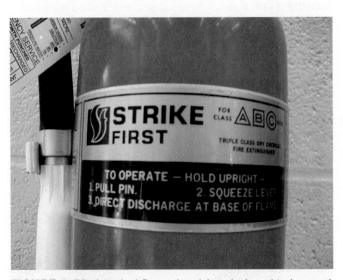

**FIGURE 1–79** A typical fire extinguisher designed to be used on Class A, B, or C fires.

**FIGURE 1–80** A CO$_2$ fire extinguisher being used on a fire set in an open drum during a demonstration at a fire training center.

## FIRE BLANKETS

**Fire blankets** are required to be available in the shop areas. If a person is on fire, a fire blanket should be removed from its storage bag and thrown over and around the victim to smother the fire. ● **SEE FIGURE 1–81** showing a typical fire blanket.

## FIRST AID AND EYE WASH STATIONS

All shop areas must be equipped with a first aid kit and an eye wash station centrally located and kept stocked with emergency supplies. ● **SEE FIGURE 1–82.**

**FIGURE 1–81** A treated wool blanket is kept in an easy-to-open wall-mounted holder and should be placed in a central location in the shop.

**FIGURE 1–82** A first aid box should be centrally located in the shop and kept stocked with the recommended supplies.

**FIRST AID KIT**    A first aid kit should include the following:

- Bandages (variety)
- Gauze pads
- Roll gauze
- Iodine swab sticks
- Antibiotic ointment
- Hydrocortisone cream
- Burn gel packets
- Eye wash solution
- Scissors
- Tweezers
- Gloves
- First aid guide

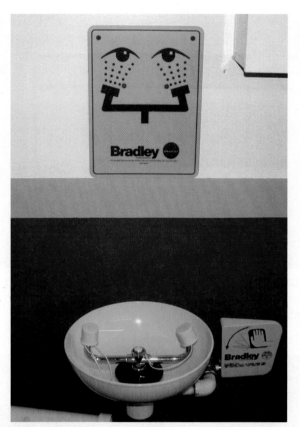

**FIGURE 1–83** A typical eye wash station. Often a thorough flushing of the eyes with water is the first and often the best treatment in the event of eye contamination.

Every shop should have a person trained in first aid. If there is an accident, call for help immediately.

**EYE WASH STATION**    An **eye wash station** should be centrally located and used whenever any liquid or chemical gets into the eyes. If such an emergency does occur, keep eyes in a constant stream of water and call for professional assistance. ● **SEE FIGURE 1–83.**

# HYBRID ELECTRIC VEHICLE SAFETY ISSUES

**Hybrid electric vehicles (HEVs)** use a high-voltage battery pack and an electric motor(s) to help propel the vehicle. ● **SEE FIGURE 1–84** for an example of a typical warning label on a hybrid electric vehicle. The gasoline or diesel engine also is equipped with a generator or a combination starter and an integrated starter generator (ISG) or integrated starter alternator (ISA).

**FIGURE 1–84** A warning label on a Honda hybrid warns that a person can be killed because of the high-voltage circuits under the cover.

**FIGURE 1–85** The high-voltage disconnect switch is in the trunk area on a Toyota Prius. Insulated rubber lineman's gloves should be worn when removing this plug. (Courtesy of Tony Martin)

To safely work around a hybrid electric vehicle, the high-voltage (HV) battery and circuits should be shut off following these steps:

**STEP 1** Turn off the ignition key (if equipped) and remove the key from the ignition switch. (This will shut off all high-voltage circuits if the relay[s] is [are] working correctly.)

**STEP 2** Disconnect the high-voltage circuits.

**TOYOTA PRIUS** The cutoff switch is located in the trunk. To gain access, remove three clips holding the upper left portion of the trunk side cover. To disconnect the high-voltage system, pull the orange handled plug while wearing insulated rubber lineman's gloves. ● **SEE FIGURE 1–85.**

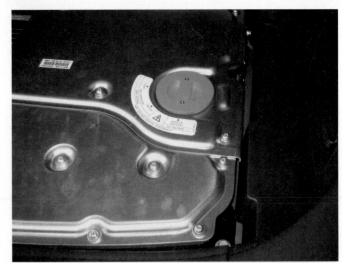

**FIGURE 1–86** The high-voltage shutoff switch on a Ford Escape hybrid. The switch is located under the carpet at the rear of the vehicle.

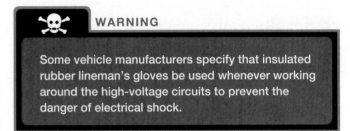

**WARNING**

Some vehicle manufacturers specify that insulated rubber lineman's gloves be used whenever working around the high-voltage circuits to prevent the danger of electrical shock.

**FORD ESCAPE/MERCURY MARINER** Ford and Mercury specify that the following steps should be included when working with the high-voltage (HV) systems of a hybrid vehicle:

- Four orange cones are to be placed at the four corners of the vehicle to create a buffer zone.
- High-voltage insulated gloves are to be worn with an outer leather glove to protect the inner rubber glove from possible damage.
- The service technician should also wear a face shield, and a fiberglass hook should be in the area and used to move a technician in the event of electrocution.

The high-voltage shutoff switch is located in the rear of the vehicle under the right-side carpet. ● **SEE FIGURE 1–86.** Rotate the handle to the "service shipping" position, lift it out to disable the high-voltage circuit, and wait 5 minutes before removing high-voltage cables.

**HONDA CIVIC** To totally disable the high-voltage system on a Honda Civic, remove the main fuse (labeled number 1) from the driver's-side underhood fuse panel. This should be all that is necessary to shut off the high-voltage circuit. If this is not possible, then remove the rear seat cushion and seat back. Remove the metal switch cover labeled "up" and remove the red locking cover. Move the "battery module switch" down to disable the high-voltage system.

**FIGURE 1–87** The shutoff switch on a GM parallel hybrid truck is green because this system uses 42 volts instead of higher, possibly fatal voltages used in other hybrid vehicles.

## CHEVROLET SILVERADO/GMC SIERRA PICKUP TRUCK

The high-voltage shutoff switch is located under the rear passenger seat. Remove the cover marked "energy storage box" and turn the green service disconnect switch to the horizontal position to turn off the high-voltage circuits. ● **SEE FIGURE 1–87.**

**WARNING**

Do not touch any orange wiring or component without following the vehicle manufacturer's procedures and wearing the specified personal protective equipment.

# HOISTING THE VEHICLE

**1** The first step in hoisting a vehicle is to properly align the vehicle in the center of the stall.

**2** Most vehicles will be correctly positioned when the left front tire is centered on the tire pad.

**3** The arms can be moved in and out and most pads can be rotated to allow for many different types of vehicle construction.

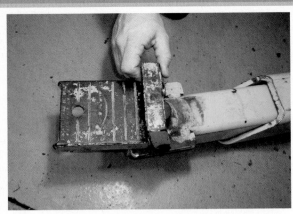

**4** Most lifts are equipped with short pad extensions that are often necessary to use to allow the pad to contact the frame of a vehicle without causing the arm of the lift to hit and damage parts of the body.

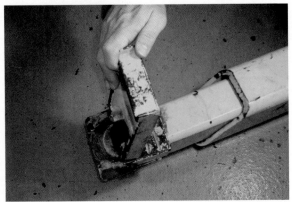

**5** Tall pad extensions can also be used to gain access to the frame of a vehicle. This position is needed to safely hoist many pickup trucks, vans, and sport utility vehicles.

**6** An additional extension may be necessary to hoist a truck or van equipped with running boards to give the necessary clearance.

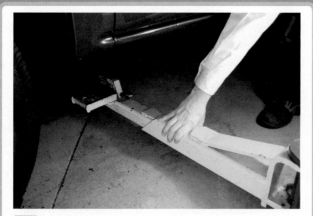

**7** Position the pads under the vehicle under the recommended locations.

**8** After being sure all pads are correctly positioned, use the electromechanical controls to raise the vehicle.

**9** With the vehicle raised 1 foot (30 cm) off the ground, push down on the vehicle to check to see if it is stable on the pads. If the vehicle rocks, lower the vehicle and reset the pads. The vehicle can be raised to any desired working level. Be sure the safety is engaged before working on or under the vehicle.

**10** If raising a vehicle without a frame, place the flat pads under the pinch weld seam to spread the load. If additional clearance is necessary, the pads can be raised as shown.

**11** When the service work is completed, the hoist should be raised slightly and the safety released before using the hydraulic lever to lower the vehicle.

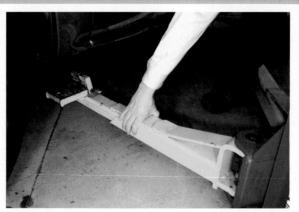

**12** After lowering the vehicle, be sure all arms of the lift are moved out of the way before driving the vehicle out of the work stall.

## SUMMARY

1. Bolts, studs, and nuts are commonly used as fasteners in the chassis. The sizes for fractional and metric threads are different and are not interchangeable. The grade is the rating of the strength of a fastener.

2. Whenever a vehicle is raised above the ground, it must be supported at a substantial section of the body or frame.

3. Wrenches are available in open end, box end, and combination open and box end.

4. An adjustable wrench should be used only where the proper size is not available.

5. Line wrenches are also called flare-nut wrenches, fitting wrenches, or tube-nut wrenches and are used to remove fuel or refrigerant lines.

6. Sockets are rotated by a ratchet or breaker bar, also called a flex handle.

7. Torque wrenches measure the amount of torque applied to a fastener.

8. Screwdriver types include straight blade (flat tip) and Phillips.

9. Hammers and mallets come in a variety of sizes and weights.

10. Pliers are a useful tool and are available in many different types, including slip-joint, multigroove, linesman's, diagonal, needle-nose, and locking pliers.

11. Other common hand tools include snap-ring pliers, files, cutters, punches, chisels, and hacksaws.

12. Hybrid electric vehicles should be depowered if any of the high-voltage components are going to be serviced.

## REVIEW QUESTIONS

1. List three precautions that must be taken whenever hoisting (lifting) a vehicle.

2. Describe how to determine the grade of a fastener, including how the markings differ between fractional and metric bolts.

3. List four items that are personal protective equipment (PPE).

4. List the types of fire extinguishers and their usage.

5. Why are wrenches offset 15 degrees?

6. What are the other names for a line wrench?

7. What are the standard automotive drive sizes for sockets?

8. Which type of screwdriver requires the use of a hammer or mallet?

9. What is inside a dead-blow hammer?

10. What type of cutter is available in left and right cutters?

## CHAPTER QUIZ

1. The correct location for the pads when hoisting or jacking the vehicle can often be found in the _____.
   a. Service manual
   b. Shop manual
   c. Owner's manual
   d. All of the above

2. For the best working position, the work should be _____.
   a. At neck or head level
   b. At knee or ankle level
   c. Overhead by about 1 foot
   d. At chest or elbow level

3. A high-strength bolt is identified by _____.
   a. A UNC symbol
   b. Lines on the head
   c. Strength letter codes
   d. The coarse threads

4. A fastener that uses threads on both ends is called a _____.
   a. Cap screw
   b. Stud
   c. Machine screw
   d. Crest fastener

5. When working with hand tools, always _____.
   a. Push the wrench—don't pull toward you
   b. Pull a wrench—don't push a wrench away from you

6. The proper term for Channel Locks® is _____.
   a. Vise Grips®
   b. Crescent wrench
   c. Locking pliers
   d. Multigroove adjustable pliers

7. The proper term for Vise Grips® is _____.
   a. Locking pliers
   b. Slip-joint pliers
   c. Side cuts
   d. Multigroove adjustable pliers

8. Two technicians are discussing torque wrenches. Technician A says that a torque wrench is capable of tightening a fastener with more torque than a conventional breaker bar or ratchet. Technician B says that a torque wrench should be calibrated regularly for the most accurate results. Which technician is correct?
   a. Technician A only
   b. Technician B only
   c. Both Technicians A and B
   d. Neither Technician A nor B

9. What type of screwdriver should be used if there is very limited space above the head of the fastener?
   a. Offset screwdriver
   b. Standard screwdriver
   c. Impact screwdriver
   d. Robertson screwdriver

10. What type of hammer is plastic coated, has a metal casing inside, and is filled with small lead balls?
   a. Dead-blow hammer
   b. Soft-blow hammer
   c. Sledgehammer
   d. Plastic hammer

# chapter 2

# ENVIRONMENTAL AND HAZARDOUS MATERIALS

**LEARNING OBJECTIVES:** **After studying this chapter, the reader should be able to:** • List the characteristics used by federal and state laws to determine if a material is hazardous. • Discuss asbestos hazards and asbestos handling guidelines. • Explain the storage and disposal of brake fluid, used oil, coolants, lead-acid batteries, used tires, and air-conditioning refrigerant oil. • Discuss the characteristics of hazardous solvents, fuel safety and storage, and airbag handling. • Explain the Hazardous Materials Identification Guide issued by the Environmental Protection Agency (EPA).

**KEY TERMS:** Aboveground storage tank (AGST) 38 • Asbestosis 37 • BCI 41 • CAA 36 • CFR 35 • EPA 35 • Hazardous waste material 35 • HEPA vacuum 37 • Mercury 42 • MSDS 36 • OSHA 35 • RCRA 36 • Right-to-know laws 36 • Solvent 37 • Underground storage tank (UST) 38 • Used oil 38 • WHMIS 36

## HAZARDOUS WASTE

### DEFINITION OF HAZARDOUS WASTE
**Hazardous waste materials** are chemicals, or components, that the shop no longer needs. These materials pose a danger to the environment and people if disposed of in ordinary garbage cans or sewers. However, no material is considered hazardous waste until the shop has finished using it and is ready to dispose of it.

### PERSONAL PROTECTIVE EQUIPMENT (PPE)
When handling hazardous waste material, one must always wear the proper protective clothing and equipment detailed in the right-to-know laws. This includes respirator equipment. Personal injury may result from improper clothing, equipment, and procedures when handling hazardous materials.

## FEDERAL AND STATE LAWS

### OCCUPATIONAL SAFETY AND HEALTH ACT
The United States Congress passed the **Occupational Safety and Health Act (OSHA)** in 1970. This legislation was designed to assist and encourage the citizens of the United States in their efforts to ensure the following:

- Safe and healthful working conditions by providing research, information, education, and training in the field of occupational safety and health

- Safe and healthful working conditions for working men and women by authorizing enforcement of the standards developed under the act

### EPA
The **Environmental Protection Agency (EPA)** publishes a list of hazardous materials that is included in the **Code of Federal Regulations (CFR).** The EPA considers waste hazardous if it is included on the EPA list of hazardous materials, or it has one or more of the following characteristics.

- **Reactive.** Any material that reacts violently with water or other chemicals is considered hazardous.

- **Corrosive.** If a material burns the skin or dissolves metals and other materials, a technician should consider it hazardous. A pH scale is used, with the number 7 indicating neutral. Pure water has a pH of 7. Lower numbers indicate an acidic solution and higher numbers indicate a caustic solution.

- **Toxic.** Materials are hazardous if they leak one or more of eight different heavy metals in concentrations greater than 100 times the primary drinking water standard.

- **Ignitable.** A liquid is hazardous if it has a flash point below 140°F (60°C), and a solid is hazardous if it ignites spontaneously.

- **Radioactive.** Any substance that emits measurable levels of radiation is radioactive.

### RIGHT-TO-KNOW LAWS
The **right-to-know laws** state that employees have a right to know when the materials they use at work are hazardous. The right-to-know laws started with the Hazard Communication Standard published by the Occupational Safety and Health Administration (OSHA) in 1983. Under the right-to-know laws, the employer has responsibilities regarding the handling of hazardous materials by their employees. All employees must be trained about the types of hazardous materials they will encounter in the

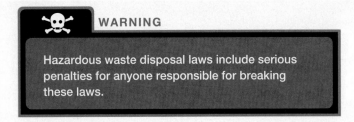
workplace. The employees must be informed about their rights under legislation regarding the handling of hazardous materials.

**MATERIAL SAFETY DATA SHEETS**  All hazardous materials must be properly labeled, and information about each hazardous material must be posted on **material safety data sheets (MSDS)**, now called simply safety data sheets (SDS), available from the manufacturer. In Canada, MSDS information is called **Workplace Hazardous Materials Information Systems (WHMIS).**

The employer has a responsibility to place MSDS information where they are easily accessible by all employees. The MSDS information provide the following information about the hazardous material: chemical name, physical characteristics, protective handling equipment, explosion/fire hazards, incompatible materials, health hazards, medical conditions aggravated by exposure, emergency and first-aid procedures, safe handling, and spill/leak procedures.

This information must be read and understood by the employee before handling the material. ● **SEE FIGURE 2–1.**

## RESOURCE CONSERVATION AND RECOVERY ACT

Federal and state laws control the disposal of hazardous waste materials and every shop employee must be familiar with these laws. Hazardous waste disposal laws include the **Resource Conservation and Recovery Act (RCRA).** This law states that hazardous material users are responsible for hazardous materials from the time they become a waste until the proper waste disposal is completed. The RCRA controls the following types of automotive waste:

- Paint and body repair products waste
- Solvents for parts and equipment cleaning
- Batteries and battery acid
- Mild acids used for metal cleaning and preparation
- Waste oil and engine coolants or antifreeze
- Air-conditioning refrigerants and oils
- Engine oil filters

**LOCKOUT/TAGOUT**  According to OSHA Title 29, Code of Federal Regulations (CPR), part 1910.147, machinery must be locked out to prevent injury to employees when maintenance or repair work is being performed. Any piece of equipment that should not be used must be tagged and the electrical power disconnected to prevent it from being used. Always read, understand, and follow all safety warning tags. ● **SEE FIGURE 2–2.**

**FIGURE 2–1** Safety data sheets (SDS), formerly known as material safety data sheets (MSDS), should be readily available for use by anyone in the area who may come into contact with hazardous materials.

**CLEAN AIR ACT**  Air-conditioning (A/C) systems and refrigerants are regulated by the **Clean Air Act (CAA),** Title VI, Section 609. Technician certification and service equipment is also regulated. Any technician working on automotive A/C systems must be certified. A/C refrigerants must not be released or vented into the atmosphere, and used refrigerants must be recovered.

## ASBESTOS HAZARDS

Friction materials such as brake and clutch linings often contain asbestos. While asbestos has been eliminated from most original equipment friction materials, the automotive service technician cannot know whether or not the vehicle being serviced is or is not equipped with friction materials containing asbestos. It is important that all friction materials be handled as if they do contain asbestos.

Asbestos exposure can cause scar tissue to form in the lungs. This condition is called **asbestosis.** It gradually causes increasing shortness of breath, and the scarring to the lungs is permanent.

Even low exposures to asbestos can cause *mesothelioma,* a type of fatal cancer of the lining of the chest or abdominal cavity. Asbestos exposure can also increase the risk of *lung cancer* as well as cancer of the voice box, stomach, and large intestine. It usually takes 15 to 30 years or more for cancer or asbestos lung scarring to show up after exposure. Scientists call this the *latency period.*

Government agencies recommend that asbestos exposure should be eliminated or controlled to the lowest level possible. These agencies have developed recommendations and standards that the automotive service technician and

**FIGURE 2–2** Tag identify that the power has been removed and service work is being done.

**FIGURE 2–3** All brakes should be moistened with water or solvent to help prevent brake dust from becoming airborne.

equipment manufacturer should follow. These U.S. federal agencies include the National Institute for Occupational Safety and Health (NIOSH), the Occupational Safety and Health Administration (OSHA), and the Environmental Protection Agency (EPA).

### ASBESTOS OSHA STANDARDS

The OSHA has established three levels of asbestos exposure. Any vehicle service establishment that does either brake or clutch work must limit employee exposure to asbestos to less than 0.2 fiber per cubic centimeter (cc) as determined by an air sample.

If the level of exposure to employees is greater than specified, corrective measures must be performed and a large fine may be imposed.

**NOTE: Research has found that worn asbestos fibers such as those from automotive brakes or clutches may not be as hazardous as first believed. Worn asbestos fibers do not have sharp flared ends that can latch onto tissue, but rather are worn down to a dust form that resembles talc. Grinding or sawing operations on unworn brake shoes or clutch discs *will* contain *harmful* asbestos fibers. To limit health damage, always use proper handling procedures while working around any component that may contain asbestos.**

### ASBESTOS EPA REGULATIONS

The federal Environmental Protection Agency (EPA) has established procedures for the removal and disposal of asbestos. The EPA procedures require that products containing asbestos be "wetted" to prevent the asbestos fibers from becoming airborne. According to the EPA, asbestos-containing materials can be disposed of as regular waste. Only when asbestos becomes airborne is it considered to be hazardous.

### ASBESTOS HANDLING GUIDELINES

The air in the shop area can be tested by a testing laboratory, but this can be expensive. Tests have determined that asbestos levels can easily be kept below the recommended levels by using a liquid, like water, or a special vacuum.

**NOTE: Even though asbestos is being removed from brake and clutch lining materials, the service technician cannot tell whether or not the old brake pads, shoes, or clutch discs contain asbestos. Therefore, to be safe, the technician should assume that all brake pads, shoes, or clutch discs contain asbestos.**

**HEPA VACUUM.** A special **high-efficiency particulate air (HEPA) vacuum** system has been proven to be effective in keeping asbestos exposure levels below 0.1 fiber per cubic centimeter.

**SOLVENT SPRAY.** Many technicians use an aerosol can of brake cleaning solvent to wet the brake dust and prevent it from becoming airborne. A **solvent** is a liquid that is used to dissolve dirt, grime, or solid particles. Commercial brake cleaners are available that use a concentrated cleaner that is mixed with water. ● **SEE FIGURE 2–3.** The waste liquid is filtered and, when dry, the filter can be disposed of as solid waste.

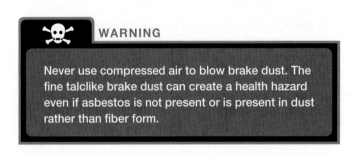

☠ WARNING

Never use compressed air to blow brake dust. The fine talclike brake dust can create a health hazard even if asbestos is not present or is present in dust rather than fiber form.

**DISPOSAL OF BRAKE DUST AND BRAKE SHOES.** The hazard of asbestos occurs when asbestos fibers are airborne. Once the asbestos has been wetted down, it is then considered to be solid waste rather than hazardous waste. Old brake shoes and pads should be enclosed, preferably in a plastic bag, to help prevent any of the brake material from becoming airborne. *Always follow current federal and local laws concerning disposal of all waste.*

## USED BRAKE FLUID

Most brake fluid is made from polyglycol, is water soluble, and can be considered hazardous if it has absorbed metals from the brake system.

### STORAGE AND DISPOSAL OF BRAKE FLUID

- Collect brake fluid in a container clearly marked to indicate that it is designated for that purpose.
- If the waste brake fluid is hazardous, be sure to manage it appropriately and use only an authorized waste receiver for its disposal.
- If the waste brake fluid is nonhazardous (such as old but unused), determine from your local solid waste collection provider what should be done for its proper disposal.
- Do not mix brake fluid with used engine oil.
- Do not pour brake fluid down drains or onto the ground.
- Recycle brake fluid through a registered recycler.

## USED OIL

**Used oil** is any petroleum-based or synthetic oil that has been used. During normal use, impurities such as dirt, metal scrapings, water, or chemicals can get mixed in with the oil. Eventually, this used oil must be replaced with virgin or re-refined oil. The EPA's used oil management standards include a three-pronged approach to determine if a substance meets the definition of *used oil.* To meet the EPA's definition of used oil, a substance must meet each of the following three criteria:

- **Origin.** The first criterion for identifying used oil is based on the oil's origin. Used oil must have been refined from crude oil or made from synthetic materials. Animal and vegetable oils are excluded from the EPA's definition of used oil.
- **Use.** The second criterion is based on whether and how the oil is used. Oils used as lubricants, hydraulic fluids, heat transfer fluids, and for other similar purposes are considered used oil. The EPA's definition also excludes products used as cleaning agents, as well as certain petroleum-derived products like antifreeze and kerosene.
- **Contaminants.** The third criterion is based on whether or not the oil is contaminated with either physical or chemical impurities. In other words, to meet the EPA's definition, used oil must become contaminated as a result of being used. This aspect of the EPA's definition includes residues and contaminants generated from handling, storing, and processing used oil.

NOTE: The release of only one gallon of used oil (a typical oil change) can make a million gallons of fresh water undrinkable.

**FIGURE 2–4** A typical aboveground oil storage tank.

If used oil is dumped down the drain and enters a sewage treatment plant, concentrations as small as 50 to 100 PPM (parts per million) in the waste water can foul sewage treatment processes. Never mix a listed hazardous waste, gasoline, waste water, halogenated solvent, antifreeze, or an unknown waste material with used oil. Adding any of these substances will cause the used oil to become contaminated, which classifies it as hazardous waste.

**STORAGE AND DISPOSAL OF USED OIL**   Once oil has been used, it can be collected, recycled, and used over and over again. An estimated 380 million gallons of used oil are recycled each year. Recycled used oil can sometimes be used again for the same job or can take on a completely different task. For example, used engine oil can be re-refined and sold at the store as engine oil or processed for furnace fuel oil. After collecting used oil in an appropriate container such as a 55-gallon steel drum, the material must be disposed of in one of two ways:

- Shipped offsite for recycling
- Burned in an onsite or offsite EPA-approved heater for energy recovery

Used oil must be stored in compliance with an existing **underground storage tank (UST)** or an **aboveground storage tank (AGST)** standard, or kept in separate containers. ● **SEE FIGURE 2–4.** Containers are portable receptacles, such as a 55-gallon steel drum.

**KEEP USED OIL STORAGE DRUMS IN GOOD CONDITION.**   This means that they should be covered, secured from vandals, properly labeled, and maintained in compliance with local fire codes. Frequent inspections for leaks, corrosion, and spillage are an essential part of container maintenance.

**NEVER STORE USED OIL IN ANYTHING OTHER THAN TANKS AND STORAGE CONTAINERS.**   Used oil may also be stored in units that are permitted to store regulated hazardous waste.

**USED OIL FILTER DISPOSAL REGULATIONS.** Used oil filters contain used engine oil that may be hazardous. Before an oil filter is placed into the trash or sent to be recycled, it must be drained using one of the following hot-draining methods approved by the EPA:

- Puncture the filter antidrainback valve or filter dome end and hot-drain for at least 12 hours
- Hot-drain and crushing
- Dismantling and hot draining
- Any other hot-draining method, which will remove all the used oil from the filter

*After the oil has been drained from the oil filter,* the filter housing can be disposed of in any of the following ways:

- Sent for recycling
- Picked up by a service contract company
- Disposed of in regular trash

## SOLVENTS

The major sources of chemical danger are liquid and aerosol brake cleaning fluids that contain chlorinated hydrocarbon solvents. Several other chemicals that do not deplete the ozone, such as heptane, hexane, and xylene, are now being used in nonchlorinated brake cleaning solvents. Some manufacturers are also producing solvents they describe as environmentally responsible, which are biodegradable and noncarcinogenic (noncancer causing).

There is no specific standard for physical contact with chlorinated hydrocarbon solvents or the chemicals replacing them. All contact should be avoided whenever possible. The law requires an employer to provide appropriate protective equipment and ensure proper work practices by an employee handling these chemicals.

### EFFECTS OF CHEMICAL POISONING  The effects of exposure to chlorinated hydrocarbon and other types of solvents can take many forms. Short-term exposure at low levels can cause symptoms such as:

- Headache
- Nausea

**FIGURE 2–5** Washing hands and removing jewelry are two important safety habits all service technicians should practice.

- Drowsiness
- Dizziness
- Lack of coordination
- Unconsciousness

It may also cause irritation of the eyes, nose, and throat and flushing of the face and neck. Short-term exposure to higher concentrations can cause liver damage with symptoms such as yellow jaundice or dark urine. Liver damage may not become evident until several weeks after the exposure.

### HAZARDOUS SOLVENTS AND REGULATORY STATUS
Most solvents are classified as hazardous wastes. Other characteristics of solvents include the following:

- Solvents with flash points below 60°C are considered flammable and, like gasoline, are federally regulated by the Department of Transportation (DOT).
- Solvents and oils with flash points above 60°C are considered combustible and, like engine oil, are also regulated by the DOT. All flammable items must be stored in a fireproof container. ● **SEE FIGURE 2–6.**

It is the responsibility of the repair shop to determine if its spent solvent is hazardous waste. Solvent reclaimers are available that clean and restore the solvent so it lasts indefinitely.

✚ SAFETY TIP

**Hand Safety**
Service technicians should wash their hands with soap and water after handling engine oil or differential or transmission fluids or wear protective rubber gloves. Another safety hint is that the service technician should not wear watches, rings, or other jewelry that could come in contact with electrical or moving parts of a vehicle. ● **SEE FIGURE 2–5.**

? FREQUENTLY ASKED QUESTION

**How can you tell if a solvent is hazardous?**
If a solvent or any of the ingredients of a product contains "fluor" or "chlor" then it is likely to be hazardous. Check the instructions on the label for proper use and disposal procedures.

FIGURE 2–6 Typical fireproof flammable storage cabinet.

FIGURE 2–7 Using a water-based cleaning system helps reduce the hazards from using strong chemicals.

**USED SOLVENTS** Used or spent solvents are liquid materials that have been generated as waste and may contain xylene, methanol, ethyl ether, and methyl isobutyl ketone (MIBK). These materials must be stored in OSHA-approved safety containers with the lids or caps closed tightly. Additional requirements include the following:

- Containers should be clearly labeled "Hazardous Waste" and the date the material was first placed into the storage receptacle should be noted.

- Labeling is not required for solvents being used in a parts washer.

- Used solvents will not be counted toward a facility's monthly output of hazardous waste if the vendor under contract removes the material.

- Used solvents may be disposed of by recycling with a local vendor, such as SafetyKleen®, to have the used solvent removed according to specific terms in the vendor agreement.

- Use aqueous-based (nonsolvent) cleaning systems to help avoid the problems associated with chemical solvents. ● **SEE FIGURE 2–7.**

FIGURE 2–8 Used antifreeze coolant should be kept separate and stored in a leakproof container until it can be recycled or disposed of according to federal, state, and local laws. Note that the storage barrel is placed inside another container to catch any coolant that may spill out of the inside barrel.

## COOLANT DISPOSAL

Coolant is a mixture of antifreeze and water. New antifreeze is not considered to be hazardous even though it can cause death if ingested. Used antifreeze may be hazardous due to dissolved metals from the engine and other components of the cooling system. These metals can include iron, steel, aluminum, copper, brass, and lead (from older radiators and

heater cores). Coolant should be disposed of in one of the following ways:

- Coolant should be recycled either onsite or offsite.

- Used coolant should be stored in a sealed and labeled container. ● **SEE FIGURE 2–8.**

- Used coolant can often be disposed of into municipal sewers with a permit. Check with local authorities and obtain a permit before discharging used coolant into sanitary sewers.

# LEAD-ACID BATTERY WASTE

About 70 million spent lead-acid batteries are generated each year in the United States alone. Lead is classified as a toxic metal and the acid used in lead-acid batteries is highly corrosive. The vast majority (95% to 98%) of these batteries are recycled through lead reclamation operations and secondary lead smelters for use in the manufacture of new batteries.

**BATTERY DISPOSAL** Used lead-acid batteries must be reclaimed or recycled in order to be exempt from hazardous waste regulations. Leaking batteries must be stored and transported as hazardous waste. Some states have more strict regulations that require special handling procedures and transportation. According to the **Battery Council International (BCI),** battery laws usually include the following rules:

1. Lead-acid battery disposal is prohibited in landfills or incinerators. Batteries are required to be delivered to a battery retailer, wholesaler, recycling center, or lead smelter.

2. All retailers of automotive batteries are required to post a sign that displays the universal recycling symbol and indicates the retailer's specific requirements for accepting used batteries.

3. Battery electrolyte contains sulfuric acid, which is a very corrosive substance capable of causing serious personal injury, such as skin burns and eye damage. In addition, the battery plates contain lead, which is highly poisonous. For this reason, disposing of batteries improperly can cause environmental contamination and lead to severe health problems.

**BATTERY HANDLING AND STORAGE** Batteries, whether new or used, should be kept indoors if possible. The storage location should be an area specifically designated for battery storage and must be well ventilated (to the outside). If outdoor storage is the only alternative, a sheltered and secured area with acid-resistant secondary containment is strongly recommended. It is also advisable that acid-resistant secondary containment be used for indoor storage. In addition, batteries should be placed on acid-resistant pallets and never stacked.

# FUEL SAFETY AND STORAGE

Gasoline is a very explosive liquid. The expanding vapors that come from gasoline are extremely dangerous. These vapors are present even in cold temperatures. Vapors formed in gasoline tanks on many vehicles are controlled, but vapors from gasoline storage may escape from the can, resulting in a hazardous situation. Therefore, place gasoline storage containers in a

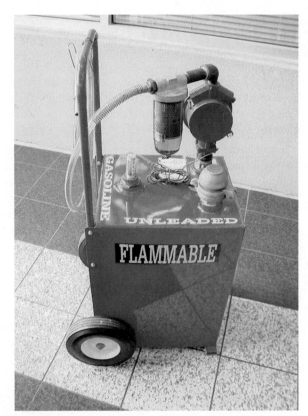

**FIGURE 2–9** This red gasoline container holds about 30 gallons of gasoline and is used to fill vehicles used for training.

well-ventilated space. Although diesel fuel is not as volatile as gasoline, the same basic rules apply to diesel fuel and gasoline storage. These rules include the following:

1. Use storage cans that have a flash-arresting screen at the outlet. These screens prevent external ignition sources from igniting the gasoline within the can when someone pours the gasoline or diesel fuel.

2. Use only a red approved gasoline container to allow for proper hazardous substance identification. ● **SEE FIGURE 2–9.**

3. Do not fill gasoline containers completely full. Always leave the level of gasoline at least 1 inch from the top of the container. This action allows expansion of the gasoline at higher temperatures. If gasoline containers are completely full, the gasoline will expand when the temperature increases. This expansion forces gasoline from the can and creates a dangerous spill. If gasoline or diesel fuel containers must be stored, place them in a designated storage locker or facility.

4. Never leave gasoline containers open, except while filling or pouring gasoline from the container.

5. Never use gasoline as a cleaning agent.

6. Always connect a ground strap to containers when filling or transferring fuel or other flammable products from one container to another to prevent static electricity that could result in explosion and fire. These ground wires prevent the buildup of a static electric charge that could result in a spark and disastrous explosion.

## AIRBAG HANDLING

Airbag modules are pyrotechnic devices that can be ignited if exposed to an electrical charge or if the body of the vehicle is subjected to a shock. Airbag safety should include the following precautions:

1. Disarm the airbag(s) if you will be working in the area where a discharged bag could make contact with any part of your body. Consult service information for the exact procedure to follow for the vehicle being serviced. The usual procedure is to deploy the airbag using a 12-volt power supply, such as a jump start box, using long wires to connect to the module to ensure a safe deployment.

2. Do not expose an airbag to extreme heat or fire.

3. Always carry an airbag pointing away from your body.

4. Place an airbag module facing upward.

5. Always follow the manufacturer's recommended procedure for airbag disposal or recycling, including the proper packaging to use during shipment.

6. Wear protective gloves if handling a deployed airbag.

7. Always wash your hands or body well if exposed to a deployed airbag. The chemicals involved can cause skin irritation and possible rash development.

## USED TIRE DISPOSAL

Used tires are an environmental concern because of several reasons, including the following:

1. In a landfill, they tend to "float" up through the other trash and rise to the surface.

2. The inside of tires traps and holds rainwater, which is a breeding ground for mosquitoes. Mosquito-borne diseases include encephalitis and dengue fever.

3. Used tires present a fire hazard and, when burned, create a large amount of black smoke that contaminates the air.

Used tires should be disposed of in one of the following ways:

1. Used tires can be reused until the end of their useful life.

2. Tires can be retreaded.

3. Tires can be recycled or shredded for use in asphalt.

4. Derimmed tires can be sent to a landfill (most landfill operators will shred the tires because it is illegal in many states to landfill whole tires).

5. Tires can be burned in cement kilns or other power plants where the smoke can be controlled.

6. A registered scrap tire handler should be used to transport tires for disposal or recycling.

**FIGURE 2–10** Air-conditioning refrigerant oil must be kept separated from other oils because it contains traces of refrigerant and must be treated as hazardous waste.

## AIR-CONDITIONING REFRIGERANT OIL DISPOSAL

Air-conditioning refrigerant oil contains dissolved refrigerant and is therefore considered to be hazardous waste. This oil must be kept separated from other waste oil or the entire amount of oil must be treated as hazardous. Used refrigerant oil must be sent to a licensed hazardous waste disposal company for recycling or disposal. ● **SEE FIGURE 2–10.**

**WASTE CHART** All automotive service facilities create some waste and, while most of it is handled properly, it is important that all hazardous and nonhazardous waste be accounted for and properly disposed. ● **SEE CHART 2–1** for a list of typical wastes generated at automotive shops, plus a checklist for keeping track of how these wastes are handled.

 **TECH TIP**

**Remove Components That Contain Mercury**

Some vehicles have a placard near the driver's side door that lists the components that contain the heavy metal, mercury. **Mercury** can be absorbed through the skin and is a heavy metal that, once absorbed by the body, does not leave. ● **SEE FIGURE 2–11.**

These components should be removed from the vehicle before the rest of the body is sent to be recycled to help prevent releasing mercury into the environment.

| WASTE STREAM | TYPICAL CATEGORY IF NOT MIXED WITH OTHER HAZARDOUS WASTE | IF DISPOSED IN LANDFILL AND NOT MIXED WITH A HAZARDOUS WASTE | IF RECYCLED |
|---|---|---|---|
| Used oil | Used oil | Hazardous waste | Used oil |
| Used oil filters | Nonhazardous solid waste, if completely drained | Nonhazardous solid waste, if completely drained | Used oil, if not drained |
| Used transmission fluid | Used oil | Hazardous waste | Used oil |
| Used brake fluid | Used oil | Hazardous waste | Used oil |
| Used antifreeze | Depends on characterization | Depends on characterization | Depends on characterization |
| Used solvents | Hazardous waste | Hazardous waste | Hazardous waste |
| Used citric solvents | Nonhazardous solid waste | Nonhazardous solid waste | Hazardous waste |
| Lead-acid automotive batteries | Not a solid waste if returned to supplier | Hazardous waste | Hazardous waste |
| Shop rags used for oil | Used oil | Depends on used oil characterization | Used oil |
| Shop rags used for solvent or gasoline spills | Hazardous waste | Hazardous waste | Hazardous waste |
| Oil spill absorbent material | Used oil | Depends on used oil characterization | Used oil |
| Spill material for solvent and gasoline | Hazardous waste | Hazardous waste | Hazardous waste |
| Catalytic converter | Not a solid waste if returned to supplier | Nonhazardous solid waste | Nonhazardous solid waste |
| Spilled or unused fuels | Hazardous waste | Hazardous waste | Hazardous waste |
| Spilled or unusable paints and thinners | Hazardous waste | Hazardous waste | Hazardous waste |
| Used tires | Nonhazardous solid waste | Nonhazardous solid waste | Nonhazardous solid waste |

**CHART 2-1**

Typical wastes generated at auto repair shops and typical category (Hazardous or Nonhazardous) by Disposal method.

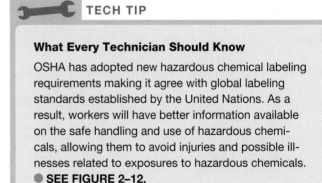

FIGURE 2-11 Placard near driver's door, including what devices in the vehicle contain mercury.

🔧 TECH TIP

**What Every Technician Should Know**

OSHA has adopted new hazardous chemical labeling requirements making it agree with global labeling standards established by the United Nations. As a result, workers will have better information available on the safe handling and use of hazardous chemicals, allowing them to avoid injuries and possible illnesses related to exposures to hazardous chemicals.
● SEE FIGURE 2-12.

| Health Hazard | Flame | Exclamation Mark |
|---|---|---|
| | | |
|  • Carcinogen<br>• Mutagenicity<br>• Reproductive Toxicity<br>• Respiratory Sensitizer<br>• Target Organ Toxicity<br>• Aspiration Toxicity |  • Flammables<br>• Pyrophorics<br>• Self-Heating<br>• Emits Flammable Gas<br>• Self-Reactives<br>• Organic Peroxides |  • Irritant (Skin and Eye)<br>• Skin Sensitizer<br>• Acute Toxicity<br>• Narcotic Effects<br>• Respiratory Tract Irritant<br>• Hazardous to Ozone Layer<br>  (Non-Mandatory) |
| **Gas Cylinder** | **Corrosion** | **Exploding Bomb** |
| | | |
|  • Gases Under Pressure |  • Skin Corrosion/Burns<br>• Eye Damage<br>• Corrosive to Metals |  • Explosives<br>• Self-Reactives<br>• Organic Peroxides |
| **Flame Over Circle** | **Environment**<br>**(Non-mandatory)** | **Skull and Crossbones** |
| | | |
|  • Oxidizers |  • Aquatic Toxicity |  • Acute Toxicity (fatal or toxic) |

**FIGURE 2–12** The OSHA global hazardous materials labels.

## SUMMARY

1. Hazardous materials include common automotive chemicals, liquids, and lubricants, especially those whose ingredients contain *chlor* or *fluor* in their name.

2. Right-to-know laws require that all workers have access to material safety data sheets (MSDS).

3. Asbestos fibers should be avoided and removed according to current laws and regulations.

4. Used engine oil contains metals worn from parts and should be handled and disposed of properly.

5. Solvents represent a serious health risk and should be avoided as much as possible.

6. Coolant should be disposed of properly or recycled.

7. Batteries are considered to be hazardous waste and should be discarded to a recycling facility.

## REVIEW QUESTIONS

1. List five common automotive chemicals or products that may be considered hazardous materials.

2. List five precautions to which every technician should adhere when working with automotive products and chemicals.

## CHAPTER QUIZ

1. Hazardous materials include all of the following *except* _____.
   - a. Engine oil
   - b. Asbestos
   - c. Water
   - d. Brake cleaner

2. To determine if a product or substance being used is hazardous, consult _____.
   - a. A dictionary
   - b. An MSDS
   - c. SAE standards
   - d. EPA guidelines

3. Exposure to asbestos dust can cause what condition?
   - a. Asbestosis
   - b. Mesothelioma
   - c. Lung cancer
   - d. All of the above are possible

4. Wetted asbestos dust is considered to be _____.
   - a. Solid waste
   - b. Hazardous waste
   - c. Toxic
   - d. Poisonous

5. An oil filter should be hot drained for how long before disposing of the filter?
   - a. 30 to 60 minutes
   - b. 4 hours
   - c. 8 hours
   - d. 12 hours

6. Used engine oil should be disposed of by all *except* which of the following methods?
   - a. Disposed of in regular trash
   - b. Shipped offsite for recycling
   - c. Burned onsite in a waste oil-approved heater
   - d. Burned offsite in a waste oil-approved heater

7. All of the following are the proper ways to dispose of a drained oil filter *except* _____.
   - a. Sent for recycling
   - b. Picked up by a service contract company
   - c. Disposed of in regular trash
   - d. Considered to be hazardous waste and disposed of accordingly

8. Which act or organization regulates air-conditioning refrigerant?
   - a. Clean Air Act (CAA)
   - b. MSDS
   - c. WHMIS
   - d. Code of Federal Regulations (CFR)

9. Gasoline should be stored in approved containers that include what color(s)?
   - a. A red container with yellow lettering
   - b. A red container
   - c. A yellow container
   - d. A yellow container with red lettering

10. What automotive devices may contain mercury?
   - a. Rear seat video displays
   - b. Navigation displays
   - c. HID headlights
   - d. All of the above

# chapter 3

# GASOLINE ENGINE OPERATION AND SPECIFICATIONS

**LEARNING OBJECTIVES:** **After studying this chapter, the reader should be able to:** • Discuss engine construction and torque and power of an engine. • Identify the various engine parts and systems. • Explain the four-stroke cycle engine operation. • Discuss engine classification and construction. • Discuss engine bore and stroke measurements. • Define compression ratio, torque, and horsepower.

**KEY TERMS:** Block 46 • Bore 54 • Bottom dead center (BDC) 49 • Boxer 49 • Cam-in-block design 51 • Camshaft 51 • Combustion 46 • Combustion chamber 46 • Compression ratio (CR) 55 • Connecting rod 49 • Crankshaft 49 • Cycle 49 • Cylinder 49 • Displacement 55 • Double overhead camshaft (DOHC) 51 • Exhaust valve 49 • External combustion engine 46 • Four-stroke cycle 49 • Intake valve 49 • Internal combustion engine 46 • Mechanical force 46 • Mechanical power 46 • Naturally aspirated 53 • Nonprincipal end 53 • Oil galleries 48 • Overhead valve (OHV) 51 • Pancake 49 • Piston stroke 49 • Principal end 53 • Pushrod engine 51 • Rotary engine 52 • Single overhead camshaft (SOHC) 51 • Stroke 54 • Supercharger 53 • Top dead center (TDC) 49 • Turbocharger 53 • Wankel engine 52

## PURPOSE AND FUNCTION

The purpose and function of an engine is to convert the heat energy of burning fuel into mechanical energy. In a typical vehicle, mechanical energy is then used to perform the following:

- Propel the vehicle
- Power the air-conditioning system and power steering
- Produce electrical power for use throughout the vehicle

## ENERGY AND POWER

Engines use energy to produce power. The chemical energy in fuel is converted to heat energy by the burning of the fuel at a controlled rate. This process is called **combustion.** If engine combustion occurs within the power chamber, the engine is called an **internal combustion engine.**

**NOTE: An external combustion engine** burns fuel outside of the engine itself, such as a steam engine.

Engines used in automobiles are internal combustion heat engines. They convert the chemical energy of the gasoline into heat within a power chamber that is called a **combustion chamber.** Heat energy released in the combustion chamber raises the temperature of the combustion gases within the chamber. The increase in gas temperature causes the pressure of the gases to increase. The pressure developed within the combustion chamber is applied to the head of a piston to produce a usable **mechanical force,** which is then converted into useful **mechanical power.**

## ENGINE CONSTRUCTION OVERVIEW

**BLOCK** All automotive and truck engines are constructed using a solid frame, called a **block.** A block is constructed of cast iron or aluminum and provides the foundation for most of the engine components and systems. The block is cast and then machined to very close tolerances to allow other parts to be installed.

**ROTATING ASSEMBLY** Pistons are installed in the block and move up and down during engine operation. Pistons are connected to connecting rods, which connect the pistons to the crankshaft. The crankshaft converts the up-and-down motion of the piston to rotary motion, which is then transmitted to the drive wheels and propels the vehicle. ● **SEE FIGURE 3–1.**

**CYLINDER HEADS** All engines use a cylinder head to seal the top of the cylinders, which are in the engine block. The cylinder head on overhead valve (OHV) engines contain both intake

FIGURE 3–1 The rotating assembly for a V-8 engine that has eight pistons and connecting rods and one crankshaft.

FIGURE 3–2 A cylinder head with four valves per cylinder: two intake valves (larger) and two exhaust valves (smaller).

FIGURE 3–3 The coolant temperature is controlled by the thermostat, which opens and allows coolant to flow to the radiator when the temperature reaches the rating temperature of the thermostat.

valves that allow air and fuel into the cylinder and exhaust valves, which allow the hot gases left over to escape from the engine. Cylinder heads are constructed of cast iron or aluminum and are then machined for the valves and other valve-related components. ● SEE FIGURE 3–2.

# ENGINE PARTS AND SYSTEMS

### INTAKE AND EXHAUST MANIFOLDS
Air and fuel enter the engine through an intake manifold and exit the engine through the exhaust manifold. Intake manifolds operate cooler than exhaust manifolds and are therefore constructed of nylon-reinforced plastic or aluminum. Exhaust manifolds must be able to withstand hot exhaust gases, so most are constructed from cast iron or steel tubing.

### COOLING SYSTEM
All engines must have a cooling system to control engine temperatures. While some older engines were air cooled, all current production passenger vehicle engines are cooled by circulating antifreeze coolant through passages in the block and cylinder head. The coolant picks up the heat from the engine and after the thermostat opens, the water pump circulates the coolant through the radiator where the excess heat is released to the outside air, cooling the coolant. The coolant is continuously circulated through the cooling system and the temperature is controlled by the thermostat. ● SEE FIGURE 3–3.

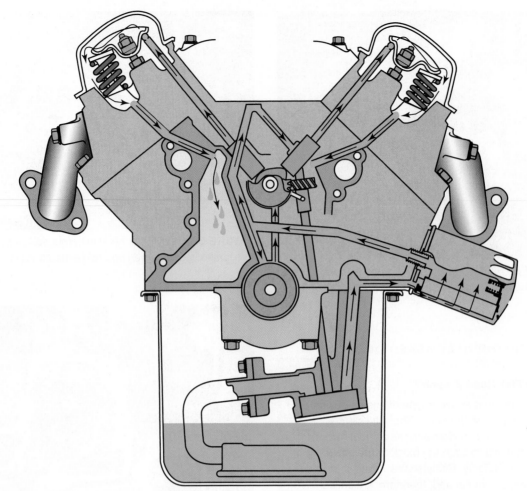

**FIGURE 3–4** A typical lubrication system, showing the oil pan, oil pump, oil filter, and oil passages.

**LUBRICATION SYSTEM** All engines contain moving and sliding parts that must be kept lubricated to reduce wear and friction. The oil pan, bolted to the bottom of the engine block, holds 4 to 7 quarts (4 to 7 liters) of oil. An oil pump, which is driven by the engine, forces the oil through the oil filter and then into passages in the crankshaft and block. These passages are called **oil galleries.** The oil is also forced up to the valves and then falls down through openings in the cylinder head and block, then back into the oil pan. ● **SEE FIGURE 3–4:**

**FUEL SYSTEM AND IGNITION SYSTEM** All engines require both a fuel system to supply fuel to the cylinders and an ignition system to ignite the air–fuel mixture in the cylinders. The fuel system includes the following components:

- Fuel tank, where fuel is stored and where most fuel pumps are located
- Fuel filter and lines, which transfer the fuel for the fuel tank to the engine
- Fuel injectors, which spray fuel into the intake manifold or directly into the cylinder, depending on the type of system used

The ignition system is designed to take 12 volts from the battery and convert it to 5,000 to 40,000 volts needed to jump the gap of a spark plug. Spark plugs are threaded into the cylinder head of each cylinder, and when the spark occurs, it ignites the air–fuel mixture in the cylinder, creating pressure and forcing the piston down in the cylinder. The following components are part of the ignition system:

- **Spark plugs.** Provide an air gap inside the cylinder where a spark occurs to start combustion
- **Sensor(s).** Includes crankshaft position (CKP) and camshaft position (CMP) sensors, used by the powertrain control module (PCM) to trigger the ignition coil(s) and the fuel injectors
- **Ignition coils.** Increase battery voltage to 5,000 to 40,000 volts
- **Ignition control module (ICM).** Controls when the spark plug fires
- **Associated wiring.** Electrically connects the battery, ICM, coil, and spark plugs

# FOUR-STROKE CYCLE OPERATION

**PRINCIPLES** The first **four-stroke cycle** engine was developed by a German engineer, Nickolaus Otto, in 1876. Most automotive engines use the four-stroke cycle of events. The process begins by the starter motor rotating the engine until combustion takes place. The four-stroke cycle is repeated for each cylinder of the engine. ● **SEE FIGURE 3–5.**

A piston that moves up and down, or reciprocates, in a **cylinder** can be seen in Figure 3–5. The piston is attached to a **crankshaft** with a **connecting rod.** This arrangement allows the piston to reciprocate (move up and down) in the cylinder as the crankshaft rotates. ● **SEE FIGURE 3–6.**

**OPERATION** Engine cycles are identified by the number of piston strokes required to complete the cycle. A **piston stroke** is a one-way piston movement either from top to bottom or bottom to top of the cylinder. During one stroke, the crankshaft rotates 180 degrees (1/2 revolution). A **cycle** is a complete series of events that continually repeats. Most automobile engines use a four-stroke cycle:

- **Intake stroke.** The **intake valve** is open and the piston inside the cylinder travels downward, drawing a mixture of air and fuel into the cylinder. The crankshaft rotates 180 degrees from **top dead center (TDC)** to **bottom dead center (BDC)** and the camshaft rotates 90 degrees.

- **Compression stroke.** As the engine continues to rotate, the intake valve closes and the piston moves upward in the cylinder, compressing the air–fuel mixture. The crankshaft rotates 180 degrees from bottom dead center (BDC) to top dead center (TDC) and the camshaft rotates 90 degrees.

- **Power stroke.** When the piston gets near the top of the cylinder, the spark at the spark plug ignites the air–fuel mixture, which forces the piston downward. The crankshaft rotates 180 degrees from top dead center (TDC) to bottom dead center (BDC) and the camshaft rotates 90 degrees. The combustion pressure developed in the combustion chamber at the correct time will push the piston downward to rotate the crankshaft.

- **Exhaust stroke.** The engine continues to rotate, and the piston again moves upward in the cylinder. The exhaust valve opens, and the piston forces the residual burned gases out of the **exhaust valve** and into the exhaust manifold and exhaust system. The crankshaft rotates 180 degrees from bottom dead center (BDC) to top dead center (TDC) and the camshaft rotates 90 degrees.

This sequence repeats as the engine rotates. To stop the engine, the electricity to the ignition system is shut off by the ignition switch, which stops the spark to the spark plugs.

**THE 720-DEGREE CYCLE** Each cycle (four strokes) of events requires that the engine crankshaft make two complete revolutions, or 720 degrees (360 degrees $\times$ 2 = 720 degrees). Each stroke of the cycle requires that the crankshaft rotate 180 degrees. The greater the number of cylinders, the closer together the power strokes of the individual cylinders will occur. The number of degrees that the crankshaft rotates between power strokes can be expressed as an angle. To find the angle between cylinders of an engine, divide the number of cylinders into 720 degrees:

Angle with 3 cylinders: 720/3 = 240 degrees

Angle with 4 cylinders: 720/4 = 180 degrees

Angle with 5 cylinders: 720/5 = 144 degrees

Angle with 6 cylinders: 720/6 = 120 degrees

Angle with 8 cylinders: 720/8 = 90 degrees

Angle with 10 cylinders: 720/10 = 72 degrees

This means that in a 4-cylinder engine, a power stroke occurs at every 180 degrees of the crankshaft rotation (every 1/2 rotation). A V-8 is a much smoother operating engine because a power stroke occurs twice as often (every 90 degrees of crankshaft rotation).

# ENGINE CLASSIFICATION AND CONSTRUCTION

Engines are classified by several characteristics, including:

- **Number of strokes.** Most automotive engines use the four-stroke cycle.

- **Cylinder arrangement.** An engine with more cylinders is smoother operating because the power pulses produced by the power strokes are more closely spaced. An inline engine places all cylinders in a straight line. The 4-, 5-, and 6-cylinder engines are commonly manufactured inline engines. A V-type engine, such as a V-6 or V-8, has the number of cylinders split and built into a V shape. ● **SEE FIGURE 3–7.** Horizontally opposed 4- and 6-cylinder engines have two banks of cylinders that are horizontal, resulting in a low engine. This style of engine is used in Porsche and Subaru engines, and is often called the **boxer** or **pancake** engine design. ● **SEE FIGURE 3–8.**

- **Longitudinal and transverse mounting.** Engines may be mounted either parallel with the length of the vehicle (longitudinally) or crosswise (transversely). ● **SEE FIGURES 3–9 AND 3–10.** The same engine may be mounted in various vehicles in either direction.

**NOTE: Although it might be possible to mount an engine in different vehicles both longitudinally and transversely, the engine component parts may *not* be interchangeable. Differences can include different engine blocks and crankshafts, as well as different water pumps.**

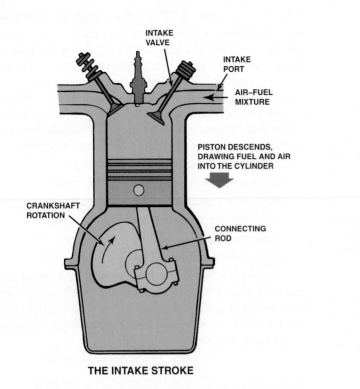

**THE INTAKE STROKE**

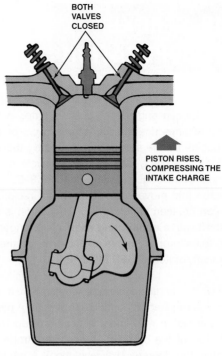

**THE COMPRESSION STROKE**

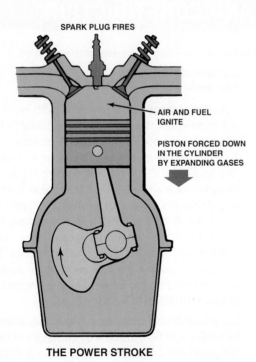

**THE POWER STROKE**

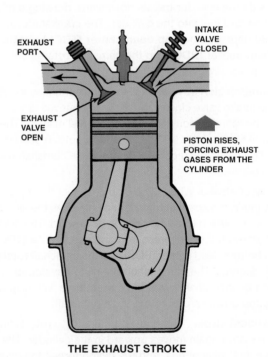

**THE EXHAUST STROKE**

**FIGURE 3–5** The downward movement of the piston draws the air–fuel mixture into the cylinder through the intake valve on the intake stroke. On the compression stroke, the mixture is compressed by the upward movement of the piston with both valves closed. Ignition occurs at the beginning of the power stroke, and combustion drives the piston downward to produce power. On the exhaust stroke, the upward-moving piston forces the burned gases out the open exhaust valve.

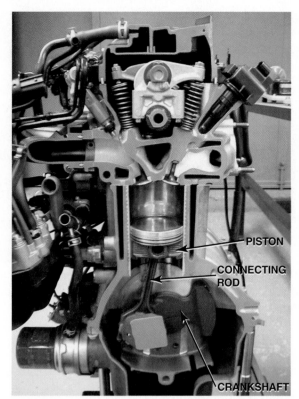

FIGURE 3–6 Cutaway of an engine showing the cylinder, piston, connecting rod, and crankshaft.

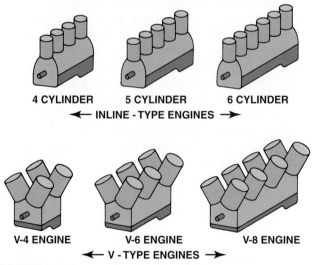

4 CYLINDER    5 CYLINDER    6 CYLINDER

← INLINE - TYPE ENGINES →

V-4 ENGINE    V-6 ENGINE    V-8 ENGINE

← V - TYPE ENGINES →

FIGURE 3–7 Automotive engine cylinder arrangements.

- **Valve and camshaft number and location.** The number of valves per cylinder and the number and location of camshafts are major factors in engine operation. A typical older-model engine uses one intake valve and one exhaust valve per cylinder. Many newer engines use two intake and two exhaust valves per cylinder. The valves are opened by a **camshaft.** Some engines use one camshaft for the intake valves and a separate camshaft for the exhaust valves. When the camshaft is located in the

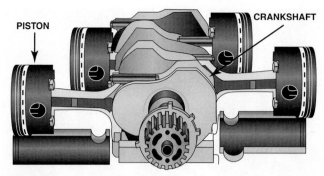

FIGURE 3–8 A horizontally opposed engine design helps to lower the vehicle's center of gravity.

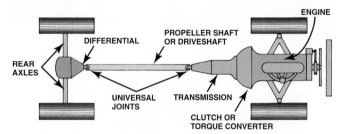

FIGURE 3–9 A longitudinally mounted engine drives the rear wheels through a transmission, driveshaft, and differential assembly.

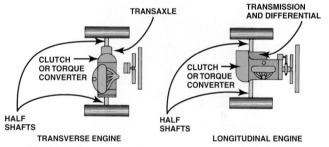

TRANSVERSE ENGINE    LONGITUDINAL ENGINE

FIGURE 3–10 Two types of front-engine, front-wheel drive mountings. A transaxle is used in most front-wheel vehicles and the longitudinal engine is used by Audi and some other manufacturers.

block, the valves are operated by lifters, pushrods, and rocker arms.

This type of engine can be called:

- A **pushrod engine**
- **Cam-in-block design**
- **Overhead valve (OHV),** because an overhead valve engine has the valves located in the cylinder head
  ● **SEE FIGURE 3–11.**

When one overhead camshaft is used, the design is called a **single overhead camshaft (SOHC)** design. When two overhead camshafts are used, the design is called a **double overhead camshaft (DOHC)** design.
  ● **SEE FIGURES 3–12 AND 3–13.**

FIGURE 3–11 Cutaway of an overhead valve (OHV) V-8 engine showing the lifters, pushrods, roller rocker arms, and valves.

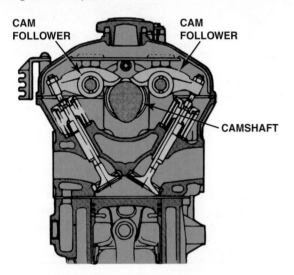

SINGLE OVERHEAD CAMSHAFT

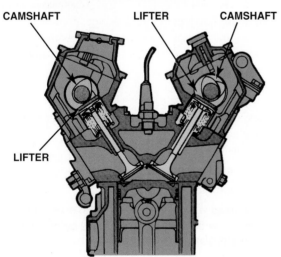

DOUBLE OVERHEAD CAMSHAFT

FIGURE 3–12 SOHC engines usually require additional components, such as a rocker arm, to operate all of the valves. DOHC engines often operate the valves directly.

FIGURE 3–13 A DOHC engine uses a camshaft for the intake valves and a separate camshaft for the exhaust valves in each cylinder head.

**?** FREQUENTLY ASKED QUESTION

**What Is a Rotary Engine?**

A successful alternative engine design is the **rotary engine,** also called the **Wankel engine** after its inventor, Felix Heinrich Wankel (1902–1988), a German inventor. The Mazda RX-7 and RX-8 represent the only long-term use of the rotary engine. The rotating combustion chamber engine runs very smoothly, and it produces high power for its size and weight.

The basic rotating combustion chamber engine has a triangular rotor turning in a housing. The housing is in the shape of a geometric figure called a two-lobed epitrochoid. A seal on each corner, or apex, of the rotor is in constant contact with the housing, so the rotor must turn with an eccentric motion. This means that the center of the rotor moves around the center of the engine. The eccentric motion can be seen in ●**FIGURE 3–14.**

**NOTE: A V-type engine uses two banks or rows of cylinders. An SOHC design, therefore, uses two camshafts but only one camshaft per bank (row) of cylinders. A DOHC V-6, therefore, has four camshafts, two for each bank.**

■ **Type of fuel.** Most engines operate on gasoline, whereas some engines are designed to operate on ethanol (E85), methanol (M85), natural gas, propane, or diesel fuel.

■ **Cooling method.** Most engines are liquid cooled, but some older models were air cooled. Air-cooled engines, such as the original VW Beetle, could not meet exhaust emission standards.

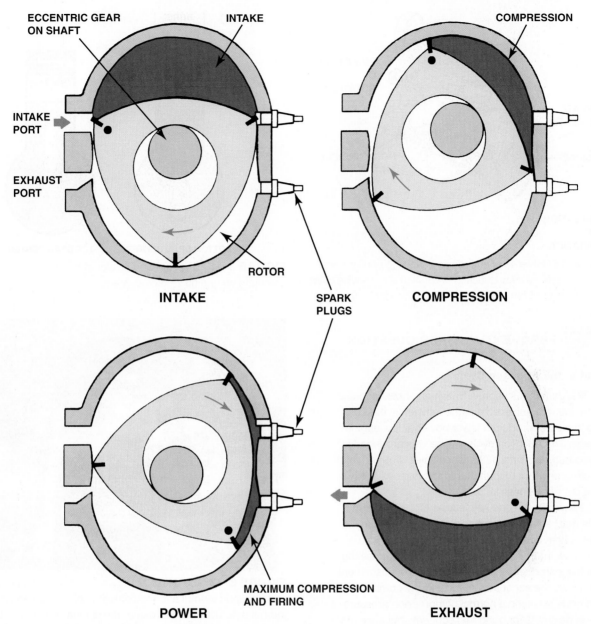

**ECCENTRIC GEAR ON SHAFT**
**INTAKE**
**INTAKE PORT**
**EXHAUST PORT**
**ROTOR**
**INTAKE**
**SPARK PLUGS**
**COMPRESSION**
**COMPRESSION**
**POWER**
**MAXIMUM COMPRESSION AND FIRING**
**EXHAUST**

**FIGURE 3–14** A rotary engine operates on the four-stroke cycle but uses a rotor instead of a piston and crankshaft to achieve intake, compression, power, and exhaust strokes.

- **Type of induction pressure.** If atmospheric air pressure is used to force the air–fuel mixture into the cylinders, the engine is called **naturally aspirated.** Some engines use a **turbocharger** or **supercharger** to force the air–fuel mixture into the cylinder for even greater power.

**ENGINE ROTATION DIRECTION** The SAE standard for automotive engine rotation is counterclockwise (CCW) as viewed from the flywheel end (clockwise as viewed from the front of the engine). The flywheel end of the engine is the end to

which the power is applied to drive the vehicle. This is called the **principal end** of the engine. The **nonprincipal end** of the engine is opposite the principal end and is generally referred to as the *front* of the engine, where the accessory belts are used. ● **SEE FIGURE 3–15.**

Therefore, in most rear-wheel-drive vehicles, the engine is mounted longitudinally with the principal end at the rear of the engine. Most transversely mounted engines also adhere to the same standard for direction of rotation. Many Honda engines, and some marine applications, may differ from this standard.

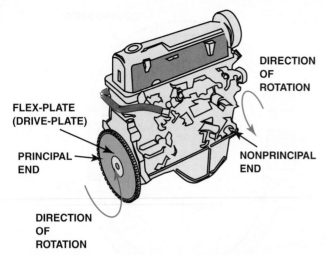

FIGURE 3–15 Inline 4-cylinder engine showing principal and nonprincipal ends. Normal direction of rotation is clockwise (CW) as viewed from the front or accessory belt (nonprincipal) end.

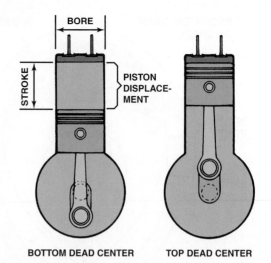

FIGURE 3–16 The bore and stroke of pistons are used to calculate an engine's displacement.

? FREQUENTLY ASKED QUESTION

**What Is the Atkinson Cycle?**

In 1882, James Atkinson, a British engineer, invented an engine that achieved a higher efficiency than the Otto cycle but produced lower power at low engine speeds. The Atkinson cycle engine was produced in limited numbers until 1890, when sales dropped, and the company that manufactured the engines finally went out of business in 1893.

However, the one key feature of the Atkinson cycle that remains in use today is that the intake valve is held open longer than normal to allow a reverse flow into the intake manifold. This reduces the effective compression ratio and engine displacement and allows the expansion to exceed the compression ratio while retaining a normal compression pressure. This is desirable for good fuel economy because the compression ratio in a spark ignition engine is limited by the octane rating of the fuel used, while a high expansion delivers a longer power stroke and reduces the heat wasted in the exhaust. This increases the efficiency of the engine because more work is being achieved. The Atkinson cycle engine design is commonly used in hybrid electric vehicles.

# ENGINE MEASUREMENT

**BORE** The diameter of a cylinder is called the **bore.** The larger the bore, the greater the area on which the gases have to work. Pressure is measured in units, such as pounds per square inch (PSI). The greater the area (in square inches), the higher the force exerted by the pistons to rotate the crankshaft. ● SEE FIGURE 3–16.

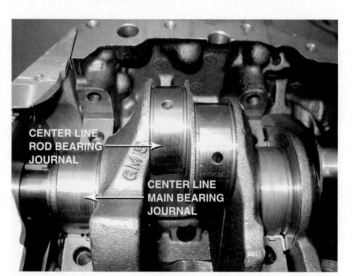

FIGURE 3–17 The distance between the centerline of the main bearing journal and the centerline of the connecting rod journal determines the stroke of the engine. This photo is a little unusual because it shows a V-6 with a splayed crankshaft used to even out the impulses on a 90-degree, V-6 engine design.

**STROKE** The **stroke** of an engine is the distance the piston travels from top dead center (TDC) to bottom dead center (BDC). This distance is determined by the throw of the crankshaft. The throw is the distance from the centerline of the crankshaft to the centerline of the crankshaft rod journal. The throw is one-half of the stroke. ● SEE FIGURE 3–17.

The longer this distance is, the greater the amount of air–fuel mixture that can be drawn into the cylinder. The more air–fuel mixture inside the cylinder, the more force will result when the mixture is ignited.

NOTE: Changing the connecting rod length does *not* change the stroke of an engine. Changing the connecting rod only changes the position of the piston in the cylinder. Only the crankshaft determines the stroke of an engine.

## DISPLACEMENT

Engine size is described as displacement. **Displacement** is the cubic inch (cu. in.) or cubic centimeter (cc) volume displaced or how much air is moved by all of the pistons. A liter (L) is equal to 1,000 cubic centimeters; therefore, most engines today are identified by their displacement in liters:

1 L = 1,000 cc

1 L = 61 cu. in.

1 cu. in. = 16.4 cc

## CONVERSION

- To convert cubic inches to liters, divide cubic inches by 61.02:

$$\text{Liters} = \frac{\text{Cubic inches}}{61.02}$$

- To convert liters into cubic inches, multiply by 61.02:

$$\text{Cubic inches} = \text{Liters} \times 61.02$$

## CALCULATING CUBIC INCH DISPLACEMENT

The formula to calculate the displacement of an engine is basically the formula for determining the volume of a cylinder multiplied by the number of cylinders.

The formula is:

$$\text{Cubic inch displacement} = \pi \text{ (pi)} \times R^2 \times \text{Stroke} \times \text{Number of cylinders}$$

R = Radius of the cylinder or one-half of the bore.

The $\pi R^2$ part is the formula for the area of a circle.

Applying the formula to a 6-cylinder engine:

- Bore = 4.000 inches
- Stroke = 3.000 inches
- $\pi$ = 3.14
- R = 2 inches
- $R^2$ = 4 ($2^2$ or $2 \times 2$)

Cubic inches = 3.14 × 4 ($R^2$) × 3 (stroke) × 6 (number of cylinders).

Cubic inches = 226 cubic inches

Because 1 cubic inch equals 16.4 cubic centimeters, this engine displacement equals 3,706 cubic centimeters or, rounded to 3,700 cubic centimeters, 3.7 liters. ● **SEE CHART 3–1** for an example of engine sizes for a variety of bore and stroke measurements.

## ENGINE SIZE CONVERSION

Many vehicle manufacturers will round the displacement so the calculated cubic inch displacement may not agree with the published displacement value. ● **SEE CHART 3–2.**

### TECH TIP

#### How Fast Can an Engine Rotate?

Most passenger vehicle engines are designed to rotate at low speed for the following reasons:

- Maximum efficiency is achieved at low engine speed. A diesel engine used in a large ship, for example, will rotate at about 100 RPM for maximum efficiency.
- Piston ring friction is the highest point of friction in the engine. The slower the engine speed, the less loss to friction from the piston rings.

However, horsepower is what is needed to get a vehicle down the road quickly. Horsepower is torque times engine speed divided by 5,252. Therefore, a high engine speed usually indicates a high horsepower. For example, a Formula 1 race car is limited to 2.4 liter V-8 but uses a 1.6 inches (40 mm) stroke. This extremely short stroke means that the engine can easily achieve the upper limit allowed by the rules of 18000 RPM while producing over 700 horsepower.

The larger the engine, the more power the engine is capable of producing. Several sayings are often quoted about engine size:

"There is no substitute for cubic inches."

"There is no replacement for displacement."

Although a large engine generally uses more fuel, making an engine larger is often the easiest way to increase power.

# COMPRESSION RATIO

**DEFINITION** Compression ratio (CR) is the ratio of the difference in the cylinder volume when the piston is at the bottom of the stroke to the volume in the cylinder above the piston when the piston is at the top of the stroke. The compression ratio of an engine is an important consideration when rebuilding or repairing an engine. ● **SEE FIGURE 3–18.**

| If Compression Is Lower | If Compression Is Higher |
|---|---|
| Lower power | Higher power possible |
| Poorer fuel economy | Better fuel economy possible |
| Easier engine cranking | Harder to crank engine, especially when hot |
| More advanced ignition timing possible without spark knock (detonation) | Less ignition timing required to prevent spark knock (detonation) |

| V-8 ENGINE | | | | | |
|---|---|---|---|---|---|
| Stroke | 3.50 | 3.75 | 3.875 | 4.00 | 4.125 |
| Bore | Cubic Inches | Cubic Inches | Cubic Inches | Cubic Inches | Cubic Inches |
| 3.00 | 199 | 212 | 219 | 226 | 233 |
| 3.125 | 214 | 229 | 237 | 244 | 252 |
| 3.250 | 232 | 249 | 257 | 265 | 274 |
| 3.375 | 251 | 269 | 277 | 286 | 295 |
| 3.500 | 269 | 288 | 298 | 308 | 317 |
| 3.625 | 288 | 309 | 319 | 330 | 339 |
| 3.750 | 309 | 332 | 343 | 354 | 365 |
| 3.875 | 331 | 354 | 366 | 378 | 390 |
| 4.00 | 352 | 377 | 389 | 402 | 414 |
| 4.125 | 373 | 399 | 413 | 426 | 439 |
| 6-CYLINDER ENGINE | | | | | |
| Stroke | 3.50 | 3.75 | 3.875 | 4.00 | 4.125 |
| Bore | Cubic Inches | Cubic Inches | Cubic Inches | Cubic Inches | Cubic Inches |
| 3.00 | 148 | 159 | 164 | 169 | 175 |
| 3.125 | 161 | 172 | 178 | 184 | 190 |
| 3.250 | 174 | 186 | 193 | 199 | 205 |
| 3.375 | 188 | 201 | 208 | 215 | 222 |
| 3.500 | 202 | 216 | 223 | 228 | 238 |
| 3.625 | 216 | 232 | 239 | 247 | 255 |
| 3.750 | 232 | 249 | 257 | 265 | 273 |
| 3.875 | 248 | 266 | 275 | 283 | 292 |
| 4.00 | 264 | 283 | 292 | 301 | 311 |
| 4.125 | 280 | 299 | 309 | 319 | 329 |
| 4-CYLINDER ENGINE | | | | | |
| Stroke | 3.50 | 3.75 | 3.875 | 4.00 | 4.125 |
| Bore | Cubic Inches | Cubic Inches | Cubic Inches | Cubic Inches | Cubic Inches |
| 3.00 | 99 | 106 | 110 | 113 | 117 |
| 3.125 | 107 | 115 | 119 | 123 | 126 |
| 3.250 | 116 | 124 | 129 | 133 | 137 |
| 3.375 | 125 | 134 | 139 | 143 | 148 |
| 3.500 | 135 | 144 | 149 | 152 | 159 |
| 3.625 | 144 | 158 | 160 | 165 | 170 |
| 3.750 | 155 | 166 | 171 | 177 | 182 |
| 3.875 | 165 | 177 | 183 | 189 | 195 |
| 4.00 | 176 | 188 | 195 | 201 | 207 |
| 4.125 | 186 | 200 | 206 | 213 | 220 |

**CHART 3-1**

To find the cubic inch displacement, find the bore that is closest to the actual value, then go across to the closest stroke value.

| LITERS TO CUBIC INCHES | | | | | |
|---|---|---|---|---|---|
| Liters | Cubic Inches | Liters | Cubic Inches | Liters | Cubic Inches |
| 1.0 | 61 | 3.2 | 196 | 5.4 | 330 |
| 1.3 | 79 | 3.3 | 200 / 201 | 5.7 | 350 |
| 1.4 | 85 | 3.4 | 204 | 5.8 | 351 |
| 1.5 | 91 | 3.5 | 215 | 6.0 | 366 / 368 |
| 1.6 | 97 / 98 | 3.7 | 225 | 6.1 | 370 |
| 1.7 | 105 | 3.8 | 229 / 231 / 232 | 6.2 | 381 |
| 1.8 | 107 / 110 / 112 | 3.9 | 239 / 240 | 6.4 | 389 / 390 / 391 |
| 1.9 | 116 | 4.0 | 241 / 244 | 6.5 | 396 |
| 2.0 | 121 / 122 | 4.1 | 250 / 252 | 6.6 | 400 |
| 2.1 | 128 | 4.2 | 255 / 258 | 6.9 | 420 |
| 2.2 | 132 / 133 / 134 / 135 | 4.3 | 260 / 262 / 265 | 7.0 | 425 / 427 / 428 / 429 |
| 2.3 | 138 / 140 | 4.4 | 267 | 7.2 | 440 |
| 2.4 | 149 | 4.5 | 273 | 7.3 | 445 |
| 2.5 | 150 / 153 | 4.6 | 280 / 281 | 7.4 | 454 |
| 2.6 | 156 / 159 | 4.8 | 292 | 7.5 | 460 |
| 2.8 | 171 / 173 | 4.9 | 300 / 301 | 7.8 | 475 / 477 |
| 2.9 | 177 | 5.0 | 302 / 304 / 305 / 307 | 8.0 | 488 |
| 3.0 | 181 / 182 / 183 | 5.2 | 318 | 8.8 | 534 |
| 3.1 | 191 | 5.3 | 327 | | |

**CHART 3-2**

Liters to cubic inches is often not exact and can result in representing several different engine sizes based on their advertised size in liters.

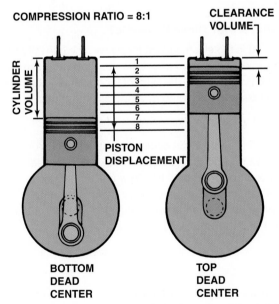

**FIGURE 3–18** Compression ratio is the ratio of the total cylinder volume (when the piston is at the bottom of its stroke) to the clearance volume (when the piston is at the top of its stroke).

**CALCULATING COMPRESSION RATIO**   The compression ratio (CR) calculation uses the formula:

$$CR = \frac{\text{Volume in cylinder with piston at bottom of cylinder}}{\text{Volume in cylinder with piston at top center}}$$

● **SEE FIGURE 3–19.**

**For example:** What is the compression ratio of an engine with 50.3 cu. in. displacement in one cylinder and a combustion chamber volume of 6.7 cu. in.?

$$CR = \frac{50 + 6.7 \text{ cu. in}}{6.7 \text{ cu. in}} = \frac{57.0}{6.7} = 8.5$$

**CHANGING COMPRESSION RATIO**   Any time an engine is modified, the compression ratio should be checked to make sure it is either the same as it was originally or has been changed to match the diesel compression ratio. Factors that can affect compression ratio include:

- **Head gasket thickness.** A thicker than stock gasket will decrease the compression ratio and a thinner than stock gasket will increase the compression ratio.

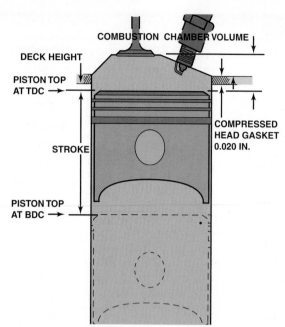

FIGURE 3–19 Combustion chamber volume is the volume above the piston with the piston is at top dead center.

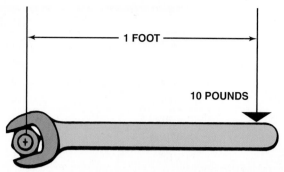

FIGURE 3–20 Torque is a twisting force equal to the distance from the pivot point times the force applied expressed in units called pound-feet (lb-ft) or newton-meters (N-m).

- **Increasing the cylinder size.** If the bore or stroke is increased, a greater amount of air will be compressed into the combustion chamber, which will increase the compression ratio.

## TORQUE AND HORSEPOWER

**DEFINITION OF TORQUE** *Torque* is the term used to describe a rotating force that may or may not result in motion. Torque is measured as the amount of force multiplied by the length of the lever through which it acts. If you use a 1 ft long wrench to apply 10 pounds (lb) of force to the end of the wrench to turn a bolt, then you are exerting 10 pound-feet (lb-ft) of torque. ●**SEE FIGURE 3–20.**

Torque is the twisting force measured at the end of the crankshaft and measured on a dynamometer. Engine torque is always expressed at a specific engine speed (RPM) or range of engine speeds where the torque is at the maximum. For example, an engine may be listed as producing 275 lb-ft @ 2400 RPM.

The metric unit for torque is newton-meters, because the newton is the metric unit for force and the distance is expressed in meters.

1 pound-foot = 1.3558 newton-meters
1 newton-meter = 0.7376 pound-foot

**DEFINITION OF POWER** The term *power* means the rate of doing work. Power equals work divided by time. Work is achieved when a certain amount of mass (weight) is moved a

**FREQUENTLY ASKED QUESTION**

**Is Torque ft-lb or lb-ft?**

The definition of torque is a force (lb) applied to an object times the distance from that object (ft). Therefore, based on the definition of the term, torque should be:

lb-ft (a force times a distance)

Newton-meter (N-m) (a force times a distance)

However, torque is commonly labeled, even on some torque wrenches, as ft-lb.

**TECH TIP**

**What's with These Kilowatts?**

A watt is the electrical unit for *power*, the capacity to do work. It is named after a Scottish inventor, James Watt (1736–1819). The symbol for power is P. Electrical power is calculated as amperes times volts:

$$P \text{ (power)} = I \text{ (amperes)} \times E \text{ (volts)}$$

Engine power is commonly rated in watts or kilowatts (1,000 watts equal 1 kilowatt), because 1 horsepower is equal to 746 watts. For example, a 200 horsepower engine can be rated in the metric system as having the power equal to 149,200 watts or 149.2 kilowatts (kW).

certain distance by a force. If the object is moved in 10 seconds or 10 minutes does not make a difference in the amount of work accomplished, but it does affect the amount of power needed. Power is expressed in units of foot-pounds per minute and power also includes the engine speed (RPM) where the maximum power is achieved. For example, an engine may be listed as producing 280 hp @ 4400 RPM.

**HORSEPOWER AND ALTITUDE** Because the density of the air is lower at high altitude, the power that a normal engine can develop is greatly reduced at high altitude. According to SAE conversion factors, a nonsupercharged or nonturbocharged engine loses about 3% of its power for every 1,000 ft (300 m) of altitude.

Therefore, an engine that develops 200 brake horsepower at sea level will produce only about 116 brake horsepower at the top of Pike's Peak in Colorado at 14,110 ft (4,300 m) (3% × 14 – 42%). Supercharged and turbocharged engines are not as greatly affected by altitude as normally aspirated engines, which are those engines that breathe air at normal atmospheric pressure.

## SUMMARY

1. The four strokes of the four-stroke cycle are intake, compression, power, and exhaust.

2. Engines are classified by number and arrangement of cylinders and by number and location of valves and camshafts, as well as by type of mounting, fuel used, cooling method, and type of air induction.

3. Most engines rotate clockwise as viewed from the front (accessory) end of the engine. The SAE standard is counterclockwise as viewed from the principal (flywheel) end of the engine.

4. Engine size is called displacement and represents the volume displaced by all of the pistons.

## REVIEW QUESTIONS

1. What are the strokes of a four-stroke cycle?

2. If an engine at sea level produces 100 hp, how many horsepower would it develop at 6,000 ft of altitude?

## CHAPTER QUIZ

1. All overhead valve engines _____.
   a. Use an overhead camshaft
   b. Have the valves located in the cylinder head
   c. Operate by the two-stroke cycle
   d. Use the camshaft to close the valves

2. An SOHC V-8 engine has how many camshafts?
   a. One
   b. Two
   c. Three
   d. Four

3. The coolant flow through the radiator is controlled by the _____.
   a. Size of the passages in the block
   b. Thermostat
   c. Cooling fan(s)
   d. Water pump

4. Torque is expressed in units of _____.
   a. Pound-feet
   b. Foot-pounds
   c. Foot-pounds per minute
   d. Pound-feet per second

5. Horsepower is expressed in units of _____.
   a. Pound-feet
   b. Foot-pounds
   c. Foot-pounds per minute
   d. Pound-feet per second

6. A normally aspirated automobile engine loses about _____ power per 1,000 ft of altitude.
   a. 1%
   b. 3%
   c. 5%
   d. 6%

7. One cylinder of an automotive four-stroke cycle engine completes a cycle every _____.
   a. 90 degrees
   b. 180 degrees
   c. 360 degrees
   d. 720 degrees

8. How many rotations of the crankshaft are required to complete each stroke of a four-stroke cycle engine?
   a. One-fourth
   b. One-half
   c. One
   d. Two

9. A rotating force is called _____.
   a. Horsepower
   b. Torque
   c. Combustion pressure
   d. Eccentric movement

10. Technician A says that a crankshaft determines the stroke of an engine. Technician B says that the length of the connecting rod determines the stroke of an engine. Which technician is correct?
    a. Technician A only
    b. Technician B only
    c. Both Technicians A and B
    d. Neither Technician A nor B

**LEARNING OBJECTIVES:** **After studying this chapter, the reader should be able to:** • List the characteristics of diesel engines. • Describe the operation of fuel tanks, lift pumps, and injection pumps. • Understand how the hydraulic electronic unit injection system works. • Discuss the purpose and function of glow plugs, diesel fuel heaters, engine-driven vacuum pumps, diesel injector nozzles, and accelerator pedal position sensors. • Explain the purpose and function of diesel engine turbochargers. • Discuss the purpose and function of the exhaust gas recirculation system and diesel oxidation catalysts. • Define diesel particulate matter, and discuss the function of diesel exhaust particulate filters. • Describe selective catalytic reduction and diesel exhaust fluid (DEF). • Explain compression testing, glow plug resistance balance test, injector pop testing, and diesel emission testing including diesel exhaust smoke diagnosis.

**KEY TERMS:** Diesel exhaust fluid (DEF) 74 • Diesel exhaust particulate filter (DPF) 71 • Diesel oxidation catalyst (DOC) 71 • Differential pressure sensor (DPS) 72 • Direct injection (DI) 62 • Glow plug 67 • Heat of compression 60 • High-pressure common rail (HPCR) 64 • Hydraulic electronic unit injection (HEUI) 64 • Indirect injection (IDI) 62 • Injection pump 60 • Lift pump 63 • Opacity 78 • Particulate matter (PM) 71 • Pop tester 77 • Regeneration 72 • Selective catalytic reduction (SCR) 74 • Soot 71 • Urea 74 • Water–fuel separator 63

## DIESEL ENGINES

**FUNDAMENTALS** In 1892, a German engineer named Rudolf Diesel perfected the compression ignition engine that bears his name. The diesel engine uses heat created by compression to ignite the fuel, so it requires no spark ignition system.

The diesel engine requires compression ratios of 16:1 and higher. Incoming air is compressed until its temperature reaches about 1,000°F (540°C). This is called **heat of compression.** As the piston reaches the top of its compression stroke, fuel is injected into the cylinder, where it is ignited by the hot air. ● **SEE FIGURE 4–1.**

As the fuel burns, it expands and produces power. Because of the very high compression and torque output of a diesel engine, it is made heavier and stronger than the same size gasoline-powered engine.

A diesel engine uses a fuel system with a precision **injection pump** and individual fuel injectors. The pump delivers fuel to the injectors at a high pressure and at timed intervals. Each injector sprays fuel into the combustion chamber at the precise moment required for efficient combustion. ● **SEE FIGURE 4–2.**

### ADVANTAGES AND DISADVANTAGES
A diesel engine has several advantages compared to a similar size gasoline-powered engine, including:

1. More torque output
2. Greater fuel economy
3. Long service life

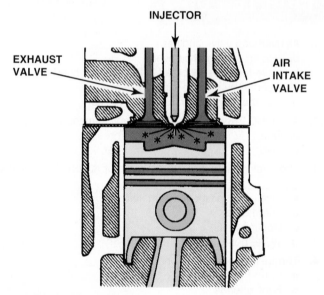

**FIGURE 4–1** Diesel combustion occurs when fuel is injected into the hot, highly compressed air in the cylinder.

A diesel engine has several disadvantages compared to a similar size gasoline-powered engine, including:

1. Engine noise, especially when cold and/or at idle speed
2. Exhaust smell
3. Cold weather startability
4. Vacuum pump that is needed to supply the vacuum needs of the heat, ventilation, and air-conditioning system

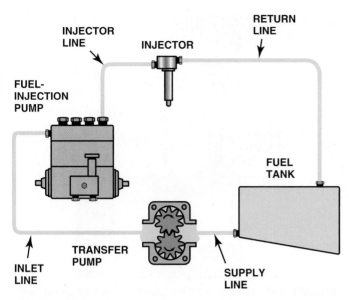

**FIGURE 4–2** A typical injector pump type of automotive diesel fuel-injection system.

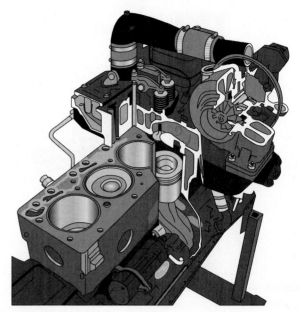

**FIGURE 4–3** A Cummins diesel engine as found in a Dodge pickup truck. A high-pressure pump (up to 30,000 PSI) is used to supply diesel fuel to this common rail, which has tubes running to each injector. Note the thick cylinder walls and heavy-duty construction.

| SYSTEM OR COMPONENT | DIESEL ENGINE | GASOLINE ENGINE |
|---|---|---|
| Block | Cast iron and heavy (● SEE FIGURE 4–3.) | Cast iron or aluminum and as light as possible |
| Cylinder head | Cast iron or aluminum | Cast iron or aluminum |
| Compression ratio | 17:1–25:1 | 8:1–12:1 |
| Peak engine speed | 2000–2500 RPM | 5000–8000 RPM |
| Pistons | Aluminum with combustion pockets and heavy-duty connecting rods (● SEE FIGURE 4–4.) | Aluminum, usually flat top or with valve relief but no combustion pockets |

**CHART 4–1**

Comparison between a typical gasoline and a diesel engine.

**FIGURE 4–4** A rod/piston assembly from a 5.9 liter Cummins diesel engine used in a Dodge pickup truck.

5. Heavier than a gasoline engine
6. Fuel availability
7. Extra cost compared to a gasoline engine

**CONSTRUCTION**  Diesel engines must be constructed heavier than gasoline engines because of the tremendous pressures that are created in the cylinders during operation. ● **SEE CHART 4–1.** The torque output of a diesel engine is often double or more than the same size gasoline-powered engines.

**AIR–FUEL RATIOS**  In a diesel engine, air is not controlled by a throttle as in a gasoline engine. Instead, the amount of fuel injected is varied to control power and speed. The air–fuel mixture

of a diesel can vary from as lean as 85:1 at idle to as rich as 20:1 at full load. This higher air–fuel ratio and the increased compression pressures make the diesel more fuel efficient than a gasoline engine, in part because diesel engines do not suffer from throttling losses. Throttling losses involve the power needed in a gasoline engine to draw air past a closed or partially closed throttle.

In a gasoline engine, the speed and power are controlled by the throttle valve, which controls the amount of air entering the engine. Adding more fuel to the cylinders of a gasoline engine without adding more air (oxygen) will not increase the speed or power of the engine. In a diesel engine, speed

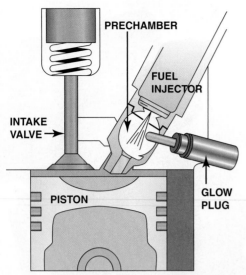

**FIGURE 4–5** An indirect injection diesel engine uses a prechamber and a glow plug.

and power are not controlled by the amount of air entering the cylinders because the engine air intake is always wide open. Therefore, the engine always has enough oxygen to burn the fuel in the cylinder and will increase speed (and power) when additional fuel is supplied.

**NOTE: Many newer diesel engines are equipped with a throttle valve. This valve is used by the emission control system and is not designed to control the speed of the engine.**

**INDIRECT AND DIRECT INJECTION** In an **indirect injection** (abbreviated **IDI**) diesel engine, fuel is injected into a small prechamber, which is connected to the cylinder by a narrow opening. The initial combustion takes place in this prechamber. This has the effect of slowing the rate of combustion, which tends to reduce noise. ● **SEE FIGURE 4–5.**

All indirect diesel injection engines require the use of a glow plug, which is an electrical heater that helps start the combustion process.

In a **direct injection** (abbreviated **DI**) diesel engine, fuel is injected directly into the cylinder. The piston incorporates a depression where initial combustion takes place. Direct injection diesel engines are generally more efficient than indirect injection engines, but have a tendency to produce greater amounts of noise. ● **SEE FIGURE 4–6.**

While some direct injection diesel engines use glow plugs to help cold starting and to reduce emissions, many direct injection diesel engines do not use glow plugs.

**DIESEL FUEL IGNITION** Ignition occurs in a diesel engine by injecting fuel into the air charge, which has been heated by compression to a temperature greater than the ignition point of the fuel or about 1,000°F (538°C). The chemical reaction of burning the fuel creates heat, which causes the gases to expand, forcing the piston to rotate the crankshaft. A four-stroke diesel engine requires two rotations of the crankshaft to complete one cycle.

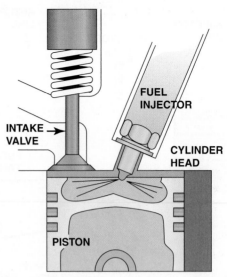

**FIGURE 4–6** A direct injection diesel engine injects the fuel directly into the combustion chamber. Many designs do not use a glow plug.

- On the intake stroke, the piston passes TDC, the intake valve(s) opens, and filtered air enters the cylinder, while the exhaust valve(s) remains open for a few degrees to allow all of the exhaust gases to escape from the previous combustion event.

- On the compression stroke, after the piston passes BDC, the intake valve(s) closes and the piston travels up to TDC (completion of the first crankshaft rotation).

- On the power stroke, the piston nears TDC on the compression stroke and diesel fuel is injected into the cylinder by the injectors. The ignition of the fuel does not start immediately but the heat of compression starts the combustion phases in the cylinder. During this power stroke, the piston passes TDC and the expanding gases force the piston down, rotating the crankshaft.

- On the exhaust stroke, as the piston passes BDC, the exhaust valve(s) opens and the exhaust gases start to flow out of the cylinder. This continues as the piston travels up to TDC, pumping the spent gases out of the cylinder. At TDC, the second crankshaft rotation is complete.

## THREE PHASES OF COMBUSTION

There are three distinct phases or parts to the combustion in a diesel engine:

1. **Ignition delay.** Near the end of the compression stroke, fuel injection begins, but ignition does not begin immediately. This period is called *ignition delay.*

2. **Rapid combustion.** This phase of combustion occurs when the fuel first starts to burn, creating a sudden rise in

cylinder pressure. It is this sudden and rapid rise in combustion chamber pressure that causes the characteristic diesel engine knock.

3. **Controlled combustion.** After the rapid combustion occurs, the rest of the fuel in the combustion chamber begins to burn and injection continues. This process occurs in an area near the injector that contains fuel surrounded by air. This fuel burns as it mixes with the air.

# FUEL TANK AND LIFT PUMP

**PARTS INVOLVED** A fuel tank used on a vehicle equipped with a diesel engine differs from the one used with a gasoline engine in the following ways:

- The filler neck is larger for diesel fuel. The nozzle size is 15/16 inch (24 mm) instead of 13/16 inch (21 mm) for gasoline filler necks. Truck stop diesel nozzles for large over-the-road trucks are usually larger, 1.25 inches or 1.5 inches (32 mm or 38 mm) to allow for faster fueling of large-capacity fuel tanks.

- There are no evaporative emission control devices or a charcoal (carbon) canister. Diesel fuel is not as volatile as gasoline, and therefore diesel vehicles do not have evaporative emission control devices.

The diesel fuel is usually drawn from the fuel tank by a separate pump, called a **lift pump,** and delivers the fuel to the injection pump. Between the fuel tank and the lift pump is a **water–fuel separator.** Water is heavier than diesel fuel and sinks to the bottom of the separator. Part of normal routine maintenance on a vehicle equipped with a diesel engine is to drain the water from the water–fuel separator. A float is often used inside the separator, which is connected to a warning light on the dash that lights if the water reaches a level where it needs to be drained. The water separator is often part of the fuel filter assembly. Both the fuel filter and the water separator are common maintenance items.

**NOTE: Water can cause corrosive damage and wear to diesel engine parts because it is not a good lubricant. Water cannot be atomized by a diesel fuel injector nozzle and will often "blow out" the nozzle tip.**

Many diesel engines also use a *fuel temperature sensor.* The computer uses this information to adjust fuel delivery based on the density of the fuel. ● **SEE FIGURE 4–7.**

# INJECTION PUMP

**NEED FOR HIGH-PRESSURE FUEL PUMP** A diesel engine injection pump is used to increase the pressure of the diesel fuel from very low values from the lift pump to the extremely high pressures needed for injection.

**FIGURE 4–7** A fuel temperature sensor is being tested using an ice bath.

**FIGURE 4–8** A typical distributor-type diesel injection pump showing the pump, lines, and fuel filter.

- The lift pump is a *low-pressure, high-volume pump.*
- The high-pressure injection pump is a *high-pressure, low-volume pump.*

Injection pumps are usually driven by a gear off the camshaft at the front of the engine. As the injection pump shaft rotates, the diesel fuel is fed from a fill port to a high-pressure chamber. If a distributor-type injection pump is used, the fuel is forced out of the injection port to the correct injector nozzle through the high-pressure line. ● **SEE FIGURE 4–8.**

**NOTE: Because of the very tight tolerances in a diesel engine, the smallest amount of dirt can cause excessive damage to the engine and to the fuel-injection system.**

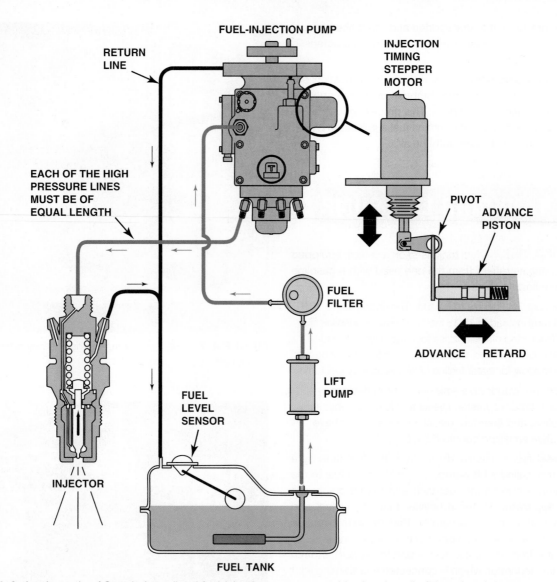

**FUEL-INJECTION PUMP**

**RETURN LINE**

**INJECTION TIMING STEPPER MOTOR**

**EACH OF THE HIGH PRESSURE LINES MUST BE OF EQUAL LENGTH**

**PIVOT**

**ADVANCE PISTON**

**FUEL FILTER**

**ADVANCE    RETARD**

**LIFT PUMP**

**FUEL LEVEL SENSOR**

**INJECTOR**

**FUEL TANK**

**FIGURE 4–9** A schematic of Standadyne diesel fuel-injection pump assembly showing all of the related components.

## DISTRIBUTOR INJECTION PUMP
A distributor diesel injection pump is a high-pressure pump assembly with lines leading to each individual injector. The high-pressure lines between the distributor and the injectors must be the exact same length to ensure proper injection timing. The high-pressure fuel causes the injectors to open. Because of the internal friction of the lines, there is a slight delay before fuel pressure opens the injector nozzle. The injection pump itself creates the injection advance needed for engine speeds above idle often by using a stepper motor attached to the advance piston, and the fuel is then discharged into the lines. ● **SEE FIGURE 4–9.**

**NOTE: The lines expand to some extent during an injection event. This is how timing checks are performed. The pulsing of the injector line is picked up by a probe used to detect the injection event similar to a timing light used to detect a spark on a gasoline engine.**

## HIGH-PRESSURE COMMON RAIL
Newer diesel engines use a fuel delivery system referred to as a **high-pressure common rail (HPCR)** design. Diesel fuel under high pressure, over

20,000 PSI (138,000 kPa), is applied to the injectors, which are opened by a solenoid controlled by the computer. Because the injectors are computer controlled, the combustion process can be precisely controlled to provide maximum engine efficiency with the lowest possible noise and exhaust emissions. ● **SEE FIGURE 4–10.**

## HEUI SYSTEM

**PRINCIPLES OF OPERATION**    Ford 7.3, 6.0, and 6.4 liter (and Navistar) diesels use a system called a **hydraulic electronic unit injection** system, or **HEUI** system. The components used include:

- High-pressure engine oil pump and reservoir
- Pressure regulator for the engine oil
- Passages in the cylinder head for flow of fuel to the injectors

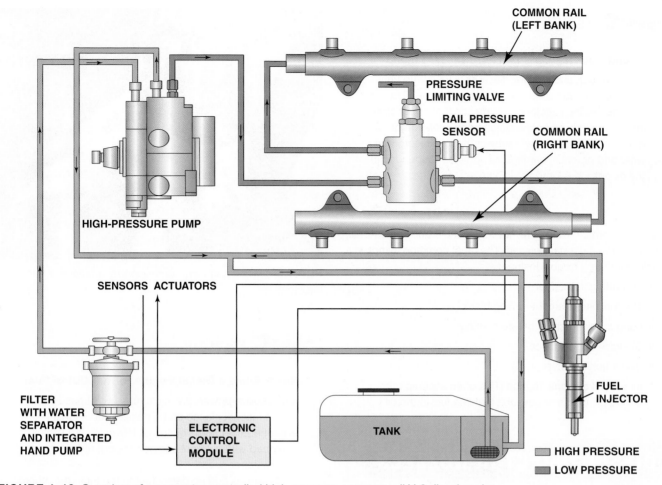

FIGURE 4–10 Overview of a computer-controlled high-pressure common rail V-8 diesel engine.

**OPERATION** The engine oil is pressurized to provide an opening pressure strong enough to overcome the fuel pressure when the solenoid is commanded to open by the PCM. The system functions as follows:

- Fuel is drawn from the tank by the tandem fuel pump, which circulates fuel at low pressure through the fuel filter/water separator/fuel heater bowl, and then fuel is directed back to the fuel pump, where fuel is pumped at high pressure into the cylinder head fuel galleries.

- The injectors, which are hydraulically actuated by engine oil pressure from the high-pressure oil pump, are then fired by the powertrain control module (PCM). The control system for the fuel injectors is the PCM, and the injectors are fired based on sensor inputs received by the PCM.
  ●**SEE FIGURE 4–11.**

HEUI injectors rely on O-rings to keep fuel and oil from mixing or escaping, causing performance problems or engine damage. HEUI injectors use five O-rings. The three external O-rings should be replaced with updated O-rings if they fail.

FIGURE 4–11 A HEUI injector from a Ford PowerStroke diesel engine. The O-ring grooves indicate the location of the O-rings that seal the fuel section of the injector from coolant and from the engine oil.

## TECH TIP

### Change Oil Regularly in a Ford Diesel Engine

Ford 7.3, 6.0, and 6.4 liter diesel engines pump unfiltered oil from the sump to the high-pressure oil pump and then to the injectors. This means that not changing oil regularly can contribute to accumulation of dirt in the engine and will subject the fuel injectors to wear and potential damage as particles suspended in the oil get forced into the injectors.

**FIGURE 4–12** Typical computer-controlled diesel engine fuel injectors.

The two internal O-rings are not replaceable and if these fail, the injector(s) must be replaced. The most common symptoms of injector O-ring trouble include:

- Oil getting in the fuel
- The fuel filter element turning black
- Long cranking times before starting
- Sluggish performance
- Reduction in power
- Increased oil consumption (This often accompanies O-ring problems or any fault that lets fuel in the oil.)

## DIESEL INJECTOR NOZZLES

**PARTS INVOLVED** Diesel injector nozzles are spring-loaded closed valves that spray fuel directly into the combustion chamber or precombustion chamber when the injector is opened. Injector nozzles are threaded or clamped into the cylinder head, one for each cylinder, and are replaceable as an assembly.

The tip of the injector nozzle has many holes to deliver an atomized spray of diesel fuel into the cylinder. Parts of a diesel injector nozzle include:

- **Heat shield.** This is the outer shell of the injector nozzle and may have external threads where it seals in the cylinder head.
- **Injector body.** This is the inner part of the nozzle and contains the injector needle valve and spring, and threads into the outer heat shield.
- **Diesel injector needle valve.** This precision machined valve and the tip of the needle seal against the injector body when it is closed. When the valve is open, diesel fuel is sprayed into the combustion chamber. This passage is controlled by a computer-controlled solenoid on diesel engines equipped with computer-controlled injection.
- **Injector pressure chamber.** The pressure chamber is a machined cavity in the injector body around the tip of the injector needle. Injection pump pressure forces fuel into this chamber, forcing the needle valve open.

## TECH TIP

### Never Allow a Diesel Engine to Run Out of Fuel

If a gasoline-powered vehicle runs out of gasoline, it is an inconvenience and a possible additional expense to get some gasoline. However, if a vehicle equipped with a diesel engine runs out of fuel, it can be a major concern.

Besides adding diesel fuel to the tank, the other problem is getting all of the air out of the pump, lines, and injectors so the engine will operate correctly.

The procedure usually involves cranking the engine long enough to get liquid diesel fuel back into the system, but at the same time keeping cranking time short enough to avoid overheating the starter. Consult service information for the exact service procedure if the diesel engine is run out of fuel.

**NOTE: Some diesel engines, such as the General Motors Duramax V-8, are equipped with a priming pump located under the hood on top of the fuel filter. Pushing down and releasing the priming pump with a vent valve open will purge any trapped air from the system. Always follow the vehicle manufacturer's instructions.**

**DIESEL INJECTOR NOZZLE OPERATION** The electric solenoid attached to the injector nozzle is computer controlled and opens to allow fuel to flow into the injector pressure chamber. ● **SEE FIGURE 4–12.**

The fuel flows down through a fuel passage in the injector body and into the pressure chamber. The high fuel pressure in the pressure chamber forces the needle valve upward, compressing the needle valve return spring and forcing the needle valve open.

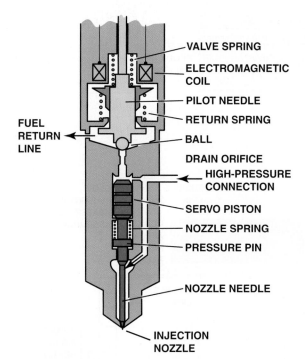

FIGURE 4–13 A Duramax injector showing all the internal parts.

VALVE SPRING
ELECTROMAGNETIC COIL
PILOT NEEDLE
RETURN SPRING
FUEL RETURN LINE
BALL
DRAIN ORIFICE
HIGH-PRESSURE CONNECTION
SERVO PISTON
NOZZLE SPRING
PRESSURE PIN
NOZZLE NEEDLE
INJECTION NOZZLE

When the needle valve opens, diesel fuel is discharged into the combustion chamber in a hollow cone spray pattern.

Any fuel that leaks past the needle valve returns to the fuel tank through a return passage and line. ● SEE FIGURE 4–13.

## GLOW PLUGS

**PURPOSE AND FUNCTION** Glow plugs are always used in diesel engines equipped with a precombustion chamber and may be used in direct injection diesel engines to aid starting. A **glow plug** is a heating element that uses 12 volts from the battery and aids in the starting of a cold engine by providing heat to help the fuel to ignite. ● SEE FIGURE 4–14.

As the temperature of the glow plug increases, the resistance of the heating element inside increases, thereby reducing the current in amperes needed by the glow plugs.

**OPERATION** Most glow plugs used in newer vehicles are controlled by the Powertrain Control Module, which monitors coolant temperature and intake air temperature. The glow plugs are turned on or pulsed on or off depending on the temperature of the engine. The PCM will also keep the glow plug turned on after the engine starts, to reduce white exhaust smoke (unburned fuel) and to improve idle quality after starting. ● SEE FIGURE 4–15.

The "wait to start" lamp (if equipped) will light when the engine and the outside temperatures are low to allow time for the glow plugs to get hot.

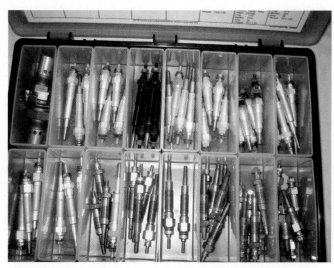

FIGURE 4–14 A glow plug assortment showing the various types and sizes of glow plugs used. Always use the specified glow plugs.

 **FREQUENTLY ASKED QUESTION**

**How Can You Tell if Gasoline Has Been Added to the Diesel Fuel by Mistake?**

If gasoline has been accidentally added to diesel fuel and is burned in a diesel engine, the result can be very damaging to the engine. The gasoline can ignite faster than diesel fuel, which would tend to increase the temperature of combustion. This high temperature can harm injectors and glow plugs, as well as pistons, head gaskets, and other major diesel engine components. If contaminated fuel is suspected, first smell the fuel at the filler neck. If the fuel smells like gasoline, then the tank should be drained and refilled with diesel fuel. If the smell test does not indicate a gasoline or any rancid smell, then test a sample for proper specific gravity.

NOTE: Diesel fuel designed for on-road use should be green. Red diesel fuel (non-taxed) should be found only in off-road or farm equipment.

**HEATED INLET AIR** Some diesel engines, such as the Dodge Cummins and the General Motors 6.6 liter Duramax V-8, use an electrical heater wire to warm the intake air to help in cold weather starting and running. ● SEE FIGURE 4–16.

## ENGINE-DRIVEN VACUUM PUMP

Because a diesel engine is unthrottled, it creates very little vacuum in the intake manifold. Several engine and vehicle components operate using vacuum, such as the exhaust gas

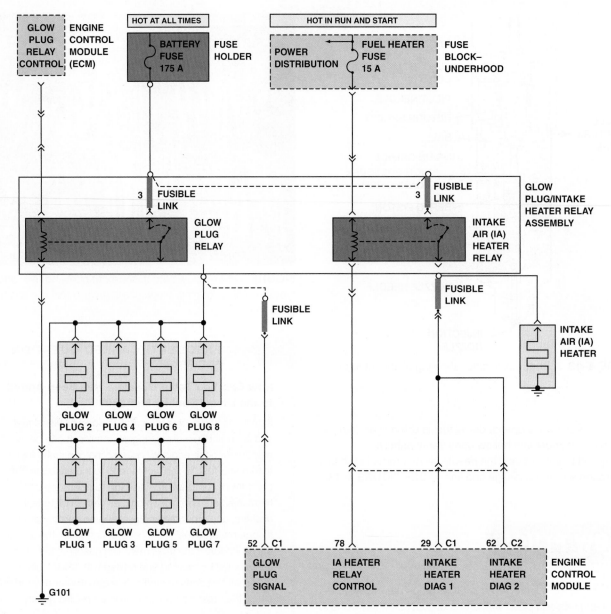

**FIGURE 4–15** A schematic of a typical glow plug circuit. Notice that the glow plug relay and intake air heater relay are both computer controlled.

recirculation (EGR) valve and the heating and ventilation blend and air doors. Most diesels used in cars and light trucks are equipped with an engine-driven vacuum pump to supply the vacuum for these components.

## DIESEL FUEL HEATERS

Diesel fuel heaters help prevent power loss and stalling in cold weather. The heater is placed in the fuel line between the tank and the primary filter. Some coolant heaters are thermostatically controlled, which allows fuel to bypass the heater once it has reached operating temperature.

## ACCELERATOR PEDAL POSITION SENSOR

Some light-truck diesel engines are equipped with an electronic throttle to control the amount of fuel injected into the engine. Because a diesel engine does not use a throttle in the air intake, the only way to control engine speed is by controlling the amount of fuel being injected into the cylinders. Instead of a mechanical link from the accelerator pedal to the diesel injection pump, a throttle-by-wire system uses an accelerator pedal position (APP) sensor. To ensure safety, it consists of three separate sensors that change in voltage as the accelerator pedal is depressed. ● **SEE FIGURE 4–17.**

FIGURE 4–16 A wire-wound electric heater is used to warm the intake air on some diesel engines.

FIGURE 4–18 A Cummins diesel turbocharger is used to increase the power and torque of the engine.

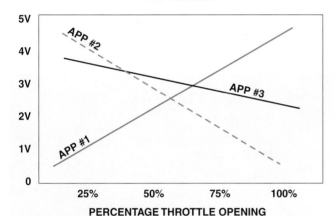

FIGURE 4–17 A typical accelerator pedal position (APP) sensor uses three different sensors in one package with each creating a different voltage as the accelerator is moved.

The computer checks for errors by comparing the voltage output of each of the three sensors inside the APP and compares them to what they should be if there are no faults. If an error is detected, the engine and vehicle speed are often reduced.

# DIESEL ENGINE TURBOCHARGERS

**TURBOCHARGED DIESELS** A turbocharger greatly increases engine power by pumping additional compressed air into the combustion chambers. This allows a greater quantity of fuel to be burned in the cylinders, resulting in greater power output. In a turbocharger, the turbine wheel spins as exhaust gas flows out of the engine and drives the turbine blades. The turbine spins the

compressor wheel at the opposite end of the turbine shaft, pumping air into the intake system. ● SEE FIGURE 4–18.

**AIR CHARGE COOLER** The first component in a typical turbocharger system is an air filter through which ambient air passes before entering the compressor. The air is compressed, which raises its density (mass/unit volume). All currently produced light-duty diesels use an air charge cooler whose purpose is to cool the compressed air to further raise the air density. Cooler air entering the engine means more power can be produced by the engine. ● SEE FIGURE 4–19.

**VARIABLE TURBOCHARGER** A variable turbocharger is used on many diesel engines for boost control. Boost pressure is controlled independent of engine speed and a wastegate is not needed. The adjustable vanes mount to a unison ring that allows the vanes to move. As the position of the unison ring rotates, the vanes change angle. The vanes are opened to minimize flow at the turbine and exhaust back pressure at low engine speeds. To increase turbine speed, the vanes are closed. The velocity of the exhaust gases increases, as does the speed of the turbine. The unison ring is connected to a cam that is positioned by a rack-and-pinion gear. The turbocharger's vane position actuator solenoid connects to a hydraulic piston, which moves the rack to rotate the pinion gear and cam. ● SEE FIGURE 4–20.

The turbocharger vane position control solenoid valve is used to advance the unison ring's relationship to the turbine and thereby articulate the vanes. This solenoid actuates a spool valve that applies oil pressure to either side of a piston. Oil flow has three modes: apply, hold, and release.

- *Apply* moves the vanes toward a closed position.
- *Hold* maintains the vanes in a fixed position.
- *Release* moves the vanes toward the open position.

The turbocharger vane position actuation is controlled by the ECM, which can change turbine boost efficiency independent of

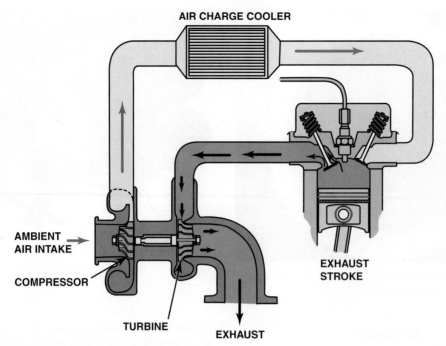

FIGURE 4–19 An air charge cooler is used to cool the compressed air.

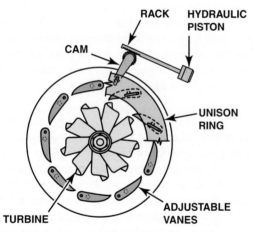

FIGURE 4–20 A variable vane turbocharger allows the boost to be controlled without the need of a wastegate.

engine speed. The ECM provides a control signal to the valve solenoid along with a low-side reference. A pulse-width-modulated signal from the ECM moves the valve to the desired position.

FIGURE 4–21 A cutaway showing the exhaust cooler. The cooler the exhaust is, the more effective it is in controlling $NO_x$ emissions.

## EXHAUST GAS RECIRCULATION

The EGR system recycles some exhaust gas back into the intake stream to cool combustion, which reduces oxides of nitrogen ($NO_x$) emissions. The EGR system includes:

- Plumbing that carries some exhaust gas from the turbocharger exhaust inlet to the intake ports

- EGR control valve
- Stainless steel cooling element used to cool the exhaust gases (● SEE FIGURE 4–21.)

The EGR valve is PCM controlled and often uses a DC stepper motor and worm gear to move the valve stem open. The gear is not attached to the valve and can only force it open. Return spring force closes the valve. The EGR valve and sensor assembly is a five-wire design. The PCM uses the position sensor to verify that valve action is as commanded.

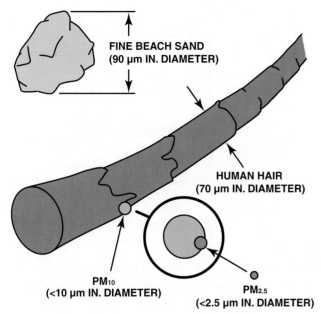

FINE BEACH SAND
(90 μm IN. DIAMETER)

HUMAN HAIR
(70 μm IN. DIAMETER)

PM₁₀
(<10 μm IN. DIAMETER)

PM₂.₅
(<2.5 μm IN. DIAMETER)

**FIGURE 4–22** Relative size of particulate matter to a human hair.

**?** FREQUENTLY ASKED QUESTION

**What Is the Big Deal for the Need to Control Very Small Soot Particles?**

For many years soot or particulate matter (PM) was thought to be less of a health concern than exhaust emissions from gasoline engines. It was felt that the soot could simply fall to the ground without causing any noticeable harm to people or the environment. However, it was discovered that the small soot particulates when breathed in are not expelled from the lungs like larger particles but instead get trapped in the deep areas of the lungs where they accumulate.

# DIESEL PARTICULATE MATTER

**PARTICULATE MATTER STANDARDS** **Particulate matter (PM),** also called **soot,** refers to tiny particles of solid or semisolid material suspended in the atmosphere. This includes particles between 0.1 micron and 50 microns in diameter. The heavier particles, larger than 50 microns, typically tend to settle out quickly due to gravity. Particulates are generally categorized as follows:

- **Total suspended particulate (TSP).** Refers to all particles between 0.1 and 50 microns. Up until 1987, the Environmental Protection Agency (EPA) standard for particulates was based on levels of TSP.

- **PM10.** Refers to particulate matter of 10 microns or less (approximately 1/6 the diameter of a human hair). EPA has a standard for particles based on levels of PM10.

- **PM2.5.** Refers to particulate matter of 2.5 microns or less (approximately 1/20 the diameter of a human hair), also called "fine" particles. In July 1997, the EPA approved a standard for PM2.5. ● **SEE FIGURE 4–22.**

**SOOT CATEGORIES** In general, soot particles produced by diesel combustion fall into the following categories.

- **Fine.** Less than 2.5 microns

- **Ultrafine.** Less than 0.1 micron, and make up 80% to 95% of soot

# DIESEL OXIDATION CATALYST

**PURPOSE AND FUNCTION** **Diesel oxidation catalysts (DOC)** are used in all light-duty diesel engines, since 2007. They consist of a flow-through honeycomb-style substrate structure that is wash coated with a layer of catalyst materials, similar to those used in a gasoline engine catalytic converter. These materials include the precious metals platinum and palladium, as well as other base metal catalysts.

Catalysts chemically react with exhaust gas to convert harmful nitrogen oxide into nitrogen dioxide, and to oxidize absorbed hydrocarbons. The chemical reaction acts as a combustor for the unburned fuel that is characteristic of diesel compression ignition. The main function of the DOC is to start a regeneration event by converting the fuel-rich exhaust gases to heat.

The DOC also reduces:

- Carbon monoxide (CO)

- Hydrocarbons (HC)

- Odor-causing compounds such as aldehydes and sulfur ● **SEE FIGURE 4–23.**

# DIESEL EXHAUST PARTICULATE FILTER

**PURPOSE AND FUNCTION** **Diesel exhaust particulate filters (DPFs)** are used in all light-duty diesel vehicles, since 2007, to meet the exhaust emissions standards. The heated exhaust gas from the DOC flows into the DPF, which captures diesel exhaust gas particulates (soot) to prevent them from being released into the atmosphere. This is done by forcing

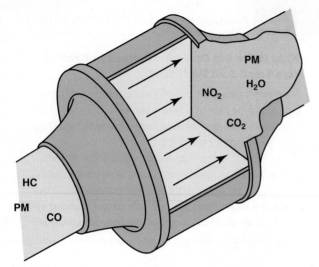

**FIGURE 4–23** Chemical reaction within the DOC.

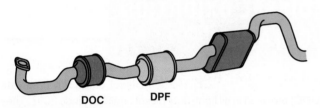

**FIGURE 4–24** Aftertreatment of diesel exhaust is handled by the DOC and DPF.

the exhaust through a porous cell which has a silicon carbide substrate with honeycomb-cell-type channels that trap the soot. The main difference between the DPF and a typical catalyst filter is that the entrance to every other cell channel in the DPF substrate is blocked at one end. So instead of flowing directly through the channels, the exhaust gas is forced through the porous walls of the blocked channels and exits through the adjacent open-ended channels. This type of filter is also referred to as a "wall-flow" filter. ● **SEE FIGURE 4–24.**

**OPERATION** Soot particulates in the gas remain trapped on the DPF channel walls where, over time, the trapped particulate matter will begin to clog the filter. The filter must therefore be purged periodically to remove accumulated soot particles. The process of purging soot from the DPF is described as **regeneration.** When the temperature of the exhaust gas is increased, the heat incinerates the soot particles trapped in the filter and is effectively renewed. ● **SEE FIGURE 4–25.**

**EXHAUST GAS TEMPERATURE SENSORS** The following two exhaust gas temperature sensors are used to help the PCM control the DPF.

- EGT sensor 1 is positioned between the DOC and the DPF where it can measure the temperature of the exhaust gas entering the DPF.
- EGT sensor 2 measures the temperature of the exhaust gas stream immediately after it exits the DPF.

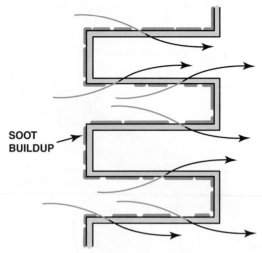

**FIGURE 4–25** The soot is trapped in the passages of the DPF. The exhaust has to flow through the sides of the trap and exit.

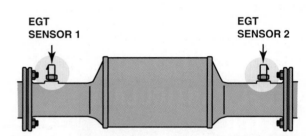

**FIGURE 4–26** EGT 1 and EGT 2 are used by the PCM to help control aftertreatment.

The powertrain control module monitors the signals from the EGT sensors as part of its calibrations to control DPF regeneration. Proper exhaust gas temperatures at the inlet of the DPF are crucial for proper operation and for starting the regeneration process. Too high a temperature at the DPF will cause the DPF substrate to melt or crack. Regeneration will be terminated at temperatures above 1,470°F (800°C). With too low a temperature, self-regeneration will not fully complete the soot-burning process. ● **SEE FIGURE 4–26.**

**DPF DIFFERENTIAL PRESSURE SENSOR** The DPF **differential pressure sensor (DPS)** has two pressure sample lines.

- One line is attached before the DPF.
- The other is located after the DPF.

The exact location of the DPS varies by vehicle model type such as medium duty, pickup, or van. By measuring the exhaust supply (upstream) pressure from the DOC, and the post DPF (downstream) pressure, the PCM can determine differential pressure, also called "delta" pressure, across the DPF. Data from the DPF differential pressure sensor is used by the PCM to calibrate for controlling DPF exhaust system operation.

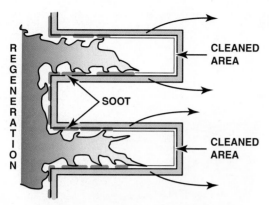

**FIGURE 4-27** Regeneration burns the soot and renews the DPF.

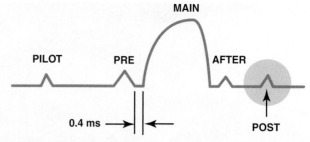

**FIGURE 4-28** The postinjection pulse occurs to create the heat needed for regeneration.

**DIESEL PARTICULATE FILTER REGENERATION** The primary reason for soot removal is to prevent the buildup of exhaust back pressure. Excessive back pressure increases fuel consumption, reduces power output, and can potentially cause engine damage. Several factors can trigger the diesel PCM to perform regeneration, including:

- Distance since last DPF regeneration
- Fuel used since last DPF regeneration
- Engine run time since last DPF regeneration
- Exhaust differential pressure across the DPF

**DPF REGENERATION PROCESS** A number of engine components are required to function together for the regeneration process to be performed, as follows:

1. PCM controls that impact DPF regeneration include late post injections, engine speed, and adjusting fuel pressure.

2. Adding late post injection pulses provides the engine with additional fuel to be oxidized in the DOC, which increases exhaust temperatures entering the DPF to 900°F (500°C) or higher. ● **SEE FIGURE 4-27.**

3. The intake air valve acts as a restrictor that reduces air entry to the engine, which increases engine operating temperature.

4. The intake air heater may also be activated to warm intake air during regeneration.

**TYPES OF DPF REGENERATION** DPF regeneration can be initiated in a number of ways, depending on the vehicle application and operating circumstances. The two main regeneration types are as follows:

- **Passive regeneration.** During normal vehicle operation when driving conditions produce sufficient load and exhaust temperatures, passive DPF regeneration may occur. This passive regeneration occurs without input from the PCM or the driver. A passive regeneration may typically occur while the vehicle is being driven at highway speed or towing a trailer.

 **FREQUENTLY ASKED QUESTION**

**Will the Postinjection Pulses Reduce Fuel Economy?**

Maybe. Due to the added fuel-injection pulses and late fuel-injection timing, an increase in fuel consumption may be noticed on the driver information center (DIC) during the regeneration time period. The overall fuel economy may or may not be affected by the regeneration depending on the how the vehicle is driven and how often regeneration is being performed. ● **SEE FIGURE 4-28.**

- **Active regeneration.** Active regeneration is commanded by the PCM when it determines that the DPF requires it to remove excess soot buildup and conditions for filter regeneration have been met. Active regeneration is usually not noticeable to the driver. The vehicle needs to be driven at speeds above 30 mph for approximately 20 to 30 minutes to complete a full regeneration. During regeneration, the exhaust gases reach temperatures above 1,000°F (550°C). Active regeneration is usually not noticeable to the driver.

 **WARNING**

Tailpipe outlet exhaust temperature will be greater than 572°F (300°C) during service regeneration. To help prevent personal injury or property damage from fire or burns, keep vehicle exhaust away from any object and people.

**ASH LOADING** Regeneration will not burn off ash. Only the particulate matter (PM) is burned off during regeneration. Ash is a noncombustible by-product from normal oil consumption. Ash accumulation in the DPF will eventually cause a restriction in the particulate filter. To service an ash-loaded DPF, the DPF will need to be removed from the vehicle and cleaned or replaced. Low ash content engine oil (API CJ-4) is required for vehicles with the DPF system. The CJ-4 rated oil is limited to 1% ash content.

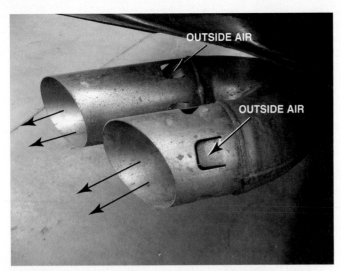

FIGURE 4–29 The exhaust is split into two outlets and has slits to help draw outside air in as the exhaust leaves the tailpipe. The end result is cooler exhaust gases exiting the tailpipe.

 **FREQUENTLY ASKED QUESTION**

**What Is an Exhaust Air Cooler?**

An exhaust air cooler is simply a section of tailpipe that has slits for air to enter. As hot exhaust rushes past the gap, outside air is drawn into the area which reduces the exhaust discharge temperature. The cooler significantly lowers exhaust temperature at the tailpipe from about 800°F (430°C) to approximately 500°F (270°C). ● SEE FIGURE 4–29.

# SELECTIVE CATALYTIC REDUCTION

**PURPOSE AND FUNCTION** Selective catalytic reduction (SCR) is a method used to reduce $NO_x$ emissions by injecting urea into the exhaust stream. Instead of using large amounts of exhaust gas recirculation (EGR), the SCR system uses urea. **Urea** is used as a nitrogen fertilizer. It is colorless, odorless, and nontoxic. Urea is called **diesel exhaust fluid (DEF)** in North America and AdBlue in Europe. ● SEE FIGURE 4–30.

The urea is injected into the catalyst where it sets off a chemical reaction that converts nitrogen oxides ($NO_x$) into nitrogen ($N_2$) and water ($H_2O$). Vehicle manufacturers size the onboard urea storage tank so that it needs to be refilled at about each scheduled oil change, or every 7,500 miles (12,000 km). A warning light alerts the driver when the urea level needs to be refilled. If the warning light is ignored and the diesel exhaust fluid is not refilled, current EPA regulations require that the operation of the engine be restricted and may not start unless the fluid is refilled. This regulation is designed to prevent the engine from

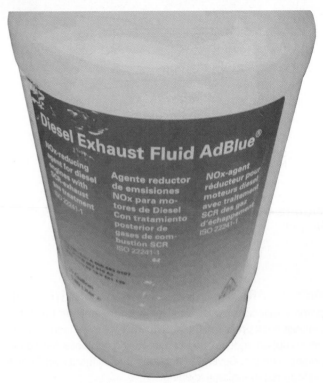

FIGURE 4–30 Diesel exhaust fluid cost $4 to $8 a gallon and is housed in a separate container that holds from 5 to 10 gallons, or enough to last until the next scheduled oil change in most diesel vehicles that use SCR.

being operated without the fluid, which, if not, would greatly increase exhaust emissions. ● SEE FIGURE 4–31.

**ADVANTAGES OF SCR** Using urea injection instead of large amounts of EGR results in the following advantages:

- Potential higher engine power output for the same size engine
- Reduced $NO_x$ emissions up to 90%
- Reduced HC and CO emissions up to 50%
- Reduced particulate matter (PM) by 50%

**DISADVANTAGES OF SCR** Using urea injection instead of large amounts of EGR results in the following disadvantages:

- Onboard storage tank required for the urea
- Difficult to find local sources of urea
- Increased costs to the vehicle owner due to having to refill the urea storage tank

# DIESEL EXHAUST SMOKE DIAGNOSIS

Although some exhaust smoke is considered normal operation for many diesel engines, especially older units, the cause of excessive exhaust smoke should be diagnosed and repaired.

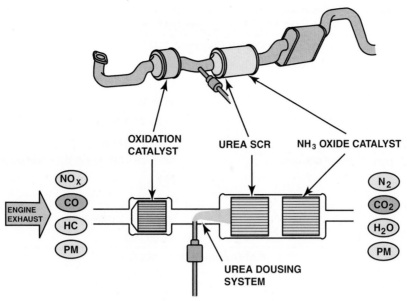

**FIGURE 4–31** Urea (diesel exhaust fluid) injection is used to reduce $NO_x$ exhaust emissions. It is injected after the diesel oxidation catalyst (DOC) and before the diesel particulate filter (DPF) on this 6.7 liter Ford diesel engine.

**BLACK SMOKE** Black exhaust smoke is caused by incomplete combustion because of a lack of air or a fault in the injection system that could cause an excessive amount of fuel in the cylinders. Items that should be checked include the following:

- Fuel specific gravity (API gravity)
- Injector balance test to locate faulty injectors using a scan tool
- Proper operation of the engine coolant temperature (ECT) sensor
- Proper operation of the fuel rail pressure (FRP) sensor
- Restrictions in the intake or turbocharger
- Engine oil usage

**WHITE SMOKE** White exhaust smoke occurs most often during cold engine starts because the smoke is usually condensed fuel droplets. White exhaust smoke is also an indication of cylinder misfire on a warm engine. The most common causes of white exhaust smoke include:

- Inoperative glow plugs
- Low engine compression
- Incorrect injector spray pattern
- Coolant leak into the combustion chamber

**GRAY OR BLUE SMOKE** Blue exhaust smoke is usually due to oil consumption caused by worn piston rings, scored cylinder walls, or defective valve stem seals. Gray or blue smoke can also be caused by a defective injector(s) or defective injector O-rings.

## DIESEL PERFORMANCE DIAGNOSIS

Always start the diagnosis of a diesel engine concern by checking the oil. Higher than normal oil level can indicate that diesel fuel has leaked into the oil. Diesel engines can be diagnosed using a scan tool in most cases, because most of the pressure sensors values can be displayed. Common faults include:

- Hard starting
- No start
- Extended cranking before starting
- Low power

Using a scan tool, check the sensor values in ● **CHART 4–2.** to help pin down the source of the problem. Also check the minimum pressures that are required to start the engine if a no-start condition is being diagnosed. ● **SEE FIGURE 4–32.**

## COMPRESSION TESTING

A compression test is fundamental for determining the mechanical condition of a diesel engine. Worn piston rings can cause low power and excessive exhaust smoke. To test the compression on a diesel engine, the following will have to be done:

- Remove the glow plug (if equipped) or the injector.
- Use a diesel compression gauge, as the compression is too high to use a gasoline engine compression gauge.

A diesel engine should produce at least 300 PSI (2,068 kPa) of compression pressure and all cylinders should be within 50 PSI (345 kPa) of each other. ● **SEE FIGURE 4–33.**

## DIESEL TROUBLESHOOTING CHART

### 5.9 Dodge Cummins 2003–2008

| | |
|---|---|
| Low-pressure pump | 8–12 PSI |
| Pump amperes | 4 A |
| Pump volume | 45 oz. in 30 sec. |
| High-pressure pump | 5,000–23,000 PSI |
| Idle PSI | 5,600–5,700 PSI |
| Electronic Fuel Control (EFC) maximum fuel pressure | Disconnect EFC to achieve maximum pressure |
| Injector volts | 90 V |
| Injector amperes | 20 A |
| Glow plug amperes | 60–80 A × 2 (120–160 A) |
| **Minimum PSI to start** | **5,000 PSI** |

### GM Duramax 2001–2008

| | |
|---|---|
| Low-pressure pump vacuum | 2–10 in. Hg |
| Pump amperes | NA |
| Pump volume | NA |
| High-pressure pump | 5 K-2.3 K-2.6 K PSI |
| Idle PSI | 5,000–6,000 PSI (30–40 MPa) |
| Fuel Rail Pressure Regulator (FRPR) maximum fuel pressure | Disconnect to achieve maximum pressure |
| Injector volts | 48 V or 93 V |
| Injector amperes | 20 A |
| Glow plug amperes | 160 A |
| **Minimum to start** | **1,500 PSI (10 MPa)** |

### Sprinter 2.7 2002–2006

| | |
|---|---|
| Low-pressure pump | 6–51 PSI |
| High-pressure pump | 800–23,000 PSI |
| Idle PSI | 4,900 PSI |
| Fuel Rail Pressure Control (FRPC) maximum fuel pressure | Apply power and ground to FRPC to achieve maximum pressure |
| Injector volts | 80 V |
| Injector amperes | 20 A |
| Glow plug amperes | 17 A each (85–95 A total) |
| **Minimum to start** | **3,200 PSI (1–1.2 V to start)** |

### 6.0 Powerstroke 2003–2008

| | |
|---|---|
| Low-pressure pump | 50–60 PSI |
| High-pressure pump | 500–4,000 PSI |
| Idle PSI | 500 PSI+ |
| Injection Pressure Regulator (IPR) maximum fuel pressure | Apply power and ground to IPR |
| Injector volts | 48 V |
| Injector amperes | 20 A |
| Glow plug amperes | 20–25 A each (160–200 A total) |
| **Minimum to start** | **500 PSI (0.85 V)** |

**CHART 4–2**

The values can be obtained by using a scan tool and basic test equipment. An inductive ammeter can be used to measure the glow plug current draw. Always follow the vehicle manufacturer's recommended procedures.

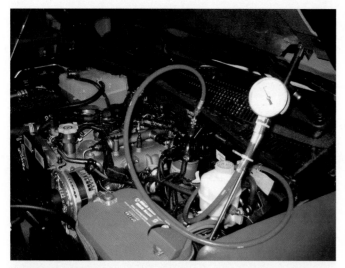

FIGURE 4–32 A pressure gauge checking the fuel pressure from the lift pump on a Cummins 6.7 liter diesel.

FIGURE 4–33 A compression gauge that is designed for the higher compression rate of a diesel engine should be used when checking the compression.

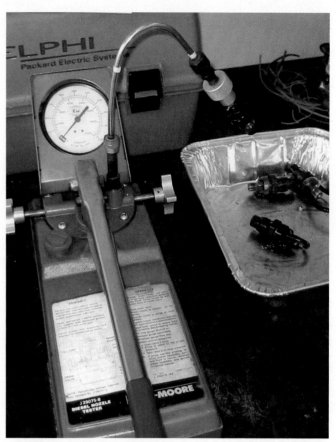

FIGURE 4–34 A typical pop tester used to check the spray pattern of a diesel engine injector.

2. With the wires still removed from the glow plugs, start the engine.

3. Allow the engine to run for several minutes to allow the combustion inside the cylinder to warm the glow plugs.

4. Turn off the engine and then measure and record the resistance of the glow plugs.

5. The resistance of all glow plugs should be higher than at the beginning of the test. A glow plug that is in a cylinder that is not firing correctly will not increase in resistance as much as the others.

6. Another test is to measure exhaust manifold temperature at each exhaust port using an infrared thermometer or a pyrometer. Misfiring cylinders will run cold.

## GLOW PLUG RESISTANCE BALANCE TEST

Glow plugs increase in resistance as their temperature increases. All glow plugs should have about the same resistance when checked with an ohmmeter. A similar test of the resistance of the glow plugs can be used to detect a weak cylinder. This test is particularly helpful on a diesel engine that is not computer controlled. To test for even cylinder balance using glow plug resistance, perform the following on a warm engine:

1. Unplug, measure, and record the resistance of all glow plugs.

## INJECTOR POP TESTING

A **pop tester** is a device used for checking a diesel injector nozzle for proper spray pattern. The handle is depressed and pop-off pressure is displayed on the gauge. ● SEE FIGURE 4–34.

The spray pattern should be a hollow cone, but will vary depending on design. The nozzle should also be tested for leakage (dripping of the nozzle) while under pressure. If the spray pattern is not correct, then cleaning, repairing, or replacing the injector nozzle may be necessary.

### Always Use Cardboard to Check for High-Pressure Leaks

If diesel fuel is found on the engine, a high-pressure leak could be present. When checking for such a leak, wear protective clothing including safety glasses, a face shield, gloves, and a long-sleeved shirt. Then use a piece of cardboard to locate the high-pressure leak. When a Duramax diesel is running, the pressure in the common rail and injector tubes can reach over 20,000 PSI. At these pressures, the diesel fuel is atomized and cannot be seen but can penetrate the skin and cause personal injury. A leak will be shown as a dark area on the cardboard. When a leak is found, shut off the engine and find the exact location of the leak without the engine running.

**CAUTION: Sometimes a leak can actually cut through the cardboard, so use extreme care.**

## DIESEL EMISSION TESTING

**OPACITY TEST**   The most common diesel exhaust emission test used in state or local testing programs is called the opacity test. **Opacity** means the percentage of light that is blocked by the exhaust smoke:

- A 0% opacity means that the exhaust has no visible smoke and does not block light from a beam projected through the exhaust smoke.

- A 100% opacity means that the exhaust is so dark that it completely blocks light from a beam projected through the exhaust smoke.

- A 50% opacity means that the exhaust blocks half of the light from a beam projected through the exhaust smoke. ● **SEE CHART 4–3.**

**SNAP ACCELERATION TEST**   In a snap acceleration test, the vehicle is held stationary, with wheel chocks in place and brakes released as the engine is rapidly accelerated to high idle, with the transmission in neutral while smoke emissions are measured. This test is conducted a minimum of six times and the three most consistent measurements are averaged for a final score.

**ROLLING ACCELERATION TEST**   Vehicles with a manual transmission are rapidly accelerated in low gear from an idle speed to a maximum governed RPM while the smoke emissions are measured.

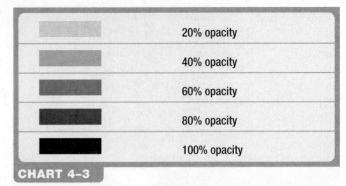

| | |
|---|---|
| | 20% opacity |
| | 40% opacity |
| | 60% opacity |
| | 80% opacity |
| | 100% opacity |

**CHART 4–3**

An opacity test is sometimes used during a state emission test on diesel engines.

**FIGURE 4–35** The letters on the side of this injector on a Cummins 6.7 liter diesel indicate the calibration number for the injector.

TECH TIP

### Do Not Switch Injectors

In the past, it was common practice to switch diesel fuel injectors from one cylinder to another when diagnosing a dead cylinder problem. However, most high-pressure common rail systems used in new diesels utilize precisely calibrated injectors that should not be mixed up during service. Each injector has its own calibration number. ● **SEE FIGURE 4–35.**

**STALL ACCELERATION TEST**   Vehicles with automatic transmissions are held in a stationary position with the parking brake and service brakes applied while the transmission is placed in "drive." The accelerator is depressed and held momentarily while smoke emissions are measured.

The standards for diesels vary according to the type of vehicle and other factors, but usually include a 40% opacity or less.

## SUMMARY

1. A diesel engine uses heat of compression to ignite the diesel fuel when it is injected into the compressed air in the combustion chamber.

2. There are two basic designs of combustion chambers used in diesel engines. Indirect injection (IDI) uses a pre-combustion chamber, whereas direct injection (DI) occurs directly into the combustion chamber.

3. The three phases of diesel combustion include:
   **a.** Ignition delay
   **b.** Rapid combustion
   **c.** Controlled combustion

4. The typical diesel engine fuel system consists of the fuel tank, lift pump, water–fuel separator, and fuel filter.

5. The engine-driven injection pump supplies high-pressure diesel fuel to the injectors.

6. The two most common types of fuel injection used in diesel engines are:
   **a.** Distributor-type injection pump
   **b.** Common rail design where all of the injectors are fed from the same fuel supply from a rail under high pressure

7. Injector nozzles are opened either by the high-pressure pulse from the distributor pump or electrically by the computer on a common rail design.

8. Glow plugs are used to help start a cold diesel engine and help prevent excessive white smoke during warm-up.

9. Emissions are controlled on newer diesel engines by using a diesel oxidation catalytic converter, a diesel exhaust particulate filter, exhaust gas recirculation, and a selective catalytic reduction system.

10. Diesel engines can be tested using a scan tool, as well as measuring the glow plug resistance or compression reading, to determine a weak or nonfunctioning cylinder.

## REVIEW QUESTIONS

1. What is the difference between direct injection and indirect injection?

2. What are the three phases of diesel ignition?

3. What are the two most commonly used types of diesel injection systems?

4. Why are glow plugs kept working after the engine starts?

5. What exhaust aftertreatment is needed to achieve exhaust emission standards for vehicles 2007 and newer?

6. What are the advantages and disadvantages of SCR?

## CHAPTER QUIZ

1. How is diesel fuel ignited in a warm diesel engine?
   **a.** Glow plugs
   **b.** Heat of compression
   **c.** Spark plugs
   **d.** Distributorless ignition system

2. Which type of diesel injection produces less noise?
   **a.** Indirect injection (IDI)
   **b.** Common rail
   **c.** Direct injection
   **d.** Distributor injection

3. Which diesel injection system requires the use of a glow plug?
   **a.** Indirect injection (IDI)
   **b.** High-pressure common rail
   **c.** Direct injection
   **d.** Distributor injection

4. The three phases of diesel ignition include _____.
   **a.** Glow plug ignition, fast burn, slow burn
   **b.** Slow burn, fast burn, slow burn
   **c.** Ignition delay, rapid combustion, controlled combustion
   **d.** Glow plug ignition, ignition delay, controlled combustion

5. What fuel system component is used in a vehicle equipped with a diesel engine that is seldom used on the same vehicle when it is equipped with a gasoline engine?
   **a.** Fuel filter
   **b.** Fuel supply line
   **c.** Fuel return line
   **d.** Water–fuel separator

6. The diesel injection pump is usually driven by a _____.
   **a.** Gear off the camshaft
   **b.** Belt off the crankshaft
   **c.** Shaft drive off the crankshaft
   **d.** Chain drive off the camshaft

7. Which diesel system supplies high-pressure diesel fuel to all of the injectors all of the time?
   a. Distributor
   b. Inline
   c. High-pressure common rail
   d. Rotary

8. Glow plugs should have high resistance when _____ and lower resistance when _____.
   a. Cold/warm
   b. Warm/cold
   c. Wet/dry
   d. Dry/wet

9. Technician A says that glow plugs are used to help start a diesel engine and are shut off as soon as the engine starts. Technician B says that the glow plugs are turned off as soon as a flame is detected in the combustion chamber. Which technician is correct?
   a. Technician A only
   b. Technician B only
   c. Both Technicians A and B
   d. Neither Technician A nor B

10. What part should be removed to test cylinder compression on a diesel engine?
    a. Injector
    b. Intake valve rocker arm and stud
    c. Glow plug
    d. Glow plug or injector

# chapter 5
# GASOLINE

**LEARNING OBJECTIVES:** **After studying this chapter, the reader should be able to:** • Discuss the chemical composition and the process of refining gasoline. • Explain how driveability is affected by volatility. • Understand the process of gasoline combustion and the means of avoiding abnormal combustion. • Describe gasoline additives, reformulated gasoline, and gasoline blending. • Discuss how to test gasoline for alcohol content and the general gasoline purchase and use recommendations.

**KEY TERMS:** Air–fuel ratio 85 • Antiknock index (AKI) 86 • American Society for Testing and Materials (ASTM) 83 • British thermal unit (BTU) 84 • Catalytic cracking 82 • Cracking 82 • Detonation 85 • Distillation 81 • Distillation curve 83 • Driveability index (DI) 83 • E10 88 • Ethanol 88 • Fungible 82 • Gasoline 81 • Hydrocracking 82 • Octane rating 85 • Oxygenated fuels 88 • Petroleum 81 • Ping 85 • Reformulated gasoline (RFG) 89 • Reid vapor pressure (RVP) 83 • Spark knock 85 • Stoichiometric 85 • Tetraethyl lead (TEL) 86 • Vapor lock 83 • Volatility 83 • World Wide Fuel Charter (WWFC) 91

## GASOLINE

**DEFINITION** **Gasoline** is a term used to describe a complex mixture of various hydrocarbons refined from crude petroleum oil for use as a fuel in engines. Gasoline and air burns in the cylinder of the engine and produces heat and pressure, which is converted to rotary motion inside the engine and eventually powers the drive wheels of a vehicle. When combustion occurs, carbon dioxide and water are produced if the process is perfect and all of the air and all of the fuel are consumed in the process.

**CHEMICAL COMPOSITION** Gasoline is a combination of hydrocarbon molecules that have between five and 12 carbon atoms. The names of these various hydrocarbons are based on the number of carbon atoms and include:

- **Methane**—one carbon atom
- **Ethane**—two carbon atoms
- **Propane**—three carbon atoms
- **Butane**—four carbon atoms
- **Pentane**—five carbon atoms
- **Hexane**—six carbon atoms
- **Heptane**—seven carbon atoms (Used to test octane rating—has an octane rating of zero)
- **Octane**—eight carbon atoms (A type of octane is used as a basis for antiknock rating)

## REFINING

**TYPES OF CRUDE OIL** Refining is a complex combination of interdependent processing units that can separate crude oil into useful products such as gasoline and diesel fuel. As it comes out of the ground, **petroleum** (meaning "rock oil") crude can be as thin and light colored as apple cider or as thick and black as melted tar. A barrel of crude oil is 42 gallons, not 55 gallons as commonly used for industrial barrels. Typical terms used to describe the type of crude oil include:

- Thin crude oil has a high American Petroleum Institute (API) gravity, and therefore, is called *high-gravity* crude.
- Thick crude oil is called *low-gravity* crude. High-gravity-type crude contains more natural gasoline and its lower sulfur and nitrogen content makes it easier to refine.
- Low-sulfur crude oil is also known as "sweet" crude.
- High-sulfur crude oil is also known as "sour" crude.

**DISTILLATION** In the late 1800s, crude was separated into different products by boiling in a process called **distillation.** Distillation works because crude oil is composed of hydrocarbons with a broad range of boiling points.

In a distillation column, the vapor of the lowest-boiling hydrocarbons, propane and butane, rises to the top. The straight-run gasoline (also called naphtha), kerosene, and diesel fuel cuts are drawn off at successively lower positions in the column.

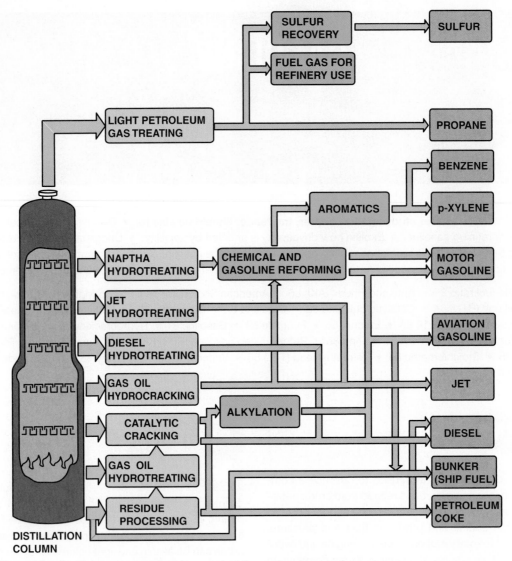

**FIGURE 5–1** The crude oil refining process showing most of the major steps and processes.

**CRACKING** **Cracking** is the process where hydrocarbons with higher boiling points could be broken down (cracked) into lower-boiling hydrocarbons by treating them to very high temperatures. This process, called *thermal cracking*, was used to increase gasoline production starting in 1913.

Instead of high heat, today cracking is performed using a catalyst and is called **catalytic cracking.** A catalyst is a material that speeds up or otherwise facilitates a chemical reaction without undergoing a permanent chemical change itself. Catalytic cracking produces gasoline of higher quality than thermal cracking.

**Hydrocracking** is similar to catalytic cracking in that it uses a catalyst, but the catalyst is in a hydrogen atmosphere. Hydrocracking can break down hydrocarbons that are resistant to catalytic cracking alone, and it is used to produce diesel fuel rather than gasoline.

Other types of refining processes include:

- Reforming
- Alkylation
- Isomerization
- Hydrotreating
- Desulfurization

● **SEE FIGURE 5–1.**

**SHIPPING** The gasoline is transported to regional storage facilities by railway tank car or by pipeline. In the pipeline method, all gasoline from many refiners is often sent through the same pipeline and can become mixed. All gasoline is said to be **fungible,** meaning that it is capable of being interchanged because each grade is created to specification so there is no reason to keep the different gasoline brands separated except for grade. Regular grade, mid-grade, and premium grades are separated in the pipeline and the additives are added at the regional storage facilities and then shipped by truck to individual gas stations.

**FIGURE 5–2** A gasoline testing kit, including an insulated container where water at 100°F is used to heat a container holding a small sample of gasoline. The reading on the pressure gauge is the Reid vapor pressure (RVP).

# VOLATILITY

**DEFINITION OF VOLATILITY** **Volatility** describes how easily the gasoline evaporates (forms a vapor). The definition of volatility assumes that the vapors will remain in the fuel tank or fuel line and will cause a certain pressure based on the temperature of the fuel.

**REID VAPOR PRESSURE (RVP)** **Reid vapor pressure (RVP)** is the pressure of the vapor above the fuel when the fuel is at 100°F (38°C). Increased vapor pressure permits the engine to start in cold weather. Gasoline without air will not burn. Gasoline must be vaporized (mixed with air) to burn in an engine. ● **SEE FIGURE 5–2.**

**SEASONAL BLENDING** Cold temperatures reduce the normal vaporization of gasoline; therefore, winter-blended gasoline is specially formulated to vaporize at lower temperatures for proper starting and driveability at low ambient temperatures. The **American Society for Testing and Materials (ASTM)** standards for winter-blend gasoline allow volatility of up to 15 pounds per square inch (PSI) RVP.

At warm ambient temperatures, gasoline vaporizes easily. However, the fuel system (fuel pump, carburetor, fuel-injector nozzles, etc.) is designed to operate with liquid gasoline. The volatility of summer-grade gasoline should be about 7.0 PSI RVP. According to ASTM standards, the maximum RVP should be 10.5 PSI for summer-blend gasoline.

**DISTILLATION CURVE** Besides Reid vapor pressure, another method of classifying gasoline volatility is the **distillation curve**. A curve on a graph is created by plotting the temperature at which the various percentage of the fuel evaporates. A typical distillation curve is shown in ● **FIGURE 5–3.**

 **FREQUENTLY ASKED QUESTION**

**Why Do I Get Lower Gas Mileage in the Winter?**

Several factors cause the engine to use more fuel in the winter than in the summer, including:

- Gasoline that is blended for use in cold climates is designed for ease of starting and contains fewer heavy molecules, which contribute to fuel economy. The heat content of winter gasoline is lower than summer-blended gasoline.
- In cold temperatures, all lubricants are stiff, causing more resistance. These lubricants include the engine oil, as well as the transmission and differential gear lubricants.
- Heat from the engine is radiated into the outside air more rapidly when the temperature is cold, resulting in longer run time until the engine has reached normal operating temperature.
- Road conditions, such as ice and snow, can cause tire slippage or additional drag on the vehicle.

**DRIVEABILITY INDEX** A distillation curve shows how much of a gasoline evaporates at what temperature range. To predict cold-weather driveability, an index was created called the **driveability index,** also called the *distillation index*, and abbreviated **DI.**

The DI was developed using the temperature for the evaporated percentage of 10% (labeled T10), 50% (labeled T50), and 90% (labeled T90). The formula for DI is:

$$DI = 1.5 \times T10 + 3 \times T50 + T90$$

The total DI is a temperature and usually ranges from 1,000°F to 1,200°F. The lower values of DI generally result in good cold-start and warm-up performance. A high DI number is less volatile than a low DI number.

**NOTE: Most premium-grade gasoline has a higher (worse) DI than regular-grade or midgrade gasoline, which could cause poor cold-weather driveability. Vehicles designed to operate on premium-grade gasoline are programmed to handle the higher DI, but engines designed to operate on regular-grade gasoline may not be able to provide acceptable cold-weather driveability.**

**VOLATILITY-RELATED PROBLEMS** At higher temperatures, liquid gasoline can easily vaporize, which can cause **vapor lock.** Vapor lock is a *lean* condition caused by vaporized fuel in the fuel system. This vaporized fuel takes up space normally occupied by liquid fuel. Bubbles that form in the fuel cause vapor lock, preventing proper operation of the fuel-injection system.

Heat causes some fuel to evaporate, thereby causing bubbles. Sharp bends cause the fuel to be restricted at the bend. When the fuel flows past the bend, the fuel can expand to fill the space after the bend. This expansion drops the pressure, and

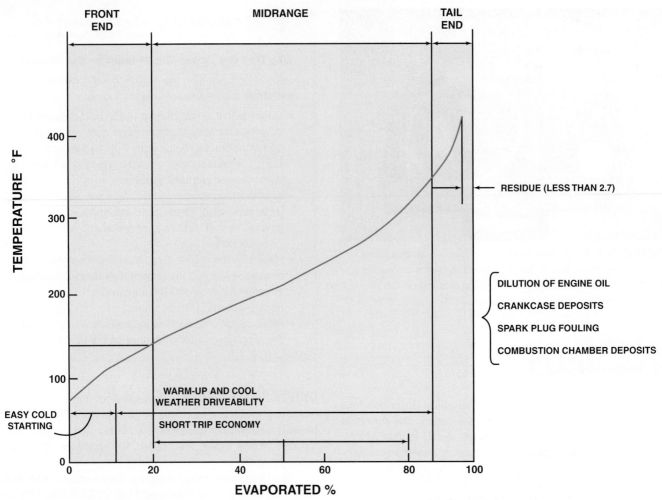

**FIGURE 5–3** A typical distillation curve. Heavier molecules evaporate at higher temperatures and contain more heat energy for power, whereas the lighter molecules evaporate easier for starting.

bubbles form in the fuel lines. When the fuel is full of bubbles, the engine is not being supplied with enough fuel and the engine runs lean. A lean engine will stumble during acceleration, will run rough, and may stall. Warm weather and alcohol-blended fuels both tend to increase vapor lock and engine performance problems.

If winter-blend gasoline (or high-RVP fuel) is used in an engine during warm weather, the following problems may occur:

1. Rough idle
2. Stalling
3. Hesitation on acceleration
4. Surging

# GASOLINE COMBUSTION PROCESS

**CHEMICAL REACTIONS** The combustion process involves the chemical combination of oxygen ($O_2$) from the air (about 21% of the atmosphere) with the hydrogen and carbon from the fuel. In a gasoline engine, a spark starts the combustion process, which takes about 3 ms (0.003 sec) to be completed inside the cylinder of an engine. The chemical reaction that takes place can be summarized as follows: hydrogen (H) plus carbon (C) plus oxygen ($O_2$) plus nitrogen (N) plus spark equals heat plus water ($H_2O$) plus carbon monoxide (CO) (if incomplete combustion) plus carbon dioxide ($CO_2$) plus hydrocarbons (HC) plus oxides of nitrogen ($NO_x$) plus many other chemicals. In an equation format it looks like this:

$$H + C + O_2 + N + Spark = Heat + CO_2 + HC + NO_x$$

**HEAT ENERGY** The heat produced by the combustion process is measured in **British thermal units (BTUs).** One BTU is the amount of heat required to raise one pound of water one Fahrenheit degree. The metric unit of heat is the *calorie* (cal). One calorie is the amount of heat required to raise the temperature of one gram (g) of water one Celsius degree:

Gasoline—About 114,000 BTUs per gallon

**AIR–FUEL RATIOS** Fuel burns best when the intake system turns it into a fine spray and mixes it with air before sending it into the cylinders. In fuel-injected engines, the fuel becomes a spray and mixes with the air in the intake manifold. There is a direct relationship between engine airflow and fuel requirements; this is called the **air–fuel ratio.**

The air–fuel ratio is the proportion by weight of air and gasoline that the injection system mixes as needed for engine combustion. The mixtures, with which an engine can operate without stalling, range from 8 to 1 to 18.5 to 1. ● **SEE FIGURE 5–4.**

These ratios are usually stated by weight, such as:

- 8 parts of air by weight combined with 1 part of gasoline by weight (8:1), which is the richest mixture that an engine can tolerate and still fire reliably.

- 18.5 parts of air mixed with 1 part of gasoline (18.5:1), which is the leanest practical ratio. Richer or leaner air–fuel ratios cause the engine to misfire badly or not run at all.

**STOICHIOMETRIC AIR–FUEL RATIO** The ideal mixture or ratio at which all of the fuel combines with all of the oxygen in the air and burns completely is called the **stoichiometric** ratio, a chemically perfect combination. In theory, this ratio for gasoline is an air–fuel mixture of 14.7 to 1. ● **SEE FIGURE 5–5.**

In reality, the exact ratio at which perfect mixture and combustion occurs depends on the molecular structure of gasoline, which can vary. The stoichiometric ratio is a compromise between maximum power and maximum economy.

## NORMAL AND ABNORMAL COMBUSTION

The **octane rating** of gasoline is the measure of its antiknock properties. *Engine knock* (also called **detonation, spark knock,** or **ping**) is a metallic noise an engine makes, usually during acceleration, resulting from abnormal or uncontrolled combustion inside the cylinder.

Normal combustion occurs smoothly and progresses across the combustion chamber from the point of ignition. ● **SEE FIGURE 5–6.**

Normal flame-front combustion travels between 45 and 90 mph (72 and 145 km/h). The speed of the flame front

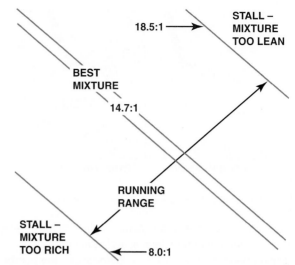

**FIGURE 5–4** An engine will not run if the air–fuel mixture is either too rich or too lean.

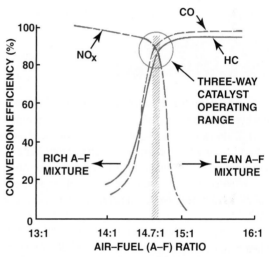

**FIGURE 5–5** With a three-way catalytic converter, emission control is most efficient with an air–fuel ratio between 14.65 to 1 and 14.75 to 1.

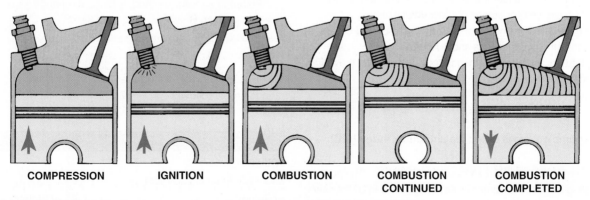

**FIGURE 5–6** Normal combustion is a smooth, controlled burning of the air–fuel mixture.

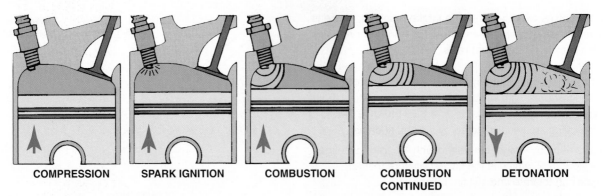

| COMPRESSION | SPARK IGNITION | COMBUSTION | COMBUSTION CONTINUED | DETONATION |

**FIGURE 5–7** Detonation is a secondary ignition of the air–fuel mixture. It is also called spark knock or pinging.

depends on the air–fuel ratio, combustion chamber design (determining amount of turbulence), and temperature.

During periods of spark knock (detonation), the combustion speed increases by up to 10 times to near the speed of sound. The increased combustion speed also causes increased temperatures and pressures, which can damage pistons, gaskets, and cylinder heads. ● **SEE FIGURE 5–7.**

One of the first additives used in gasoline was **tetraethyl lead (TEL).** TEL was added to gasoline in the early 1920s to reduce the tendency to knock. It was often called ethyl or high-test gasoline.

# OCTANE RATING

The antiknock standard or basis of comparison is the knock-resistant hydrocarbon isooctane, chemically called trimethylpentane ($C_8H_{18}$), also known as 2-2-4 trimethylpentane. If a gasoline tested had the exact same antiknock characteristics as isooctane, it was rated as 100-octane gasoline. If the gasoline tested had only 85% of the antiknock properties of isooctane, it was rated as 85 octane. Remember, octane rating is only a comparison test.

The two basic methods used to rate gasoline for antiknock properties (octane rating) are the *research method* and the *motor method*. Each uses a model of the special cooperative fuel research (CFR) single-cylinder engine. The research method and the motor method vary as to temperature of air, spark advance, and other parameters. The research method typically results in readings that are 6 to 10 points higher than those of the motor method. For example, a fuel with a research octane number (RON) of 93 might have a motor octane number (MON) of 85.

The octane rating posted on pumps in the United States is the average of the two methods and is referred to as (R + M) ÷ 2, meaning that, for the fuel used in the previous example, the rating posted on the pumps would be

$$\frac{RON + MON}{2} = \frac{93 + 85}{2} = 89$$

The pump octane is called the **antiknock index (AKI).**

**GASOLINE GRADES AND OCTANE NUMBER** The posted octane rating on gasoline pumps is the rating achieved by the average of the research and the motor methods. ● **SEE FIGURE 5–8.**

**FIGURE 5–8** A pump showing regular with a pump octane of 87, plus rated at 89, and premium rated at 93. These ratings can vary with brand as well as in different parts of the country.

 **FREQUENTLY ASKED QUESTION**

**What Grade of Gasoline Does the EPA Use When Testing Engines?**

Due to the various grades and additives used in commercial fuel, the government (EPA) uses a liquid called indolene. Indolene has a research octane number of 96.5 and a motor method octane rating of 88, which results in an R + M ÷ 2 rating of 92.25.

Except in high-altitude areas, the grades and octane ratings are as follows:

| Grades | Octane rating |
| --- | --- |
| Regular | 87 |
| Midgrade (also called Plus) | 89 |
| Premium | 91 or higher |

**FIGURE 5–9** The posted octane rating in most high-altitude areas shows regular at 85 instead of the usual 87.

 **TECH TIP**

### Horsepower and Fuel Flow

To produce 1 hp, the engine must be supplied with 0.50 lb of fuel per hour (lb/hr). Fuel injectors are rated in pounds per hour. For example, a V-8 engine equipped with 25 lb/hr fuel injectors could produce 50 hp per cylinder (per injector) or 400 hp. Even if the cylinder head or block is modified to produce more horsepower, the limiting factor may be the injector flow rate.

The following are flow rates and resulting horsepower for a V-8 engine:

30 lb/hr: 60 hp per cylinder or 480 hp
35 lb/hr: 70 hp per cylinder or 560 hp
40 lb/hr: 80 hp per cylinder or 640 hp

Of course, injector flow rate is only one of many variables that affect power output. Installing larger injectors without other major engine modification could decrease engine output and drastically increase exhaust emissions.

# HIGH-ALTITUDE OCTANE REQUIREMENTS

As the altitude increases, atmospheric pressure drops. The air is less dense because a pound of air takes more volume. The octane rating of fuel does not need to be as high because the engine cannot take in as much air. This process will reduce the combustion (compression) pressures inside the engine. In mountainous areas, gasoline (R + M) ÷ 2 octane ratings are two or more numbers lower than normal (according to the SAE, about one octane number lower per 1,000 ft or 300 m in altitude). 
● **SEE FIGURE 5–9.**

A secondary reason for the lowered octane requirement of engines running at higher altitudes is the normal enrichment of the air–fuel ratio and lower engine vacuum with the decreased air density. Some problems, therefore, may occur when driving out of high-altitude areas into lower-altitude areas where the octane rating must be higher. Most computerized engine control systems can compensate for changes in altitude and modify air–fuel ratio and ignition timing for best operation.

Because the combustion burn rate slows at high altitude, the ignition (spark) timing can be advanced to improve power. The amount of timing advance can be about 1 degree per 1,000 ft over 5,000 ft. Therefore, if driving at 8,000 ft of altitude, the ignition timing can be advanced 3 degrees.

High altitude also allows fuel to evaporate more easily. The volatility of fuel should be reduced at higher altitudes to prevent vapor from forming in sections of the fuel system, which can cause driveability and stalling problems. The extra heat generated in climbing to higher altitudes plus the lower atmospheric pressure at higher altitudes combine to cause vapor lock problems as the vehicle goes to higher altitudes.

# GASOLINE ADDITIVES

**DYE** Dye is usually added to gasoline at the distributor to help identify the grade and/or brand of fuel. In many countries, fuels are required to be colored using a fuel-soluble dye. In the United States and Canada, diesel fuel used for off-road use and not taxed is required to be dyed red for identification. Gasoline sold for off-road use in Canada is dyed purple.

**OCTANE IMPROVER ADDITIVES** When gasoline companies, under federal EPA regulations, removed tetraethyl lead from gasoline, other methods were developed to help maintain the antiknock properties of gasoline. Octane improvers (enhancers) can be grouped into three broad categories:

1. Aromatic hydrocarbons (hydrocarbons containing the benzene ring) such as xylene and toluene
2. Alcohols such as ethanol (ethyl alcohol), methanol (methyl alcohol), and tertiary butyl alcohol (TBA)
3. Metallic compounds such as methylcyclopentadienyl manganese tricarbonyl (MMT)

**NOTE: MMT has been proven to be harmful to catalytic converters and can cause spark plug fouling. However, MMT is currently one of the active ingredients commonly found in octane improvers available to the public and in some gasoline sold in Canada. If an octane boost additive has been used that contains MMT, the spark plug porcelain will be rust colored around the tip.**

Propane and butane, which are volatile by-products of the refinery process, are also often added to gasoline as octane improvers. The increase in volatility caused by the added propane and butane often leads to hot-weather driveability problems.

### Can Regular-Grade Gasoline Be Used If Premium Is the Recommended Grade?

Maybe. It is usually possible to use regular-grade or midgrade (plus) gasoline in most newer vehicles without danger of damage to the engine. Most vehicles built since the 1990s are equipped with at least one knock sensor. If a lower octane gasoline than specified is used, the engine ignition timing setting will usually cause the engine to spark knock, also called detonation or ping. This spark knock is detected by the knock sensor(s), which sends a signal to the computer. The computer then retards the ignition timing until the spark knock stops.

**NOTE: Some scan tools will show the "estimated octane rating" of the fuel being used, which is based on knock sensor activity.**

As a result of this spark timing retardation, the engine torque is reduced. While this reduction in power is seldom noticed, it will reduce fuel economy, often by 4 to 5 miles per gallon. If premium gasoline is then used, the PCM will gradually permit the engine to operate at the more advanced ignition timing setting. Therefore, it may take several tanks of premium gasoline to restore normal fuel economy. For best overall performance, use the grade of gasoline recommended by the vehicle manufacturer.

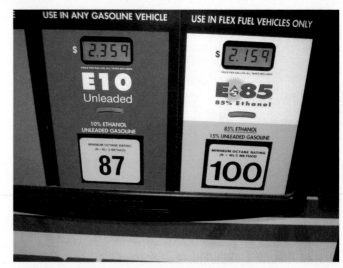

**FIGURE 5–10** This refueling pump indicates that the gasoline is blended with 10% ethanol (ethyl alcohol) and can be used in any gasoline vehicle. E85 contains 85% ethanol and can be used only in vehicles specifically designed to use it.

### What Is Meant by "Phase Separation?"

All alcohols absorb water, and the alcohol–water mixture can separate from the gasoline and sink to the bottom of the fuel tank. This process is called *phase separation*. To help avoid engine performance problems, try to keep at least a quarter tank of fuel at all times, especially during seasons when there is a wide temperature span between daytime highs and nighttime lows. These conditions can cause moisture to accumulate in the fuel tank as a result of condensation of the moisture in the air.

**OXYGENATED FUEL ADDITIVES** Oxygenated fuels contain oxygen in the molecule of the fuel itself. Examples of oxygenated fuels include methanol, ethanol, methyl tertiary butyl ether (MTBE), tertiary-amyl methyl ether (TAME), and ethyl tertiary butyl ether (ETBE).

Oxygenated fuels are commonly used in high-altitude areas to reduce carbon monoxide (CO) emissions. The extra oxygen in the fuel itself is used to convert harmful CO into carbon dioxide ($CO_2$). The extra oxygen in the fuel helps ensure that there is enough oxygen to convert all CO into $CO_2$ during the combustion process in the engine or catalytic converter.

**METHYL TERTIARY BUTYL ETHER (MTBE).** MTBE is manufactured by means of the chemical reaction of methanol and isobutylene. Unlike methanol, MTBE does not increase the volatility of the fuel and is not as sensitive to water as are other alcohols. The maximum allowable volume level, according to the EPA, is 15% but is currently being phased out because of health concerns, as well as MTBE contamination of drinking water if spilled from storage tanks.

**TERTIARY-AMYL METHYL ETHER.** Tertiary-amyl methyl ether (TAME) contains an oxygen atom bonded to two carbon atoms and is added to gasoline to provide oxygen to the fuel. It is slightly soluble in water, very soluble in ethers and alcohol, and soluble in most organic solvents including hydrocarbons.

**ETHYL TERTIARY BUTYL ETHER.** ETBE is derived from ethanol. The maximum allowable volume level is 17.2%. The use of ETBE is the cause of much of the odor from the exhaust of vehicles using reformulated gasoline.

**ETHANOL.** Ethanol, also called *ethyl alcohol* is drinkable alcohol and is usually made from grain. Adding 10% ethanol (ethyl alcohol or grain alcohol) increases the (R + M) ÷ 2 octane rating by three points. The alcohol added to the base gasoline, however, also raises the volatility of the fuel about 0.5 PSI. Most automobile manufacturers permit up to 10% ethanol if driveability problems are not experienced.

The oxygen content of a 10% blend of ethanol in gasoline, called **E10**, is 3.5% oxygen by weight. ● **SEE FIGURE 5–10.**

Keeping the fuel tank full reduces the amount of air and moisture in the tank. ● **SEE FIGURE 5–11.**

**FIGURE 5–11** A container with gasoline containing alcohol. Notice the separation line where the alcohol–water mixture separated from the gasoline and sank to the bottom.

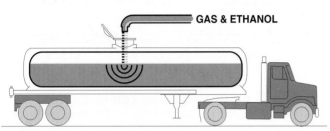

**FIGURE 5–12** In-line blending is the most accurate method for blending ethanol with gasoline because computers are used to calculate the correct ratio.

## GASOLINE BLENDING

Gasoline additives, such as ethanol and dyes, are usually added to the fuel at the distributor. Adding ethanol to gasoline is a way to add oxygen to the fuel itself. Gasoline containing an addition that has oxygen is called *oxygenated fuel*. There are three basic methods used to blend ethanol with gasoline to create E10 (10% ethanol, 90% gasoline):

1. **In-line blending.** Gasoline and ethanol are mixed in a storage tank or in the tank of a transport truck while it is being filled. Because the quantities of each can be accurately measured, this method is most likely to produce a well-mixed blend of ethanol and gasoline. ● SEE FIGURE 5–12.

2. **Sequential blending.** This method is usually performed at the wholesale terminal and involves adding a measured amount of ethanol to a tank truck followed by a measured amount of gasoline. ● SEE FIGURE 5–13.

3. **Splash blending.** Splash blending can be done at the retail outlet or distributor and involves separate purchases of ethanol and gasoline. In a typical case, a distributor can purchase gasoline, and then drive to another supplier and purchase ethanol. The ethanol is then added (splashed) into the tank of gasoline. This method is the least-accurate method of blending and can result in ethanol concentration for E10 that should be 10% to range from 5% to over 20% in some cases. ● SEE FIGURE 5–14.

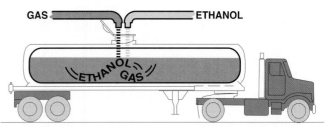

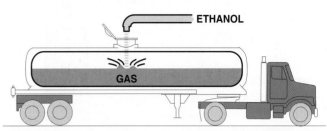

**FIGURE 5–13** Sequential blending uses a computer to calculate the correct ratio as well as the prescribed order in which the products are loaded.

**FIGURE 5–14** Splash blending occurs when the ethanol is added to a tanker with gasoline and is mixed as the truck travels to the retail outlet.

---

**?  FREQUENTLY ASKED QUESTION**

**Is Water Heavier Than Gasoline?**

Yes. Water weighs about 8 pounds per gallon, whereas gasoline weighs about 6 pounds per gallon. The density as measured by specific gravity includes:

Water = 1.000 (the baseline for specific gravity)

Gasoline = 0.730 to 0.760

This means that any water that gets into the fuel tank will sink to the bottom.

---

## REFORMULATED GASOLINE

**Reformulated gasoline (RFG)** is manufactured to help reduce emissions. The gasoline refiners reformulate gasoline by using additives that contain at least 2% oxygen by weight and reducing the additive benzene to a maximum of 1% by volume. Two other major changes done at the refineries are as follows:

1. **Reduce light compounds.** Refineries eliminate butane, pentane, and propane, which have a low boiling point and evaporate easily. These unburned hydrocarbons are released into the atmosphere during refueling and through the fuel tank vent system, contributing to smog formation. Therefore, reducing the light compounds from gasoline helps reduce evaporative emissions.

2. **Reduce heavy compounds.** Refineries eliminate heavy compounds with high boiling points such as aromatics and olefins. The purpose of this reduction is to reduce the amount of unburned hydrocarbons that enter the catalytic

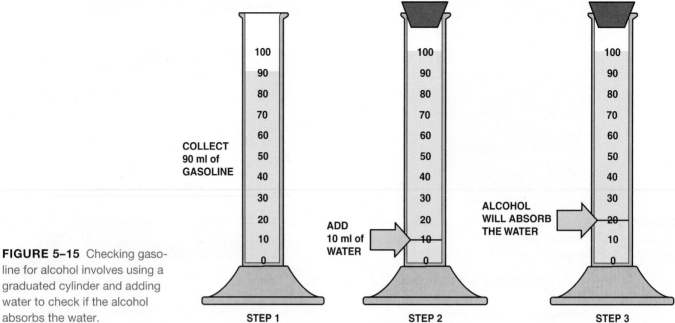

**FIGURE 5–15** Checking gasoline for alcohol involves using a graduated cylinder and adding water to check if the alcohol absorbs the water.

COLLECT 90 ml of GASOLINE

STEP 1

ADD 10 ml of WATER

STEP 2

ALCOHOL WILL ABSORB THE WATER

STEP 3

converter, which makes the converter more efficient, thereby reducing emissions.

Because many of the heavy compounds are eliminated, a drop in fuel economy of about 1 mpg has been reported in areas where reformulated gasoline is being used. Formaldehyde is formed when RFG is burned, and the vehicle exhaust has a unique smell when reformulated gasoline is used.

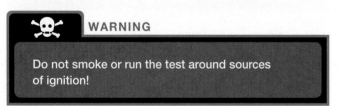

# TESTING GASOLINE FOR ALCOHOL CONTENT

Take the following steps when testing gasoline for alcohol content:

**☠ WARNING**

Do not smoke or run the test around sources of ignition!

1. Pour suspect gasoline into a graduated cylinder.
2. Carefully fill the graduated cylinder to the 90-mL mark.
3. Add 10 mL of water to the graduated cylinder by counting the number of drops from an eyedropper.
4. Put the stopper in the cylinder and shake vigorously for 1 minute. Relieve built-up pressure by occasionally removing the stopper. Alcohol dissolves in water and will drop to the bottom of the cylinder.
5. Place the cylinder on a flat surface and let it stand for 2 minutes.

**❓ FREQUENTLY ASKED QUESTION**

**How Does Alcohol Content in the Gasoline Affect Engine Operation?**

In most cases, the use of gasoline containing 10% or less of ethanol (ethyl alcohol) has little or no effect on engine operation. However, because the addition of 10% ethanol raises the volatility of the fuel slightly, occasional rough idle or stalling may be noticed, especially during warm weather. The rough idle and stalling may also be noticeable after the engine is started, driven, then stopped for a short time. Engine heat can vaporize the alcohol-enhanced fuel causing bubbles to form in the fuel system. These bubbles in the fuel prevent the proper operation of the fuel injection system and result in a hesitation during acceleration, rough idle, or in severe cases repeated stalling until all the bubbles have been forced through the fuel system, replaced by cooler fuel from the fuel tank.

6. Take a reading near the bottom of the cylinder at the boundary between the two liquids.
7. For percentage of alcohol in gasoline, subtract 10 from the reading.

For example,

**The reading is 20 mL: 20 − 10 = 10% alcohol**

If the increase in volume is 0.2% or less, it may be assumed that the test gasoline contains no alcohol. ● **SEE FIGURE 5–15.** Alcohol content can also be checked using an electronic tester. See the step-by-step sequence at the end of the chapter.

**FIGURE 5–16** Many vehicle manufacturers include warning labels to avoid E15 (15% ethanol and 85% gasoline) as well as E85 (85% ethanol and 15% gasoline) on vehicles that are not designated a flex fuel vehicle (FFV).

 **FREQUENTLY ASKED QUESTION**

**What Is "Top-Tier" Gasoline?**

Top-tier gasoline is gasoline that has specific standards for quality, including enough detergent to keep all intake valves clean. Four automobile manufacturers, including BMW, General Motors, Honda, and Toyota, developed the standards. Top-tier gasoline exceeds the quality standards developed by the **World Wide Fuel Char-ter (WWFC)** that was established in 2002 by vehicle and engine manufacturers. The gasoline companies that agreed to make fuel that matches or exceeds the standards as a top-tier fuel include ChevronTexaco and ConocoPhillips. For more information, visit http://www.toptiergas.com. ● **SEE FIGURE 5–16.**

# GENERAL GASOLINE RECOMMENDATIONS

The fuel used by an engine is a major expense in the operation cost of the vehicle. The proper operation of the engine depends on clean fuel of the proper octane rating and vapor pressure for the atmospheric conditions.

 **TECH TIP**

**The Sniff Test**

Problems can occur with stale gasoline from which the lighter parts of the gasoline have evaporated. Stale gasoline usually results in a no-start situation. If stale gasoline is suspected, sniff it. If it smells rancid, replace it with fresh gasoline.

**NOTE: If storing a vehicle, boat, or lawnmower over the winter, put some gasoline stabilizer into the gasoline to reduce the evaporation and separation that can occur during storage. Gas-oline stabilizer is easily available at lawnmower repair shops or marinas.**

To help ensure proper engine operation and keep fuel costs to a minimum, follow these guidelines:

1. Purchase fuel from a busy station to help ensure that it is fresh and less likely to be contaminated with water or moisture.

2. Keep the fuel tank above one-quarter full, especially during seasons in which the temperature rises and falls by more than 20°F between daytime highs and nighttime lows. This helps to reduce condensed moisture in the fuel tank and could prevent gas line freeze-up in cold weather.

   **NOTE: Gas line freeze-up occurs when the water in the gasoline freezes and forms an ice blockage in the fuel line.**

3. Do not purchase fuel with a higher octane rating than is nec-essary. Most newer engines are equipped with a detonation (knock) sensor that signals the vehicle computer to retard the ignition timing when spark knock occurs. Therefore, an operating difference may not be noticeable to the driver when using a low-octane fuel, except for a decrease in power and fuel economy. In other words, the engine with a knock sen-sor will tend to operate knock free on regular fuel, even if premium, higher-octane fuel is specified. Using premium fuel may result in more power and greater fuel economy. The in-crease in fuel economy, however, would have to be substan-tial to justify the increased cost of high-octane premium fuel. Some drivers find a good compromise by using midgrade (plus) fuel to benefit from the engine power and fuel economy gains without the cost of using premium fuel all the time.

4. Avoid using gasoline with alcohol in warm weather, even though many alcohol blends do not affect engine driveabil-ity. If warm-engine stumble, stalling, or rough idle occurs, change brands of gasoline.

5. Do not purchase fuel from a retail outlet when a tanker truck is filling the underground tanks. During the refill-ing procedure, dirt, rust, and water may be stirred up in the underground tanks. This undesirable material may be pumped into your vehicle's fuel tank.

**Why Should I Keep the Fuel Gauge Above One-Quarter Tank?**

The fuel pickup inside the fuel tank can help keep water from being drawn into the fuel system unless water is all that is left at the bottom of the tank. Over time, moisture in the air inside the fuel tank can condense, causing liquid water to drop to the bottom of the fuel tank (water is heavier than gasoline—about 8 lb per gallon for water and about 6 lb per gallon for gasoline). If alcohol-blended gasoline is used, the alcohol can absorb the water and the alcohol–water combination can be burned inside the engine. However, when water combines with alcohol, a separation layer occurs between the gasoline at the top of the tank and the alcohol–water combination at the bottom. When the fuel level is low, the fuel pump will draw from this concentrated level of alcohol and water. Because alcohol and water do not burn as well as pure gasoline, severe driveability problems can occur such as stalling, rough idle, hard starting, and missing.

**Do Not Overfill the Fuel Tank**

Gasoline fuel tanks have an expansion volume area at the top. The volume of this expansion area is equal to 10% to 15% of the volume of the tank. This area is normally not filled with gasoline, but rather is designed to provide a place for the gasoline to expand into, if the vehicle is parked in the hot sun and the gasoline expands. This prevents raw gasoline from escaping from the fuel system. A small restriction is usually present to control the amount of air and vapors that can escape the tank and flow to the charcoal canister.

This volume area could be filled with gasoline if the fuel is slowly pumped into the tank. Since it can hold an extra 10% (2 gallons in a 20-gallon tank), some people deliberately try to fill the tank completely. When this expansion volume is filled, liquid fuel (rather than vapors) can be drawn into the charcoal canister. When the purge valve opens, liquid fuel can be drawn into the engine, causing an excessively rich air–fuel mixture. Not only can this liquid fuel harm vapor recovery parts, but overfilling the gas tank could also cause the vehicle to fail an exhaust emission test, particularly during an enhanced test when the tank could be purged while on the rollers.

**FIGURE 5–17** Many gasoline service stations have signs posted warning customers to place plastic fuel containers on the ground while filling. If placed in a trunk or pickup truck bed equipped with a plastic liner, static electricity could build up during fueling and discharge from the container to the metal nozzle, creating a spark and possible explosion. Some service stations have warning signs not to use cell phones while fueling to help avoid the possibility of an accidental spark creating a fire hazard.

6. Do not overfill the gas tank. After the nozzle clicks off, add just enough fuel to round up to the next dime. Adding additional gasoline will cause the excess to be drawn into the charcoal canister. This can lead to engine flooding and excessive exhaust emissions.

7. Be careful when filling gasoline containers. Always fill a gas can on the ground to help prevent the possibility of static electricity buildup during the refueling process. ● **SEE FIGURE 5–17.**

**1** A fuel composition tester (SPX Kent-Moore J-44175) is the recommended tool, by General Motors, to use to test the alcohol content of gasoline.

**2** This battery-powered tester uses light-emitting diodes (LEDs), meter lead terminals, and two small openings for the fuel sample.

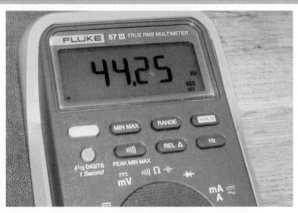

**3** The first step is to verify the proper operation of the tester by measuring the air frequency by selecting AC hertz on the meter. The air frequency should be between 35 and 48 Hz.

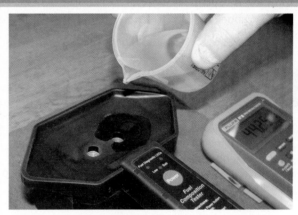

**4** After verifying that the tester is capable of correctly reading the air frequency, gasoline is poured into the testing cell of the tool.

**5** Record the AC frequency as shown on the meter and subtract 50 from the reading. (e.g., 60.50 − 50.00 = 10.5). This number (10.5) is the percentage of alcohol in the gasoline sample.

**6** Adding additional amounts of ethyl alcohol (ethanol) increases the frequency reading.

## SUMMARY

1. Gasoline is a complex blend of hydrocarbons. Gasoline is blended for seasonal usage to achieve the correct volatility for easy starting and maximum fuel economy under all driving conditions.

2. Winter-blend fuel used in a vehicle during warm weather can cause a rough idle and stalling because of its higher Reid vapor pressure (RVP).

3. Abnormal combustion (also called detonation or spark knock) increases both the temperature and the pressure inside the combustion chamber.

4. Most regular-grade gasoline today, using the (R + M) ÷ 2 rating method, is 87 octane; midgrade (plus) is 89; and premium grade is 91 or higher.

5. Oxygenated fuels contain oxygen to lower CO exhaust emissions.

6. Gasoline should always be purchased from a busy station, and the tank should not be overfilled.

## REVIEW QUESTIONS

1. What is the difference between summer-blend and winter-blend gasoline?

2. What is Reid vapor pressure?

3. What is vapor lock?

4. What does the (R + M) ÷ 2 gasoline pump octane rating indicate?

5. What are the octane improvers that may be used during the refining process?

6. What is stoichiometric?

## CHAPTER QUIZ

1. Winter-blend gasoline _____.
   a. Vaporizes more easily than summer-blend gasoline
   b. Has a higher RVP
   c. Can cause engine driveability problems if used during warm weather
   d. All of the above

2. Vapor lock can occur _____.
   a. As a result of excessive heat near fuel lines
   b. If a fuel line is restricted
   c. During both a and b
   d. During neither a nor b

3. Technician A says that spark knock, ping, and detonation are different names for abnormal combustion. Technician B says that any abnormal combustion raises the temperature and pressure inside the combustion chamber and can cause severe engine damage. Which technician is correct?
   a. Technician A only
   b. Technician B only
   c. Both Technicians A and B
   d. Neither Technician A nor B

4. Technician A says that the research octane number is higher than the motor octane number. Technician B says that the octane rating posted on fuel pumps is an average of the two ratings. Which technician is correct?
   a. Technician A only
   b. Technician B only
   c. Both Technicians A and B
   d. Neither Technician A nor B

5. Technician A says that in going to high altitudes, engines produce lower power. Technician B says that most engine control systems can compensate the air–fuel mixture for changes in altitude. Which technician is correct?
   a. Technician A only
   b. Technician B only
   c. Both Technicians A and B
   d. Neither Technician A nor B

6. Which method of blending ethanol with gasoline is the most accurate?
   a. In-line
   b. Sequential
   c. Splash
   d. All of the above are equally accurate methods

7. What can be used to measure the alcohol content in gasoline?
   a. Graduated cylinder
   b. Electronic tester
   c. Scan tool
   d. Either a or b

8. To avoid problems with the variation of gasoline, all government testing uses _____ as a fuel during testing procedures.
   a. MTBE (methyl tertiary butyl ether)
   b. Indolene
   c. Xylene
   d. TBA (tertiary butyl alcohol)

9. Avoid topping off the fuel tank because _____.
   a. It can saturate the charcoal canister
   b. The extra fuel simply spills onto the ground
   c. The extra fuel increases vehicle weight and reduces performance
   d. The extra fuel goes into the expansion area of the tank and is not used by the engine

10. Using ethanol-enhanced or reformulated gasoline can result in reduced fuel economy.
    a. True
    b. False

# chapter 6

# ALTERNATIVE FUELS

**LEARNING OBJECTIVES:** **After studying this chapter, the reader should be able to:** • Discuss the alternatives to gasoline.
• Understand how alternative fuels affect driveability. • Explain how alternative fuels reduce CO exhaust emissions.
• List the safety precautions to be taken when working with alternative fuels.

**KEY TERMS:** AFV 97 • Anhydrous ethanol 96 • Biomass 101 • Cellulose ethanol 96 • Cellulosic biomass 96
• Coal to liquid (CTL) 105 • Compressed natural gas (CNG) 102 • E85 96 • Ethanol 95 • Ethyl alcohol 95 • FFV 97
• Fischer–Tropsch 104 • Flex fuels 97 • FTD 105 • Fuel compensation sensor 97 • Gas to liquid (GTL) 105 • Grain
alcohol 95 • Liquefied petroleum gas (LPG) 101 • LP-gas 101 • M85 101 • Methanol 100 • Methanol to gasoline (MTG) 105 •
NGV 102 • Propane 101 • Switchgrass 96 • Syncrude 105 • Syn-gas 101 • Synthetic fuel 104 • Underground coal gasification
(UCG) 105 • V-FFV 98 • Variable fuel sensor 97

## ETHANOL

**ETHANOL TERMINOLOGY** **Ethanol** is also called **ethyl alcohol** or **grain alcohol,** because it is usually made from grain and is the type of alcohol found in alcoholic drinks such as beer, wine, and distilled spirits like whiskey. Ethanol is composed of two carbon atoms and six hydrogen atoms with one added oxygen atom. ● **SEE FIGURE 6–1.**

**ETHANOL PRODUCTION** Conventional ethanol is derived from grains, such as corn, wheat, or soybeans. Corn, for example, is converted to ethanol through either a dry or wet milling process. In dry milling operations, liquefied cornstarch is produced by heating cornmeal with water and enzymes. A second enzyme converts the liquefied starch to sugars, which are fermented by yeast into ethanol and carbon dioxide. Wet milling operations separate the fiber, germ (oil), and protein from the starch before it is fermented into ethanol.

The majority of the ethanol in the United States is made from:

■ Corn
■ Grain
■ Sorghum
■ Wheat
■ Barley
■ Potatoes

In Brazil, the world's largest ethanol producer, it is made from sugarcane. Ethanol can be made by the dry mill process

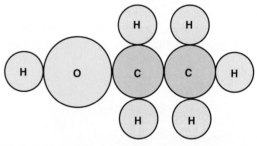

**FIGURE 6–1** The ethanol molecule showing two carbon atoms, six hydrogen atoms, and one oxygen atom.

 **FREQUENTLY ASKED QUESTION**

**Does Ethanol Production Harm the Environment?**

The production of ethanol is referred to as being carbon neutral because the amount of $CO_2$ released during production is equal to the amount of $CO_2$ that would be released if the corn or other products were left to decay.

in which the starch portion of the corn is fermented into sugar and then distilled into alcohol.

The major steps in the dry mill process include:

1. **Milling.** The feedstock passes through a hammer mill that turns it into a fine powder called *meal*.

2. **Liquefaction.** The meal is mixed with water and then passed through cookers where the starch is liquefied.

Heat is applied at this stage to enable liquefaction. Cookers use a high-temperature stage of about 250°F to 300°F (120°C to 150°C) to reduce bacteria levels and then a lower temperature of about 200°F (95°C) for a holding period.

3. **Saccharification.** The mash from the cookers is cooled and a secondary enzyme is added to convert the liquefied starch to fermentable sugars (dextrose).

4. **Fermentation.** Yeast is added to the mash to ferment the sugars to ethanol and carbon dioxide.

5. **Distillation.** The fermented mash, now called beer, contains about 10% alcohol plus all the nonfermentable solids from the corn and yeast cells. The mash is pumped to the continuous-flow, distillation system where the alcohol is removed from the solids and the water. The alcohol leaves the top of the final column at about 96% strength, and the residue mash, called *silage*, is transferred from the base of the column to the co-product processing area.

6. **Dehydration.** The alcohol from the top of the column passes through a dehydration system where the remaining water will be removed. The alcohol product at this stage is called **anhydrous ethanol** (pure, no more than 0.5% water).

7. **Denaturing.** Ethanol that will be used for fuel must be denatured, or made unfit for human consumption, with a small amount of gasoline (2% to 5%), methanol, or denatonium benzoate. This is done at the ethanol plant.

# CELLULOSE ETHANOL

**TERMINOLOGY**    **Cellulose ethanol** can be produced from a wide variety of cellulose biomass feedstock, including:

- Agricultural plant wastes (corn stalks, cereal straws)
- Plant wastes from industrial processes (sawdust, paper pulp)
- Energy crops grown specifically for fuel production.

These nongrain products are often referred to as **cellulosic biomass.** Cellulosic biomass is composed of cellulose and lignin, with smaller amounts of proteins, lipids (fats, waxes, and oils), and ash. About two-thirds of cellulosic materials are present as cellulose, with lignin making up the bulk of the remaining dry mass.

**REFINING CELLULOSE BIOMASS**    As with grains, processing cellulose biomass involves extracting fermentable sugars from the feedstock. But the sugars in cellulose are locked in complex carbohydrates called polysaccharides (long chains of simple sugars). Separating these complex structures into fermentable sugars is needed to achieve the efficient and economic production of cellulose ethanol.

 **FREQUENTLY ASKED QUESTION**

**What Is Switchgrass?**

**Switchgrass** (*Panicum virgatum*) can be used to make ethanol and is a summer perennial grass that is native to North America. It is a natural component of the tall-grass prairie, which covered most of the Great Plains, but was also found on the prairie soils in the Black Belt of Alabama and Mississippi. Switchgrass is resistant to many pests and plant diseases, and is capable of producing high yields with very low applications of fertilizer. This means that the need for agricultural chemicals to grow switchgrass is relatively low. Switchgrass is also very tolerant of poor soils, flooding, and drought, which are widespread agricultural problems in the Southeast.

There are two main types of switchgrass:

- **Upland types**—usually grow 5 to 6 feet tall
- **Lowland types**—grow up to 12 feet tall and are typically found on heavy soils in bottomland sites

Better energy efficiency is gained because less energy is used to produce ethanol from switchgrass.

Two processing options are employed to produce fermentable sugars from cellulose biomass:

- Acid hydrolysis is used to break down the complex carbohydrates into simple sugars.
- Enzymes are employed to convert the cellulose biomass to fermentable sugars. The final step involves microbial fermentation, yielding ethanol and carbon dioxide.

**NOTE: Cellulose ethanol production substitutes biomass for fossil fuels. The greenhouse gases produced by the combustion of biomass are offset by the $CO_2$ absorbed by the biomass as it grows in the field.**

# E85

**WHAT IS E85?**    Vehicle manufacturers have available vehicles that are capable of operating on gasoline plus ethanol or a combination of gasoline and ethanol called **E85.** E85 is composed of 85% ethanol and 15% gasoline.

Pure ethanol has an octane rating of about 113. E85, which contains 35% oxygen by weight, has an octane rating of about 100 to 105. This compares to a regular unleaded gasoline, which has a rating of 87. ● **SEE FIGURE 6–2.**

**NOTE: The octane rating of E85 depends on the exact percent of ethanol used, which can vary from 81% to 85%. It also depends on the octane rating of the gasoline used to make E85.**

**FIGURE 6–2** Some retail stations offer a variety of fuel choices, such as this station in Ohio where E10 and E85 are available.

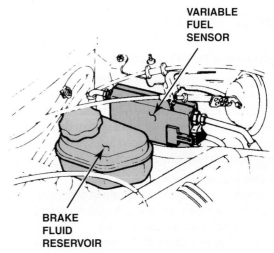

**FIGURE 6–3** The location of the variable fuel sensor can vary, depending on the make and model of vehicle, but it is always in the fuel line between the fuel tank and the fuel injectors.

**HEAT ENERGY OF E85** E85 has less heat energy than gasoline.

**Gasoline = 114,000 BTUs per gallon**

**E85 = 87,000 BTUs per gallon**

This means that the fuel economy is reduced by 20% to 30% if E85 is used instead of gasoline.

**Example:** A Chevrolet Tahoe 5.3-liter V-8 with an automatic transmission has an EPA rating of 15 mpg in the city and 20 mpg on the highway when using gasoline. If this same vehicle was fueled with E85, the EPA fuel economy rating drops to 11 mpg in the city and 15 mpg on the highway.

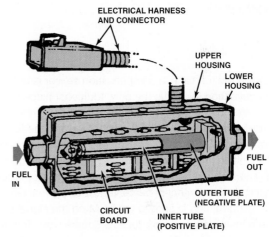

**FIGURE 6–4** A cutaway view of a typical variable fuel sensor.

## ALTERNATIVE-FUEL VEHICLES

The 15% gasoline in this blend helps the engine start, especially in cold weather. Vehicles equipped with this capability are commonly referred to as **alternative-fuel vehicles (AFVs)**, **flex fuels,** and **flexible fuel vehicles,** or **FFVs.** Using E85 in a flex-fuel vehicle can result in a power increase of about 5%. For example, an engine rated at 200 hp using gasoline or E10 could produce 210 hp if using E85.

**NOTE: E85 may test as containing less than 85% ethanol if tested in cold climates because it is often blended according to outside temperature. A lower percentage of ethanol with a slightly higher percentage of gasoline helps engines start in cold climates.**

These vehicles are equipped with an electronic sensor in the fuel supply line that detects the presence and percentage of ethanol. The PCM then adjusts the fuel injector on-time and ignition timing to match the needs of the fuel being used.

E85 contains less heat energy, and therefore will use more fuel, but the benefits include a lower cost of the fuel and the environmental benefit associated with using an oxygenated fuel.

General Motors, Ford, Chrysler, Mazda, and Honda are a few of the manufacturers offering E85 compatible vehicles. E85 vehicles use fuel system parts designed to withstand the additional alcohol content, modified driveability programs that adjust fuel delivery and timing to compensate for the various percentages of ethanol fuel, and a **fuel compensation sensor** that measures both the percentage of ethanol blend and the temperature of the fuel. This sensor is also called a **variable fuel sensor.** ● **SEE FIGURES 6–3 AND 6–4.**

**E85 FUEL SYSTEM REQUIREMENTS** Most E85 vehicles are very similar to non-E85 vehicles. Fuel system components may be redesigned to withstand the effects of higher concentrations of ethanol. In addition, since the stoichiometric point for ethanol is 9:1 instead of 14.7:1 as for gasoline, the air–fuel mixture has to be adjusted for the percentage of ethanol present in the fuel tank. In order to determine this percentage

**Purchase a Flex-Fuel Vehicle**

If purchasing a new or used vehicle, try to find a flex-fuel vehicle. Even though you may not want to use E85, a flex-fuel vehicle has a more robust fuel system than a conventional fuel system designed for gasoline or E10. The enhanced fuel system components and materials usually include:

- Stainless steel fuel rail
- Graphite commutator bars instead of copper in the fuel pump motor (ethanol can oxidize into acetic acid, which can corrode copper)
- Diamond-like carbon (DLC) corrosion-resistant fuel injectors
- Alcohol-resistant O-rings and hoses

The cost of a flex-fuel vehicle compared with the same vehicle designed to operate on gasoline is a no-cost or a low-cost option.

**FIGURE 6–5** A pump for E85 (85% ethanol and 15% gasoline). E85 is available in more locations every year.

of ethanol in the fuel tank, a compensation sensor is used. The fuel compensation sensor is the only additional piece of hardware required on some E85 vehicles. The fuel compensation sensor provides both the ethanol percentage and the fuel temperature to the PCM. The PCM uses this information to adjust both the ignition timing and the quantity of fuel delivered to the engine. The fuel compensation sensor uses a microprocessor to measure both the ethanol percentage and the fuel temperature. This information is sent to the PCM on the signal circuit. The compensation sensor produces a square wave frequency and pulse width signal. The normal frequency range of the fuel compensation sensor is 50 hertz, which represents 0% ethanol and 150 hertz, which represents 100% ethanol. The pulse width of the signal varies from 1 millisecond to 5 milliseconds. One millisecond would represent a fuel temperature of −40°F (−40°C), and 5 milliseconds would represent a fuel temperature of 257°F (125°C). Since the PCM knows both the fuel temperature and the ethanol percentage of the fuel, it can adjust fuel quantity and ignition timing for optimum performance and emissions.

The benefits of E85 vehicles are less pollution, less $CO_2$ production, and less dependence on oil. ● **SEE FIGURE 6–5.**

Ethanol-fueled vehicles generally produce the same pollutants as gasoline vehicles; however, they produce less CO and $CO_2$ emissions. While $CO_2$ is not considered a pollutant, it is thought to lead to global warming and is called a greenhouse gas.

**FLEX-FUEL VEHICLE IDENTIFICATION** Flexible fuel vehicles (FFVs) can be identified by:

- Emblems on the side, front, and/or rear of the vehicle
- Yellow fuel cap showing E85/gasoline (● **SEE FIGURE 6–6**)

**FIGURE 6–6** A flex-fuel vehicle often has a yellow gas cap, which is labeled E85/gasoline.

**How Does a Sensorless Flex-Fuel System Work?**

Many General Motors flex-fuel vehicles do not use a fuel compensation sensor and instead use the oxygen sensor to detect the presence of the lean mixture and the extra oxygen in the fuel.

The powertrain control module (PCM) then adjusts the injector pulse-width and the ignition timing to optimize engine operation to the use of E85. This type of vehicle is called a **virtual flexible fuel vehicle**, abbreviated **V-FFV**. The virtual flexible fuel vehicle can operate on pure gasoline or blends up to 85% ethanol.

**Ford Motor Company**
**VEHICLE EMISSION CONTROL INFORMATION**

Conforms to regulations: 2011 MY FFV

**U.S. EPA:** T2B5 LDV
**OBD:** F II        **Fuel:** Gasoline/Ethanol

**California:** Not for sale in states with *California emissions.*
**OBD:** N/A        **Fuel:** N/A

| TWC/HO2S/EGR/SFI/HAFS | No adjustments needed. |
|---|---|

3.0L Group: BFMXV03.0VEG
Evap: BFMXR0155GAV

BW7E-9C485-
**U K F**

**FIGURE 6–7** This flexible fuel vehicle (FFV) vehicle emission control information (VECI) sticker located under the hood indicates that it can operate on either gasoline or ethanol.

- Vehicle emission control information (VECI) label under the hood (● **SEE FIGURE 6–7**)
- Vehicle identification number (VIN)

Vehicles that are flexible fuel include:

### Chrysler

**2004+**
- 4.7L Dodge Ram Pickup 1500 Series
- 2.7L Dodge Stratus Sedan
- 2.7L Chrysler Sebring Sedan
- 3.3L Caravan and Grand Caravan SE

**2003–2004**
- 2.7L Dodge Stratus Sedan
- 2.7L Chrysler Sebring Sedan

**2003**
- 3.3L Dodge Cargo Minivan

**2000–2003**
- 3.3L Chrysler Voyager Minivan
- 3.3L Dodge Caravan Minivan 3.3L Chrysler Town and Country Minivan

**1998–1999**
- 3.3L Dodge Caravan Minivan
- 3.3L Plymouth Voyager Minivan
- 3.3L Chrysler Town & Country Minivan

### Ford Motor Company
*Ford offers the flex fuel capability as an option on select vehicles—see the owner's manual.

**2004+**
- 4.0L Explorer Sport Trac
- 4.0L Explorer (4-door)
- 3.0L Taurus Sedan and Wagon

**2002–2004**
- 4.0L Explorer (4-door)
- 3.0L Taurus Sedan and Wagon

**2002–2003**
- 3.0L Supercab Ranger Pickup 2WD

**2001**
- 3.0L Supercab Ranger Pickup 2WD
- 3.0L Taurus LX, SE, and SES Sedan

**1999–2000**
- 3.0L Ranger Pickup 4WD and 2WD

### General Motors
*Select vehicles only—see your owner's manual.

**2005+**
- 5.3L Vortec-Engine Avalanche
- 5.3L Vortec-Engine Police Package Tahoe

**2003–2005**
- 5.3L V8 Chevy Silverado* and GMC Sierra* Half-Ton Pickups 2WD and 4WD
- 5.3L Vortec-Engine Suburban, Tahoe, Yukon, and Yukon XL

**2002**
- 5.3L V8 Chevy Silverado* and GMC Sierra* Half-Ton Pickups 2WD and 4WD
- 5.3L Vortec-Engine Suburban, Tahoe, Yukon, and Yukon XL
- 2.2L Chevy S10 Pickup 2WD
- 2.2L Sonoma GMC Pickup 2WD

**2000–2001**
- 2.2L Chevy S10 Pickup 2WD
- 2.2L GMC Sonoma Pickup 2WD

### Isuzu

**2000–2001**
- 2.2L Hombre Pickup 2WD

Mazda

**1999–2003**
- 3.0L Selected B3000 Pickups

### Mercedes-Benz

**2005+**
- 2.6L C240 Luxury Sedan and Wagon

**2003**
- 3.2L C320 Sport Sedan and Wagon

### Mercury
**2002–2004**
- 4.0L Selected Mountaineers

**2000–2004**
- 3.0L Selected Sables

### Nissan

**2005+**
- 5.6L DOHC V8 Engine

*Select vehicles only—see the owner's manual or VECI sticker under the hood.

### Avoid Resetting Fuel Compensation

Starting in 2006, General Motors vehicles designed to operate on E85 do not use a fuel compensation sensor, but instead use the oxygen sensor and refueling information to calculate the percentage of ethanol in the fuel. The PCM uses the fuel level sensor to sense that fuel has been added and starts to determine the resulting ethanol content by using the oxygen sensor. However, if a service technician were to reset fuel compensation by clearing long-term fuel trim, the PCM starts the calculation based on base fuel, which is gasoline with less than or equal to 10% ethanol (E10). If the fuel tank has E85, then the fuel compensation cannot be determined unless the tank is drained and refilled with base fuel. Therefore, avoid resetting the fuel compensation setting unless it is known that the fuel tank contains gasoline or E10 only.

## HOW TO READ A VEHICLE IDENTIFICATION NUMBER

The vehicle identification number (VIN) is required by federal regulation to contain specific information about the vehicle. The following chart shows the character in the eighth position of the VIN number from Ford Motor Company, General Motors, and Chrysler that designates their vehicles as flexible fuel vehicles.

### Ford Motor Company

| Vehicle | 8th Character |
| --- | --- |
| Ford Crown Victoria | V |
| Ford F-150 | V |
| Ford Explorer | K |
| Ford Ranger | V |
| Ford Taurus | 2 |
| Lincoln Town Car | V |
| Mercury Mountaineer | K |
| Mercury Sable | 2 |
| Mercury Grand Marquis | V |

### General Motors

| Vehicle | 8th Character |
| --- | --- |
| Chevrolet Avalanche | Z |
| Chevrolet Impala | K |
| Chevrolet Monte Carlo | K |
| Chevrolet S-10 Pickup | 5 |
| Chevrolet Sierra | Z |
| Chevrolet Suburban | Z |
| Chevrolet Tahoe | Z |
| GMC Yukon and Yukon XL | Z |
| GMC Silverado | Z |
| GMC Sonoma | 5 |

### How Long Can Oxygenated Fuel Be Stored Before All of the Oxygen Escapes?

The oxygen in oxygenated fuels, such as E10 and E85, is not in a gaseous state like the $CO_2$ in soft drinks. The oxygen is part of the molecule of ethanol or other oxygenates and does not bubble out of the fuel. Oxygenated fuels, just like any fuel, have a shelf life of about 90 days.

### Chrysler

| Vehicle | 8th Character |
| --- | --- |
| Chrysler Sebring | T |
| Chrysler Town & Country | E, G or 3 |
| Dodge Caravan | E, G or 3 |
| Dodge Cargo Minivan | E, G or 3 |
| Dodge Durango | P |
| Dodge Ram | P |
| Dodge Stratus | T |
| Plymouth Voyageur | E, G or 3 |

### Mazda

| Vehicle | 8th Character |
| --- | --- |
| B3000 Pickup | V |

### Nissan

| Vehicle | 4th Character |
| --- | --- |
| Titan | B |

### Mercedes Benz

Check owner's manual or the VECI sticker under the hood.

NOTE: For additional information on E85 and for the location of E85 stations in your area, visit www.e85fuel.com.

## METHANOL

**METHANOL TERMINOLOGY** **Methanol,** also known as *methyl alcohol, wood alcohol,* or *methyl hydrate,* is a chemical compound formula that includes one carbon atom and four hydrogen atoms and one oxygen. ● **SEE FIGURE 6–8.**

Methanol is a light, volatile, colorless, tasteless, flammable, poisonous liquid with a very faint odor. It is used as an antifreeze, a solvent, and a fuel. Methanol burns in air, forming $CO_2$ (carbon dioxide) and $H_2O$ (water). A methanol flame is almost colorless. Because of its poisonous properties, methanol is also

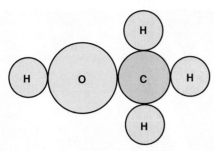

**FIGURE 6–8** The molecular structure of methanol showing the one carbon atom, four hydrogen atoms, and one oxygen atom.

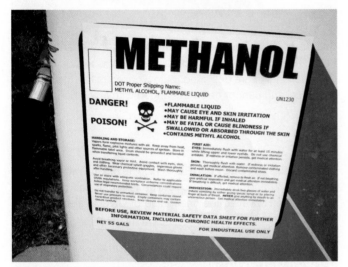

**FIGURE 6–9** Sign on methanol pump shows that methyl alcohol is a poison and can cause skin irritation and other personal injury. Methanol is used in industry as well as being a fuel.

used to denature ethanol. Methanol is often called wood alcohol because it was once produced chiefly as a by-product of the destructive distillation of wood. ● **SEE FIGURE 6–9.**

**PRODUCTION OF METHANOL** The biggest source of methanol in the United States is coal. Using a simple reaction between coal and steam, a gas mixture called **syn-gas** (*synthesis gas*) is formed. The components of this mixture are carbon monoxide and hydrogen, which, through an additional chemical reaction, are converted to methanol.

Natural gas can also be used to create methanol and is re-formed or converted to synthesis gas, which is later made into methanol.

Biomass can be converted to synthesis gas by a process called partial oxidation, and later converted to methanol. **Biomass** is organic material, such as:

- Urban wood wastes
- Primary mill residues
- Forest residues
- Agricultural residues
- Dedicated energy crops (e.g., sugarcane and sugar beets) that can be made into fuel

**FIGURE 6–10** Propane fuel storage tank in the trunk of a Ford taxi.

Electricity can be used to convert water into hydrogen, which is then reacted with carbon dioxide to produce methanol.

Methanol is toxic and can cause blindness and death. It can enter the body by ingestion, inhalation, or absorption through the skin. Dangerous doses will build up if a person is regularly exposed to fumes or handles liquid without skin protection. If methanol has been ingested, a doctor should be contacted immediately. The usual fatal dose is 4 fl oz (100 to 125 mL).

**M85** Some flexible fuel vehicles are designed to operate on 85% methanol and 15% gasoline called **M85.** Methanol is very corrosive and requires that the fuel system components be constructed of stainless steel and other alcohol-resistant rubber and plastic components. The heat content of M85 is about 60% of that of gasoline.

## PROPANE

**Propane** is the most widely used of all of the alternative fuels. Propane is normally a gas but is easily compressed into a liquid and stored in inexpensive containers. When sold as a fuel, it is also known as **liquefied petroleum gas (LPG)** or **LP-gas** because the propane is often mixed with about 10% of other gases such as butane, propylene, butylenes, and mercaptan to give the colorless and odorless propane a smell. Propane is nontoxic, but if inhaled can cause asphyxiation through lack of oxygen. Propane is heavier than air and lays near the floor if released into the atmosphere. Propane is commonly used in forklifts and other equipment used inside warehouses and factories because the exhaust from the engine using propane is not harmful. Propane is a by-product of petroleum refining of natural gas. In order to liquefy the fuel, it is stored in strong tanks at about 300 PSI (2,000 kPa). The heating value of propane is less than that of gasoline; therefore, more is required, which reduces the fuel economy. ● **SEE FIGURE 6–10.**

**FIGURE 6–11** The blue sticker on the rear of this vehicle indicates that it is designed to use compressed natural gas.

**FIGURE 6–12** A CNG storage tank from a Honda Civic GX shown with the fixture used to support it while it is being removed or installed in the vehicle. Honda specifies that three technicians be used to remove or install the tank through the rear door of the vehicle due to the size and weight of the tank.

# COMPRESSED NATURAL GAS

**CNG VEHICLE DESIGN**    Another alternative fuel that is often used in fleet vehicles is **compressed natural gas,** or **CNG,** and vehicles using this fuel are often referred to as **natural gas vehicles (NGVs).** Look for the blue CNG label on vehicles designed to operate on compressed natural gas. ● **SEE FIGURE 6–11.**

Natural gas has to be compressed to about 3,000 PSI (20,000 kPa) or more, so that the weight and the cost of the storage container is a major factor when it comes to preparing a vehicle to run on CNG. The tanks needed for CNG are typically constructed of 0.5-inch-thick (3 mm) aluminum reinforced

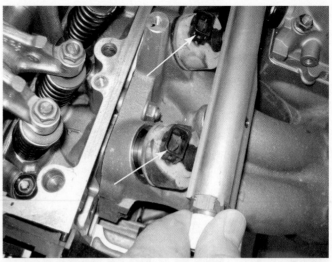

**FIGURE 6–13** The fuel injectors used on this Honda Civic GX CNG engine are designed to flow gaseous fuel instead of liquid fuel and cannot be interchanged with any other type of injector.

with fiberglass. ● **SEE FIGURE 6–12.** The octane rating of CNG is about 130 and the cost per gallon is about half of the cost of gasoline. However, the heat value of CNG is also less, and therefore more is required to produce the same power and the miles per gallon is less.

**CNG COMPOSITION**    Compressed natural gas is made up of a blend of:

- Methane
- Propane
- Ethane
- N-butane
- Carbon dioxide
- Nitrogen

Once it is processed, it is at least 93% methane. Natural gas is nontoxic, odorless, and colorless in its natural state. It is odorized during processing, using ethyl mercaptan ("skunk"), to allow for easy leak detection. Natural gas is lighter than air and will rise when released into the air. Since CNG is already a vapor, it does not need heat to vaporize before it will burn, which improves cold start-up and results in lower emissions during cold operation. However, because it is already in a gaseous state, it does displace some of the air charge in the intake manifold. This leads to about a 10% reduction in engine power as compared to an engine operating on gasoline. Natural gas also burns slower than gasoline; therefore, the ignition timing must be advanced more when the vehicle operates on natural gas. The stoichiometric ratio, the point at which all the air and fuel is used or burned is 16.5:1 compared to 14.7:1 for gasoline. This means that more air is required to burn one pound of natural gas than is required to burn one pound of gasoline. ● **SEE FIGURE 6–13.**

**What Is the Amount of CNG Equal to in Gasoline?**

To achieve the amount of energy of one gallon of gasoline, 122 cubic feet of compressed natural gas (CNG) is needed. While the octane rating of CNG is much higher than gasoline (130 octane), using CNG instead of gasoline in the same engine would result in a reduction of 10% to 20% of power due to the lower heat energy that is released when CNG is burned in the engine.

The CNG engine is designed to include:

- Increased compression ratio
- Strong pistons and connecting rods
- Heat-resistant valves
- Fuel injectors designed for gaseous fuel instead of liquid fuel

**CNG FUEL SYSTEMS**    When completely filled, the CNG tank has 3,600 PSI of pressure in the tank. When the ignition is turned on, the alternate fuel electronic control unit activates the high-pressure lock-off, which allows high-pressure gas to pass to the high-pressure regulator. The high-pressure regulator reduces the high-pressure CNG to approximately 170 PSI and sends it to the low-pressure lock-off. The low-pressure lock-off is also controlled by the alternate fuel electronic control unit and is activated at the same time that the high-pressure lock-off is activated. From the low-pressure lock-off, the CNG is directed to the low-pressure regulator. This is a two-stage regulator that first reduces the pressure to approximately 4 to 6 PSI in the first stage and then to 4.5 to 7 inches of water in the second stage. Twenty-eight inches of water is equal to 1 PSI, therefore, the final pressure of the natural gas entering the engine is very low. From here, the low-pressure gas is delivered to the gas mass sensor/mixture control valve. This valve controls the air–fuel mixture. The CNG gas distributor adapter then delivers the gas to the intake stream.

CNG vehicles are designed for fleet use that usually have their own refueling capabilities. One of the drawbacks to using CNG is the time that it takes to refuel a vehicle. The ideal method of refueling is the slow fill method. The slow filling method compresses the natural gas as the tank is being fueled. This method ensures that the tank will receive a full charge of CNG; however, this method can take three to five hours to accomplish. If more than one vehicle needs filling, the facility will need multiple CNG compressors to refuel the vehicles.

There are three commonly used CNG refilling station pressures:

P24—2,400 PSI

P30—3,000 PSI

P36—3,600 PSI

**FIGURE 6–14** This CNG pump is capable of supplying compressed natural gas at either 3,000 PSI or 3,600 PSI. The price per gallon is higher for the higher pressure.

Try to find and use a station with the highest refilling pressure. Filling at lower pressures will result in less compressed natural gas being installed in the storage tank, thereby reducing the driving range. ● **SEE FIGURE 6–14.**

The fast fill method uses CNG that is already compressed. However, as the CNG tank is filled rapidly, the internal temperature of the tank will rise, which causes a rise in tank pressure. Once the temperature drops in the CNG tank, the pressure in the tank also drops, resulting in an incomplete charge in the CNG tank. This refueling method may take only about five minutes; however, it will result in an incomplete charge to the CNG tank, reducing the driving range.

## LIQUEFIED NATURAL GAS

Natural gas can be turned into a liquid if cooled to below −260°F (−127°C). The natural gas condenses into a liquid at normal atmospheric pressure and the volume is reduced by about 600 times. This means that the natural gas can be more efficiently transported over long distances where no pipelines are present when liquefied.

Because the temperature of liquefied natural gas (LNG) must be kept low, it is only practical for use in short haul trucks where they can be refueled from a central location.

## P-SERIES FUELS

P-series alternative fuel is patented by Princeton University and is a non-petroleum- or natural gas-based fuel suitable for use in flexible fuel vehicles or any vehicle designed to operate on E85 (85% ethanol, 15% gasoline). P-series fuel is recognized by the United States Department of Energy as being an alternative fuel,

### What Is a Tri-Fuel Vehicle?

In Brazil, most vehicles are designed to operate on ethanol or gasoline or any combination of the two. In this South American country, ethanol is made from sugarcane, is commonly available, and is lower in price than gasoline. Compressed natural gas (CNG) is also being made available so many vehicle manufacturers in Brazil, such as General Motors and Ford, are equipping vehicles to be capable of using gasoline, ethanol, or CNG. These vehicles are called tri-fuel vehicles.

but is not yet available to the public. P-series fuels are blends of the following:

- Ethanol (ethyl alcohol)
- Methyltetrahydrofuron, abbreviated MTHF
- Natural gas liquids, such as pentanes
- Butane

The ethanol and MTHF are produced from renewable feedstocks, such as corn, waste paper, biomass, agricultural waste, and wood waste (scraps and sawdust). The components used in P-type fuel can be varied to produce regular grade, premium grade, or fuel suitable for cold climates.
● **SEE CHART 6–1** for the percentages of the ingredients based on fuel grade.

● **SEE CHART 6–2** for a comparison of the most frequently used alternative fuels.

# SYNTHETIC FUELS

**Synthetic fuels** can be made from a variety of products, using several different processes. Synthetic fuel must, however, make these alternatives practical only when conventional petroleum products are either very expensive or not available.

**FISCHER–TROPSCH** Synthetic fuels were first developed using the **Fischer–Tropsch** method and have been in use since the 1920s to convert coal, natural gas, and other fossil fuel products into a fuel that is high in quality and clean-burning. The

### COMPOSITION OF P-SERIES FUELS (BY VOLUME)

| COMPONENT | REGULAR GRADE (%) | PREMIUM GRADE (%) | COLD WEATHER |
|---|---|---|---|
| Pentanes plus | 32.5 | 27.5 | 16.0 |
| MTHF | 32.5 | 17.5 | 26.0 |
| Ethanol | 35.0 | 55.0 | 47.0 |
| Butane | 0.0 | 0.0 | 11.0 |

**CHART 6–1**

P-series fuel varies in composition, depending on the octane rating and temperature.

### ALTERNATIVE FUEL COMPARISON CHART

| CHARACTERISTIC | PROPANE | CNG | METHANOL | ETHANOL | REGULAR UNLEADED GAS |
|---|---|---|---|---|---|
| Octane | 104 | 130 | 100 | 100 | 87–93 |
| BTU per gallon | 91,000 | N.A. | 70,000 | 83,000 | 114,000–125,000 |
| Gallon equivalent | 1.15 | 122 cubic feet— 1 gallon of gasoline | 1.8 | 1.5 | 1 |
| On-board fuel storage | Liquid | Gas | Liquid | Liquid | Liquid |
| Miles/gallon as compared to gas | 85% | N.A. | 55% | 70% | 100% |
| Relative tank size required to yield driving range equivalent to gas | Tank is 1.25 times larger | Tank is 3.5 times larger | Tank is 1.8 times larger | Tank is 1.5 times larger | |
| Pressure | 200 PSI | 3,000-3,600 PSI | N.A. | N.A. | N.A. |
| Cold weather capability | Good | Good | Poor | Poor | Good |
| Vehicle power | 5%-10% power loss | 10%-20% power loss | 4% power increase | 5% power increase | Standard |
| Toxicity | Nontoxic | Nontoxic | Highly toxic | Toxic | Toxic |
| Corrosiveness | Noncorrosive | Noncorrosive | Corrosive | Corrosive | Minimally corrosive |
| Source | Natural gas/ petroleum refining | Natural gas/ crude oil | Natural gas/coal | Sugar and starch crops/biomass | Crude oil |

**CHART 6–2**

The characteristics of alternative fuels compared to regular unleaded gasoline shows that all have advantages and disadvantages.

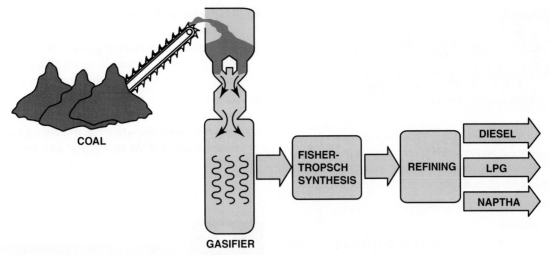

**FIGURE 6–15** A Fischer–Tropsch processing plant is able to produce a variety of fuels from coal.

process for producing Fischer–Tropsch fuels was patented by two German scientists, Franz Fischer and Hans Tropsch, during World War I. The Fischer–Tropsch method uses carbon monoxide and hydrogen (the same synthesis gas used to produce hydrogen fuel) to convert coal and other hydrocarbons to liquid fuels in a process similar to hydrogenation, another method for hydrocarbon conversion. The process using natural gas, also called **gas-to-liquid (GTL)** technology, uses a catalyst, usually iron or cobalt, and incorporates steam re-forming to give off the by-products of carbon dioxide, hydrogen, and carbon monoxide. ● **SEE FIGURE 6–15.**

Whereas traditional fuels emit environmentally harmful particulates and chemicals, namely sulfur compounds, Fischer–Tropsch fuels combust with no soot or odors and emit only low levels of toxins. Fischer–Tropsch fuels can also be blended with traditional transportation fuels with little equipment modification, as they use the same engine and equipment technology as traditional fuels.

The fuels contain a very low sulfur and aromatic content and they produce virtually no particulate emissions. Researchers also expect reductions in hydrocarbon and carbon monoxide emissions. Fischer–Tropsch fuels do not differ in fuel performance from gasoline and diesel. At present, Fischer–Tropsch fuels are very expensive to produce on a large scale, although research is under way to lower processing costs. Diesel fuel created using the Fischer–Tropsch diesel **(FTD)** process is often called *GTL diesel*. GTL diesel can also be combined with petroleum diesel to produce a GTL blend. This fuel product is currently being sold in Europe and plans are in place to introduce it in North America.

### COAL TO LIQUID (CTL)

Coal is very abundant in the United States and coal can be converted to a liquid fuel through a process called **coal to liquid (CTL).** The huge cost is the main obstacle to these plants. The need to invest $1.4 billion per plant before it can make product is the reason no one has built a CTL plant yet in the United States. Investors need to be convinced that the cost of oil is going to remain high in order to get them to commit this kind of money.

A large plant might be able to produce 120,000 barrels of liquid fuel a day and would consume about 50,000 tons of coal per day. However, such a plant would create about 6,000 tons

of $CO_2$ per day. These $CO_2$ emissions, which could contribute to global warming, and the cost involved make CTL a technology that is not likely to expand.

Two procedures can be used to convert coal-to-liquid fuel:

1. **Direct.** In the direct method, coal is broken down to create liquid products. First the coal is reacted with hydrogen $(H_2)$ at high temperatures and pressure with a catalyst. This process creates a synthetic crude, called **syncrude,** which is then refined to produce gasoline or diesel fuel.

2. **Indirect.** In the indirect method, coal is first turned into a gas and the molecules are reassembled to create the desired product. This process involves turning coal into a gas called syngas. The syngas is then converted into liquid, using the Fischer–Tropsch (FT) process.

Russia has been using CTL by injecting air into the underground coal seams. Ignition is provided and the resulting gases are trapped and converted to liquid gasoline and diesel fuel through the Fischer–Tropsch process. This underground method is called **underground coal gasification (UCG).**

### METHANOL TO GASOLINE

Exxon Mobil has developed a process for converting methanol (methyl alcohol) into gasoline in a process called **methanol-to-gasoline (MTG).** The MTG process was discovered by accident when a gasoline additive made from methanol was being created. The process instead created olefins (alkenes), paraffins (alkenes), and aromatic compounds, which in combination are known as gasoline. The process uses a catalyst and is currently being produced in New Zealand.

### FUTURE OF SYNTHETIC FUELS

Producing gasoline and diesel fuels by other methods besides refining from crude oil has usually been more expensive. With the increasing cost of crude oil, alternative methods are now becoming economically feasible. Whether or not the diesel fuel or gasoline is created from coal, natural gas, or methanol, or created by refining crude oil, the transportation and service pumps are already in place. Compared to using compressed natural gas or other similar alternative fuels, synthetic fuels represent the lowest cost.

# SAFETY PROCEDURES WHEN WORKING WITH ALTERNATIVE FUELS

All fuels are flammable and many are explosive under certain conditions. Whenever working around compressed gases of any kind (CNG, LNG, propane, or LPG), always wear personal protective equipment (PPE), including at least the following items:

1. Safety glasses and/or face shield.

2. Protective gloves.

3. Long-sleeved shirt and pants to help protect bare skin from the freezing effects of gases under pressure in the event that the pressure is lost.

4. If any fuel gets on the skin, the area should be washed immediately.

5. If fuel spills on clothing, change into clean clothing as soon as possible.

6. If fuel spills on a painted surface, flush the surface with water and air dry. If simply wiped off with a dry cloth, the paint surface could be permanently damaged.

7. As with any fuel-burning vehicle, always vent the exhaust to the outside. If methanol fuel is used, the exhaust contains *formaldehyde*, which has a sharp odor and can cause severe burning of the eyes, nose, and throat.

 **WARNING**

Do not smoke or have an open flame in the area when working around or refueling any vehicle.

## SUMMARY

1. Flexible fuel vehicles (FFVs) are designed to operate on gasoline or gasoline-ethanol blends up to 85% ethanol (E85).

2. Ethanol can be made from grain, such as corn, or from cellulosic biomass, such as switchgrass.

3. E85 has fewer BTUs of energy per gallon compared with gasoline and will therefore provide lower fuel economy.

4. Older flexible fuel vehicles used a fuel compensation sensor but newer models use the oxygen sensor to calculate the percentage of ethanol in the fuel being burned.

5. Methanol is also called methyl alcohol or wood alcohol and, while it can be made from wood, it is mostly made from natural gas.

6. Propane is the most widely used alternative fuel. Propane is also called liquefied petroleum gas (LPG).

7. Compressed natural gas (CNG) is available for refilling in several pressures, including 2,400 PSI, 3,000 PSI, and 3,600 PSI.

8. P-series fuel is recognized by the United States Department of Energy as being an alternative fuel. P-series fuel is a non-petroleum-based fuel suitable for use in a flexible fuel vehicle. However, P-series fuel is not commercially available.

9. Synthetic fuels are usually made using the Fischer–Tropsch method to convert coal or natural gas into gasoline and diesel fuel.

10. Safety procedures when working around alternative fuel include wearing the necessary personal protective equipment (PPE), including safety glasses and protective gloves.

## REVIEW QUESTIONS

1. Ethanol is also known by what other terms?

2. The majority of ethanol in the United States is made from what farm products?

3. How is a flexible fuel vehicle identified?

4. Methanol is also known by what other terms?

5. What other gases are often mixed with propane?

6. Why is it desirable to fill a compressed natural gas (CNG) vehicle with the highest pressure available?

7. P-series fuel is made of what products?

8. The Fischer–Tropsch method can be used to change what into gasoline?

1. Ethanol can be produced from what products?
   a. Switchgrass
   c. Sugarcane
   b. Corn
   d. Any of the above

2. E85 means that the fuel is made from _____.
   a. 85% gasoline, 15% ethanol
   b. 85% ethanol, 15% gasoline
   c. Ethanol that has 15% water
   d. Pure ethyl alcohol

3. A flex-fuel vehicle can be identified by _____.
   a. Emblems on the side, front, and/or rear of the vehicle
   b. VECI
   c. VIN
   d. Any of the above

4. Methanol is also called _____.
   a. Methyl alcohol
   c. Methyl hydrate
   b. Wood alcohol
   d. All of the above

5. Which alcohol is dangerous (toxic)?
   a. Methanol
   b. Ethanol
   c. Both ethanol and methanol
   d. Neither ethanol nor methanol

6. Which is the most widely used alternative fuel?
   a. E85
   c. CNG
   b. Propane
   d. M85

7. Liquefied petroleum gas (LPG) is also called _____.
   a. E85
   c. Propane
   b. M85
   d. P-series fuel

8. How much compressed natural gas (CNG) does it require to achieve the energy of one gallon of gasoline?
   a. 130 cubic feet
   c. 105 cubic feet
   b. 122 cubic feet
   d. 91 cubic feet

9. When refueling a CNG vehicle, why is it recommended that the tank be filled to a high pressure?
   a. The range of the vehicle is increased
   b. The cost of the fuel is lower
   c. Less of the fuel is lost to evaporation
   d. Both a and c

10. Producing liquid fuel from coal or natural gas usually uses which process?
    a. Syncrude
    c. Fischer–Tropsch
    b. P-series
    d. Methanol to gasoline (MTG)

# chapter 7

# DIESEL AND BIODIESEL FUELS

**LEARNING OBJECTIVES:** **After studying this chapter, the reader should be able to:** • Discuss the specifications of diesel fuel. • Discuss API gravity. • List the advantages and disadvantages of biodiesel. • Discuss E-diesel.

**KEY TERMS:** API gravity 109 • ASTM 108 • B20 111 • Biodiesel 110 • Cetane number 108 • Cloud point 108 • Diesohol 112 • E-diesel 112 • Petrodiesel 111 • PPO 111 • SVO 111 • UCO 111 • ULSD 110 • WVO 111

## DIESEL FUEL

**FEATURES OF DIESEL FUEL** Diesel fuel must meet an entirely different set of standards than gasoline. Diesel fuel contains 12% more heat energy than the same amount of gasoline. The fuel in a diesel engine is not ignited with a spark, but is ignited by the heat generated by high compression. The pressure of compression (400 to 700 PSI or 2,800 to 4,800 kPa) generates temperatures of 1,200°F to 1,600°F (700°C to 900°C), which speeds the preflame reaction to start the ignition of fuel injected into the cylinder.

**DIESEL FUEL REQUIREMENTS** All diesel fuel must have the following characteristics:

- **Cleanliness.** It is imperative that the fuel used in a diesel engine be clean and free from water. Unlike the case with gasoline engines, the fuel is the lubricant and coolant for the diesel injector pump and injectors. Good-quality diesel fuel contains additives such as oxidation inhibitors, detergents, dispersants, rust preventatives, and metal deactivators.

- **Low-temperature fluidity.** Diesel fuel must be able to flow freely at all expected ambient temperatures. One specification for diesel fuel is its "pour point," which is the temperature below which the fuel would stop flowing.

- **Cloud point** is another concern with diesel fuel at lower temperatures. **Cloud point** is the low-temperature point when the waxes present in most diesel fuels tend to form crystals that can clog the fuel filter. Most diesel fuel suppliers distribute fuel with the proper pour point and cloud point for the climate conditions of the area.

**CETANE NUMBER** The cetane number for diesel fuel is the opposite of the octane number for gasoline. The **cetane number** is a measure of the ease with which the fuel can be ignited.

The cetane rating of the fuel determines, to a great extent, its ability to start the engine at low temperatures and to provide smooth warm-up and even combustion. The cetane rating of diesel fuel should be between 45 and 50. The higher the cetane rating, the more easily the fuel is ignited.

**SULFUR CONTENT** The sulfur content of diesel fuel is very important to the life of the engine. Sulfur in the fuel creates sulfuric acid during the combustion process, which can damage engine components and cause piston ring wear. Federal regulations are getting extremely tight on sulfur content to less than 15 parts per million (ppm). High-sulfur fuel contributes to acid rain.

**DIESEL FUEL COLOR** Diesel fuel intended for use on the streets and highways is clear or green in color. Diesel fuel to be used on farms and off-road use is dyed red because it is not taxed for on road use. ● **SEE FIGURE 7–1.**

**GRADES OF DIESEL FUEL** **American Society for Testing Materials (ASTM)** also classifies diesel fuel by volatility (boiling range) into the following grades:

**GRADE #1** This grade of diesel fuel has the lowest boiling point and the lowest cloud and pour points, as well as a lower BTU content—less heat per pound of fuel. As a result, grade #1 is suitable for use during low-temperature (winter) operation. Grade #1 produces less heat per pound of fuel compared to grade #2 and may be specified for use in diesel engines involved in frequent changes in load and speed, such as those found in city buses and delivery trucks.

**GRADE #2** This grade has a higher boiling point, cloud point, and pour point as compared with grade #1. It is usually specified where constant speed and high loads are encountered, such as in long-haul trucking and automotive diesel applications. Most diesel is Grade #2.

(a)

(b)

**FIGURE 7–1** (a) Regular diesel fuel on the left has a clear or greenish tint, whereas fuel for off-road use is tinted red for identification. (b) A fuel pump in a farming area that clearly states the red diesel fuel is for off-road use only.

## DIESEL FUEL SPECIFIC GRAVITY TESTING

The density of diesel fuel should be tested whenever there is a driveability concern. The density or specific gravity of diesel fuel is measured in units of **API gravity.** API gravity is an arbitrary scale expressing the gravity or density of liquid petroleum products devised jointly by the American Petroleum Institute and the National Bureau of Standards. The measuring scale is calibrated in terms of degrees API. Oil with the least specific gravity has the highest API gravity. The formula for determining API gravity is as follows:

$$\text{Degrees API gravity} = (141.5 \div \text{specific gravity at } 60°F) - 131.5$$

The normal API gravity for #1 diesel fuel is 39 to 44 (typically 40). The normal API gravity for #2 diesel fuel is 30 to 39 (typically 35). A hydrometer calibrated in API gravity units should be used to test diesel fuel. ● **SEE FIGURE 7–2.**

**FIGURE 7–2** Testing the API viscosity of a diesel fuel sample using a hydrometer.

 **FREQUENTLY ASKED QUESTION**

### How Can You Tell If Gasoline Has Been Added to the Diesel Fuel by Mistake?

If gasoline has been accidentally added to diesel fuel and is burned in a diesel engine, the result can be very damaging to the engine. The gasoline can ignite faster than diesel fuel, which would tend to increase the temperature of combustion. This high temperature can harm injectors and glow plugs, as well as pistons, head gaskets, and other major diesel engine components. If contaminated fuel is suspected, first smell the fuel at the filler neck. If the fuel smells like gasoline, then the tank should be drained and refilled with diesel fuel. If the smell test does not indicate a gasoline smell (or any rancid smell), then test a sample for proper API gravity.

**NOTE: Diesel fuel designed for on-road use should be green in color. Red diesel fuel (high sulfur) should be found only in off-road or farm equipment.**

● **SEE CHART 7–1** for a comparison among specific gravity, weight density, pounds per gallon, and API gravity of diesel fuel.

## DIESEL FUEL HEATERS

Diesel fuel heaters, either coolant or electric, help prevent power loss and stalling in cold weather. The heater is placed in the fuel line between the tank and the primary filter. Some coolant heaters are thermostatically controlled, which allows fuel to bypass the heater once it has reached operating temperature. ● **SEE FIGURE 7–3.**

| API GRAVITY COMPARISON CHART | | | |
|---|---|---|---|
| *Values for API Scale Oil* | | | |
| API GRAVITY SCALE | SPECIFIC GRAVITY | WEIGHT DENSITY, LB/FT | POUNDS PER GALLON |
| 0 | | | |
| 2 | | | |
| 4 | | | |
| 6 | | | |
| 8 | | | |
| 10 | 1.0000 | 62.36 | 8.337 |
| 12 | 0.9861 | 61.50 | 8.221 |
| 14 | 0.9725 | 60.65 | 8.108 |
| 16 | 0.9593 | 59.83 | 7.998 |
| 18 | 0.9465 | 59.03 | 7.891 |
| 20 | 0.9340 | 58.25 | 7.787 |
| 22 | 0.9218 | 57.87 | 7.736 |
| 24 | 0.9100 | 56.75 | 7.587 |
| 26 | 0.8984 | 56.03 | 7.490 |
| 28 | 0.8871 | 55.32 | 7.396 |
| 30 | 0.8762 | 54.64 | 7.305 |
| 32 | 0.8654 | 53.97 | 7.215 |
| 34 | 0.8550 | 53.32 | 7.128 |
| 36 | 0.8448 | 52.69 | 7.043 |
| 38 | 0.8348 | 51.06 | 6.960 |
| 40 | 0.8251 | 50.96 | 6.879 |
| 42 | 0.8155 | 50.86 | 6.799 |
| 44 | 0.8030 | 50.28 | 6.722 |
| 46 | 0.7972 | 49.72 | 6.646 |
| 48 | 0.7883 | 49.16 | 6.572 |
| 50 | 0.7796 | 48.62 | 6.499 |
| 52 | 0.7711 | 48.09 | 6.429 |
| 54 | 0.7628 | 47.57 | 6.359 |
| 56 | 0.7547 | 47.07 | 6.292 |
| 58 | 0.7467 | 46.57 | 6.225 |
| 60 | 0.7389 | 46.08 | 6.160 |
| 62 | 0.7313 | 45.61 | 6.097 |
| 64 | 0.7238 | 45.14 | 6.034 |
| 66 | 0.7165 | 44.68 | 5.973 |
| 68 | 0.7093 | 44.23 | 5.913 |
| 70 | 0.7022 | 43.79 | 5.854 |
| 72 | 0.6953 | 43.36 | 5.797 |
| 74 | 0.6886 | 42.94 | 5.741 |
| 76 | 0.6819 | 42.53 | 5.685 |
| 78 | 0.6754 | 41.12 | 5.631 |
| 80 | 0.6690 | 41.72 | 5.577 |
| 82 | 0.6628 | 41.33 | 5.526 |
| 84 | 0.6566 | 40.95 | 5.474 |
| 86 | 0.6506 | 40.57 | 5.424 |
| 88 | 0.6446 | 40.20 | 5.374 |
| 90 | 0.6388 | 39.84 | 5.326 |
| 92 | 0.6331 | 39.48 | 5.278 |
| 94 | 0.6275 | 39.13 | 5.231 |
| 96 | 0.6220 | 38.79 | 5.186 |
| 98 | 0.6116 | 38.45 | 5.141 |
| 100 | 0.6112 | 38.12 | 5.096 |

**CHART 7–1**

The API gravity scale is based on the specific gravity of the fuel.

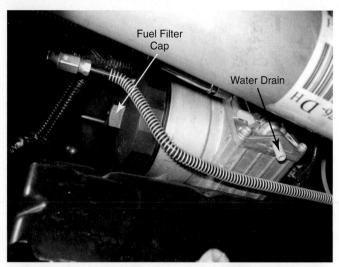

**FIGURE 7–3** A fuel heater is part of the fuel filter and water separator located on the frame rail of a Ford pickup truck equipped with a PowerStroke 6.0 liter V-8 diesel engine.

**ULTRA-LOW-SULFUR DIESEL FUEL** Diesel fuel is used in diesel engines and is usually readily available throughout the United States, Canada, and Europe, where many more cars are equipped with diesel engines. Diesel engines manufactured to 2007 or newer standards must use ultra-low-sulfur diesel fuel containing less than 15 ppm of sulfur compared to the older, low-sulfur specification of 500 ppm. The purpose of the lower sulfur amount in diesel fuel is to reduce emissions of sulfur oxides ($SO_x$) and particulate matter (PM) from heavy-duty highway engines and vehicles that use diesel fuel. The emission controls used on 2007 and newer diesel engines require the use of **ultra-low-sulfur diesel (ULSD)** for reliable operation.

Ultra-low-sulfur diesel (ULSD) will eventually replace the current highway diesel fuel, low-sulfur diesel, which can have as much as 500 ppm of sulfur. ULSD is required for use in all model year 2007 and newer vehicles equipped with advanced emission control systems. ULSD looks lighter in color and has less smell than other diesel fuel.

## BIODIESEL

**DEFINITION OF BIODIESEL** **Biodiesel** is a domestically produced, renewable fuel that can be manufactured from vegetable oils, animal fats, or recycled restaurant greases. Biodiesel is safe, biodegradable, and reduces serious air pollutants such as particulate matter (PM), carbon monoxide, and hydrocarbons. Biodiesel is defined as mono-alkyl esters of long-chain fatty acids derived from vegetable oils or animal fats which conform to ASTM D6751 specifications for use in diesel engines. Biodiesel refers to the pure fuel before blending with diesel fuel. ● **SEE FIGURE 7–4.**

**FIGURE 7–4** A pump decal indicating that the biodiesel fuel is ultra-low-sulfur diesel (ULSD) and must be used in 2007 and newer diesel vehicles.

**BIODIESEL BLENDS** Biodiesel blends are denoted as "BXX" with "XX" representing the percentage of biodiesel contained in the blend (i.e., **B20** is 20% biodiesel, 80% petroleum diesel). Blends of 20% biodiesel with 80% petroleum diesel (B20) can generally be used in unmodified diesel engines; however, users should consult their OEM and engine warranty statement. Biodiesel can also be used in its pure form (B100), but it may require certain engine modifications to avoid maintenance and performance problems and may not be suitable for wintertime use. Most diesel engine or vehicle manufacturers of diesel vehicles allow the use of B5 (5% biodiesel). For example, Cummins, used in Dodge trucks, allows the use of B20 only if the optional extra fuel filter has been installed. Users should consult their engine warranty statement for more information on fuel blends of greater than 20% biodiesel.

In general, B20 costs 30 to 40 cents more per gallon than conventional diesel. Although biodiesel costs more than regular diesel fuel, often called **petrodiesel,** fleet managers can make the switch to alternative fuels without purchasing new vehicles, acquiring new spare parts inventories, rebuilding refueling stations, or hiring new service technicians.

**FEATURES OF BIODIESEL** Biodiesel has the following characteristics:

1. Purchasing biodiesel in bulk quantities decreases the cost of fuel.
2. Biodiesel maintains similar horsepower, torque, and fuel economy.
3. Biodiesel has a higher cetane number than conventional diesel, which increases the engine's performance.
4. It is nontoxic, which makes it safe to handle, transport, and store. Maintenance requirements for B20 vehicles and petrodiesel vehicles are the same.
5. Biodiesel acts as a lubricant and this can add to the life of the fuel system components.

### I Thought Biodiesel Was Vegetable Oil?

Biodiesel is vegetable oil with the glycerin component removed by means of reacting the vegetable oil with a catalyst. The resulting hydrocarbon esters are 16 to 18 carbon atoms in length, almost identical to the petroleum diesel fuel atoms. This allows the use of biodiesel fuel in a diesel engine with no modifications needed. Biodiesel-powered vehicles do not *need* a second fuel tank, whereas vegetable-oil-powered vehicles do. There are three main types of fuel used in diesel engines. These are:

- Petroleum diesel, a fossil hydrocarbon with a carbon chain length of about 16 carbon atoms.
- Biodiesel, a hydrocarbon with a carbon chain length of 16 to 18 carbon atoms.
- Vegetable oil is a triglyceride with a glycerin component joining three hydrocarbon chains of 16 to 18 carbon atoms each, called **straight vegetable oil (SVO)**. Other terms used when describing vegetable oil include:
  - **Pure plant oil (PPO)**—a term most often used in Europe to describe SVO
  - **Waste vegetable oil (WVO)**—this oil could include animal or fish oils from cooking
  - **Used cooking oil (UCO)**—a term used when the oil may or may not be pure vegetable oil

Vegetable oil is not liquid enough at common ambient temperatures for use in a diesel engine fuel delivery system designed for the lower-viscosity petroleum diesel fuel. Vegetable oil needs to be heated to obtain a similar viscosity to biodiesel and petroleum diesel. This means that a heat source needs to be provided before the fuel can be used in a diesel engine. This is achieved by starting on petroleum diesel or biodiesel fuel until the engine heat can be used to sufficiently warm a tank containing the vegetable oil. It also requires purging the fuel system of vegetable oil with petroleum diesel or biodiesel fuel prior to stopping the engine to avoid the vegetable oil's thickening and solidifying in the fuel system away from the heated tank. The use of vegetable oil in its natural state does, however, eliminate the need to remove the glycerin component. Many vehicle and diesel engine fuel system suppliers permit the use of biodiesel fuel that is certified as meeting testing standards. None permit the use of vegetable oil in its natural state.

**NOTE:** For additional information on biodiesel and the locations where it can be purchased, visit www.biodiesel.org.

# E-DIESEL FUEL

**DEFINITION OF E-DIESEL** **E-diesel,** also called **diesohol** outside of the United States, is standard No. 2 diesel fuel that contains up to 15% ethanol. While E-diesel can have up to 15% ethanol by volume, typical blend levels are from 8% to 10%.

**CETANE RATING OF E-DIESEL** The higher the cetane number, the shorter the delay between injection and ignition. Normal diesel fuel has a cetane number of about 50. Adding 15% ethanol lowers the cetane number. To increase the cetane number back to that of conventional diesel fuel, a cetane-enhancing additive is added to E-diesel. The additive used to increase the cetane rating of E-diesel is ethylhexylnitrate or ditertbutyl peroxide.

E-diesel has better cold-flow properties than conventional diesel. The heat content of E-diesel is about 6% less than conventional diesel, but the particulate matter (PM) emissions are reduced by as much as 40%, 20% less carbon monoxide, and a 5% reduction in oxides of nitrogen (NOx).

Currently, E-diesel is considered to be experimental and can be used legally in off-road applications or in mass-transit buses with EPA approval. For additional information, visit www.e-diesel.org.

## SUMMARY

1. Diesel fuel produces 12% more heat energy than the same amount of gasoline.
2. Diesel fuel requirements include cleanliness, low-temperature fluidity, and proper cetane rating.
3. Emission control devices used on 2007 and newer engines require the use of ultra-low-sulfur diesel (ULSD) that has less than 15 ppm of sulfur.
4. The density of diesel fuel is measured in a unit called API gravity.
5. The cetane rating of diesel fuel is a measure of the ease with which the fuel can be ignited.
6. Biodiesel is the blend of vegetable-based liquid with regular diesel fuel. Most diesel engine manufacturers allow the use of a 5% blend, called B20 without any changes to the fuel system or engine.
7. E-diesel is a blend of ethanol with diesel fuel up to 15% ethanol by volume.

## REVIEW QUESTIONS

1. What is meant by the cloud point?
2. What is ultra-low-sulfur diesel?
3. Biodiesel blends are identified by what designation?

## CHAPTER QUIZ

1. What color is diesel fuel dyed if it is for off-road use only?
   a. Red
   b. Green
   c. Blue
   d. Yellow

2. What clogs fuel filters when the temperature is low on a vehicle that uses diesel fuel?
   a. Alcohol
   b. Sulfur
   c. Wax
   d. Cetane

3. The specific gravity of diesel fuel is measured in what units?
   a. Hydrometer units
   b. API gravity
   c. Grade number
   d. Cetane number

4. What rating of diesel fuel indicates how well a diesel engine will start?
   a. Specific gravity rating
   b. Sulfur content
   c. Cloud point
   d. Cetane rating

5. Ultra-low-sulfur diesel fuel has how much sulfur content?
   a. 15 ppm
   b. 50 ppm
   c. 500 ppm
   d. 1500 ppm

6. E-diesel is diesel fuel with what additive?
   a. Methanol
   b. Sulfur
   c. Ethanol
   d. Vegetable oil

7. Biodiesel is regular diesel fuel with vegetable oil added.
   a. True
   b. False

8. B20 biodiesel has how much regular diesel fuel?
   a. 20%
   b. 40%
   c. 80%
   d. 100%

9. Most diesel fuel is what grade?
   a. Grade #1
   b. Grade #2
   c. Grade #3
   d. Grade #4

10. Most manufacturers of vehicles equipped with diesel engines allow what type of biodiesel?
    a. B100
    b. B80
    c. B20
    d. B5

# INTAKE AND EXHAUST SYSTEMS

**LEARNING OBJECTIVES: After studying this chapter, the reader should be able to:** • Discuss air intake filtration. • Explain throttle-body injection and port fuel-injection intake manifolds. • Discuss exhaust gas recirculation passages and exhaust manifolds. • Understand the purpose and function of mufflers.

**KEY TERMS:** EGR 118 • Hangers 120 • Helmholtz resonator 115 • Micron 113 • Plenum 117

## AIR INTAKE FILTRATION

**NEED FOR AIR FILTERING** Gasoline must be mixed with air to form a combustible mixture. Air movement into an engine occurs due to low pressure (vacuum) being created in the engine. ●**SEE FIGURE 8–1.**

Air contains dirt and other materials that cannot be allowed to reach the engine. Just as fuel filters are used to clean impurities from gasoline, an air cleaner and filter are used to remove contaminants from the air. The three main jobs of the air cleaner and filter include:

1. Clean the air before it is mixed with fuel
2. Silence air intake noise
3. Act as a flame arrester in case of a backfire

The automotive engine uses about 9,000 gallons (34,000 liters) of air for every gallon of gasoline burned at an air–fuel ratio of 14.7:1 by weight. Without proper filtering of the air before it enters the engine, dust and dirt in the air can seriously damage engine parts and shorten engine life.

Abrasive particles can cause wear any place inside the engine where two surfaces move against each other, such as piston rings against the cylinder wall. The dirt particles then pass by the piston rings and into the crankcase. From the crankcase, the particles circulate throughout the engine in the oil. Large amounts of abrasive particles in the oil can damage other moving engine parts.

The filter that cleans the intake air is in a two-piece air cleaner housing made of either:

- Stamped steel
- Composite (usually nylon reinforced plastic) materials

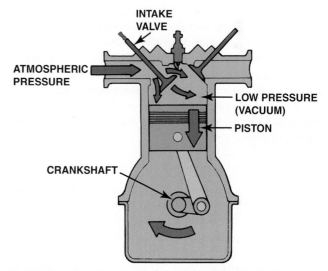

**FIGURE 8–1** Downward movement of the piston lowers the air pressure inside the combustion chamber. The pressure differential between the atmosphere and the inside of the engine forces air into the engine.

**AIR FILTER ELEMENTS** The paper air filter element is the most common type of filter. It is made of a chemically treated paper stock that contains tiny passages in the fibers. These passages form an indirect path for the airflow to follow. The airflow passes through several fiber surfaces, each of which traps microscopic particles of dust, dirt, and carbon. Most air filters are capable of trapping dirt and other particles larger than 10 to 25 microns in size. One **micron** is equal to 0.000039 inches.

**NOTE: A person can only see objects that are 40 microns or larger in size. A human hair is about 50 microns in diameter.**

●**SEE FIGURE 8–2.**

**FIGURE 8–2** Dust and dirt in the air are trapped in the air filter so they do not enter the engine. A restricted air filter reduces the amount of air that can enter the engine thereby reducing engine power and performance.

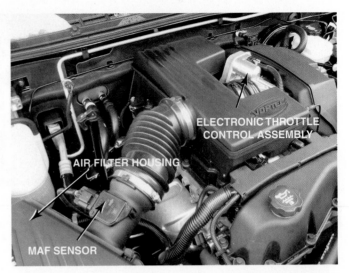

**FIGURE 8–3** Most air filter housings are located on the side of the engine compartment and use flexible rubber hose to direct the airflow into the throttle body of the engine.

**FILTER REPLACEMENT** Manufacturers recommend cleaning or replacing the air filter element at periodic intervals, usually listed in terms of distance driven or months of service. The distance and time intervals are based on so-called normal driving. More frequent air filter replacement is necessary when the vehicle is driven under dusty, dirty, or other severe conditions.

It is best to replace a filter element before it becomes too dirty to be effective. A dirty air filter that passes contaminants can cause engine wear.

**REMOTELY MOUNTED AIR FILTERS AND DUCTS** Air cleaner and duct design depend on a number of factors such as the size, shape, and location of other engine compartment components, as well as the vehicle body structure.

Port fuel-injection systems generally use a horizontally mounted throttle body.

Some systems also have a mass airflow (MAF) sensor between the throttle body and the air cleaner. Because placing the air cleaner housing next to the throttle body would cause engine and vehicle design problems, it is more efficient to use this remote air cleaner placement. ● SEE FIGURE 8–3.

Turbocharged engines present a similar problem. The air cleaner connects to the air inlet elbow at the turbocharger. However, the tremendous heat generated by the turbocharger makes it impractical to place the air cleaner housing too close to the turbocharger. Remote air cleaners are connected to the turbocharger air inlet elbow or fuel-injection throttle body by composite ducting that is usually retained by clamps. The ducting used may be rigid or flexible, but all connections must be airtight.

**AIR FILTER RESTRICTION INDICATOR** Some vehicles, especially pickup trucks that are often driven in dusty conditions, are equipped with an air filter restriction indicator. The purpose of this device is to give a visual warning when the air filter is restricted and needs to be replaced. The device operates

**FIGURE 8–4** A typical air filter restriction indicator used on a General Motors truck engine. The indicator turns red when it detects enough restriction to require a filter replacement.

by detecting the slight drop in pressure that occurs when an air filter is restricted. The calibration before the red warning bar or "replace air filter" message appears varies, but is usually:

■ 15 to 20 inch of water (inch $H_2O$) for gasoline engines
■ 20 to 30 inch of water (inch $H_2O$) for diesel engines

The unit of inches of water is used to measure the difference in air pressure before and after the air filter. The unit is very small, because 28 inch of water is equal to a pound per square inch (PSI).

Some air filter restriction indicators, especially on diesel engines, include an electrical switch used to light a dash-mounted warning lamp when the air filter needs to be replaced. ● SEE FIGURE 8–4.

(a)

(b)

**FIGURE 8–5** (a) Note the discovery as the air filter housing was opened during service on a Pontiac. The nuts were obviously deposited by squirrels (or some other animal). (b) Not only was the housing filled with nuts, but also this air filter was extremely dirty, indicating that this vehicle had not been serviced for a long time.

 **TECH TIP**

**Always Check the Air Filter**

Always inspect the air filter and the air intake system carefully during routine service. Debris or objects deposited by animals can cause a restriction to the airflow and can reduce engine performance. ● **SEE FIGURE 8–5.**

# THROTTLE-BODY INJECTION INTAKE MANIFOLDS

**TERMINOLOGY** The *intake manifold* is also called an *inlet manifold.* Smooth engine operation can occur only when each combustion chamber produces the same pressure as every

**FIGURE 8–6** A resonance tube, called a Helmholtz resonator, is used on the intake duct between the air filter and the throttle body to reduce air intake noise during engine acceleration.

**?** **FREQUENTLY ASKED QUESTION**

**What Does this Tube Do?**

What is the purpose of the odd-shape tube attached to the inlet duct between the air filter and the throttle body, as seen in ● **FIGURE 8–6?**

The tube shape is designed to dampen out certain resonant frequencies that can occur at specific engine speeds. The length and shape of this tube are designed to absorb shock waves that are created in the air intake system and to provide a reservoir for the air that will then be released into the airstream during cycles of lower pressure. This resonance tube is often called a **Helmholtz resonator,** named for the discoverer of the relationship between shape and value of frequency, Herman L. F. von Helmholtz (1821–1894) of the University of Hönizsberg in East Prussia. The overall effect of these resonance tubes is to reduce the noise of the air entering the engine.

other chamber in the engine. For this to be achieved, each cylinder must receive an intake charge exactly like the charge going into the other cylinders in quality and quantity. The charges must have the same physical properties and the same air–fuel mixture.

A throttle-body fuel injector forces finely divided droplets of liquid fuel into the incoming air to form a combustible air–fuel mixture. ● **SEE FIGURE 8–7** for an example of a typical throttle-body injection (TBI) unit.

**INTAKE AIR SPEEDS** These droplets start to evaporate as soon as they leave the throttle-body injector nozzles. *The droplets stay in the charge as long as the charge flows at high velocities.* At maximum engine speed, these velocities may reach 300 ft per second. Separation of the droplets from the charge as it passes through the manifold occurs when the velocity drops

**FIGURE 8–7** A throttle-body injection (TBI) unit used on a GM V-6 engine.

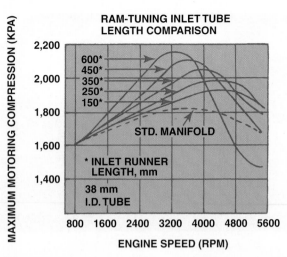

**FIGURE 8–9** The graph shows the effect of sonic tuning of the intake manifold runners. The longer runners increase the torque peak and move it to a lower RPM. The 600 mm intake runner is about 24 inch long.

**FIGURE 8–8** Heavy fuel droplets separate as they flow around an abrupt bend in an intake manifold.

below 50 ft per second. Intake charge velocities at idle speeds are often below this value. When separation occurs—at low engine speeds—extra fuel must be supplied to the charge in order to have a combustible mixture reach the combustion chamber.

Manifold sizes and shapes represent a compromise.

- They must have a cross section large enough to allow charge flow for maximum power.

- The cross section must be small enough that the flow velocities of the charge will be high enough to keep the fuel droplets in suspension. This is required so that equal mixtures reach each cylinder. Manifold cross-sectional size is one reason why engines designed especially for racing will not run at low engine speeds.

- Racing manifolds must be large enough to reach maximum horsepower. This size, however, allows the charge to move slowly, and the fuel will separate from the charge at low engine speeds. Fuel separation leads to poor accelerator response. ● **SEE FIGURE 8–8.**

Standard passenger vehicle engines are primarily designed for economy during light-load, partial-throttle operation. Their manifolds, therefore, have a much smaller cross-sectional area than do those of racing engines. This small size will help keep flow velocities of the charge high throughout the normal operating speed range of the engine.

# PORT FUEL-INJECTION INTAKE MANIFOLDS

**TERMINOLOGY** The size and shape of port fuel-injected engine intake manifolds can be optimized because the only thing in the manifold is air. The fuel injector is located in the intake manifold about 3 to 4 inches (70 to 100 mm) from the intake valve. Therefore, the runner length and shape are designed for tuning only. There is no need to keep an air–fuel mixture thoroughly mixed (homogenized) throughout its trip from the TBI unit to the intake valve. Intake manifold runners are tuned to improve engine performance.

- Long runners build low-RPM torque.

- Shorter runners provide maximum high-RPM power.

   ● **SEE FIGURES 8–9 AND 8–10.**

**VARIABLE INTAKES** Some engines with four valve heads utilize a dual or variable intake runner design. At lower engine speeds, long intake runners provide low-speed torque. At higher engine speeds, shorter intake runners are opened by means of a computer-controlled valve to increase high-speed power.

Many intake manifolds are designed to provide both short runners best for higher engine speed power and longer runners best for lower engine speed torque. The valve(s) that control the flow of air through the passages of the intake manifold are computer controlled. ● **SEE FIGURES 8–11 AND 8–12.**

**PLASTIC INTAKE MANIFOLDS** Most intake manifolds are made from thermoplastic molded from fiberglass-reinforced nylon by either casting or by injection molding. Some manifolds are molded in two parts and bonded together. Plastic intake manifolds are lighter than aluminum manifolds and can better insulate engine heat from the fuel injectors.

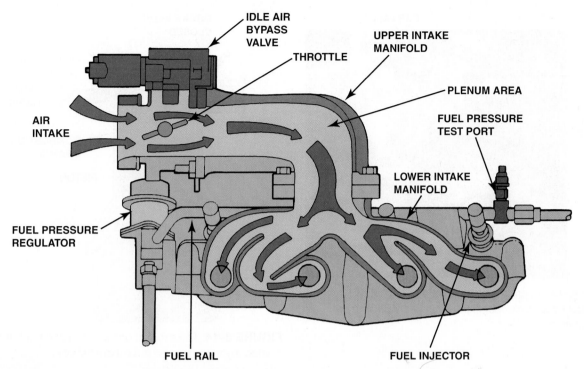

FIGURE 8–10 Airflow through the large diameter upper intake manifold is distributed to smaller diameter individual runners in the lower manifold in this two-piece manifold design.

FIGURE 8–11 The air flowing into the engine can be directed through long or short runners for best performance and fuel economy.

Plastic intake manifolds have smoother interior surfaces than do other types of manifolds, resulting in greater airflow.

## UPPER AND LOWER INTAKE MANIFOLDS
Many intake manifolds are constructed in two parts.

- A lower section attaches to the cylinder heads and includes passages from the intake ports.

- An upper manifold, usually called the **plenum,** connects to the lower unit and includes the long passages needed to help provide the ram effect that helps the engine deliver maximum torque at low engine speeds. The throttle body attaches to the upper intake.

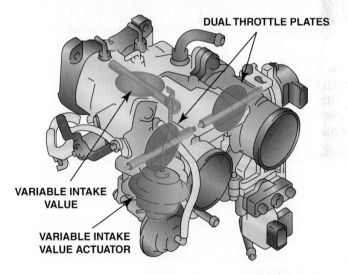

FIGURE 8–12 A variable intake manifold is used to change the length of the intake runner to enhance engine performance. At lower engine speed, a longer intake runner results in improved engine torque whereas at higher engine speeds, a shorter intake runner helps the engine produce high-speed power. The actuator can be vacuum or electric depending on the make and model of vehicle.

The use of a two-part intake manifold allows for easier manufacturing as well as assembly, but can create additional locations for leaks.

If the lower intake manifold gasket leaks, not only could a vacuum leak occur affecting the operation of the engine, but a coolant leak or an oil leak can also occur if the manifold has coolant flowing through it. A leak at the gasket(s) of the upper intake manifold usually results in a vacuum (air) leak only.

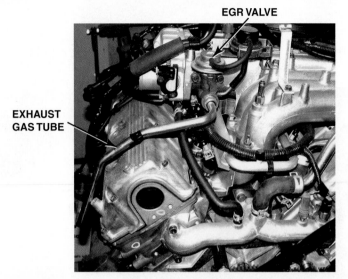

FIGURE 8–13 A typical long exhaust gas line used to cool the exhaust gases before being recirculated back into the intake manifold.

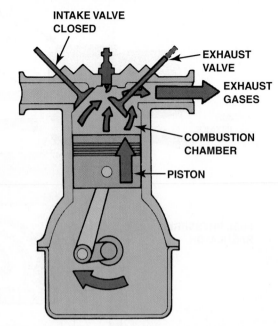

FIGURE 8–14 The exhaust gases are pushed out of the cylinder by the piston on the exhaust stroke.

## EXHAUST GAS RECIRCULATION PASSAGES

**PURPOSE AND FUNCTION** To reduce the emission of oxides of nitrogen (NOx), engines have been equipped with **exhaust gas recirculation (EGR)** valves. From 1973 until recently, they were used on almost all vehicles. Most EGR valves are mounted on the intake manifold. Because of the efficiency of computer-controlled fuel injection, some newer engines do not require an EGR system to meet emission standards. These engines' variable valve timing to close the exhaust valve sooner than normal, trapping some exhaust in the cylinder, is an alternative to using an EGR valve.

On engines with EGR systems, the EGR valve opens at speeds above idle on a warm engine. When open, the valve allows a small portion of the exhaust gas (5% to 10%) to enter the intake manifold.

The EGR system has some means of interconnecting of the exhaust and intake manifolds. The EGR valve controls the gas flow through the passages.

- On V-type engines, the intake manifold crossover is used as a source of exhaust gas for the EGR system. A cast passage connects the exhaust crossover to the EGR valve.
- On inline-type engines, an external tube is generally used to carry exhaust gas to the EGR valve.

**EXHAUST GAS COOLERS** The exhaust gases are more effective in reducing oxide of nitrogen (NOx) emissions if the exhaust is cooled before being drawn into the cylinders. This tube is often designed to be long so that the exhaust gas is cooled before it enters the EGR valve. ● **SEE FIGURE 8–13.**

## EXHAUST MANIFOLDS

**PURPOSE AND FUNCTION** The exhaust manifold is designed to collect high-temperature spent gases from the individual head exhaust ports and direct them into a single outlet connected to the exhaust system. ● **SEE FIGURE 8–14.**

The hot gases are sent to an exhaust pipe, then to a catalytic converter, to the muffler, to a resonator, and on to the tailpipe, where they are vented to the atmosphere. The exhaust system is designed to meet the following needs.

- Provide the least possible amount of restriction or backpressure
- Keep the exhaust noise at a minimum

Exhaust gas temperature will vary according to the power produced by the engine. The manifold must be designed to operate at both engine idle and continuous full power. Under full-power conditions, the exhaust manifold can become red-hot, causing a great deal of expansion.

The temperature of an exhaust manifold can exceed 1,500°F (815°C).

**CONSTRUCTION** Most exhaust manifolds are made from the following:

- Cast iron
- Steel tubing

During vehicle operation, manifold temperatures usually reach the high-temperature extremes. The manifold is bolted to the head in a way that will allow expansion and contraction. In some cases, hollow-headed bolts are used to maintain

FIGURE 8–15 This exhaust manifold (red area) is equipped with a heat shield to help retain heat and reduce exhaust emissions.

FIGURE 8–16 Many exhaust manifolds are constructed of steel tubing and are free flowing to improve engine performance.

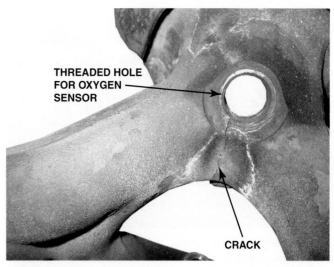

FIGURE 8–17 A crack in an exhaust manifold is often not visible because a heat shield usually covers the area. A crack in the exhaust manifold upstream of the oxygen sensor can fool the sensor and affect engine operation.

**? FREQUENTLY ASKED QUESTION**

**How Can a Cracked Exhaust Manifold Affect Engine Performance?**

Cracks in an exhaust manifold will not only allow exhaust gases to escape and cause noise, but also allow air to enter the exhaust manifold. ● SEE FIGURE 8–17.

Exhaust flows from the cylinders as individual puffs or pressure pulses. Behind each of these pressure pulses, a low pressure (below atmospheric pressure) is created. Outside air at atmospheric pressure is then drawn into the exhaust manifold through the crack. This outside air contains 21% oxygen and is measured by the oxygen sensor (O2S). The air passing the O2S signals the engine computer that the engine is operating too lean (excess oxygen) and the computer, not knowing that the lean indicator is false, adds additional fuel to the engine. The result is that the engine will be operating richer (more fuel than normal) and spark plugs could become fouled by fuel, causing poor engine operation.

a gas-tight seal while still allowing normal expansion and contraction.

Many exhaust manifolds have heat shields to help keep exhaust heat off the spark plug wires and to help keep the heat from escaping to improve exhaust emissions. ● SEE FIGURE 8–15.

Exhaust systems are especially designed for the engine-chassis combination. The exhaust system length, pipe size, and silencer are designed, where possible, to make use of the tuning effect within the exhaust system. Tuning occurs when the exhaust pulses from the cylinders are emptied into the manifold between the pulses of other cylinders. ● SEE FIGURE 8–16.

**EXHAUST MANIFOLD GASKETS** Exhaust heat will expand the manifold more than it will expand the head. The heat causes the exhaust manifold to slide on the sealing surface of the head. The heat also causes thermal stress. When the manifold is removed from the engine for service, the stress is relieved, which may cause the manifold to warp slightly. Exhaust manifold gaskets are included in gasket sets to seal slightly warped exhaust manifolds. These gaskets *should* be used, even if the engine did not originally use exhaust manifold gaskets. When an exhaust manifold gasket has facing on one side only, put the facing side against the head and put the manifold against the perforated metal core. The manifold can slide on the metal of the gasket just as it slid on the sealing surface of the head.

FIGURE 8–18 Typical exhaust manifold gaskets. Note how they are laminated to allow the exhaust manifold to expand and contract due to heating and cooling.

Gaskets are used on new engines with tubing- or header-type exhaust manifolds. They may have several layers of steel for high-temperature sealing. The layers are spot welded together. Some are embossed where special sealing is needed. ● SEE FIGURE 8–18.

Many new engines do not use gaskets with cast exhaust manifolds. The flat surface of the new cast-iron exhaust manifold fits tightly against the flat surface of the new head.

# MUFFLERS

**PURPOSE AND FUNCTION** When the exhaust valve opens, it rapidly releases high-pressure gas. This sends a strong air pressure wave through the atmosphere inside the exhaust system, which produces a sound we call an explosion. It is the same sound produced when the high-pressure gases from burned gunpowder are released from a gun. In an engine, the pulses are released one after another. The explosions come so fast that they blend together in a steady roar.

EXHAUST MANIFOLD SPREADER TOOL

FIGURE 8–19 An exhaust manifold spreader tool is absolutely necessary when reinstalling exhaust manifolds. When they are removed from the engine, the manifolds tend to warp slightly even though the engine is allowed to cool before being removed. The spreader tool allows the technician to line up the bolt holes without harming the manifold.

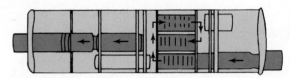

FIGURE 8–20 Exhaust gases expand and cool as they travel through passages in the muffler.

Sound is air vibration. When the vibrations are large, the sound is loud. The muffler catches the large bursts of high-pressure exhaust gas from the cylinder, smoothing out the pressure pulses and allowing them to be released at an even and constant rate. It does this through the use of perforated tubes within the muffler chamber. The smooth-flowing gases are released to the tailpipe. In this way, the muffler silences engine exhaust noise. ● SEE FIGURE 8–20.

**CONSTRUCTION** Most mufflers have a larger inlet diameter than outlet diameter. As the exhaust enters the muffler, it expands and cools. The cooler exhaust is denser and occupies less volume. The diameter of the outlet of the muffler and the diameter of the tailpipe can be reduced with no decrease in efficiency.

Sometimes resonators are used in the exhaust system and the catalytic converter also acts as a muffler. They provide additional expansion space at critical points in the exhaust system to smooth out the exhaust gas flow.

The tailpipe carries the exhaust gases from the muffler to the air, away from the vehicle. In most cases, the tailpipe exit is at the rear of the vehicle, below the rear bumper. In some cases, the exhaust is released at the side of the vehicle, just ahead of or just behind the rear wheel.

The muffler and tailpipe are supported with brackets, called **hangers,** which help to isolate the exhaust noise from

**FIGURE 8–21** A hole in the muffler allows condensed water to escape.

**FIGURE 8–22** A high-performance aftermarket air filter often can increase airflow into the engine for more power.

 **FREQUENTLY ASKED QUESTION**

**Why Is There a Hole in My Muffler?**

Many mufflers are equipped with a small hole in the lower rear part to drain accumulated water. About 1 gallon of water is produced in the form of steam for each gallon of gasoline burned. The water is formed when gasoline is burned in the cylinder. Water consists of two molecules of hydrogen and one of oxygen ($H_2O$). The hydrogen (H) comes from the fuel and the oxygen (O) comes from the air. During combustion, the hydrogen from the fuel combines with some of the oxygen in the air to form water vapor. The water vapor condenses on the cooler surfaces of the exhaust system, especially in the muffler, until the vehicle has been driven long enough to fully warm the exhaust above the boiling point of water (212°F [100°C]). ● **SEE FIGURE 8–21.**

 **HIGH-PERFORMANCE TIP**

**More Airflow = More Power**

One of the most popular high-performance modifications is to replace the factory exhaust system with a low-restriction design and to replace the original air filter and air filter housing with a low-restriction unit, as shown in ● **FIGURE 8–22.**

The installation of an aftermarket air filter not only increases power, but also increases air induction noise, which many drivers prefer. The aftermarket filter housing, however, may not be able to effectively prevent water from being drawn into the engine if the vehicle is traveling through deep water.

Almost every modification that increases performance has a negative effect on some other part of the vehicle, or else the manufacturer would include the change at the factory. Use an aftermarket intake system with caution as it can sometimes affect the emissions and driveablity.

the rest of the vehicle. The types of exhaust system hangers include:

■ Rubberized fabric with metal ends that hold the muffler and tailpipe in position so that they do not touch any metal part, to isolate the exhaust noise from the rest of the vehicle

■ Rubber material that looks like large rubber bands, which slip over the hooks on the exhaust system and the hooks attached to the body of the vehicle

## SUMMARY

1. All air entering an engine must be filtered.
2. Engines that use throttle-body injection units are equipped with intake manifolds that keep the airflow speed through the manifold at 50 to 300 ft per second.
3. Most intake manifolds have an EGR valve that regulates the amount of recirculated exhaust that enters the engine to reduce NOx emissions.
4. Exhaust manifolds can be made from cast iron or steel tubing.
5. The exhaust system also contains a catalytic converter, exhaust pipes, and muffler. The entire exhaust system is supported by rubber hangers that isolate the noise and vibration of the exhaust from the rest of the vehicle.

1. Why is it necessary to have intake charge velocities of about 50 ft per second?
2. Why can port fuel-injected engines use larger (and longer) intake manifolds and still operate at low engine speed?
3. What is a tuned runner in an intake manifold?
4. How does a muffler quiet exhaust noise?

## CHAPTER QUIZ

1. Intake charge velocity has to be _____ to prevent fuel droplet separation.
   a. 25 ft per second
   b. 50 ft per second
   c. 100 ft per second
   d. 300 ft per second

2. The air filter restriction indicator uses what to detect when it signals to replace the filter?
   a. Number of hours of engine operation
   b. Number of miles or vehicle travel
   c. The amount of light that can past through the filter
   d. The amount of restriction measured in inches of water

3. Why are the EGR gases cooled before entering the engine on some engines?
   a. Cool exhaust gas is more effective at controlling NOx emissions
   b. To help prevent the exhaust from slowing down
   c. To prevent damage to the intake valve
   d. To prevent heating the air–fuel mixture in the cylinder

4. The air–fuel mixture flows through the intake manifold on what type of system?
   a. Port fuel-injection systems
   b. Throttle-body fuel-injection systems
   c. Both a port-injected and throttle-body injected engine
   d. Any fuel-injected engine

5. Air filters can remove particles and dirt as small as _____.
   a. 5 to 10 microns
   b. 10 to 25 microns
   c. 30 to 40 microns
   d. 40 to 50 microns

6. Why do many port fuel-injected engines use long intake manifold runners?
   a. To reduce exhaust emissions
   b. To heat the incoming air
   c. To increase high-RPM power
   d. To increase low-RPM torque

7. Exhaust passages are included in some intake manifolds. Technician A says that the exhaust passages are used for exhaust gas recirculation (EGR) systems. Technician B says that the upper intake is often called the plenum. Which technician is correct?
   a. Technician A only
   b. Technician B only
   c. Both Technicians A and B
   d. Neither Technician A nor B

8. The upper portion of a two-part intake manifold is often called the _____.
   a. Housing
   b. Lower part
   c. Plenum
   d. Vacuum chamber

9. Technician A says that a cracked exhaust manifold can affect engine operation. Technician B says that a leaking lower intake manifold gasket could cause a vacuum leak. Which technician is correct?
   a. Technician A only
   b. Technician B only
   c. Both Technicians A and B
   d. Neither Technician A nor B

10. Technician A says that some intake manifolds are plastic. Technician B says that some intake manifolds are constructed in two parts or sections: upper and lower. Which technician is correct?
    a. Technician A only
    b. Technician B only
    c. Both Technicians A and B
    d. Neither Technician A nor B

# TURBOCHARGING AND SUPERCHARGING

**LEARNING OBJECTIVES:** **After studying this chapter, the reader should be able to:** • Discuss airflow requirements and volumetric efficiency of engines. • Understand forced induction principles. • List the advantages and disadvantages of superchargers. • Explain the purpose and function of turbochargers. • Explain boost control and turbocharger failures. • Describe the purpose and function of a nitrous oxide system.

**KEY TERMS:** Boost 123 • BOV 130 • Bypass valve 126 • CBV 130 • Dry system 132 • Dump valve 130 • Forced induction systems 124 • Intercooler 129 • Naturally (normally) aspirated 123 • Nitrous oxide ($N_2O$) 132 • Positive displacement 126 • Power adder 132 • Roots supercharger 126 • Supercharger 125 • Turbocharger 127 • Turbo lag 128 • Vent valve 130 • Volumetric efficiency 123 • Wastegate 129 • Wet system 132

## INTRODUCTION

### AIRFLOW REQUIREMENTS
Naturally aspirated engines with throttle plates use atmospheric pressure to push an air–fuel mixture into the combustion chamber vacuum created by the downstroke of a piston. The mixture is then compressed before ignition to increase the force of the burning, expanding gases. The greater the compression of the air–fuel mixture, the higher the engine power output resulting from combustion.

A four-stroke engine can take in only so much air, and how much fuel it needs for proper combustion depends on how much air it takes in. Engineers calculate engine airflow requirements using three factors:

1. Engine displacement
2. Engine revolutions per minute (RPM)
3. Volumetric efficiency

### VOLUMETRIC EFFICIENCY
**Volumetric efficiency** is a measure of how well an engine breathes. It is a comparison of the actual volume of air–fuel mixture drawn into an engine to the theoretical maximum volume that could be drawn in. Volumetric efficiency is expressed as a percentage. If the engine takes in the airflow volume slowly, a cylinder might fill to capacity. It takes a definite amount of time for the airflow to pass through all the curves of the intake manifold and valve port. Therefore, volumetric efficiency decreases as engine speed increases because of the shorter amount of time for the cylinders to be filled with air during the intake stroke. At high speed, it may drop to as low as 50%.

The average stock gasoline engine never reaches 100% volumetric efficiency. A new engine is about 85% efficient.

A race engine usually has 95% or better volumetric efficiency. These figures apply only to naturally aspirated engines. However, with either turbochargers or superchargers, engines can easily achieve more than 100% volumetric efficiency. Many vehicles are equipped with a supercharger or a turbocharger from the factory to increase power. ● **SEE FIGURES 9–1 AND 9–2.**

## FORCED INDUCTION PRINCIPLES

### PURPOSE AND FUNCTION
The amount of force an air–fuel charge produces when it is ignited is largely a function of the charge density. Charge density is a term used to define the amount of the air–fuel charge introduced into the cylinders. Density is the mass of a substance in a given amount of space.
● **SEE FIGURE 9–3.**

The greater the density of an air–fuel charge forced into a cylinder, the greater the force it produces when ignited, and the greater the engine power.

An engine that uses atmospheric pressure for its intake charge is called a **naturally (normally) aspirated** engine. A better way to increase air density is to use some type of air pump, such as a turbocharger or supercharger.

When air is pumped into the cylinder, the combustion chamber receives an increase of air pressure, known as **boost,** and can be measured in the following ways:

- Pounds per square inch (PSI)
- Atmospheres (ATM) (1 atmosphere is 14.7 PSI)
- Bars (1 bar is 14.7 PSI)

FIGURE 9–1 A supercharger on a Ford V-8.

FIGURE 9–2 A turbocharger on a Toyota engine.

While boost pressure increases air density, friction heats air in motion and causes an increase in temperature. This increase in temperature works in the opposite direction, decreasing air density. Because of these and other variables, an increase in pressure does not always result in greater air density.

**FORCED INDUCTION PRINCIPLES** **Forced induction systems** use an air pump to pack a denser air–fuel charge into the cylinders. Because the density of the air–fuel charge is greater, the following occurs:

- The weight of the air–fuel charge is higher.
- Power is increased because it is directly related to the weight of an air–fuel charge consumed within a given time period.

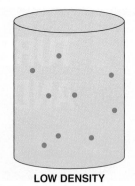 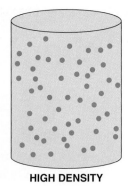

LOW DENSITY          HIGH DENSITY

FIGURE 9–3 The more air and fuel that can be packed in a cylinder, the greater the density of the air–fuel charge.

Pumping air into the intake system under pressure forces it through the bends and restrictions of the air intake system at a greater speed than it would travel under normal atmospheric pressure. This added pressure allows more air to enter the intake port before the intake valve closes. By increasing the airflow into the intake, more fuel can be mixed with the air while still maintaining the same air–fuel ratio. The denser the air–fuel charge entering the engine during its intake stroke, the greater the potential energy released during combustion. In addition to the increased power resulting from combustion, there are several other advantages of supercharging an engine including the following:

- It increases the air–fuel charge density to provide high-compression pressure when power is required but allows the engine to run on lower pressures when additional power is not required.
- The pumped air pushes the remaining exhaust from the combustion chamber during intake and exhaust valve overlap. (Overlap is when both the intake and the exhaust valves are partially open when the piston is near the top at the end of the exhaust stroke and the beginning of the intake stroke.)
- The forced airflow and removal of hot exhaust gases lowers the temperature of the cylinder head, pistons, and valves and helps extend the life of the engine.

A supercharger or turbocharger pressurizes air to greater than atmospheric pressure. The pressurization above atmospheric pressure, or boost, can be measured in the same way as atmospheric pressure. Atmospheric pressure drops as altitude increases, but boost pressure remains the same. If a supercharger develops 12 PSI (83 kPa) boost at sea level, it will develop the same amount at a 5,000-foot altitude because boost pressure is measured inside the intake manifold. ● SEE FIGURE 9–4.

**BOOST AND COMPRESSION RATIOS** Boost increases the amount of air entering the cylinder during the intake stroke. This extra air causes the effective compression ratio to be greater than the mechanical compression ratio designed into

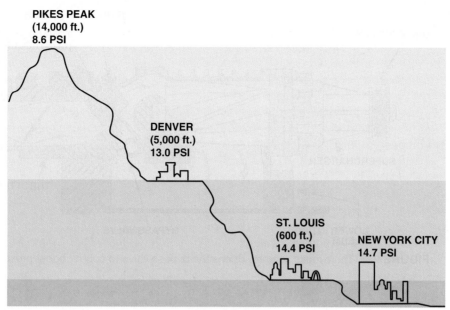

FIGURE 9–4 Atmospheric pressure decreases with increases in altitude.

| FINAL COMPRESSION RATIO CHART AT VARIOUS BOOST LEVELS | | | | | | | | | | |
|---|---|---|---|---|---|---|---|---|---|---|
| | BLOWER BOOST (PSI) | | | | | | | | | |
| Comp Ratio | 2 | 4 | 6 | 8 | 10 | 12 | 14 | 16 | 18 | 20 |
| 6.5 | 7.4 | 8.3 | 9.2 | 10 | 10.9 | 11.8 | 12.7 | 13.6 | 14.5 | 15.3 |
| 7 | 8 | 8.9 | 9.9 | 10.8 | 11.8 | 12.7 | 13.6 | 14.5 | 15.3 | 16.2 |
| 7.5 | 8.5 | 9.5 | 10.6 | 11.6 | 12.6 | 13.6 | 14.6 | 15.7 | 16.7 | 17.8 |
| 8 | 9.1 | 10.2 | 11.3 | 12.4 | 13.4 | 14.5 | 15.6 | 16.7 | 17.8 | 18.9 |
| 8.5 | 9.7 | 10.8 | 12 | 13.1 | 14.3 | 15.4 | 16.6 | 17.8 | 18.9 | 19.8 |
| 9 | 10.2 | 11.4 | 12.7 | 13.9 | 15.1 | 16.3 | 17.6 | 18.8 | 20 | 21.2 |
| 9.5 | 10.8 | 12.1 | 13.4 | 14.7 | 16 | 17.3 | 18.5 | 19.8 | 21.1 | 22.4 |
| 10 | 11.4 | 12.7 | 14.1 | 15.4 | 16.8 | 18.2 | 19.5 | 20.9 | 22.2 | 23.6 |

**CHART 9–1**

The effective compression ratio compared to the boost pressure.

the engine. The higher the boost pressure, the greater the compression ratio. This means that any engine that uses a supercharger or turbocharger must use all of the following engine components:

- Forged pistons to withstand the increased combustion pressures
- Stronger-than-normal connecting rods
- Piston oil squirters that direct a stream of oil to the underneath part of the piston to keep piston temperatures under control
- Lower compression ratio compared to a naturally aspirated engine

● **SEE CHART 9–1.**

## SUPERCHARGERS

**INTRODUCTION**  A **supercharger** is an engine-driven air pump that supplies more than the normal amount of air into the intake manifold and boosts engine torque and power. A supercharger provides an instantaneous increase in power without any delay. However, a supercharger, because it is driven by the engine, requires horsepower to operate and is not as efficient as a turbocharger.

**PARTS AND OPERATION**  Gears, shafts, chains, or belts from the crankshaft can all be used to turn the pump. This means that the air pump or supercharger pumps air in direct relation to engine speed.

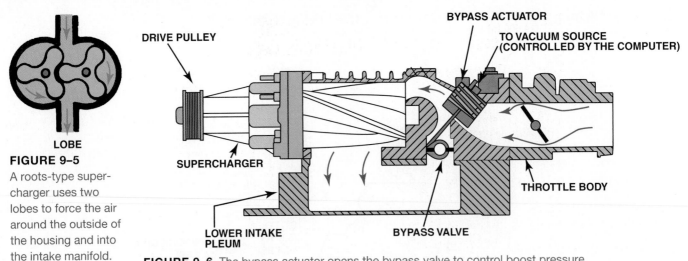

**FIGURE 9–5**

A roots-type super-charger uses two lobes to force the air around the outside of the housing and into the intake manifold.

**FIGURE 9–6** The bypass actuator opens the bypass valve to control boost pressure.

**TYPES OF SUPERCHARGERS** There are two general types of superchargers:

- **Roots type.** Named for Philander and Francis Roots, two brothers from Connersville, Indiana, the **roots supercharger** was patented in 1860 as a type of water pump to be used in mines. Later, it was used to move air and is used today on two-stroke-cycle Detroit diesel engines and other supercharged engines. The roots-type supercharger is called a **positive displacement** design because all of the air that enters is forced through the unit. Example of a roots-type supercharger includes the GMC 6-71 (used originally on GMC diesel engines that had 6 cylinders each with 71 cu. in.). Eaton used the roots design for the supercharger on the 3800 V-6 GM engine. ● **SEE FIGURE 9–5.**

- **Centrifugal supercharger.** A centrifugal supercharger is similar to a turbocharger but is mechanically driven by the engine instead of being powered by the hot exhaust gases. A centrifugal supercharger is not a positive displacement pump, and all of the air that enters is not forced through the unit. Air enters a centrifugal supercharger housing in the center and exits at the outer edges of the compressor wheels at a much higher speed because of centrifugal force. The speed of the blades has to be higher than engine speed, so a smaller pulley is used on the supercharger, compared to the engine crankshaft which overdrives the impeller through an internal gear box, achieving about seven times the speed of the engine. Examples of centrifugal superchargers include Vortech and Paxton.

**SUPERCHARGER BOOST CONTROL** Many factory installed superchargers are equipped with a **bypass valve** that allows intake air to flow directly into the intake manifold, bypassing the supercharger. The computer controls the bypass valve actuator. ● **SEE FIGURE 9–6.**

**TECH TIP**

**Faster Moves More Air**

One of the high-performance measures that can be used to increase horsepower on a supercharged engine is to install a smaller diameter pulley. The smaller the pulley diameter, the faster the supercharger will rotate and the higher the potential boost pressure will be. The change will require a shorter belt, and the extra boost could cause serious engine damage.

The airflow is directed around the supercharger whenever any of the following conditions occur:

- The boost pressure, as measured by the MAP sensor, indicates that the intake manifold pressure is reaching the predetermined boost level.
- During deceleration, to prevent excessive pressure buildup in the intake.
- Reverse gear is selected.
- When the engine is at idle speed.

**SUPERCHARGER SERVICE** Superchargers are usually lubricated with synthetic engine oil inside the unit. This oil level should be checked and replaced as specified by the vehicle or supercharger manufacturer. The drive belt should also be inspected and replaced as necessary. The air filter should be replaced regularly, and always use the filter specified for a supercharged engine. Many factory supercharger systems use a separate cooling system for the air charge cooler located under the supercharger. Check service information for the exact service procedures to follow. ● **SEE FIGURE 9–7.**

**FIGURE 9–7** A Ford supercharger cutaway display showing the roots-type blower and air charge cooler (intercooler). The air charge cooler is used to reduce the temperature of the compressed air before it enters the engine to increase the air charge density.

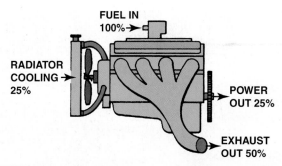

**FIGURE 9–8** A turbocharger uses some of the heat energy that would normally be wasted.

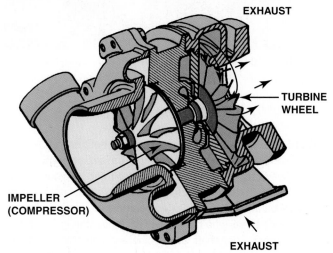

**FIGURE 9–9** A turbine wheel is turned by the expanding exhaust gases.

# TURBOCHARGERS

**INTRODUCTION** The major disadvantage of a supercharger is that it takes some of the engine power to drive the unit. In some installations, as much as 20% of the engine power is used by a mechanical supercharger. A **turbocharger** uses the heat of the exhaust to power a turbine wheel and therefore does not directly reduce engine power. In a naturally aspirated engine, about half of the heat energy contained in the fuel goes out the exhaust system. As much as 50% of the heat is lost to the exhaust system. Some of this lost energy is regained by using a turbocharger that uses the normally wasted combustion heat energy to perform useful work. Another 25% is lost through radiator cooling. Only about 25% is actually converted to mechanical power. A mechanically driven pump uses some of this mechanical output, but a turbocharger gets its energy from the exhaust gases, converting more of the fuel's heat energy into useful mechanical energy. ● **SEE FIGURE 9–8.**

**OPERATION** A turbocharger turbine looks much like a typical centrifugal pump used for supercharging.

Hot exhaust gases flow from the combustion chamber to the *turbine wheel*. The gases are heated and expanded as they leave the engine. It is not the speed of force of the exhaust gases that forces the turbine wheel to turn, as is commonly thought, but the expansion of hot gases against the turbine wheel's blades.

A turbocharger consists of two chambers connected with a center housing. The two chambers contain a turbine wheel and an *impeller* (compressor) *wheel* connected by a shaft that passes through the center housing. ● **SEE FIGURE 9–9.**

To take full advantage of the exhaust heat that provides the rotating force, a turbocharger must be positioned as close as possible to the exhaust manifold. This allows the hot exhaust to pass directly into the unit with minimal heat loss. As exhaust gas enters the turbocharger, it rotates the turbine blades. The turbine wheel and compressor wheel are on the same shaft so that they turn at the same speed. Rotation of the compressor wheel draws air in through a central inlet, and centrifugal force pumps it through an outlet at the edge of the housing. A pair of bushings in the center housing supports the turbine and compressor wheel shaft and is lubricated by engine oil. ● **SEE FIGURE 9–10.**

Both the turbine and the compressor wheels must operate with extremely close clearances to minimize possible leakage around their blades. Any leakage around the turbine blades causes a dissipation of the heat energy required for compressor rotation. Leakage around the compressor blades prevents the turbocharger from developing its full boost pressure.

**TURBOCHARGER OPERATION** When the engine is started and runs at low speed, both exhaust heat and pressure are low, and the turbine runs at a low speed (approximately

**FIGURE 9–10** The exhaust drives the turbine wheel on the left, which is connected to the impeller wheel on the right through a shaft. The bushings that support the shaft are lubricated with engine oil under pressure.

**FIGURE 9–11** Engine oil is fed to the center of the turbocharger to lubricate the bushings and returns to the oil pan through a return line.

1000 RPM). Because the compressor does not turn fast enough to develop boost pressure, air simply passes through it, and the engine works like any naturally aspirated engine. As the engine runs faster or load increases, both exhaust heat and flow increase, causing the turbine and compressor wheels to rotate faster. Since there is no brake and very little rotating resistance on the turbocharger shaft, the turbine and compressor wheels accelerate as the exhaust heat energy increases. When an engine is running at full power, the typical turbocharger rotates at speeds between 100,000 and 150,000 RPM. The turbocharger is lubricated by engine oil through an oil line to the center bushing assembly. ● **SEE FIGURE 9–11.**

Engine deceleration from full power to idle requires only a second or two because of its internal friction, pumping resistance, and drivetrain load. The turbocharger, however, has no such load on its shaft and is already turning many times faster than the engine at top speed. As a result, it can take as much as a minute or more after the engine has returned to idle speed before the turbocharger also has returned to idle. If the engine is decelerated to idle and then shut off immediately, engine lubrication stops flowing to the center housing bushings while the turbocharger is still spinning at thousands of RPM. The oil in the center housing is then subjected to extreme heat and can gradually "coke" or oxidize. The coked oil can clog passages and will reduce the life of the turbocharger.

The high rotating speeds and extremely close clearances of the turbine and compressor wheels in their housings require equally critical bushing clearances. The bushings must keep radial clearances of 0.003 to 0.006 inches (0.08 to 0.15 mm). Axial clearance (endplay) must be maintained at 0.001 to 0.003 inches (0.025 to 0.08 mm). If properly maintained, the turbocharger also is a trouble-free device. However, to prevent problems, the following conditions must be met:

- The turbocharger bushings must be constantly lubricated with clean engine oil. Turbocharged engines usually have specified oil changes at more frequent intervals than nonturbocharged engines. Always use the specified engine oil, which is likely to be vehicle specific and synthetic.

- Dirt particles and other contamination must be kept out of the intake and exhaust housings.

- Whenever a basic engine bearing (crankshaft or camshaft) has been damaged, the turbocharger must be flushed with clean engine oil after the bearing has been replaced.

- If the turbocharger is damaged, the engine oil must be drained and flushed and the oil filter replaced as part of the repair procedure.

Late-model turbochargers all have liquid-cooled center bushings to prevent heat damage. In a liquid-cooled turbocharger, engine coolant is circulated through passages cast in the center housing to draw off the excess heat. This allows the bushings to run cooler and minimizes the probability of oil coking when the engine is shut down.

**TURBOCHARGER SIZE AND RESPONSE TIME** A time lag occurs between an increase in engine speed and the increase in the speed of the turbocharger. This delay between acceleration and turbo boost is called **turbo lag.** Like any material, moving exhaust gas has inertia. Inertia also is present in the turbine and compressor wheels as well as the intake airflow. Unlike a supercharger, the turbocharger cannot supply an adequate amount of boost at low speed.

Turbocharger response time is directly related to the size of the turbine and compressor wheels. Small wheels accelerate rapidly; large wheels accelerate slowly. While small wheels would seem to have an advantage over larger ones, they may not have enough airflow capacity for an engine. To minimize turbo lag, the intake and exhaust breathing capacities of an engine must be matched to the exhaust and intake airflow capabilities of the turbocharger.

# BOOST CONTROL

**PURPOSE AND FUNCTION** Both supercharged and turbocharged systems are designed to provide a pressure greater than atmospheric pressure in the intake manifold. This increased pressure forces additional amounts of air into the combustion chamber over what would normally be forced in by atmospheric pressure. This increased charge increases engine power. The amount of "boost" (or pressure in the intake manifold) is measured in pounds per square inch (PSI), in inches of mercury (in. Hg), in bars, or in atmospheres. The following values will vary, depending on altitude and weather conditions (barometric pressure):

1 atmosphere = 14.7 PSI

1 atmosphere = 29.50 in. Hg

1 atmosphere = 1 bar

1 bar = 14.7 PSI

**BOOST CONTROL FACTORS** The higher the level of boost (pressure), the greater the horsepower output potential. However, other factors must be considered when increasing boost pressure:

1. As boost pressure increases, the temperature of the air also increases.

2. As the temperature of the air increases, combustion temperatures also increase, as does the possibility of detonation.

3. Power can be increased by cooling the compressed air after it leaves the turbocharger. *The power can be increased about 1% per 10°F by which the air is cooled.* A typical cooling device is called an **intercooler.** It is similar to a radiator, wherein outside air can pass through, cooling the pressurized heated air. An intercooler is located between the turbocharger and the intake manifold. ● **SEE FIGURE 9–12.** Some intercoolers use engine coolant to cool the hot compressed air that flows from the turbocharger to the intake.

4. As boost pressure increases, combustion temperature and pressures increase, which, if not limited, can do severe engine damage. The maximum exhaust gas temperature must be 1,550°F (840°C). Higher temperatures decrease the durability of the turbocharger *and* the engine.

**WASTEGATE** Turbochargers use exhaust gases to increase boost, which causes the engine to make more exhaust gases, which in turn increases the boost from the turbocharger. To prevent overboost and severe engine damage, most turbocharger systems use a wastegate. A **wastegate** is a valve similar to a door that can open and close. It is a bypass valve at the exhaust inlet to the turbine, which allows all of the exhaust into the

**FIGURE 9–12** The unit on top of this Subaru that looks like a radiator is the intercooler, which cools the air after it has been compressed by the turbocharger.

 **TECH TIP**

**Boost Is the Result of Restriction**

The boost pressure of a turbocharger (or supercharger) is commonly measured in pounds per square inch. If a cylinder head is restricted because of small valves and ports, the turbocharger will quickly provide boost. Boost results when the air being forced into the cylinder heads cannot flow into the cylinders fast enough and "piles up" in the intake manifold, increasing boost pressure. If an engine had large valves and ports, the turbocharger could provide a much greater *amount* of air into the engine at the same boost pressure as an identical engine with smaller valves and ports. Therefore, by increasing the size of the valves, a turbocharged or supercharged engine will be capable of producing much greater power.

turbine, or it can route part of the exhaust past the turbine to the exhaust system. If the valve is closed, all of the exhaust travels to the turbocharger. When a predetermined amount of boost pressure develops in the intake manifold, the wastegate valve is opened. As the valve opens, most of the exhaust flows directly out the exhaust system, bypassing the turbocharger. With less exhaust flowing across the vanes of the turbocharger, the turbocharger decreases in speed, and boost pressure is reduced. When the boost pressure drops, the wastegate valve closes to direct the exhaust over the turbocharger vanes to again allow the boost pressure to rise. Wastegate operation is a continuous process to control boost pressure.

**FIGURE 9–13** A wastegate is used on many turbocharged engines to control maximum boost pressure. The wastegate is controlled by a computer-controlled valve.

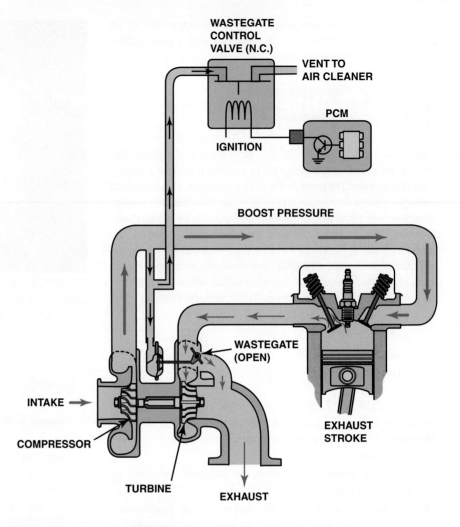

The wastegate is the pressure control valve of a turbocharger system. It is usually controlled by the engine control computer through a boost control solenoid, also called a wastegate control valve. ● **SEE FIGURE 9–13.**

**RELIEF VALVES** A wastegate controls the exhaust side of the turbocharger. A relief valve controls the intake side. A relief valve vents pressurized air from the connecting pipe between the outlet of the turbocharger and the throttle whenever the throttle is closed during boost, such as during shifts. If the pressure is not released, the turbocharger turbine wheel will slow down, creating a lag when the throttle is opened again after a shift has been completed. There are two basic types of relief valves:

1. **Compressor bypass valve (CBV).** This type of relief valve routes the pressurized air to the inlet side of the turbocharger for reuse and is quiet during operation.

2. **Blow-off valve (BOV).** Also called a **dump valve** or **vent valve,** the BOV features an adjustable spring design that keeps the valve closed until a sudden release of the throttle. The resulting pressure increase opens the valve and vents the pressurized air directly into the atmosphere. This type of relief valve is noisy in operation and creates a whooshing sound when the valve opens. ● **SEE FIGURE 9–14.**

**TECH TIP**

**If One Is Good, Two Are Better**

A turbocharger uses the exhaust from the engine to spin a turbine, which is connected to an impeller inside a turbocharger. This impeller then forces air into the engine under pressure, higher than is normally achieved without a turbocharger. The more air that can be forced into an engine, the greater the power potential. A V-type engine has two exhaust manifolds, so two small turbochargers can be used to help force greater quantities of air into an engine, as shown in ● **FIGURE 9–15.**

# TURBOCHARGER FAILURES

**SYMPTOMS OF FAILURE** When turbochargers fail to function correctly, a noticeable drop in power occurs. To restore proper operation, the turbocharger must be rebuilt,

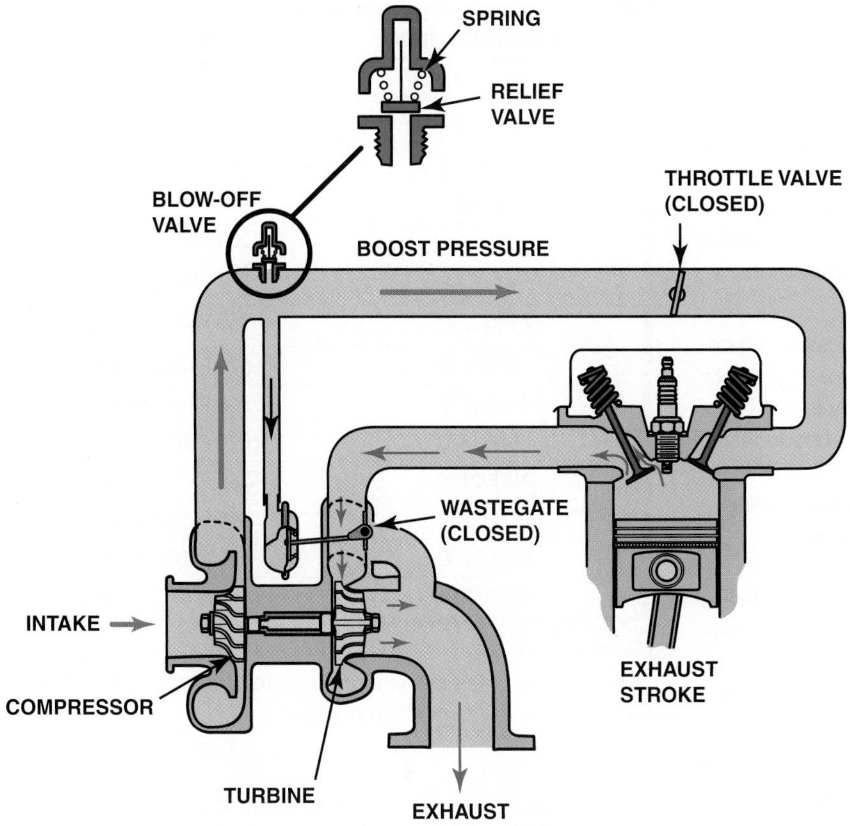

SPRING

RELIEF VALVE

BLOW-OFF VALVE

BOOST PRESSURE

THROTTLE VALVE (CLOSED)

WASTEGATE (CLOSED)

INTAKE

COMPRESSOR

TURBINE

EXHAUST

EXHAUST STROKE

**FIGURE 9–14** A blow-off valve is used in some turbocharged systems to relieve boost pressure during deceleration.

**FIGURE 9–15** A dual turbocharger system installed on a small block Chevrolet V-8 engine.

repaired, or replaced. It is not possible to simply remove the turbocharger, seal any openings, and maintain decent driveability. Bushing failure is a common cause of turbocharger failure, and replacement bushings are usually available only to rebuilders. Another common turbocharger problem is excessive and continuous oil consumption resulting in blue exhaust smoke. Turbochargers use small rings similar to piston rings

on the shaft to prevent exhaust (combustion gases) from entering the central bushings. Because there are no seals to keep oil in, excessive oil consumption is usually caused by the following:

1. Plugged positive crankcase ventilation (PCV) system, resulting in excessive crankcase pressures forcing oil into the air inlet (This failure is not related to the turbocharger, but the turbocharger is often blamed.)

2. Clogged air filter, which causes a low-pressure area in the inlet, drawing oil past the turbo shaft rings and into the intake manifold.

3. Clogged oil return (drain) line from the turbocharger to the oil pan (sump), which can cause the engine oil pressure to force oil past the turbocharger's shaft rings and into the intake *and* exhaust manifolds (Obviously, oil being forced into both the intake and exhaust would create lots of smoke.)

**PREVENTING TURBOCHARGER FAILURES** To help prevent turbocharger failures, the wise vehicle owner should follow the vehicle manufacturer's recommended routine service procedures. The most critical of these services include the following:

■ Regular oil changes (synthetic oil would be best)

■ Regular air filter replacement intervals

■ Performing any other inspections and services recommended, such as cleaning the intercooler

# NITROUS OXIDE

**INTRODUCTION** Nitrous oxide is used for racing or high-performance only and is not used from the factory on any vehicle. This system is a relatively inexpensive way to get additional power from an engine but can cause serious engine damage if not used correctly or in excess amounts or without proper precautions.

**PRINCIPLES** **Nitrous oxide (N$_2$O)** is a colorless, nonflammable gas. It was discovered by a British chemist, Joseph Priestly (1733–1804), who also discovered oxygen. Priestly found that if a person breathed in nitrous oxide, it caused light-headedness, and so the gas soon became known as *laughing gas*. Nitrous oxide was used in dentistry during tooth extractions to reduce the pain and cause the patient to forget the experience.

Nitrous oxide has two nitrogen atoms and one oxide atom. About 36% of the molecule weight is oxygen. Nitrous oxide is a manufactured gas because, even though both nitrogen and oxygen are present in our atmosphere, they are not combined into one molecule and require heat and a catalyst to be combined.

**ENGINE POWER ADDER** A **power adder** is a device or system added to an engine, such as a supercharger, turbocharger, or nitrous oxide, to increase power. When nitrous oxide is injected into an engine along with gasoline, engine power is increased. The addition of N$_2$O supplies the needed oxygen for the extra fuel. N$_2$O by itself does not burn but provides the oxygen for additional fuel that is supplied along with the N$_2$O to produce more power.

**NOTE: Nitrous oxide was used as a power adder in World War II on some fighter aircraft. Having several hundred more horsepower for a short time saved many lives.**

**PRESSURE AND TEMPERATURE** It requires about 11 pounds of pressure per degree Fahrenheit to condense nitrous oxide gas into liquid nitrous oxide. For example, at 70°F, it requires a pressure of about 770 PSI to condense N$_2$O into a liquid. To change N$_2$O from a liquid under pressure to a gas, all that is needed is to lower its pressure below the pressure it takes to cause it to become a liquid.

The temperature also affects the pressure of N$_2$O. ● **SEE CHART 9–2.**

Nitrous oxide is stored in a pressurized storage container and installed at an angle so the pickup tube is in the liquid. The front or discharge end of the storage bottle should be toward the front of the vehicle. ● **SEE FIGURE 9–16.**

**WET AND DRY SYSTEM** There are two different types of N$_2$O systems that depend on whether additional fuel (gasoline) is supplied at the same time as when the nitrous oxide is squirted:

- The **wet system** involves additional fuel being injected. It is identified as having both a red and a blue nozzle, with the red flowing gasoline and the blue flowing nitrous oxide.

| TEMPERATURE (°F/°C) | PRESSURE (PSI/KPA) |
|---|---|
| −30°F/−34°C | 67 PSI/468 kPa |
| −20°F/−29°C | 203 PSI/1,400 kPa |
| −10°F/−23°C | 240 PSI/1,655 kPa |
| 0°F/−18°C | 283 PSI/1,950 kPa |
| 10°F/−12°C | 335 PSI/2,310 kPa |
| 20°F/−7°C | 387 PSI/2,668 kPa |
| 30°F/−1°C | 460 PSI/3,172 kPa |
| 40°F/4°C | 520 PSI/3,585 kPa |
| 50°F/10°C | 590 PSI/4,068 kPa |
| 60°F/16°C | 675 PSI/4,654 kPa |
| 70°F/21°C | 760 PSI/5,240 kPa |
| 80°F/27°C | 865 PSI/5,964 kPa |
| 90°F/32°C | 985 PSI/6,792 kPa |
| 100°F/38°C | 1,120 PSI/7,722 kPa |

**CHART 9–2**

Temperature/pressure relation for nitrous oxide: The higher the temperature, the higher the pressure.

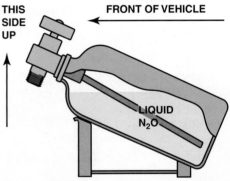

**FIGURE 9–16** Nitrous bottles have to be mounted at an angle to ensure that the pickup tube is in the liquid N$_2$O.

- In a **dry system,** such as an engine using port fuel injection, only nitrous oxide needs to be injected because the PCM can be commanded to provide more fuel when the N$_2$O is being sprayed. As a result, the intake manifold contains only air and the injected gaseous N$_2$O.

 **TECH TIP**

**Increase Bottle Pressure**

To increase the pressure of the nitrous oxide in a bottle, an electrical warming blanket can be used, as seen in ● **FIGURE 9–17.** The higher the temperature, the higher the pressure and the greater the amount of N$_2$O flow when energized.

**FIGURE 9–17** An electrical heating mat is installed on the bottle of nitrous oxide to increase the pressure of the gas inside.

**ENGINE CHANGES NEEDED FOR N₂O** If nitrous oxide is going to be used to increase horsepower more than 50 hp, the engine must be designed and built to withstand the greater heat and pressure that will occur in the combustion chambers. For example, the following items should be considered if adding a turbocharger, supercharger, or nitrous oxide system:

- Forged pistons are best able to withstand the pressure and temperature when using nitrous oxide or other power adder.

- Cylinder-to-wall clearance should be increased. Because of the greater amount of heat created by the extra fuel and N₂O injection, the piston temperature will be increased. Although using forged pistons will help, most experts recommend using increased cylinder-to-wall clearance.

- Using forged crankshaft and connecting rods.

Check the instructions from the nitrous oxide supplier for details and other suggested changes.

**CAUTION: The use of a nitrous oxide injection system can cause catastrophic engine damage. Always follow the instructions that come with the kit and be sure that all of the internal engine parts meet the standard specified to help avoid severe engine damage.**

**SYSTEM INSTALLATION AND CALIBRATION** Nitrous oxide systems are usually purchased as a kit with all of the needed components included. The kit also includes one or more sizes of nozzle(s) that are calibrated to control the flow of nitrous oxide into the intake manifold.

The sizes of the nozzles are often calibrated in horsepower that can be gained by their use. Commonly sized nozzles include the following:

- 50 hp
- 100 hp
- 150 hp

Installation of a nitrous oxide kit also includes the installation of an on-off switch and a switch on or near the throttle, which is used to activate the system only when the throttle is fully opened (WOT).

## SUMMARY

1. Volumetric efficiency is a comparison of the actual volume of air–fuel mixture drawn into the engine to the theoretical maximum volume that can be drawn into the cylinder.

2. A supercharger operates from the engine by a drive belt, and, although it consumes some engine power, it forces a greater amount of air into the cylinders for even more power.

3. There are two types of superchargers: roots type and centrifugal.

4. A turbocharger uses the normally wasted heat energy of the exhaust to turn an impeller at high speed. The impeller is linked to a turbine wheel on the same shaft and is used to force air into the engine.

5. A bypass valve is used to control the boost pressure on most factory-installed superchargers.

6. An intercooler is used on many turbocharged and some supercharged engines to reduce the temperature of air entering the engine for increased power.

7. A wastegate is used on most turbocharger systems to limit and control boost pressures, as well as a relief valve, to keep the speed of the turbine wheel from slowing down during engine deceleration.

8. Nitrous oxide injection can be used as a power adder but only with extreme caution.

1. What are the reasons why supercharging increases engine power?

2. How does the bypass valve work on a supercharged engine?

3. What are the advantages and disadvantages of supercharging?

4. What are the advantages and disadvantages of turbocharging?

5. What turbocharger control valves are needed for proper engine operation?

## CHAPTER QUIZ

1. Boost pressure is generally measured in _____.
   a. in. Hg
   b. PSI
   c. in. $H_2O$
   d. in. lb

2. Two types of superchargers include _____.
   a. Rotary and reciprocating
   b. Roots-type and centrifugal
   c. Double and single acting
   d. Turbine and piston

3. Which valve is used on a factory supercharger to limit boost?
   a. Bypass valve
   b. Wastegate
   c. Blow-off valve
   d. Air valve

4. How are most superchargers lubricated?
   a. By engine oil under pressure through lines from the engine
   b. By an internal oil reservoir
   c. By greased bearings
   d. No lubrication is needed because the incoming air cools the supercharger

5. How are most turbochargers lubricated?
   a. By engine oil under pressure through lines from the engine
   b. By an internal oil reservoir
   c. By greased bearings
   d. No lubrication is needed because the incoming air cools the supercharger

6. Two technicians are discussing the term *turbo lag.* Technician A says that it refers to the delay between when the exhaust leaves the cylinder and when it contacts the turbine blades of the turbocharger. Technician B says that it refers to the delay in boost pressure that occurs when the throttle is first opened. Which technician is correct?
   a. Technician A only
   b. Technician B only
   c. Both Technicians A and B
   d. Neither Technician A nor B

7. What is the purpose of an intercooler?
   a. To reduce the temperature of the air entering the engine
   b. To cool the turbocharger
   c. To cool the engine oil on a turbocharged engine
   d. To cool the exhaust before it enters the turbocharger

8. Which type of relief valve used on a turbocharged engine is noisy?
   a. Bypass valve
   b. BOV
   c. Dump valve
   d. Both b and c

9. Technician A says that a stuck open wastegate can cause the engine to burn oil. Technician B says that a clogged PCV system can cause the engine to burn oil. Which technician is correct?
   a. Technician A only
   b. Technician B only
   c. Both Technicians A and B
   d. Neither Technician A nor B

10. What service operation is *most* important on engines equipped with a turbocharger?
    a. Replacing the air filter regularly
    b. Replacing the fuel filter regularly
    c. Regular oil changes
    d. Regular exhaust system maintenance

# chapter 10

# ENGINE CONDITION DIAGNOSIS

**LEARNING OBJECTIVES:** **After studying this chapter, the reader should be able to:** • Discuss typical engine-related complaints and engine smoke diagnosis. • Explain the importance of visual checks. • Discuss engine noise diagnosis. • Describe oil pressure testing. • Explain cranking and running compression tests. • Describe cylinder leakage test and cylinder power balance test. • Describe vacuum testing and discuss the testing of back pressure with a vacuum gauge and a pressure gauge. • Explain the operation of dash warning lights.

**KEY TERMS:** Back pressure 147 • compression test 140 • cranking vacuum test 144 • cylinder leakage test 143 • dynamic compression test 142 • idle vacuum test 144 • inches of mercury (in. Hg) 144 • paper test 141 • power balance test 144 • restricted exhaust 146 • running compression test 142 • vacuum test 144 • wet compression test 142

If there is an engine operation problem, then the cause could be any one of many items, including the engine itself. The condition of the engine should be tested anytime the operation of the engine is not satisfactory.

## TYPICAL ENGINE-RELATED COMPLAINTS

Many driveability problems are *not* caused by engine mechanical problems. A thorough inspection and testing of the ignition and fuel systems should be performed before testing for mechanical engine problems.

Typical engine mechanical-related complaints include the following:

- Excessive oil consumption
- Engine misfiring
- Loss of power
- Smoke from the engine or exhaust
- Engine noise

## ENGINE SMOKE DIAGNOSIS

The color of engine exhaust smoke can indicate what engine problem might exist.

| Typical Exhaust Smoke Color | Possible Causes |
|---|---|
| Blue | Blue exhaust indicates that the engine is burning oil. Oil is getting into the combustion chamber either past the piston rings or past the valve stem seals. Blue smoke only after start-up is usually due to defective valve stem seals. ● **SEE FIGURE 10–1.** |
| Black | Black exhaust smoke is due to excessive fuel being burned in the combustion chamber. Typical causes include a defective or misadjusted throttle body, leaking fuel injector, or excessive fuel-pump pressure. |
| White (steam) | White smoke or steam from the exhaust is normal during cold weather and represents condensed steam. Every engine creates about 1 gallon of water for each gallon of gasoline burned. If the steam from the exhaust is excessive, then water (coolant) is getting into the combustion chamber. Typical causes include a defective cylinder head gasket, a cracked cylinder head, or in severe cases a cracked block. ● **SEE FIGURE 10–2.** |

**Note:** White smoke can also be created when automatic transmission fluid (ATF) is burned. A common source of ATF getting into the engine is through a defective vacuum modulator valve on older automatic transmissions.

ENGINE CONDITION DIAGNOSIS

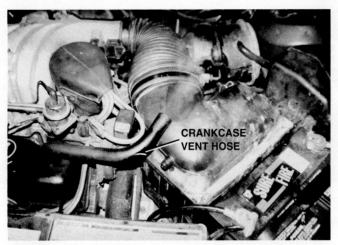

FIGURE 10–1 Blowby gases coming out of the crankcase vent hose. Excessive amounts of combustion gases flow past the piston rings and into the crankcase.

 **TECH TIP**

**Your Nose Knows**

Whenever diagnosing any vehicle, try to use all senses including the smell. Some smells and their causes include the following:

- **Gasoline.** If the exhaust smells like gasoline or unburned fuel, then a fault with the ignition system is a likely cause. Unburned fuel due to lean air–fuel mixture causing a lean misfire is also possible.
- **Sweet smell.** A coolant leak often gives off a sweet smell, especially if the leaking coolant flows onto the hot exhaust.
- **Exhaust smell.** Check for an exhaust leak, including a possible cracked exhaust manifold, which can be difficult to find because it often does not make noise.

# THE DRIVER IS YOUR BEST RESOURCE

The driver of the vehicle knows a lot about the vehicle and how it is driven. *Before* diagnosis is started, always ask the following questions:

- When did the problem first occur?
- Under what conditions does it occur?
  1. Cold or hot?
  2. Acceleration, cruise, or deceleration?
  3. How far was it driven?

After the nature and scope of the problem are determined, the complaint should be verified before further diagnostic tests are performed.

FIGURE 10–2 White steam is usually an indication of a blown (defective) cylinder head gasket that allows engine coolant to flow into the combustion chamber where it is turned to steam.

# VISUAL CHECKS

The first and most important "test" that can be performed is a careful visual inspection.

**OIL LEVEL AND CONDITION** The first area for visual inspection is oil level and condition.

1. Oil level—oil should be to the proper level
2. Oil condition
   a. Using a match or lighter, try to light the oil on the dipstick; if the oil flames up, gasoline is present in the engine oil.
   b. Drip some of the engine oil from the dipstick onto the hot exhaust manifold. If the oil bubbles or boils, there is coolant (water) in the oil.
   c. Check for grittiness by rubbing the oil between your fingers.

**COOLANT LEVEL AND CONDITION** Most mechanical engine problems are caused by overheating. The proper operation of the cooling system is critical to the life of any engine.

**NOTE: Check the coolant level in the radiator only if the radiator is cool. If the radiator is hot and the radiator cap is removed, the drop in pressure above the coolant will cause the coolant to boil immediately and can cause severe burns when the coolant explosively expands upward and outward from the radiator opening.**

1. The coolant level in the coolant recovery container should be within the limits indicated on the overflow bottle. If this level is too low or the coolant recovery container is empty, then check the level of coolant in the radiator (only when cool) and also check the operation of the pressure cap.

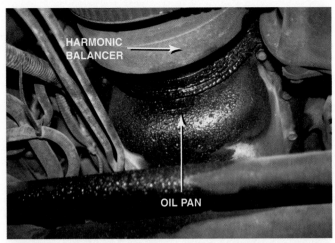

FIGURE 10–3 What looks like an oil pan gasket leak can be a rocker cover gasket leak. Always look up and look for the highest place you see oil leaking; that should be repaired first.

2. The coolant should be checked with a hydrometer for boiling and freezing temperature. This test indicates if the concentration of the antifreeze is sufficient for proper protection.

3. Pressure test the cooling system and look for leakage. Coolant leakage can often be seen around hoses or cooling system components because it will often cause the following:
   a. A grayish white stain
   b. A rusty color stain
   c. Dye stains from antifreeze (greenish or yellowish depending on the type of coolant)

4. Check for cool areas of the radiator indicating clogged sections.

5. Check operation and condition of the fan clutch, electric fan, and water pump drive belt.

## OIL LEAKS
Oil leaks can lead to severe engine damage if the resulting low oil level is not corrected. Besides causing an oily mess where the vehicle is parked, the oil leak can cause blue smoke to occur under the hood as leaking oil drips on the exhaust system. *Finding* the location of the oil leak can often be difficult. ● **SEE FIGURES 10–3 AND 10–4.** To help find the source of oil leaks, follow these steps:

**STEP 1** Clean the engine or area around the suspected oil leak. Use a high-powered hot-water spray to wash the engine. While the engine is running, spray the entire engine and the engine compartment. Avoid letting the water come into direct contact with the air inlet and ignition distributor or ignition coil(s).

**NOTE: If the engine starts to run rough or stalls when the engine gets wet, then the secondary ignition wires (spark plug wires) or distributor cap may be defective or have weak insulation. Be certain to wipe all wires and the distributor cap dry with a soft, dry cloth if the engine stalls.**

FIGURE 10–4 The transmission and flexplate (flywheel) were removed to check the exact location of this oil leak. The rear main seal and/or the oil pan gasket could be the cause of this leak.

  **TECH TIP**

### What's Leaking?
The color of the leaks observed under a vehicle can help the technician determine and correct the cause. Some leaks, such as condensate (water) from the air-conditioning system, are normal, whereas a brake fluid leak is very dangerous. The following are colors of common leaks:

| Sooty Black | Engine Oil |
| --- | --- |
| Yellow, green, blue, or orange | Antifreeze (coolant) |
| Red | Automatic transmission fluid |
| Murky brown | Brake or power steering fluid or very neglected antifreeze (coolant) |
| Clear | Air-conditioning condensate (water) (normal) |

An alternative method is to spray a degreaser on the engine, then start and run the engine until warm. Engine heat helps the degreaser penetrate the grease and dirt. Use a water hose to rinse off the engine and engine compartment.

**FIGURE 10–5** Using a black light to spot leaks after adding dye to the oil.

**FIGURE 10–6** An accessory belt tensioner. Most tensioners have a mark that indicates normal operating location. If the belt has stretched, this indicator mark will be outside of the normal range. Anything wrong with the belt or tensioner can cause noise.

 **TECH TIP**

**The Foot Powder Spray Trick**

The source of an oil or other fluid leak is often difficult to determine. A quick and easy method that works is the following. First, clean the entire area. This can best be done by using a commercially available degreaser to spray the entire area. Let it soak to loosen all accumulated oil and greasy dirt. Clean off the degreaser with a water hose. Let the area dry. Start the engine and, using spray foot powder or other aerosol powder product, spray the entire area. The leak will turn the white powder dark. The exact location of any leak can be quickly located.

**NOTE: Most oil leaks appear at the bottom of the engine due to gravity. Look for the highest, most forward location for the source of the leak.**

**STEP 2** If the oil leak is not visible or oil seems to be coming from "everywhere," use a white talcum powder. The leaking oil will show as a dark area on the white powder. See the Tech Tip, "The Foot Powder Spray Trick."

**STEP 3** Fluorescent dye can be added to the engine oil. Add about 1/2 ounce (15 cc) of dye per 5 quarts of engine oil. Start the engine and allow it to run about 10 minutes to thoroughly mix the dye throughout the engine. A black light can then be shown around every suspected oil leak location. The black light will easily show all oil leak locations because the dye will show as a bright yellow/green area. ● **SEE FIGURE 10–5.**

**NOTE: Fluorescent dye works best with clean oil.**

# ENGINE NOISE DIAGNOSIS

An engine knocking noise is often difficult to diagnose. Several items that can cause a deep engine knock include the following:

- **Valves clicking.** This can happen because of lack of oil to the lifters. This noise is most noticeable at idle when the oil pressure is the lowest.

- **Torque converter.** The attaching bolts or nuts may be loose on the flex plate. This noise is most noticeable at idle or when there is no load on the engine.

- **Cracked flex plate.** The noise of a cracked flex plate is often mistaken for a rod- or main-bearing noise.

- **Loose or defective drive belts or tensioners.** If an accessory drive belt is loose or defective, the flopping noise often sounds similar to a bearing knock. ● **SEE FIGURE 10–6.**

- **Piston pin knock.** This knocking noise is usually not affected by load on the cylinder. If the clearance is too great, a double knock noise is heard when the engine idles. If all cylinders are grounded out one at a time and the noise does not change, a defective piston pin could be the cause.

- **Piston slap.** A piston slap is usually caused by an undersized or improperly shaped piston or oversized cylinder bore. A piston slap is most noticeable when the engine is cold and tends to decrease or stop making noise as the piston expands during engine operation.

- **Timing chain noise.** An excessively loose timing chain can cause a severe knocking noise when the chain hits the timing chain cover. This noise can often sound like a rod-bearing knock.

| Typical Noises | Possible Causes |
|---|---|
| Clicking noise—like the clicking of a ballpoint pen | 1. Loose spark plug<br>2. Loose accessory mount (for air-conditioning compressor, alternator, power steering pump, etc.)<br>3. Loose rocker arm<br>4. Worn rocker arm pedestal<br>5. Fuel pump (broken mechanical fuel pump return spring)<br>6. Worn camshaft<br>7. Exhaust leak ● SEE FIGURE 10–7. |
| Clacking noise—like tapping on metal | 1. Worn piston pin<br>2. Broken piston<br>3. Excessive valve clearance<br>4. Timing chain hitting cover |
| Knock—like knocking on a door | 1. Rod bearing(s)<br>2. Main bearing(s)<br>3. Thrust bearing(s)<br>4. Loose torque converter<br>5. Cracked flex plate (drive plate) |
| Rattle—like a baby rattle | 1. Manifold heat control valve<br>2. Broken harmonic balancer<br>3. Loose accessory mounts<br>4. Loose accessory drive belt or tensioner |
| Clatter—like rolling marbles | 1. Rod bearings<br>2. Piston pin<br>3. Loose timing chain |
| Whine—like an electric motor running | 1. Alternator bearing<br>2. Drive belt<br>3. Power steering<br>4. Belt noise (accessory or timing) |
| Clunk—like a door closing | 1. Engine mount<br>2. Drive axle shaft U-joint or constant velocity (CV) joint |

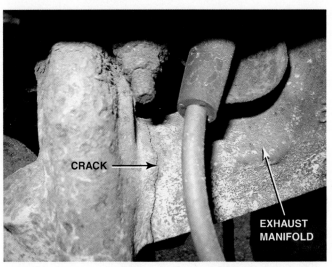

FIGURE 10–7 A cracked exhaust manifold on a Ford V-8.

CRACK

EXHAUST MANIFOLD

   TECH TIP

**Engine Noise and Cost**

A light ticking noise often heard at one-half engine speed and associated with valve train noise is a less serious problem than many deep-sounding knocking noises. Generally, the deeper the sound of the engine noise, the more the owner will have to pay for repairs. A light "tick tick tick," though often not cheap, is usually far less expensive than a deep "knock knock knock" from the engine.

## OIL PRESSURE TESTING

Proper oil pressure is very important for the operation of any engine. *Low oil pressure can cause engine wear, and engine wear can cause low oil pressure.*

If main thrust or rod bearings are worn, oil pressure is reduced because of leakage of the oil around the bearings. Oil pressure testing is usually performed with the following steps:

STEP 1  Operate the engine until normal operating temperature is achieved.

STEP 2  With the engine off, remove the oil pressure sending unit or sender, usually located near the oil filter. Thread an oil pressure gauge into the threaded hole. ● SEE FIGURE 10–8.

NOTE: An oil pressure gauge can be made from another gauge, such as an old air-conditioning gauge and a flexible brake hose. The threads are often the same as those used for the oil pressure sending unit.

- **Rod-bearing noise.** The noise from a defective rod bearing is usually load sensitive and changes in intensity as the load on the engine increases and decreases. A rod-bearing failure can often be detected by grounding out the spark plugs one cylinder at a time. If the knocking noise decreases or is eliminated when a particular cylinder is grounded (disabled), then the grounded cylinder is the one from which the noise is originating.

- **Main-bearing knock.** A main-bearing knock often cannot be isolated to a particular cylinder. The sound can vary in intensity and may disappear at times depending on engine load.

Regardless of the type of loud knocking noise, after the external causes of the knocking noise have been eliminated, the engine should be disassembled and carefully inspected to determine the exact cause.

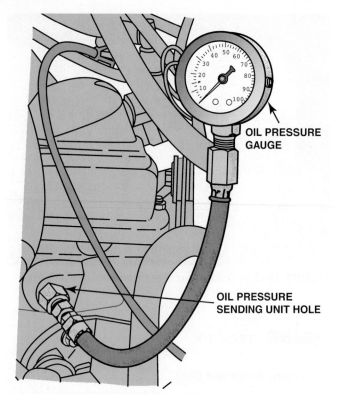

FIGURE 10–8 To measure engine oil pressure, remove the oil pressure sending (sender) unit usually located near the oil filter. Screw the pressure gauge into the oil pressure sending unit hole.

**STEP 3** Start the engine and observe the gauge. Record the oil pressure at idle and at 2500 RPM. Most vehicle manufacturers recommend a minimum oil pressure of 10 PSI per 1000 RPM. Therefore, at 2500 RPM, the oil pressure should be at least 25 PSI. Always compare your test results with the manufacturer's recommended oil pressure.

Besides engine bearing wear, other possible causes for low oil pressure include the following:

- Low oil level
- Diluted oil
- Stuck oil pressure relief valve

## OIL PRESSURE WARNING LAMP

The red oil pressure warning lamp in the dash usually lights when the oil pressure is less than 4 to 7 PSI, depending on vehicle and engine. The oil light should not be on during driving. If the oil warning lamp is on, stop the engine immediately. Always confirm oil pressure with a reliable mechanical gauge before performing engine repairs. The sending unit or circuit may be defective.

 TECH TIP

### Use the KISS Test Method

Engine testing is done to find the cause of an engine problem. All the simple things should be tested first. Just remember KISS–"keep it simple, stupid." A loose alternator belt or loose bolts on a torque converter can sound just like a lifter or rod bearing. A loose spark plug can make the engine perform as if it had a burned valve. Some simple items that can cause serious problems include the following:

### Oil Burning

- Low oil level
- Clogged PCV valve or system, causing blowby and oil to be blown into the air cleaner
- Clogged drainback passages in the cylinder head
- Dirty oil that has not been changed for a long time (Change the oil and drive for about 1,000 miles [1,600 kilometers] and change the oil and filter again.)

### Noises

- Carbon on top of the piston(s) can sound like a bad rod bearing (often called a carbon knock)
- Loose torque-to-flex plate bolts (or nuts), causing a loud knocking noise

NOTE: Often this problem will cause noise only at idle; the noise tends to disappear during driving or when the engine is under load.

- A loose and/or defective drive belt, which may cause a rod- or main-bearing knocking noise (A loose or broken mount for the generator [alternator], power steering pump, or air-conditioning compressor can also cause a knocking noise.)

## COMPRESSION TEST

An engine **compression test** is one of the fundamental engine diagnostic tests that can be performed. For smooth engine operation, all cylinders must have equal compression. An engine can lose compression by leakage of air through one or more of only three routes:

- Intake or exhaust valve
- Piston rings (or piston, if there is a hole)
- Cylinder head gasket

For best results, the engine should be warmed to normal operating temperature before testing. An accurate compression test should be performed as follows:

**STEP 1** Remove all spark plugs. This allows the engine to be cranked to an even speed. Be sure to label all spark plug wires.

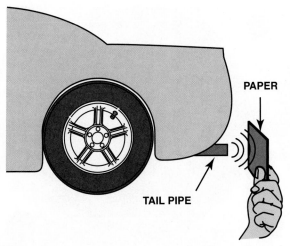

FIGURE 10–9 The paper test involves holding a piece of paper near the tailpipe of an idling engine. A good engine should produce even, outward puffs of exhaust. If the paper is sucked in toward the tailpipe, a burned valve is a possibility.

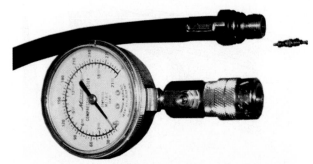

FIGURE 10–10 A two-piece compression gauge set. The threaded hose is screwed into the spark plug hole after removing the spark plug. The gauge part is then snapped onto the end of the hose.

### TECH TIP

**The Paper Test**

A soundly running engine should produce even and steady exhaust at the tailpipe. You can test this with the **paper test.** Hold a piece of paper or a 3″ × 5″ index card (even a dollar bill works) within 1 inch (25 mm) of the tailpipe with the engine running at idle. ● **SEE FIGURE 10–9.**

The paper should blow out evenly without "puffing." If the paper is drawn *toward* the tailpipe at times, the exhaust valves in one or more cylinders could be burned. Other reasons why the paper might be sucked toward the tailpipe include the following:

1. The engine could be misfiring because of a lean condition that could occur normally when the engine is cold.
2. Pulsing of the paper toward the tailpipe could also be caused by a hole in the exhaust system. If exhaust escapes through a hole in the exhaust system, air could be drawn in during the intervals between the exhaust puffs from the tailpipe to the hole in the exhaust, causing the paper to be drawn toward the tailpipe.
3. Ignition fault causing misfire.

**STEP 2** Block open the throttle. This permits the maximum amount of air to be drawn into the engine. This step also ensures consistent compression test results.

**CAUTION: Disable the ignition system by disconnecting the primary leads from the ignition coil or module or by grounding the coil wire after**

removing it from the center of the distributor cap. Also disable the fuel-injection system to prevent the squirting of fuel into the cylinder.

**STEP 3** Thread a compression gauge into one spark plug hole and crank the engine. ● **SEE FIGURE 10–10.**

Continue cranking the engine through *four* compression strokes. Each compression stroke makes a puffing sound.

**NOTE: Note the reading on the compression gauge after the first puff. This reading should be at least one-half the final reading. For example, if the final, highest reading is 150 PSI, then the reading after the first puff should be higher than 75 PSI. A low first-puff reading indicates possible weak piston rings. Release the pressure on the gauge and repeat for the other cylinders.**

**STEP 4** Record the highest readings and compare the results. Most vehicle manufacturers specify the minimum compression reading and the maximum allowable variation among cylinders. Most manufacturers specify a maximum difference of 20% between the highest reading and the lowest reading. An example follows:

| | |
|---|---|
| **If the high reading is** | **150 PSI** |
| **Subtract 20%** | **−30 PSI** |
| **Lowest allowable compression is** | **120 PSI** |

**NOTE: To make the math quick and easy, think of 10% of 150, which is 15 (move the decimal point to the left by one place). Now double it: 15 × 2 = 30. This represents 20%.**

**NOTE: During cranking, the oil pump cannot maintain normal oil pressure. Extended engine cranking, such as that which occurs during a compression test, can cause hydraulic lifters to collapse. When the engine starts, loud valve clicking noises may be heard. This should be considered normal after performing a compression test, and the noise should stop after the vehicle has been driven a short distance.**

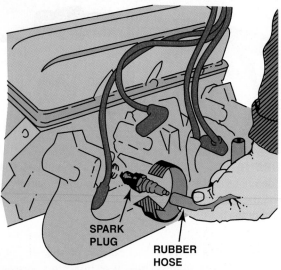

SPARK PLUG

RUBBER HOSE

**FIGURE 10–11** Use a vacuum or fuel line hose over the spark plug to install it without danger of cross-threading the cylinder head.

**FIGURE 10–12** Badly burned exhaust valve. A compression test could have detected a problem, and a cylinder leakage test (leak-down test) could have been used to determine the exact problem.

 **TECH TIP**

**The Hose Trick**

Installing spark plugs can be made easier by using a rubber hose on the end of the spark plug. The hose can be a vacuum hose, fuel line, or even an old spark plug wire end. ● **SEE FIGURE 10–11.**

The hose makes it easy to start the threads of the spark plug into the cylinder head. After starting the threads, continue to thread the spark plug for several turns. Using the hose eliminates the chance of cross-threading the plug. This is especially important when installing spark plugs in aluminum cylinder heads.

# WET COMPRESSION TEST

If the compression test reading indicates low compression on one or more cylinders, add three squirts of oil to the cylinder and retest. This is called a **wet compression test** when oil is used to help seal around the piston rings.

**CAUTION: Do not use more oil than three squirts from a hand-operated oil squirt can. Too much oil can cause a hydrostatic lock, which can damage or break pistons or connecting rods or even crack a cylinder head.**

Perform the compression test again and observe the results. If the first-puff readings greatly improve and the readings are much higher than without the oil, the cause of the low compression is worn or defective piston rings. If the compression

readings increase only slightly (or not at all), then the cause of the low compression is usually defective valves. ● **SEE FIGURE 10–12.**

**NOTE: During both the dry and wet compression tests, be sure that the battery and starting system are capable of cranking the engine at normal cranking speed.**

# RUNNING (DYNAMIC) COMPRESSION TEST

A compression test is commonly used to help determine engine condition and is usually performed with the engine cranking.

What is the RPM of a cranking engine? An engine idles at about 600 to 900 RPM, and the starter motor obviously cannot crank the engine as fast as the engine idles. Most manufacturers' specifications require the engine to crank at 80 to 250 cranking RPM. Therefore, a check of the engine's compression at cranking speed determines the condition of an engine that does not run at such low speeds.

But what should be the compression of a running engine? Some would think that the compression would be substantially higher because the valve overlap of the cam is more effective at higher engine speeds, which would tend to increase the compression.

A **running compression test,** also called a **dynamic compression test,** is a compression test done with the engine running rather than during engine cranking as is done in a regular compression test.

Actually, the compression pressure of a running engine is much *lower* than cranking compression pressure. This results from the volumetric efficiency. The engine is revolving faster, and therefore there is less *time* for air to enter the combustion

**FIGURE 10–13** A typical handheld cylinder leakage tester.

chamber. With less air to compress, the compression pressure is lower. Typically, the higher the engine RPM, the lower the running compression. For most engines, the value ranges are as follows:

- Compression during cranking:     125 to 160 PSI
- Compression at idle:     60 to 90 PSI
- Compression at 2000 RPM:     30 to 60 PSI

As with cranking compression, the running compression of all cylinders should be equal. Therefore, a problem is likely to be detected not by single compression values but by *variations* in running compression values among the cylinders. Broken valve springs, worn valve guides, bent pushrods, and worn cam lobes are some items that would be indicated by a low running compression test reading on one or more cylinders.

**PERFORMING A RUNNING COMPRESSION TEST** To perform a running compression test, remove just one spark plug at a time. With one spark plug removed from the engine, use a jumper wire to *ground* the spark plug wire to a good engine ground. This prevents possible ignition coil damage. Start the engine, push the pressure release on the gauge, and read the compression. Increase the engine speed to about 2000 RPM and push the pressure release on the gauge again. Read the gauge. Stop the engine, reinstall the spark plug, reattach the spark plug wire, and repeat the test for each of the remaining cylinders. Just like the cranking compression test, the running compression test can inform a technician of the *relative* compression of all the cylinders.

## CYLINDER LEAKAGE TEST

One of the best tests that can be used to determine engine condition is the **cylinder leakage test.** This test involves injecting air under pressure into the cylinders one at a time. The amount and location of any escaping air helps the technician determine

**FIGURE 10–14** A whistle stop used to find top dead center. Remove the spark plug and install the whistle stop, then rotate the engine by hand. When the whistle stops making a sound, the piston is at the top.

the condition of the engine. The air is injected into the cylinder through a cylinder leakage gauge into the spark plug hole. ● SEE FIGURE 10–13. To perform the cylinder leakage test, take the following steps:

**STEP 1** For best results, the engine should be at normal operating temperature (upper radiator hose hot and pressurized).

**STEP 2** The cylinder being tested must be at top dead center (TDC) of the compression stroke. ● SEE FIGURE 10–14.

> **NOTE: The greatest amount of wear occurs at the top of the cylinder because of the heat generated near the top of the cylinders. The piston ring flex also adds to the wear at the top of the cylinder.**

**STEP 3** Calibrate the cylinder leakage unit as per manufacturer's instructions.

**STEP 4** Inject air into the cylinders one at a time, rotating the engine as necessitated by firing order to test each cylinder at TDC on the compression stroke.

**STEP 5** Evaluate the results:
Less than 10% leakage: good
Less than 20% leakage: acceptable
Less than 30% leakage: poor
More than 30% leakage: definite problem

> **NOTE: If leakage seems unacceptably high, repeat the test, being certain that it is being performed correctly and that the cylinder being tested is at TDC on the compression stroke.**

**STEP 6** Check the source of air leakage.
   a. If air is heard escaping from the oil filler cap, the *piston rings* are worn or broken.
   b. If air is observed bubbling out of the radiator, there is a possible blown *head gasket* or cracked *cylinder head.*
   c. If air is heard coming from the throttle body or air inlet on fuel injection-equipped engines, there is a defective *intake valve(s).*
   d. If air is heard coming from the tailpipe, there is a defective *exhaust valve(s).*

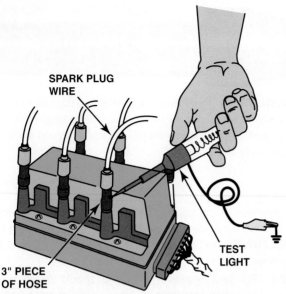

**FIGURE 10–15** Using a vacuum hose and a test light to ground one cylinder at a time on a distributorless ignition system. This works on all types of ignition systems and provides a method for grounding out one cylinder at a time without fear of damaging any component. Do not short out a cylinder for longer than 15 seconds to prevent possible damage to the catalytic converter.

# CYLINDER POWER BALANCE TEST

Most large engine analyzers and scan tools have a cylinder power balance feature. The purpose of a cylinder **power balance test** is to determine if all cylinders are contributing power equally. It determines this by shorting out one cylinder at a time. If the engine speed (RPM) does not drop as much for one cylinder as for other cylinders of the same engine, then the shorted cylinder must be weaker than the other cylinders. An example follows:

| Cylinder Number | RPM Drop When Ignition Is Shorted |
|:---:|:---:|
| 1 | 75 |
| 2 | 70 |
| 3 | 15 |
| 4 | 65 |
| 5 | 75 |
| 6 | 70 |

Cylinder #3 is the weak cylinder.

**NOTE: Most automotive test equipment uses automatic means for testing cylinder balance. Be certain to correctly identify the offending cylinder. Cylinder #3 as identified by the equipment may be the third cylinder in the firing order instead of the actual cylinder #3.**

# POWER BALANCE TEST PROCEDURE

When point-type ignition was used on all vehicles, the common method for determining which, if any, cylinder was weak was to remove a spark plug wire from one spark plug at a time while watching a tachometer and a vacuum gauge. This method is not recommended on any vehicle with any type of electronic ignition. If any of the spark plug wires are removed from a spark plug with the engine running, the ignition coil tries to supply increasing levels of voltage attempting to jump the increasing gap as the plug wires are removed. This high voltage could easily track the ignition coil, damage the ignition module, or both.

The acceptable method of canceling cylinders, which will work on all types of ignition systems, including distributorless, is to *ground* the secondary current for each cylinder. ● **SEE FIGURE 10–15.** The cylinder with the least RPM drop is the cylinder not producing its share of power.

# VACUUM TESTS

Vacuum is pressure below atmospheric pressure and is measured in **inches (or millimeters) of mercury (in. Hg).** An engine in good mechanical condition will run with high manifold vacuum. Manifold vacuum is developed by the pistons as they move down on the intake stroke to draw the charge from the throttle body and intake manifold. Air to refill the manifold comes past the throttle plate into the manifold. Vacuum will increase anytime the engine turns faster or has better cylinder sealing while the throttle plate remains in a fixed position. Manifold vacuum will decrease when the engine turns more slowly or when the cylinders no longer do an efficient job of pumping. **Vacuum tests** include testing the engine for **cranking vacuum, idle vacuum,** and vacuum at 2500 RPM.

**CRANKING VACUUM TEST** Measuring the amount of manifold vacuum during cranking is a quick and easy test to determine if the piston rings and valves are properly sealing. (For accurate results, the engine should be warm and the throttle closed.) To perform the cranking vacuum test, take the following steps:

**STEP 1** Disable the ignition or fuel injection.

**STEP 2** Connect the vacuum gauge to a manifold vacuum source.

**STEP 3** Crank the engine while observing the vacuum gauge.

Cranking vacuum should be higher than 2.5 in. Hg. (Normal cranking vacuum is 3 to 6 in. Hg.) If it is lower than 2.5 in. Hg, then the following could be the cause:

■ Too slow a cranking speed

■ Worn piston rings

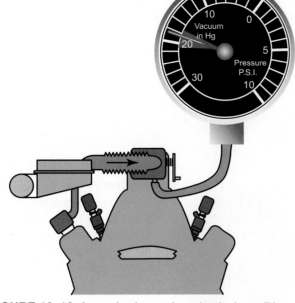

FIGURE 10–16 An engine in good mechanical condition should produce 17 to 21 in. Hg of vacuum at idle at sea level.

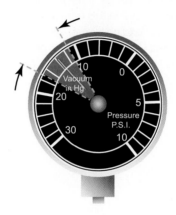

FIGURE 10–18 A gauge reading with the needle fluctuating 3 to 9 in. Hg below normal often indicates a vacuum leak in the intake system.

- Leaking valves
- Excessive amounts of air bypassing the throttle plate (This could give a false low vacuum reading. Common sources include a throttle plate partially open or a high-performance camshaft with excessive overlap.)

**IDLE VACUUM TEST**  An engine in proper condition should idle with a steady vacuum between 17 and 21 in. Hg. ● **SEE FIGURE 10–16.**

NOTE: **Engine vacuum readings vary with altitude. A reduction of 1 in. Hg per 1,000 feet (300 m) of altitude should be subtracted from the expected values if testing a vehicle above 1,000 feet (300 m).**

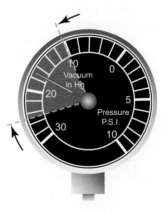

FIGURE 10–19 A leaking head gasket can cause the needle to vibrate as it moves through a range from below to above normal.

**LOW AND STEADY VACUUM**  If the vacuum is lower than normal yet the gauge reading is steady, the most common causes include the following:

- Retarded ignition timing
- Retarded cam timing (check timing chain for excessive slack or timing belt for proper installation)
    ● **SEE FIGURE 10–17.**

**FLUCTUATING VACUUM**  If the needle drops, then returns to a normal reading, then drops again, and again returns, this indicates a sticking valve. A common cause of sticking valves is lack of lubrication of the valve stems. ● **SEE FIGURES 10–18 THROUGH 10–26.** If the vacuum gauge fluctuates above and below a center point, burned valves or weak valve springs may be indicated. If the fluctuation is slow and steady, unequal fuel mixture could be the cause.

FIGURE 10–20 An oscillating needle 1 or 2 in. Hg below normal could indicate an incorrect air–fuel mixture (either too rich or too lean).

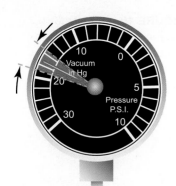

FIGURE 10–21 A rapidly vibrating needle at idle that becomes steady as engine speed is increased indicates worn valve guides.

FIGURE 10–22 If the needle drops 1 or 2 in. Hg from the normal reading, one of the engine valves is burned or not seating properly.

FIGURE 10–23 Weak valve springs will produce a normal reading at idle, but as engine speed increases, the needle will fluctuate rapidly between 12 and 24 in. Hg.

FIGURE 10–24 A steady needle reading that drops 2 or 3 in. Hg when the engine speed is increased slightly above idle indicates that the ignition timing is retarded.

FIGURE 10–25 A steady needle reading that rises 2 or 3 in. Hg when the engine speed is increased slightly above idle indicates that the ignition timing is advanced.

**NOTE: A common trick that some technicians use is to squirt some automatic transmission fluid (ATF) down the throttle body or into the air inlet of a warm engine. Often the idle quality improves, and normal vacuum gauge readings are restored. The use of ATF does create excessive exhaust smoke for a short time, but it should not harm oxygen sensors or catalytic converters.**

## EXHAUST RESTRICTION TEST

If the exhaust system is restricted, the engine will be low on power yet smooth. Common causes of **restricted exhaust** include the following:

- **Clogged catalytic converter.** Always check the ignition and fuel-injection systems for faults that could cause excessive amounts of unburned fuel to be exhausted. Excessive unburned fuel can overheat the catalytic converter and cause the beads or structure of the converter to fuse together, creating the restriction. A defective fuel delivery system could also cause excessive unburned fuel to be dumped into the converter.

- **Clogged or restricted muffler.** This can cause low power. Often a defective catalytic converter will shed particles that can clog a muffler. Broken internal baffles can also restrict exhaust flow.

FIGURE 10–26 A needle that drops to near zero when the engine is accelerated rapidly and then rises slightly to a reading below normal indicates an exhaust restriction.

- **Damaged or defective piping.** This can reduce the power of any engine. Some exhaust pipe is constructed with double walls, and the inside pipe can collapse and form a restriction that is not visible on the outside of the exhaust pipe.

FIGURE 10–27 A technician-made adapter used to test exhaust system back pressure.

FIGURE 10–28 A tester that uses a blue liquid to check for exhaust gases in the coolant, which would indicate a head gasket leak problem.

## TESTING BACK PRESSURE WITH A VACUUM GAUGE

A vacuum gauge can be used to measure manifold vacuum at a high idle (2000 to 2500 RPM). If the exhaust system is restricted, pressure increases in the exhaust system. This pressure is called **back pressure.** Manifold vacuum will drop gradually if the engine is kept at a constant speed if the exhaust is restricted.

The reason the vacuum will drop is that all exhaust leaving the engine at the higher engine speed cannot get through the restriction. After a short time (within 1 minute), the exhaust tends to "pile up" above the restriction and eventually remains in the cylinder of the engine at the end of the exhaust stroke. Therefore, at the beginning of the intake stroke, when the piston traveling downward should be lowering the pressure (raising the vacuum) in the intake manifold, the extra exhaust in the cylinder *lowers* the normal vacuum. If the exhaust restriction is severe enough, the vehicle can become undriveable because cylinder filling cannot occur except at idle.

## TESTING BACK PRESSURE WITH A PRESSURE GAUGE

Exhaust system back pressure can be measured directly by installing a pressure gauge into an exhaust opening. This can be accomplished in one of the following ways:

- **With an oxygen sensor.** Use a back pressure gauge and adapter or remove the inside of an old, discarded oxygen sensor and thread in an adapter to convert to a vacuum or pressure gauge.

**NOTE: An adapter can be easily made by inserting a metal tube or pipe. A short section of brake line works great. The pipe can be brazed to the oxygen sensor housing or it can be glued in with epoxy. An 18-millimeter compression gauge adapter can also be adapted to fit into the oxygen sensor opening.** ● **SEE FIGURE 10–27.**

- **With the exhaust gas recirculation (EGR) valve.** Remove the EGR valve and fabricate a plate to connect to a pressure gauge.

- **With the air-injection reaction (AIR) check valve.** Remove the check valve from the exhaust tubes leading down to the exhaust manifold. Use a rubber cone with a tube inside to seal against the exhaust tube. Connect the tube to a pressure gauge.

At idle, the maximum back pressure should be less than 1.5 PSI (10 kPa), and it should be less than 2.5 PSI (15 kPa) at 2500 RPM.

## DIAGNOSING HEAD GASKET FAILURE

Several items can be used to help diagnose a head gasket failure:

- **Exhaust gas analyzer.** With the radiator cap removed, place the probe from the exhaust analyzer above the radiator filler neck. If the HC reading increases, the exhaust (unburned hydrocarbons) is getting into the coolant from the combustion chamber.

- **Chemical test.** A chemical tester using blue liquid is also available. The liquid turns yellow if combustion gases are present in the coolant. ● **SEE FIGURE 10–28.**

- **Bubbles in the coolant.** Remove the coolant pump belt to prevent pump operation. Remove the radiator cap and start the engine. If bubbles appear in the coolant before it begins to boil, a defective head gasket or cracked cylinder head is indicated.

- **Excessive exhaust steam.** If excessive water or steam is observed coming from the tailpipe, this means that coolant is getting into the combustion chamber from a defective head gasket or a cracked head. If there is leakage between cylinders, the engine usually misfires and a power balancer test and/or compression test can be used to confirm the problem.

If any of the preceding indicators of head gasket failure occur, remove the cylinder head(s) and check all of the following:

1. Head gasket
2. Sealing surfaces—for warpage
3. Castings—for cracks

**NOTE:** A leaking thermal vacuum valve can cause symptoms similar to those of a defective head gasket. Most thermal vacuum valves thread into a coolant passage, and they often leak only after they get hot.

## DASH WARNING LIGHTS

Most vehicles are equipped with several dash warning lights often called "telltale" or "idiot" lights. These lights are often the only warning a driver receives that there may be engine problems. A summary of typical dash warning lights and their meanings follows.

### OIL (ENGINE) LIGHT
The red oil light indicates that the engine oil pressure is too low (usually lights when oil pressure is 4 to 7 PSI [20 to 50 kPa]). Normal oil pressure should be 10 to 60 PSI (70 to 400 kPa) or 10 PSI per 1000 engine RPM.

When this light comes on, the driver should shut off the engine immediately and check the oil level and condition for possible dilution with gasoline caused by a fuel system fault. If the oil level is okay, then there is a possible serious engine problem or a possible defective oil pressure sending (sender) unit. The automotive technician should always check the oil pressure using a reliable mechanical oil pressure gauge if low oil pressure is suspected.

**NOTE:** Some automobile manufacturers combine the dash warning lights for oil pressure and coolant temperature into one light, usually labeled "engine." Therefore, when the engine light comes on, the technician should check for possible coolant temperature and/or oil pressure problems.

### COOLANT TEMPERATURE LIGHT
Most vehicles are equipped with a coolant temperature gauge or dash warning light. The warning light may be labeled "coolant," "hot," or "temperature." If the coolant temperature warning light comes on during driving, this usually indicates that the coolant temperature is above a safe level, or above about 250°F (120°C). Normal coolant temperature should be about 200° to 220°F (90° to 105°C).

If the coolant temperature light comes on during driving, the following steps should be followed to prevent possible engine damage:

1. Turn off the air conditioning and turn on the heater. The heater will help get rid of some of the heat in the cooling system.
2. Raise the engine speed in neutral or park to increase the circulation of coolant through the radiator.
3. If possible, turn the engine off and allow it to cool (this may take over an hour).
4. Do not continue driving with the coolant temperature light on (or the gauge reading in the red warning section or above 260°F), or serious engine damage may result.

**NOTE:** If the engine does not feel or smell hot, it is possible that the problem is a faulty coolant temperature sensor or gauge.

**TECH TIP**

**Misfire Diagnosis**
If a misfire goes away with propane added to the air inlet, suspect a lean injector.

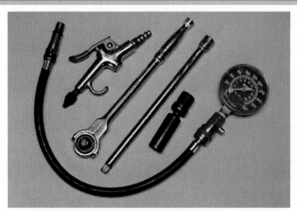

**1** The tools and equipment needed to perform a compression test include a compression gauge, an air nozzle, and the socket ratchets and extensions that may be necessary to remove the spark plugs from the engine.

**2** To prevent ignition and fuel-injection operation while the engine is being cranked, remove both the fuel-injection fuse and the ignition fuse. If the fuses cannot be removed, disconnect the wiring connectors for the injectors and the ignition system.

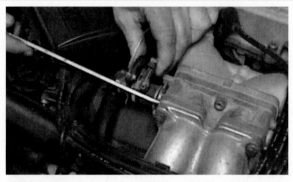

**3** Block open the throttle (and choke if the engine is equipped with a carburetor). Here a screwdriver is being used to wedge the throttle linkage open. Keeping the throttle open ensures that enough air will be drawn into the engine so that the compression test results will be accurate.

**4** Before removing the spark plugs, use an air nozzle to blow away any dirt that may be around the spark plug. This step helps prevent debris from getting into the engine when the spark plugs are removed.

**5** Remove all of the spark plugs. Be sure to mark the spark plug wires so that they can be reinstalled onto the correct spark plugs after the compression test has been performed.

**6** Select the proper adapter for the compression gauge. The threads on the adapter should match those on the spark plug.

CONTINUED E

# COMPRESSION TEST (CONTINUED)

**7** If necessary, connect a battery charger to the battery before starting the compression test. It is important that consistent cranking speed be available for each cylinder being tested.

**8** Make a note of the reading on the gauge after the first "puff," which indicates the first compression stroke that occurred on that cylinder as the engine was being rotated. If the first puff reading is low and the reading gradually increases with each puff, weak or worn piston rings may be indicated.

**9** After the engine has been cranked for four "puffs," stop cranking the engine and observe the compression gauge.

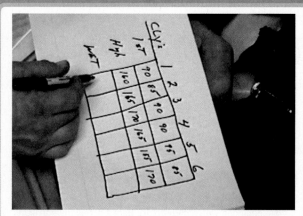

**10** Record the first puff and this final reading for each cylinder. The final readings should all be within 20% of each other.

**11** If a cylinder(s) is lower than most of the others, use an oil can and squirt two squirts of engine oil into the cylinder and repeat the compression test. This is called performing a wet compression test.

**12** If the gauge reading is now much higher than the first test results, then the cause of the low compression is due to worn or defective piston rings. The oil in the cylinder temporarily seals the rings which causes the higher reading.

## SUMMARY

1. The first step in diagnosing engine condition is to perform a thorough visual inspection, including a check of oil and coolant levels and condition.

2. Oil leaks can be found by using a white powder or a fluorescent dye and a black light.

3. Many engine-related problems make a characteristic noise.

4. A compression test can be used to test the condition of valves and piston rings.

5. A cylinder leakage test fills the cylinder with compressed air, and the gauge indicates the percentage of leakage.

6. A cylinder balance test indicates whether all cylinders are working equally.

7. Testing engine vacuum is another procedure that can help the service technician determine engine condition.

## REVIEW QUESTIONS

1. Describe the visual checks that should be performed on an engine if a mechanical malfunction is suspected.

2. List three simple items that could cause excessive oil consumption.

3. List three simple items that could cause engine noises.

4. Describe how to perform a compression test and how to determine what is wrong with an engine based on a compression test result.

5. Describe the cylinder leakage test.

6. Describe how a vacuum gauge would indicate if the valves were sticking in their guides.

7. Describe the test procedure for determining if the exhaust system is restricted (clogged) using a vacuum gauge.

## CHAPTER QUIZ

1. Technician A says that the paper test could detect a burned valve. Technician B says that a grayish white stain on the engine could be a coolant leak. Which technician is correct?
   a. Technician A only
   b. Technician B only
   c. Both Technicians A and B
   d. Neither Technician A nor B

2. Two technicians are discussing oil leaks. Technician A says that an oil leak can be found using a fluorescent dye in the oil with a black light to check for leaks. Technician B says that a white spray powder can be used to locate oil leaks. Which technician is correct?
   a. Technician A only
   b. Technician B only
   c. Both Technicians A and B
   d. Neither Technician A nor B

3. Which of the following is the *least likely* to cause an engine noise?
   a. Carbon on the pistons
   b. Cracked exhaust manifold
   c. Loose accessory drive belt
   d. Vacuum leak

4. A good engine should produce how much compression during a running (dynamic) compression test at idle?
   a. 150–200 PSI
   b. 100–150 PSI
   c. 60–90 PSI
   d. 30–60 PSI

5. A smoothly operating engine depends on _____.
   a. High compression on most cylinders
   b. Equal compression between cylinders
   c. Cylinder compression levels above 100 PSI (700 kPa) and within 70 PSI (500 kPa) of each other
   d. Compression levels below 100 PSI (700 kPa) on most cylinders

6. A good reading for a cylinder leakage test would be _____.
   a. Within 20% between cylinders
   b. All cylinders below 20% leakage
   c. All cylinders above 20% leakage
   d. All cylinders above 70% leakage and within 7% of each other

7. Technician A says that during a power balance test, the cylinder that causes the biggest RPM drop is the weak cylinder. Technician B says that if one spark plug wire is grounded out and the engine speed does not drop, a weak or dead cylinder is indicated. Which technician is correct?
   a. Technician A only
   b. Technician B only
   c. Both Technicians A and B
   d. Neither Technician A nor B

8. *Cranking* vacuum should be _____.
   a. 2.5 in. Hg or higher
   b. Over 25 in. Hg
   c. 17 to 21 in. Hg
   d. 6 to 16 in. Hg

9. Technician A says that a leaking head gasket can be tested for using a chemical tester. Technician B says that leaking head gasket can be found using an exhaust gas analyzer.
   a. Technician A only
   b. Technician B only
   c. Both Technicians A and B
   d. Neither Technician A nor B

10. The low oil pressure warning light usually comes on _____.
   a. Whenever an oil change is required
   b. Whenever oil pressure drops dangerously low (4–7 PSI)
   c. Whenever the oil filter bypass valve opens
   d. Whenever the oil filter antidrainback valve opens

# chapter
# 11

# ON-BOARD DIAGNOSIS

**LEARNING OBJECTIVES:** **After studying this chapter, the reader should be able to:** • Understand the purpose and function of on-board diagnostics generation-II (OBD-II) systems. • List the continuous and noncontinuous monitors. • Understand the information obtained from an on-board diagnostics monitor and the criteria to enable an OBD monitor. • Discuss the numbering designation of OBD-II diagnostic trouble codes. • Explain powertrain control module (PCM) tests and the modes of operation of OBD-II vehicles.

**KEY TERMS:** California Air Resources Board (CARB) 153 • Component identification (CID) 160 • Comprehensive component monitor (CCM) 154 • Diagnostic executive 154 • Enable criteria 156 • Exponentially weighted moving average (EWMA) monitor 156 • Federal Test Procedure (FTP) 154 • Freeze-frame 154 • Functionality 155 • Malfunction indicator lamp (MIL) 153 • On-board diagnosis (OBD) 153 • Parameter identification (PID) 160 • Rationality 155 • Society of Automotive Engineers (SAE) 157 • Test identification (TID) 160 • Task manager 154

## ON-BOARD DIAGNOSTICS GENERATION-II (OBD-II) SYSTEMS

**PURPOSE AND FUNCTION OF OBD II** During the 1980s, most manufacturers began equipping their vehicles with full-function control systems capable of alerting the driver of a malfunction and of allowing the technician to retrieve codes that identify circuit faults. These early diagnostic systems were meant to reduce emissions and speed up vehicle repair.

The automotive industry calls these systems **On-board Diagnostics (OBDs)**. The **California Air Resources Board (CARB)** developed the first regulation requiring manufacturers selling vehicles in that state to install OBD. OBD Generation I (OBD I) applies to all vehicles sold in California beginning with the 1988 model year. It specifies the following requirements:

1. An instrument panel warning lamp able to alert the driver of certain control system failures, now called a **malfunction indicator lamp (MIL)**. ● **SEE FIGURE 11–1.**

2. The system's ability to record and transmit diagnostic trouble codes (DTCs) for emission-related failures.

3. Electronic system monitoring of the HO2S, EGR valve, and evaporative purge solenoid. Although not U.S. EPA required, during this time most manufacturers also equipped vehicles sold outside of California with OBD I.

By failing to monitor the catalytic converter, the evaporative system for leaks, and the presence of engine misfire, OBD I

**FIGURE 11–1** A typical malfunction indicator lamp (MIL) often labeled "check engine" or "service engine soon" (SES).

did not do enough to lower automotive emissions. This led the CARB and the EPA to develop OBD Generation II (OBD II).

**OBD-II OBJECTIVES** Generally, the CARB defines an OBD-II-equipped vehicle by its ability to do the following:

1. Detect component degradation or a faulty emission-related system that prevents compliance with federal emission standards.

2. Alert the driver of needed emission-related repair or maintenance.

3. Use standardized DTCs and accept a generic scan tool.

These requirements apply to all 1996 and later model light-duty vehicles. The Clean Air Act of 1990 directed the EPA to develop new regulations for OBD. The primary purpose of OBD II is emission related, whereas the primary purpose of OBD I (1988) was to detect faults in sensors or sensor circuits. OBD-II regulations require that not only sensors be tested but also all exhaust emission control devices and that they be verified for proper operation.

All new vehicles must pass the **Federal Test Procedure (FTP)** for exhaust emissions while being tested for 1874 seconds on dynamometer rollers that simulate the urban drive cycle around downtown Los Angeles.

**NOTE: IM 240 is simply a shorter 240-second version of the 1874-second federal test procedure.**

The regulations for OBD-II vehicles state that the vehicle computer must be capable of testing for, and determining, if the exhaust emissions are within 1.5 times the FTP limits. To achieve this goal, the computer must do the following:

1. Test all exhaust emission system components for correct operation.
2. Actively operate the system and measure the results.
3. Continuously monitor all aspects of the engine operation to be certain that the exhaust emissions do not exceed 1.5 times the FTP limit.
4. Check engine operation for misfire.
5. Turn on the MIL (check engine) if the computer senses a fault in a circuit or system.
6. Record a **freeze-frame,** which is a snapshot of all key engine data at the time the DTC was set.
7. Flash the MIL if an engine misfire occurs that could damage the catalytic converter.

## DIAGNOSTIC EXECUTIVE AND TASK MANAGER

On OBD-II systems, the powertrain control module (PCM) incorporates a special segment of software. On Ford and GM systems, this software is called the **diagnostic executive.** On Chrysler systems, it is called the **task manager.** This software program is designed to manage the operation of all OBD-II monitors by controlling the sequence of steps necessary to execute the diagnostic tests and monitors.

## MONITORS

A monitor is an organized method of testing a specific part of the system. Monitors are simply tests that the computer performs to evaluate components and systems. If a component or system failure is detected while a monitor is running, a DTC will

be stored and the MIL illuminated during the second trip. The two types of monitors are continuous and noncontinuous.

**CONTINUOUS MONITORS**   As required conditions are met, continuous monitors begin to run. These continuous monitors will run for the remainder of the vehicle drive cycle. The three continuous monitors are as follows:

- **Comprehensive component monitor (CCM).** This monitor watches the sensors and actuators in the OBD-II system. Sensor values are constantly compared with known-good values stored in the PCM's memory.

    The CCM is an internal program in the PCM designed to monitor a failure in any electronic component or circuit (including emission-related and non–emission-related circuits) that provides input or output signals to the PCM. The PCM considers that an input or output signal is inoperative when a failure exists because of an open circuit or out-of-range value or if an onboard rationality check fails. If an emission-related fault is detected, the PCM will set a code and activate the MIL (requires two consecutive trips).

    Many PCM sensors and output devices are tested at key-on or immediately after engine start-up. However, some devices are tested by the CCM only after the engine meets certain engine conditions. The number of times the CCM must detect a fault before it will activate the MIL depends upon the manufacturer, but most require two consecutive trips to activate the MIL. The components tested by the CCM include the following:

    Four-wheel-drive low switch

    Brake switch

    Camshaft (CMP) and crankshaft (CKP) sensors

    Clutch switch (manual transmissions/transaxles only)

    Cruise servo switch

    Engine coolant temperature (ECT) sensor

    EVAP purge sensor or switch

    Fuel composition sensor

    Intake air temperature (IAT) sensor

    Knock sensor (KS)

    Manifold absolute pressure (MAP) sensor

    Mass airflow (MAF) sensor

    Throttle-position (TP) sensor

    Transmission temperature sensor

    Transmission turbine speed sensor

    Vacuum sensor

    Vehicle speed (VS) sensor

    EVAP canister purge and EVAP purge vent solenoid

    Idle air control (IAC)

    Ignition control system

    Transmission torque converter clutch solenoid

    Transmission shift solenoids

- **Misfire monitor.** This monitor looks at engine misfire. The PCM uses the information received from the crankshaft position sensor (CKP) to calculate the time between the edges of the reluctor as well as the rotational speed and acceleration. By comparing the acceleration of each firing event, the PCM can determine if a cylinder is not firing correctly.

**Misfire type A.** Upon detection of a misfire type A (200 revolutions), which would cause catalyst damage, the MIL will blink once per second during the actual misfire, and a DTC will be stored.

**Misfire type B.** Upon detection of a misfire type B (1,000 revolutions), which will exceed 1.5 times the EPA federal test procedure (FTP) standard or cause a vehicle to fail an inspection and maintenance tailpipe emissions test, the MIL will illuminate, and a DTC will be stored.

The DTC associated with multiple cylinder misfire for a type A or type B misfire is DTC P0300. The DTCs associated with an individual cylinder misfire for a type A or type B misfire are DTCs P0301, P0302, P0303, P0304, P0305, P0306, P0307, P0308, P0309, and P0310.

■ **Fuel trim monitor.** The PCM continuously monitors short- and long-term fuel trim. Constantly updated adaptive fuel tables are stored in long-term memory (KAM) and used by the PCM for compensation due to wear and aging of the fuel system components. The MIL will illuminate when the PCM determines the fuel trim values have reached and stayed at their limits for too long a period of time.

**NONCONTINUOUS MONITORS** Noncontinuous monitors run (at most) once per vehicle drive cycle. The noncontinuous monitors are as follows:

O2S monitor

O2S heater monitor

Catalyst monitor

EGR monitor

EVAP monitor

Secondary AIR monitor

Transmission monitor

PCV system monitor

Thermostat monitor

Once a noncontinuous monitor has run to completion, it will not be run again until the conditions are met during the next vehicle drive cycle. Also, after a noncontinuous monitor has run to completion, the readiness status on your scan tool will show "complete" or "done" for that monitor. Monitors that have not run to completion will show up on your scanner as "incomplete."

## OBD-II MONITOR INFORMATION

**COMPREHENSIVE COMPONENT MONITOR** The circuits and components covered by the comprehensive component monitor (CCM) do not include those directly monitored by another monitor.

However, OBD-II also requires that inputs from powertrain components to the PCM be tested for **rationality** and that outputs to powertrain components from the PCM be tested for **functionality.** Both inputs and outputs are to be checked electrically. Rationality checks refer to a PCM comparison of input value to values from other sensors to determine if they make sense and are normal (rational):

**Example:**

| TPS | 3 V |
| MAP | 18 in. Hg |
| RPM | 700 RPM |
| PRNDL | Park |

**NOTE: Comprehensive component monitors are continuous. Therefore, enabling conditions do not apply.**

■ Monitor runs continuously.

■ Monitor includes sensors, switches, relays, solenoids, and PCM hardware.

■ All are checked for opens, shorts-to-ground, and shorts-to-voltage.

■ Inputs are checked for rationality.

■ Outputs are checked for functionality.

■ Most are one-trip DTCs.

■ Freeze-frame is priority 3.

■ Three consecutive good trips are used to extinguish the MIL.

■ Forty warm-up cycles are required to self-erase DTC and freeze-frame.

■ Two minutes run time without reoccurrence of the fault constitutes a "good trip."

**CONTINUOUS RUNNING MONITORS** Continuous monitors run continuously and only stop if they fail and include:

■ Fuel system: rich/lean.

■ Misfire: catalyst damaging/FTP (emissions).

■ Two-trip faults (except early generation catalyst damaging misfire).

■ MIL, DTC, freeze-frame after two consecutive faults.

■ Freeze-frame is priority 2 on first trip.

■ Freeze-frame is priority 4 on maturing trip.

■ Three consecutive good trips in a similar condition window are used to extinguish the MIL.

■ Forty warm-up cycles are used to erase DTC and freeze-frame (80 to erase one-trip failure if similar conditions cannot be met).

**ONCE PER TRIP MONITORS**

■ Monitor runs once per trip, pass or fail.

■ $O_2$ response, $O_2$ heaters, EGR, purge flow EVAP leak, secondary air, catalyst.

■ Two-trip DTCs.

- MIL, DTC, freeze-frame after two consecutive faults.
- Freeze-frame is priority 1 on first trip.
- Freeze-frame is priority 3 on maturing trip.
- Three consecutive good trips are used to extinguish the MIL.
- Forty warm-up cycles are used to erase DTC and freeze-frame.

## EXPONENTIALLY WEIGHTED MOVING AVERAGE (EWMA) MONITORS

The **exponentially weighted moving average (EWMA) monitor** is a mathematical method used to determine performance. This method smooths out any variables in the readings over time and results in a running average. This method is used by some vehicle manufacturers for two monitors.

1. Catalyst monitor.
2. EGR monitor.

## ENABLING CRITERIA

With so many different tests (monitors) to run, the PCM needs an internal director to keep track of when each monitor should run. As mentioned, different manufacturers have different names for this director, such as the diagnostic executive or the task manager. Each monitor has enabling criteria. These criteria are a set of conditions that must be met before the task manager will give the go-ahead for each monitor to run. Most enabling criteria follow simple logic, such as the following:

- The task manager will not authorize the start of the O2S monitor until the engine has reached operating temperature and the system has entered closed loop.

- The task manager will not authorize the start of the EGR monitor when the engine is at idle because the EGR is always closed at this time.

Because each monitor is responsible for testing a different part of the system, the enabling criteria can differ greatly from one monitor to the next. The task manager must decide when each monitor should run, and in what order, to avoid confusion.

There may be a conflict if two monitors were to run at the same time. The results of one monitor might also be tainted if a second monitor were to run simultaneously. In such cases, the task manager decides which monitor has a higher priority. Some monitors also depend on the results of other monitors before they can run.

A monitor may be classified as pending if a failed sensor or other system fault is keeping it from running on schedule.

The task manager may suspend a monitor if the conditions are not correct to continue. For example, if the catalyst monitor is running during a road test and the PCM detects a misfire, the catalyst monitor will be suspended for the duration of the misfire.

**?**     FREQUENTLY ASKED QUESTION

### What Is a Drive Cycle?

A drive cycle is a vehicle being driven under specified speed and times that will allow all monitors to run. In other words, the powertrain control module (PCM) is looking at a series of data points representing speed and time and determines from these data points when the conditions are right to perform a monitor or a test of a component. These data points, and therefore the drive cycle, are vehicle specific and are not the same for each vehicle. Some common conditions for a drive cycle to successfully run all of the monitors include:

1. Cold start intake air temperature (IAT) and engine coolant temperature (ECT) close to each other, indicating that the engine has cooled to the temperature of the surrounding air temperature.
2. Fuel level within a certain range usually between 15% and 85%.
3. Vehicle speed within a certain speed range for a certain amount of time, usually 4 to 12 minutes.
4. Stop and idle for a certain time.

Each monitor requires its own set of parameters needed to run the test and sometimes these conditions cannot be met. For example, some evaporate emissions control (EVAP) systems require a temperature that may not be possible in winter months in a cold climatic area.

A typical universal drive cycle that works for many vehicles includes the following steps.

MIL must be off.

No DTCs present.

Fuel fill between 15% and 85%.

Cold start—Preferred = 8-hour soak at 68°F to 86°F.

Alternative = ECT below 86°F.

**STEP 1**   With the ignition off, connect scan tool.
**STEP 2**   Start engine and drive between 20 and 30 mph for 22 minutes, allowing speed to vary.
**STEP 3**   Stop and idle for 40 seconds, gradually accelerate to 55 mph.
**STEP 4**   Maintain 55 mph for 4 minutes using a steady throttle input.
**STEP 5**   Stop and idle for 30 seconds, then accelerate to 30 mph.
**STEP 6**   Maintain 30 mph for 12 minutes.
**STEP 7**   Repeat steps 4 and 5 four times.

Using scan tool, check readiness. Always check service information for the exact drive cycle conditions for the vehicle being serviced for best results.

**TRIP** A trip is defined as a key-on condition that contains the necessary conditions for a particular test to be performed followed by a key-off. These conditions are called the **enable criteria.** For example, for the EGR test to be performed, the engine must be at normal operating temperature and decelerating for a minimum amount of time. Some tests are performed when the engine is cold, whereas others require that the vehicle be cruising at a steady highway speed.

**WARM-UP CYCLE** Once a MIL is deactivated, the original code will remain in memory until 40 warm-up cycles are completed without the fault reappearing. A warm-up cycle is defined as a trip with an engine temperature increase of at least 40°F and where engine temperature reaches at least 160°F (71°C).

**MIL CONDITION: OFF** This condition indicates that the PCM has not detected any faults in an emissions-related component or system or that the MIL circuit is not working.

**MIL CONDITION: ON STEADY** This condition indicates a fault in an emissions-related component or system that could affect the vehicle emission levels.

**MIL CONDITION: FLASHING** This condition indicates a misfire or fuel control system fault that could damage the catalytic converter.

**NOTE: In a misfire condition with the MIL on steady, if the driver reaches a vehicle speed and load condition with the engine misfiring at a level that could cause catalyst damage, the MIL would start flashing. It would continue to flash until engine speed and load conditions caused the level of misfire to subside. Then the MIL would go back to the on-steady condition. This situation might result in a customer complaint of a MIL with an intermittent flashing condition.**

**MIL: OFF** The PCM will turn off the MIL if any of the following actions or conditions occur:

- The codes are cleared with a scan tool.
- Power to the PCM is removed at the battery or with the PCM power fuse for an extended period of time (may be up to several hours or longer).
- A vehicle is driven on three consecutive trips with a warm-up cycle and meets all code set conditions without the PCM detecting any faults.

The PCM will set a code if a fault is detected that could cause tailpipe emissions to exceed 1.5 times the FTP standard; however, the PCM will not deactivate the MIL until the vehicle has been driven on three consecutive trips with vehicle conditions similar to actual conditions present when the fault was detected. This is not merely three vehicle start-ups and trips. It means three trips during which certain engine operating conditions are met so that the OBD-II monitor that found the fault can run again and pass the diagnostic test.

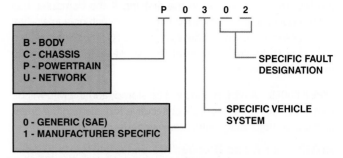

**FIGURE 11–2** OBD-II DTC identification format.

# OBD-II DTC NUMBERING DESIGNATION

A scan tool is required to retrieve DTCs from an OBD-II vehicle. Every OBD-II scan tool will be able to read all generic **Society of Automotive Engineers (SAE)** DTCs from any vehicle. ● **SEE FIGURE 11–2** for definitions and explanations of OBD alphanumeric DTCs. The diagnostic trouble codes (DTCs) are grouped into major categories, depending on the location of the fault on the system involved:

Pxxx codes—powertrain DTCs (engine, transmission-related faults)

Bxxx codes—body DTCs (accessories, interior-related faults)

Cxxx codes—chassis DTCs (suspension and steering-related faults)

Uxxx codes—network DTCs (module communication-related faults)

**DTC NUMBERING EXPLANATION** The number in the hundredth position indicates the specific vehicle system or subgroup that failed. This position should be consistent for P0xxx and P1xxx type codes. The following numbers and systems were established by SAE:

- P0100—Air metering and fuel system fault
- P0200—Fuel system (fuel injector only) fault
- P0300—Ignition system or misfire fault
- P0400—Emission control system fault
- P0500—Idle speed control, vehicle speed (VS) sensor fault
- P0600—Computer output circuit (relay, solenoid, etc.) fault
- P0700—Transaxle, transmission faults

**NOTE: The tens and ones numbers indicate the part of the system at fault.**

**TYPES OF DTCS** Not all OBD-II DTCs are of the same importance for exhaust emissions. Each type of DTC has different requirements for it to set, and the computer will turn on the MIL only for emissions-related DTCs.

**TYPE A CODES**  A type A DTC is emission related and will cause the MIL to be turned on in the first trip if the computer has detected a problem. Engine misfire or a very rich or lean air–fuel ratio, for example, would cause a type A DTC. These codes alert the driver to an emission problem that may cause damage to the catalytic converter.

**TYPE B CODES**  A type B code will be stored, and the MIL will be turned on during the second consecutive trip, alerting the driver to the fact that a diagnostic test was performed and failed.

**NOTE: Type A and B codes are emission-related codes that will cause the lighting of the malfunction indicator lamp (MIL), usually labeled "check engine" or "service engine soon."**

**TYPE C AND D CODES**  Type C and D codes are for use with non–emission-related diagnostic tests; they will cause the lighting of a "service" lamp (if the vehicle is so equipped). Type C codes are also called type C1 codes, and D codes are also called type C0 codes.

## DIAGNOSTIC TROUBLE CODE PRIORITY
CARB has also mandated that all diagnostic trouble codes (DTCs) be stored according to individual priority. DTCs with a higher priority overwrite those with a lower priority. The OBD-II System DTC Priority is listed here:

    Priority 0—Non–emission-related codes

    Priority 1—One-trip failure of two-trip fault for nonfuel, nonmisfire codes

    Priority 2—One-trip failure of two-trip fault for fuel or misfire codes

    Priority 3—Two-trip failure or matured fault of nonfuel, nonmisfire codes

    Priority 4—Two-trip failure or matured fault for fuel or misfire codes

# OBD-II FREEZE-FRAME

To assist the service technician, OBD II requires the computer to take a "snapshot" or freeze-frame of all data at the instant an emission-related DTC is set. A scan tool is required to retrieve these data.

**NOTE: Although OBD-II requires that just one freeze-frame of data be stored, the instant an emission-related DTC is set, vehicle manufacturers usually provide expanded data about the DTC beyond that required such as General Motors's *failure recorders*. However, retrieving this enhanced data usually requires the use of the vehicle-specific scan tool.**

    Freeze-frame items include the following:

- Calculated load value
- Engine speed (RPM)
- Short-term and long-term fuel trim percent
- Fuel system pressure (on some vehicles)
- Vehicle speed (mph)

- Engine coolant temperature
- Intake manifold pressure
- Closed-open-loop status
- Fault code that triggered the freeze-frame
- If a misfire code is set, identify which cylinder is misfiring

A DTC should not be cleared from the vehicle computer memory unless the fault has been corrected and the technician is so directed by the diagnostic procedure. If the problem that caused the DTC to be set has been corrected, the computer will automatically clear the DTC after 40 consecutive warm-up cycles with no further faults detected (misfire and excessively rich or lean condition codes require 80 warm-up cycles). The codes can also be erased by using a scan tool. ● **SEE CHART 11–1.**

**NOTE: Disconnecting the battery may not erase OBD-II DTCs or freeze-frame data. Most vehicle manufacturers recommend using a scan tool to erase DTCs rather than disconnecting the battery because the memory for the radio, seats, and learned engine operating parameters is lost if the battery is disconnected.**

# ENABLING CONDITIONS

These are the exact engine operating conditions required for a diagnostic monitor to run:

**Example:**

Specific RPM
Specific ECT, MAP, run time, and so on.

**PENDING**  Under some situations the PCM will not run a monitor if the MIL is illuminated and a fault is stored from another monitor. In these situations, the PCM postpones monitors pending a resolution of the original fault. The PCM does not run the test until the problem is remedied.

    For example, when the MIL is illuminated for an oxygen sensor fault, the PCM does not run the catalyst monitor until the oxygen sensor fault is remedied. Since the catalyst monitor is based on signals from the oxygen sensor, running the test would produce inaccurate results.

| MONITOR NAME | MONITOR TYPE (HOW OFTEN IT COMPLETES) | NUMBER OF FAULTS ON SEPARATE TRIPS TO SET A PENDING DTC | NUMBER OF SEPARATE CONSECUTIVE TRIPS TO LIGHT MIL, STORE A DTC | NUMBER OF TRIPS WITH NO FAULTS TO ERASE A MATURING DTC | NUMBER OF TRIPS WITH NO FAULT TO TURN THE MIL OFF | NUMBER OF WARM-UP CYCLES TO ERASE DTC AFTER MIL IS TURNED OFF |
|---|---|---|---|---|---|---|
| CCM | Continuous (when trip conditions allow it) | 1 | 2 | 1 | 3–Trips | 40 |
| Catalyst | Once per drive cycle | 1 | 3 | 1 | 3–OBD-II drive cycle | 40 |
| Misfire type A | Continuous | | 1 | | 3–Similar conditions | 80 |
| Misfire type B | Continuous | 1 | 2 | 1 | 3–Similar conditions | 80 |
| Fuel system | Continuous | 1 | 2 | 1 | 3–Similar conditions | 80 |
| Oxygen sensor | Once per trip | 1 | 2 | 1 | 3–Trips | 40 |
| EGR | Once per trip | 1 | 2 | 1 | 3–Trips | 40 |
| EVAP | Once per trip | 1 | 1 | 1 | 3–Trips | 40 |
| AIR | Once per trip | 1 | 2 | 1 | 3–Trips | 40 |

**CHART 11–1**

PCM determination of faults chart.

**CONFLICT** There are also situations when the PCM does not run a monitor if another monitor is in progress. In these situations, the effects of another monitor running could result in an erroneous failure. If this conflict is present, the monitor is not run until the conflicting condition passes. Most likely, the monitor will run later after the conflicting monitor has passed.

For example, if the fuel system monitor is in progress, the PCM does not run the EGR monitor. Since both tests monitor changes in air–fuel ratio and adaptive fuel compensation, the monitors conflict with each other.

**SUSPEND** Occasionally, the PCM may not allow a two-trip fault to mature. The PCM will suspend the maturing fault if a condition exists that may induce erroneous failure. This prevents illuminating the MIL for the wrong fault and allows more precise diagnosis.

For example, if the PCM is storing a one-trip fault for the oxygen sensor and the EGR monitor, the PCM may still run the EGR monitor but will suspend the results until the oxygen sensor monitor either passes or fails. At that point, the PCM can determine if the EGR system is actually failing or if an oxygen sensor is failing.

## PCM TESTS

**RATIONALITY TEST** While input signals to the PCM are constantly being monitored for electrical opens and shorts, they are also tested for rationality. This means that the input signal is compared against other inputs and information to see if it makes sense under the current conditions.

PCM sensor inputs that are checked for rationality include the following:

- MAP sensor
- $O_2$ sensor
- ECT
- Camshaft position sensor (CMP)
- VS sensor
- Crankshaft position sensor (CKP)
- IAT sensor
- TP sensor
- Ambient air temperature sensor
- Power steering switch
- $O_2$ sensor heater
- Engine controller
- Brake switch
- P/N switch (range switch)
- Transmission controls

**FUNCTIONALITY TEST** A functionality test refers to PCM inputs checking the operation of the outputs:

**Example:**

PCM commands IAC to increase engine speed
PCM monitors engine RPM
Functionality test fails if engine speed does not increase

PCM outputs that are checked for functionality include the following:

- EVAP canister purge solenoid
- EVAP purge vent solenoid
- Cooling fan
- Idle air control solenoid
- Ignition control system
- Transmission torque converter clutch solenoid
- Transmission shift solenoids (A, B, 1–2, etc.)

**ELECTRICAL TEST** Refers to the PCM check of both input and outputs for the following:

- Open
- Shorts
- Ground

**Example:**

ECT

Shorted high (input to PCM) above capable voltage; that is, 5-volt sensor with 12-volt input to PCM would indicate a short to voltage.

| Monitor Type | Conditions to Set DTC and Illuminate MIL | Extinguish MIL | Clear DTC Criteria | Applicable DTC |
|---|---|---|---|---|
| Continuous 1-trip monitor | (See note below) Input and output failure— rationally, functionally, electrically | 3 consecutive pass trips | 40 warm-up cycles | P0123 |

**NOTE: The number of times the comprehensive component monitor must detect a fault depends on the vehicle manufacturer. On some vehicles, the comprehensive component monitor will activate the MIL as soon as it detects a fault. On other vehicles, the comprehensive component monitor must fail two times in a row.**

- Freeze-frame captured on first-trip failure.
- Enabling conditions: Many PCM sensors and output devices are tested at key-on or immediately after engine start-up. However, some devices (ECT, idle speed control) are tested by the CCM only after the engine meets particular engine conditions.
- Pending: No pending condition
- Conflict: No conflict conditions
- Suspend: No suspend conditions

# GLOBAL OBD-II

All OBD-II vehicles must be able to display data on a global (also called *generic*) scan tool under nine different modes of operation. These modes include the following:

| Mode One | Current power train data (**parameter identification** display or **PID**) |
|---|---|
| Mode Two | Freeze-frame data |
| Mode Three | Diagnostic trouble codes |

| Mode Four | Clear and reset diagnostic trouble codes (DTCs), freeze-frame data, and readiness status monitors for noncontinuous monitors only |
|---|---|
| Mode Five | Oxygen sensor monitor test results |
| Mode Six | Onboard monitoring of test results for non-continuously monitored systems |
| Mode Seven | Onboard monitoring of test results for continuously monitored systems |
| Mode Eight | Bidirectional control of onboard systems |
| Mode Nine | Module identification |

**Mode TEN ($0A)** Permanent diagnostic trouble codes (DTCs)

The global (generic) data are used by most state emission programs. Global OBD-II displays often use hexadecimal numbers, which use 16 numbers instead of 10. The numbers 0 to 9 (zero counts as a number) make up the first 10 and then capital letters A to F complete the 16 numbers. To help identify the number as being in a hexadecimal format, a dollar sign ($) is used in front of the number or letter. See the following conversion chart:

| Decimal Number | Hexadecimal Code |
|---|---|
| 0 | $0 |
| 1 | $1 |
| 2 | $2 |
| 3 | $3 |
| 4 | $4 |
| 5 | $5 |
| 6 | $6 |
| 7 | $7 |
| 8 | $8 |
| 9 | $9 |
| 10 | $A |
| 11 | $B |
| 12 | $C |
| 13 | $D |
| 14 | $E |
| 15 | $F |

Hexadecimal coding is also used to identify tests (**test identification [TID]** and **component identification [CID]**).

 **FREQUENTLY ASKED QUESTION**

**How Can You Tell Generic from Factory?**

When using a scan tool on an OBD-II-equipped vehicle, if the display asks for make, model, and year, then the factory or enhanced part of the PCM is being accessed. If the generic or global part of the PCM is being scanned, then there is no need to know the vehicle identification details.

# DIAGNOSING PROBLEMS USING MODE SIX

Mode six information can be used to diagnose faults by following three steps:

1. Check the monitor status before starting repairs. This step will show how the system failed.
2. Look at the component or parameter that triggered the fault. This step will help pin down the root cause of the failure.
3. Look to the monitor enable criteria, which will show what it takes to fail or pass the monitor.

Many scan tools display all of the parameters and information needed so that additional mode $06 data is not needed. Many vehicle manufacturers post mode $06 information on the service information websites. This information is often free, unlike other service information. Refer to the National Automotive Service Task Force (NASTF) website for the website address of all vehicle manufacturers' service information sites (*www.NASTF.org*). Two examples include: *http://service.gm.com* (free access to mode $06 information) *www.motorcraftservice.com* (search for mode $06 free access)

## SUMMARY

1. If the MIL is on, retrieve the DTC and follow the manufacturer's recommended procedure to find the root cause of the problem.
2. All monitors must have the enable criteria achieved before a test is performed.
3. OBD-II vehicles use common generic DTCs.
4. OBD-II includes generic (SAE) as well as vehicle manufacturer-specific DTCs and data display.

## REVIEW QUESTIONS

1. What does the PCM do during a trip to test emission-related components?
2. What is the difference between a type A and type B OBD-II DTC?
3. What is the difference between a trip and a warm-up cycle?
4. What could cause the MIL to flash?

## CHAPTER QUIZ

1. A freeze-frame is generated on an OBD-II vehicle _____.
   a. When a type C or D diagnostic trouble code is set
   b. When a type A or B diagnostic trouble code is set
   c. Every other trip
   d. When the PCM detects a problem with the O2S

2. An ignition misfire or fuel mixture problem is an example of what type of DTC?
   a. Type A
   b. Type B
   c. Type C
   d. Type D

3. The comprehensive component monitor checks computer-controlled devices for _____.
   a. Opens
   b. Rationality
   c. Shorts-to-ground
   d. All of the above

4. OBD-II has been on all passenger vehicles in the United States since _____.
   a. 1986
   b. 1991
   c. 1996
   d. 2000

5. Which is a continuous monitor?
   a. Fuel system monitor
   b. EGR monitor
   c. Oxygen sensor monitor
   d. Catalyst monitor

6. DTC P0302 is a _____.
   a. Generic DTC
   b. Vehicle manufacturer–specific DTC
   c. Idle speed–related DTC
   d. Transmission/transaxle-related DTC

7. Global (generic) OBD-II contains some data in what format?
   a. Plain English
   b. Hexadecimal
   c. Roman numerals
   d. All of the above

8. By looking at the way diagnostic trouble codes are formatted, which DTC could indicate that the gas cap is loose or defective?
   a. P0221
   b. P1301
   c. P0442
   d. P1603

9. The computer will automatically clear a DTC if there are no additional detected faults after _____.
   a. Forty consecutive warm-up cycles
   b. Eighty warm-up cycles
   c. Two consecutive trips
   d. Four key-on/key-off cycles

10. A pending code is set when a fault is detected on _____.
    a. A one-trip fault item
    b. The first fault of a two-trip failure
    c. The catalytic converter efficiency
    d. Thermostat problem (too long to closed-loop status)

# chapter 12

# CAN AND NETWORK COMMUNICATIONS

**LEARNING OBJECTIVES:** **After studying this chapter, the reader should be able to:** • Explain the fundamentals of module communications and their configuration. • Explain the classifications of network communications and the communications protocols of General Motors, Ford, and Chrysler. • Explain the features of Controller Area Networks (CAN) and European BUS Communications. • Discuss how to diagnose network communication faults.

**KEY TERMS:** Breakout box (BOB) 173 • BUS 165 • CAN 165 • Chrysler Collision Detection (CCD) 169 • Class 2 165 • E & C 165 • GMLAN 165 • Keyword 165 • Multiplexing 162 • Network 162 • Node 162 • Plastic optical fiber (POF) 173 • Programmable controller interface (PCI) 170 • Protocol 165 • Serial communications interface (SCI) 170 • Serial data 162 • Splice pack 163 • Standard corporate protocol (SCP) 168 • State of health (SOH) 174 • SWCAN 166 • Terminating resistors 174 • Twisted pair 162 • UART 165 • UART-based protocol (UBP) 168

## MODULE COMMUNICATIONS AND NETWORKS

**NEED FOR NETWORK**  Since the 1990s, vehicles have used modules to control the operation of most electrical components. A typical vehicle will have 10 or more modules and they communicate with each other over data lines or hard wiring, depending on the application.

**ADVANTAGES**  Most modules are connected together in a network because of the following advantages:

- A decreased number of wires are needed, thereby saving weight and cost, as well as helping with installation at the factory and decreased complexity, making servicing easier.

- Common sensor data can be shared with those modules that may need the information, such as vehicle speed, outside air temperature, and engine coolant temperature.
  - ● **SEE FIGURE 12–1.**

## NETWORK FUNDAMENTALS

**MODULES AND NODES**  Each module, also called a **node,** must communicate to other modules. For example, if the driver depresses the window-down switch, the power window switch sends a window-down message to the body control module. The body control module then sends the request to the driver's side window module. This

module is responsible for actually performing the task by supplying power and ground to the window lift motor in the current polarity to cause the window to go down. The module also contains a circuit that monitors the current flow through the motor and will stop and/or reverse the window motor if an obstruction causes the window motor to draw more than the normal amount of current.

**TYPES OF COMMUNICATION**  The types of communications include the following:

- **Differential.** In the differential form of module communication, a difference in voltage is applied to two wires, which are twisted to help reduce electromagnetic interference (EMI). These transfer wires are called a **twisted pair.**

- **Parallel.** In the parallel type of module communication, the send and receive signals are on different wires.

- **Serial data.** The **serial data** is data transmitted over one wire by a series of rapidly changing voltage signals pulsed from low to high or from high to low.

- **Multiplexing.** The process of **multiplexing** involves the sending of multiple signals of information at the same time over a signal wire and then separating the signals at the receiving end.

This system of intercommunication of computers or processors is referred to as a **network.** ● **SEE FIGURE 12–2.**

By connecting the computers together on a communications network, they can easily share information back and forth. This multiplexing has the following advantages:

- Elimination of redundant sensors and dedicated wiring for these multiple sensors

- Reduction of the number of wires, connectors, and circuits

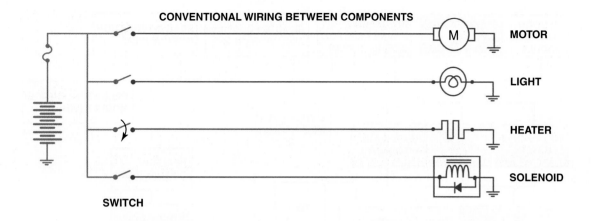

CONVENTIONAL WIRING BETWEEN COMPONENTS

MOTOR

LIGHT

HEATER

SOLENOID

SWITCH

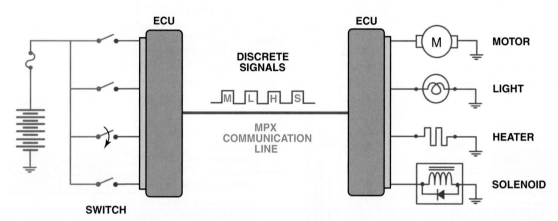

ECU

ECU

DISCRETE
SIGNALS

MPX
COMMUNICATION
LINE

MOTOR

LIGHT

HEATER

SOLENOID

SWITCH

**FIGURE 12–1** Module communications makes controlling multiple electrical devices and accessories easier by utilizing simple low-current switches to signal another module, which does the actual switching of the current to the device.

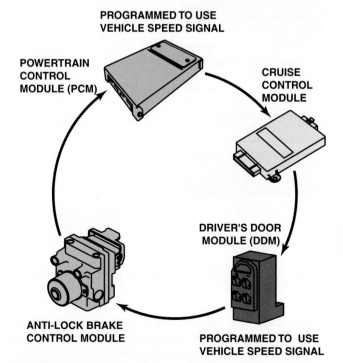

PROGRAMMED TO USE
VEHICLE SPEED SIGNAL

POWERTRAIN
CONTROL
MODULE (PCM)

CRUISE
CONTROL
MODULE

DRIVER'S DOOR
MODULE (DDM)

ANTI-LOCK BRAKE
CONTROL MODULE

PROGRAMMED TO USE
VEHICLE SPEED SIGNAL

**FIGURE 12–2** A network allows all modules to communicate with other modules.

■ Addition of more features and option content to new vehicles

■ Weight reduction due to fewer components, wires, and connectors, thereby increasing fuel economy

■ Changeable features with software upgrades versus component replacement

# MODULE COMMUNICATIONS CONFIGURATION

The three most common types of networks used on vehicles are the following:

1. **Ring link networks.** In a ring-type network, all modules are connected to each other by a serial data line (in a line) until all are connected in a ring. ● **SEE FIGURE 12–3.**

2. **Star link networks.** In a star link network, a serial data line attaches to each module, and then each is connected to a central point. This central point is called a **splice pack**, abbreviated SP, such as in "SP 306." The splice pack uses a bar to splice all of the serial lines together. Some GM vehicles use two or more splice packs to tie the modules

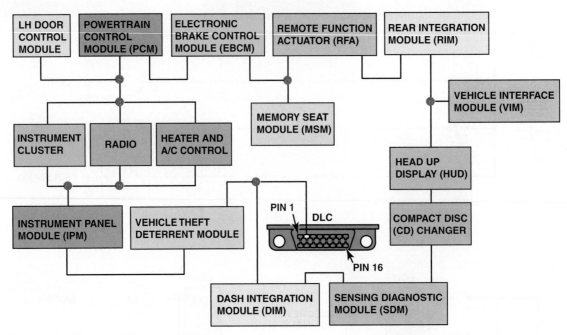

**FIGURE 12–3** A ring link network reduces the number of wires it takes to interconnect all of the modules.

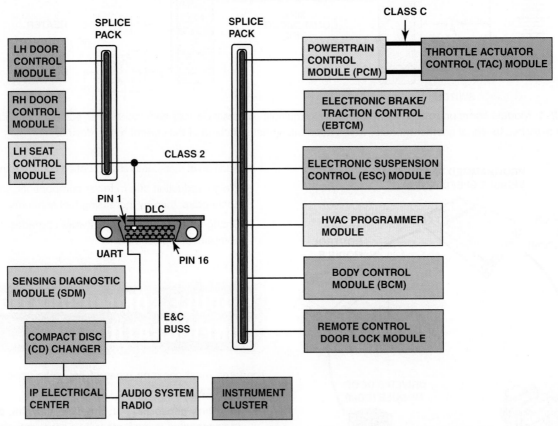

**FIGURE 12–4** In a star link network, all of the modules are connected using splice packs. Network circuits can be easily tested by separating the circuits at the splice packs.

together. When more than one splice pack is used, a serial data line connects one splice pack to the others. In most applications, the BUS bar used in each splice pack can be removed. When the BUS bar is removed, a special tool (J 42236) can be installed in place of the removed BUS bar.

Using this tool, the serial data line for each module can be isolated and tested for a possible problem. Using the special tool at the splice pack makes diagnosing this type of network easier than many others. ● **SEE FIGURE 12–4.**

3. **Ring/star hybrid.** In a ring/star network, the modules are connected using both types of network configurations. Check service information (SI) for details on how this network is connected on the vehicle being diagnosed and always follow the recommended diagnostic steps.

# NETWORK COMMUNICATIONS CLASSIFICATIONS

The Society of Automotive Engineers (SAE) standards include the following three categories of in-vehicle network communications.

**CLASS A** Low-speed networks, meaning less than 10,000 bits per second (bps, or 10 Kbs), are generally used for trip computers, entertainment, and other convenience features.

**CLASS B** Medium-speed networks, meaning 10,000 to 125,000 bps (10 to 125 Kbs), are generally used for information transfer among modules, such as instrument clusters, temperature sensor data, and other general uses.

**CLASS C** High-speed networks, meaning 125,000 to 1,000,000 bps, are generally used for real-time powertrain and vehicle dynamic control. High-speed BUS communication systems now use a **controller area network (CAN).** ● SEE FIGURE 12–5.

# GENERAL MOTORS COMMUNICATIONS PROTOCOLS

**UART** General Motors and others use UART communications for some electronic modules or systems. **UART** is a serial data communications protocol that stands for **universal asynchronous receive and transmit.** UART uses a master control module connected to one or more remote modules. The master control module is used to control message traffic on the data line by poling all of the other UART modules. The remote modules send a response message back to the master module.

UART uses a fixed pulse-width switching between 0 and 5 V. The UART data BUS operates at a baud rate of 8,192 bps. ● SEE FIGURE 12–6.

## ENTERTAINMENT AND COMFORT COMMUNICATION
The GM **entertainment and comfort (E & C)** serial data is similar to UART but uses a 0- to 12-V toggle. Like UART, the E & C serial data uses a master control module connected to other remote modules, which could include the following:

- Compact disc (CD) player
- Instrument panel (IP) electrical center
- Audio system (radio)
- Heating, ventilation, and air-conditioning (HVAC) programmer and control head
- Steering wheel controls
    - ● SEE FIGURE 12–7.

**CLASS 2 COMMUNICATIONS** Class 2 is a serial communications system that operates by toggling between 0 and 7 V at a transfer rate of 10.4 Kbs. Class 2 is used for most high-speed communications between the powertrain control module (PCM) and other control modules, plus to the scan tool. ● SEE FIGURE 12–8.

**KEYWORD COMMUNICATION** Keyword 81, 82, and 2000 serial data are also used for some module-to-module communication on GM vehicles. Keyword data BUS signals are toggled from 0 to 12 V when communicating. The voltage or the data stream is zero volts when not communicating. Keyword serial communication is used by the seat heater module and others but is not connected to the data link connector (DLC). ● SEE FIGURE 12–9.

**GMLAN** General Motors, like all vehicle manufacturers, must use high-speed serial data to communicate with scan tools on all vehicles effective with the 2008 model year. As mentioned, the standard is called controller area network (CAN), which General Motors calls **GMLAN,** which stands for **GM local area network.**

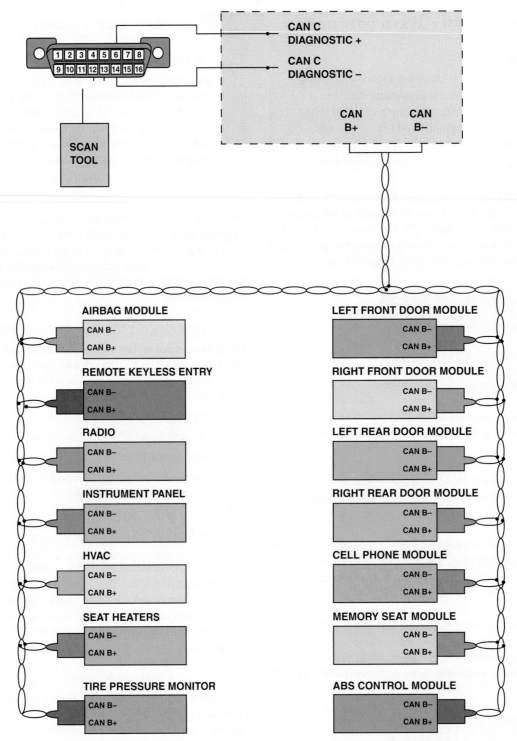

**FIGURE 12–5** A typical BUS system showing module CAN communications and twisted pairs of wire.

General Motors uses two versions of GMLAN.

- **Low-speed GMLAN.** The low-speed version is used for driver-controlled functions such as power windows and door locks. The baud rate for low-speed GMLAN is 33,300 bps. The GMLAN low-speed serial data is not connected directly to the data link connector and uses one wire. The voltage toggles between 0 and 5 V after an

initial 12-V spike, which indicates to the modules to turn on or wake up and listen for data on the line. Low-speed GMLAN is also known as **single-wire CAN,** or **SWCAN.**

- **High-speed GMLAN.** The baud rate is almost real time at 500 Kbs. This serial data method uses a two-twisted-wire circuit that is connected to the data link connector on pins 6 and 14. ● **SEE FIGURE 12–10.**

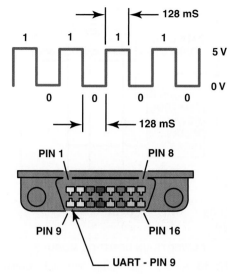

**FIGURE 12–6** UART serial data master control module is connected to the data link connector at pin 9.

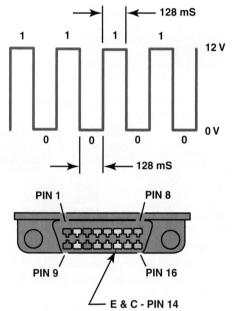

**FIGURE 12–7** The E & C serial data is connected to the data link connector (DLC) at pin 14.

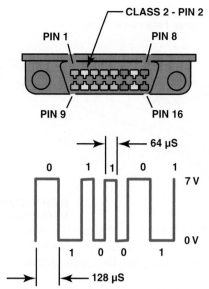

**FIGURE 12–8** Class 2 serial data communication is accessible at the data link connector (DLC) at pin 2.

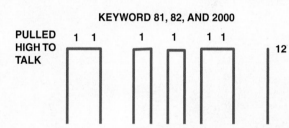

**FIGURE 12–9** Keyword 82 operates at a rate of 8,192 bps, similar to UART, and keyword 2000 operates at a baud rate of 10,400 bps (the same as a Class 2 communicator).

**?** **FREQUENTLY ASKED QUESTION**

**Why Is a Twisted Pair Used?**

A twisted pair is where two wires are twisted to prevent electromagnetic radiation from affecting the signals passing through the wires. By twisting the two wires about once every inch (9 to 16 times per foot), the interference is canceled by the adjacent wire.
● **SEE FIGURE 12–11.**

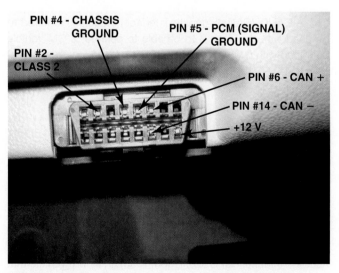

**FIGURE 12–10** GMLAN uses pins at terminals 6 and 14.

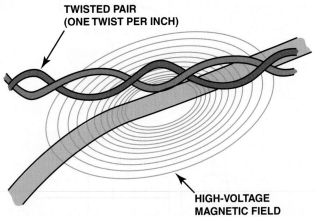

FIGURE 12–11 A twisted pair is used by several different network communications protocols to reduce interference that can be induced in the wiring from nearby electromagnetic sources.

FIGURE 12–12 A CANDi module will flash the green LED rapidly if communication is detected.

A CANDi (CAN diagnostic interface) module is required to be used with the Tech 2 to be able to connect a GM vehicle equipped with GMLAN. ● SEE FIGURE 12–12.

# FORD NETWORK COMMUNICATIONS PROTOCOLS

**STANDARD CORPORATE PROTOCOL** Only a few Fords had scan tool data accessible through the OBD-I data link connector. To identify an OBD-I (1988–1995) on a Ford vehicle that is equipped with **standard corporate protocol (SCP)** and be

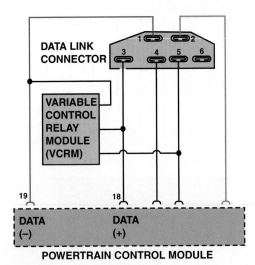

FIGURE 12–13 A Ford OBD-I diagnostic link connector showing that SCP communication uses terminals in cavities 1 (upper left) and 3 (lower left).

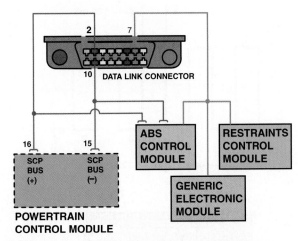

FIGURE 12–14 A scan tool can be used to check communications with the SCP BUS through terminals 2 and 10 and to the other modules connected to terminal 7 of the data link connector (DLC).

able to communicate through a scan tool, look for terminals in cavities 1 and 3 of the DLC. ● SEE FIGURE 12–13.

SCP uses the J-1850 protocol and is active with the key on. The SCP signal is from 4 V negative to 4.3 V positive, and a scan tool does not have to be connected for the signal to be detected on the terminals. OBD-II (EECV) Ford vehicles use terminals 2 (positive) and 10 (negative) of the 16-pin data link connector (DLC) for network communication, using the SCP module communications.

**UART-BASED PROTOCOL** Newer Fords use the CAN for scan tool diagnosis but still retain SCP and **UART-based protocol (UBP)** for some modules. ● SEE FIGURES 12–14 AND 12–15.

**?** FREQUENTLY ASKED QUESTION

**What Are U Codes?**

The U diagnostic trouble codes were at first "undefined" but are now network-related codes. Use the network codes to help pinpoint the circuit or module that is not working correctly.

# CHRYSLER COMMUNICATIONS PROTOCOLS

**CCD** Since the late 1980s, the **Chrysler Collision Detection (CCD)** multiplex network is used for scan tool and module communications. It is a differential-type communication and uses a twisted pair of wires. The modules connected to the network apply a bias voltage on each wire. CCD, signals are divided into plus and minus (CCD+ and CCD−), and the voltage difference does not exceed 0.02 V. The baud rate is 7,812.5 bps.

**NOTE: The "collision" in the Chrysler Collision detection BUS communications refers to the program that avoids conflicts of information exchange within the BUS and does not refer to airbags or other accident-related circuits of the vehicle.**

The circuit is active without a scan tool command. ● **SEE FIGURE 12–16.**

The modules on the CCD BUS apply a bias voltage on each wire by using termination resistors. ● **SEE FIGURE 12–17.**

The difference in voltage between CCD+ and CCD− is less than 20 mV. For example, using a digital meter with the

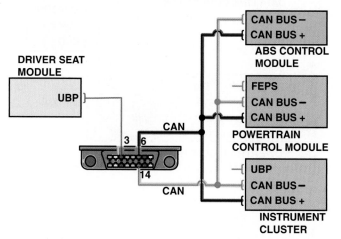

**FIGURE 12–15** Many Fords use UBP module communications along with CAN.

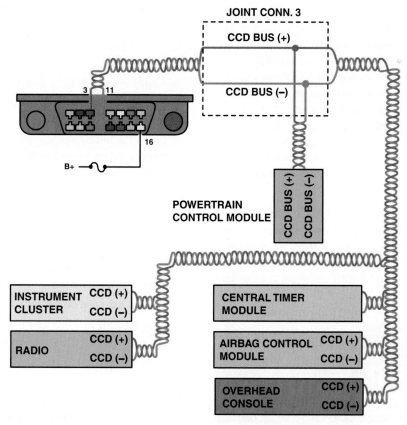

**FIGURE 12–16** CCD signals are labeled plus and minus and use a twisted pair of wires. Notice that terminals 3 and 11 of the data link connector are used to access the CCD BUS from a scan tool. Pin 16 is used to supply 12 volts to the scan tool.

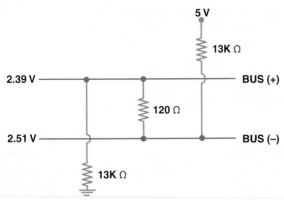

FIGURE 12–17 The differential voltage for the CCD BUS is created by using resistors in a module.

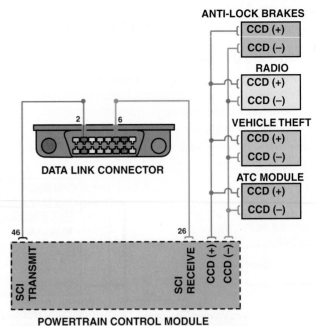

FIGURE 12–18 Many Chrysler vehicles use both SCI and CCD for module communication.

black meter lead attached to ground and the red meter lead attached at the data link connector (DLC), a normal reading could include the following:

- Terminal 3 = 2.45 volts
- Terminal 11 = 2.47 volts

This is an acceptable reading because the readings are 20 mV (0.020 volt) of each other. If both had been exactly 2.5 volts, then this could indicate that the two data lines are shorted together. The module providing the bias voltage is usually the body control module on passenger cars and the front control module on Jeeps and trucks.

**PROGRAMMABLE CONTROLLER INTERFACE**  The Chrysler **programmable controller interface (PCI)** is a one-wire communication protocol that connects at the OBD-II DLC at terminal 2. The PCI BUS is connected to all modules on the BUS in a star configuration and operates at a baud rate of 10,200 bps. The voltage signal toggles between 7.5 and 0 V. If this voltage is checked at terminal 2 of the OBD-II DLC, a voltage of about 1 V indicates the average voltage and means that the BUS is functioning and is not shorted-to-ground. PCI and CCD are often used in the same vehicle. ● **SEE FIGURE 12–18.**

**SERIAL COMMUNICATIONS INTERFACE**  Chrysler used **serial communications interface (SCI)** for most scan tool and flash reprogramming functions until it was replaced with CAN. SCI is connected at the OBD-II diagnostic link connector (DLC) at terminals 6 (SCI receive) and 7 (SCI transmit). A scan tool must be connected to test the circuit.

# CONTROLLER AREA NETWORK

**BACKGROUND**  Robert Bosch Corporation developed the CAN protocol, which was called CAN 1.2, in 1993. The CAN protocol was approved by the Environmental Protection Agency (EPA) for 2003 and newer vehicle diagnostics and a legal requirement for all vehicles by 2008. The CAN diagnostic systems use pins 6 and 14 in the standard 16 pin OBD-II (J-1962) connector. Before CAN, the scan tool protocol had been manufacturer specific.

**CAN FEATURES**  The CAN protocol offers the following features:

- Faster than other BUS communication protocols
- Cost effective because it is an easier system than others to use
- Less effected by electromagnetic interference (Data is transferred on two wires that are twisted together, called twisted pair, to help reduce EMI interference.)
- Message based rather than address based, making it easier to expand
- No wake-up needed because it is a two-wire system
- Supports up to 15 modules plus a scan tool
- Uses a 120-ohm resistor at the ends of each pair to reduce electrical noise
- Applies 2.5 volts on both wires:
  H (high) goes to 3.5 volts when active
  L (low) goes to 1.5 volts when active
  ● **SEE FIGURE 12–19.**

**CAN CLASS A, B, AND C**  There are three classes of CAN, and they operate at different speeds. The CAN A, B, and C networks can all be linked using a gateway within the same vehicle. The gateway is usually one of the many modules in the vehicle.

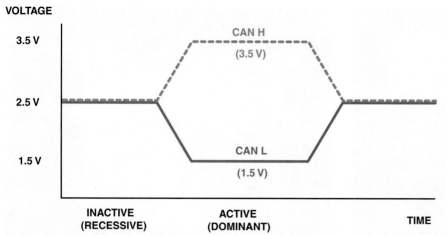

FIGURE 12–19 CAN uses a differential type of module communication where the voltage on one wire is the equal but opposite voltage on the other wire. When no communication is occurring, both wires have 2.5 volts applied. When communication is occurring, CAN H (high) goes up 1 to 3.5 volts, and CAN L (low) goes down 1 to 1.5 volts.

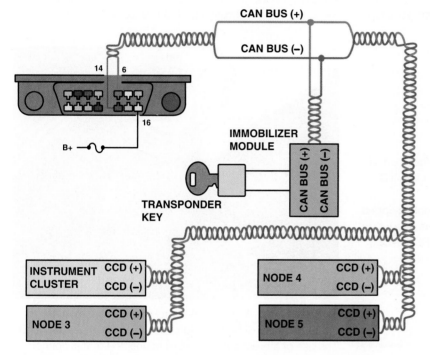

FIGURE 12–20 A typical (generic) system showing how the CAN BUS is connected to various electrical accessories and systems in the vehicle.

- **CAN A.** This class operates on only one wire at slow speeds and is therefore less expensive to build. CAN A operates a data transfer rate of 33.33 Kbs in normal mode and up to 83.33 Kbs during reprogramming mode. CAN A uses the vehicle ground as the signal return circuit.

- **CAN B.** This class operates on a two-wire network and does not use the vehicle ground as the signal return circuit. CAN B uses a data transfer rate of 95.2 Kbs. Instead, CAN B (and CAN C) uses two network wires for differential signaling. This means that the two data signal voltages are opposite to each other and used for error detection by constantly being compared. In this case, when the signal voltage at one of the CAN data wires goes high (CAN H), the other one goes low (CAN L), hence the name *differential signaling*. Differential signaling is also used for redundancy in case one of the signal wires shorts out.

- **CAN C.** This class is the highest speed CAN protocol with speeds up to 500 Kbs. Beginning with 2008 models, all vehicles sold in the United States must use CAN BUS for scan tool communications. Most vehicle manufacturers started using CAN in older models, and it is easy to determine if a vehicle is equipped with CAN. The CAN BUS communicates to the scan tool through terminals 6 and 14 of the DLC, indicating that the vehicle is equipped with CAN. ● **SEE FIGURE 12–20.**

The total voltage remains constant at all times, and the electromagnetic field effects of the two data BUS lines cancel each other out. The data BUS line is protected against received radiation and is virtually neutral in sending radiation.

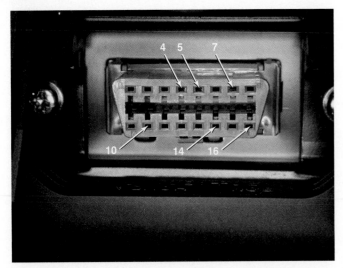

**FIGURE 12–21** A DLC from a pre-CAN Acura. It shows terminals in cavities 4, 5 (grounds), 7, 10, 14, and 16 (B+).

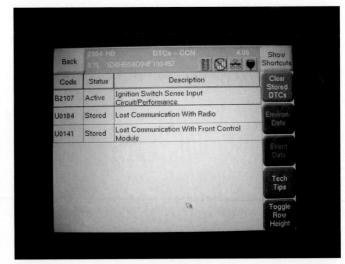

**FIGURE 12–22** A Chrysler scan tool display showing a B and two U codes, all indicating a BUS-related problem(s).

# HONDA/TOYOTA COMMUNICATIONS

The primary BUS communications on pre-CAN-equipped vehicles is ISO 9141-2 using terminals 7 and 15 at the OBD-II DLC. ● **SEE FIGURE 12–21.**

A factory scan tool or an aftermarket scan tool equipped with enhanced original equipment (OE) software is needed to access many of the BUS messages. ● **SEE FIGURE 12–22.**

# EUROPEAN BUS COMMUNICATIONS

**UNIQUE DIAGNOSTIC CONNECTOR** Many different types of module communications protocols are used on European vehicles such as Mercedes and BMW.

Most of these communication BUS messages cannot be accessed through the data link connector (DLC). To check the operation of the individual modules, a scan tool equipped with factory-type software will be needed to communicate with the module through the gateway module. ● **SEE FIGURE 12–23** for an alternative access method to the modules.

**MEDIA ORIENTED SYSTEM TRANSPORT BUS** The media-oriented system transport (MOST) BUS uses fiber optics for module-to-module communications in a ring or star configuration. This BUS system is currently being used for entertainment equipment data communications for videos, CDs, and other media systems in the vehicle.

**FIGURE 12–23** A typical 38-cavity diagnostic connector as found on many BMW and Mercedes vehicles under the hood. The use of a breakout box (BOB) connected to this connector can often be used to gain access to module BUS information.

**MOTOROLA INTERCONNECT BUS** Motorola interconnect (MI) is a single-wire serial communications protocol, using one master control module and many slave modules. Typical application of the MI BUS protocol is with power and memory mirrors, seats, windows, and headlight levelers.

**DISTRIBUTED SYSTEM INTERFACE BUS** Distributed system interface (DSI) BUS protocol was developed by Motorola and uses a two-wire serial BUS. This BUS protocol is currently being used for safety-related sensors and components.

**BOSCH-SIEMANS-TEMIC BUS** The Bosch-Siemans-Temic (BST) BUS is another system that is used for safety-related components and sensors in a vehicle, such as airbags. The BST BUS is a two-wire system and operates up to 250,000 bps.

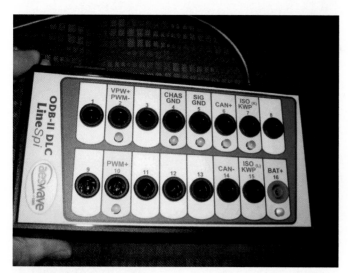

**FIGURE 12–24** A breakout box (BOB) used to access the BUS terminals while using a scan tool to activate the modules. This breakout box is equipped with LEDs that light when circuits are active.

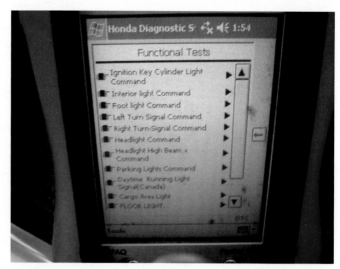

**FIGURE 12–25** This Honda scan tool allows the technician to turn on individual lights and operate individual power windows and other accessories that are connected to the BUS system.

## FREQUENTLY ASKED QUESTION

**How Do You Know What System Is Used?**

Use service information to determine which network communication protocol is used. However, because of the various systems on some vehicles, it may be easier to look at the data link connection to determine the system. All OBD-II vehicles have terminals in the following cavities:

Terminal 4: chassis ground

Terminal 5: computer (signal) ground

Terminal 16: 12 V positive

The terminals in cavities 6 and 14 mean that this vehicle is equipped with CAN as the only module communication protocol available at the DLC. To perform a test of the BUS, use a **breakout box (BOB)** to gain access to the terminals while connecting to the vehicle, using a scan tool. ● **SEE FIGURE 12–24** or a typical OBD-II connector breakout box.

**BYTEFLIGHT BUS** The byteflight BUS is used in safety critical systems, such as airbags, and uses the time division multiple access (TDMA) protocol, which operates at 10 million bps using a **plastic optical fiber (POF).**

**FLEXRAY BUS** FlexRay BUS is a version of byteflight and is a high-speed serial communication system for in-vehicle networks. FlexRay is commonly used for steer-by-wire and brake-by-wire systems.

**DOMESTIC DIGITAL BUS** The domestic digital BUS, commonly designated D2B, is an optical BUS system connecting audio, video, computer, and telephone components in a single-ring structure with a speed of up to 5,600,000 bps.

**LOCAL INTERCONNECT NETWORK BUS** Local interconnect network (LIN) is a BUS protocol used between intelligent sensors and actuators and has a BUS speed of 19,200 bps.

## NETWORK COMMUNICATIONS DIAGNOSIS

**STEPS TO FINDING A FAULT** When a network communications fault is suspected, perform the following steps:

**STEP 1** **Check everything that does and does not work.** Often accessories that do not seem to be connected can help identify which module or BUS circuit is at fault.

**STEP 2** **Perform module status test.** Use a factory-level scan tool or an aftermarket scan tool equipped with enhanced software that allows OE-like functions. Check if the components or systems can be operated through the scan tool. ● **SEE FIGURE 12–25.**

- **Ping modules.** Start the Class 2 diagnosis by using a scan tool and select *diagnostic circuit check*. If no diagnostic trouble codes (DTCs) are shown, there could be a communication problem. Select

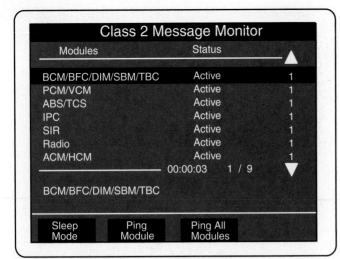

FIGURE 12–26 Modules used in a General Motors vehicle can be "pinged" using a Tech 2 scan tool.

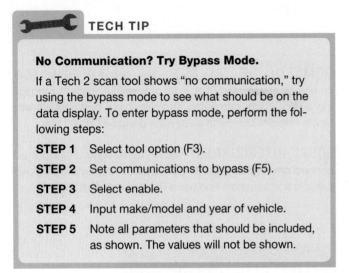

🔧 **TECH TIP**

**No Communication? Try Bypass Mode.**

If a Tech 2 scan tool shows "no communication," try using the bypass mode to see what should be on the data display. To enter bypass mode, perform the following steps:

**STEP 1** Select tool option (F3).

**STEP 2** Set communications to bypass (F5).

**STEP 3** Select enable.

**STEP 4** Input make/model and year of vehicle.

**STEP 5** Note all parameters that should be included, as shown. The values will not be shown.

message monitor, which will display the status of all of the modules on the Class 2 BUS circuit. The modules that are awake will be shown as active and the scan tool can be used to ping individual modules or command all modules. The ping command should change the status from "active" to "inactive." ● **SEE FIGURE 12–26.**

**NOTE: If an excessive parasitic draw is being diagnosed, use a scan tool to ping the modules in one way to determine if one of the modules is not going to sleep and causing the excessive battery drain.**

▪ **Check state of health.** All modules on the Class 2 BUS circuit have at least one other module responsible for reporting **state of health (SOH).** If a module fails to send a state of health message within five seconds, the companion module will set a diagnostic trouble code for the module that did not respond. The defective module is not capable of sending this message.

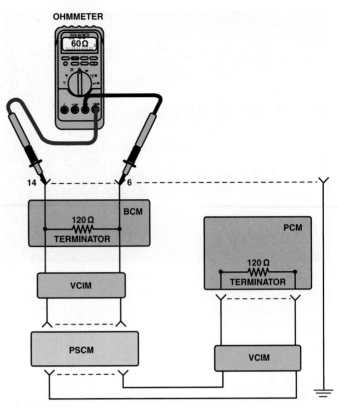

FIGURE 12–27 Checking the terminating resistors using an ohmmeter at the DLC.

**STEP 3** **Check the resistance of the terminating resistors.** Most high-speed BUS systems use resistors at each end, called **terminating resistors.** These resistors are used to help reduce interference into other systems in the vehicle. Usually two 120-ohm resistors are installed at each end and are therefore connected electrically in parallel. Two 120-ohm resistors connected in parallel would measure 60 ohms if being tested using an ohmmeter. ● **SEE FIGURE 12–27.**

**STEP 4** **Check data BUS for voltages.** Use a digital multimeter set to DC volts to monitor communications and check the BUS for proper operation. Some BUS conditions and possible causes include the following:

▪ **Signal is zero volt all of the time.** Check for short-to-ground by unplugging modules one at a time to check if one module is causing the problem.

▪ **Signal is high or 12 volts all of the time.** The BUS circuit could be shorted to 12 V. Check with the customer to see if any service or body repair work was done recently. Try unplugging each module one at a time to pin down which module is causing the communications problem.

▪ **A variable voltage usually indicates that messages are being sent and received.** CAN and Class 2 can be identified by looking at the data link connector (DLC) for a terminal in cavity number 2. Class 2 is active all of the time the ignition is on, and therefore voltage variation between 0 and 7 V can be measured using a DMM set to read DC volts. ● **SEE FIGURE 12–28.**

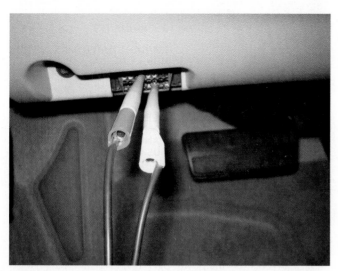

**FIGURE 12–28** Use front-probe terminals to access the data link connector. Always follow the specified back-probe and front-probe procedures as found in service information.

**STEP 5** **Use a digital storage oscilloscope to monitor the waveforms of the BUS circuit.** Using a scope on the data line terminals can show if communication is being transmitted. Typical faults and their causes include the following:

- **Normal operation.** Normal operation shows variable voltage signals on the data lines. It is impossible to know what information is being transmitted, but if there is activity with short sections of inactivity, this indicates normal data line transmission activity. ● **SEE FIGURE 12–29.**
- **High voltage.** If there is a constant high-voltage signal without any change, this indicates that the data line is shorted to voltage.
- **Zero or low voltage.** If the data line voltage is zero or almost zero and not showing any higher voltage signals, then the data line is short-to-ground.

**STEP 6** **Follow factory service information instructions to isolate the cause of the fault.** This step often involves disconnecting one module at a time to see if it is the cause of a short-to-ground or an open in the BUS circuit.

(a)

CAN BUS LOOKS GOOD

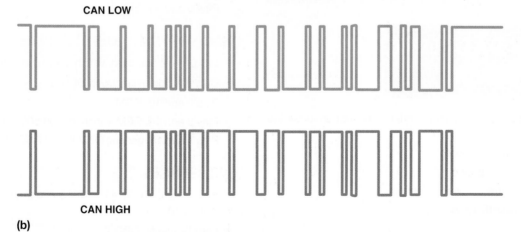

(b)

**FIGURE 12–29** (a) Data are sent in packets, so it is normal to see activity then a flat line between messages. (b) A CAN BUS should show voltages that are opposite when there is normal communications. CAN H (high) circuit should go from 2.5 volts at rest to 3.5 volts when active. The CAN L (low) circuit goes from 2.5 volts at rest to 1.5 volts when active.

**The Radio Caused No-Start Story**

A 2005 Chevrolet Cobalt did not start. A technician checked with a subscription-based helpline service and discovered that a fault with the Class 2 data circuit could prevent the engine from starting. The advisor suggested that a module should be disconnected one at a time to see if one of them was taking the data line to ground. The two most common components on the Class 2 serial data line that have been known to cause a lack of communication and become shorted-to-ground are the radio and electronic brake control module (EBCM). The first one the technician disconnected was the radio. The engine started and ran. Apparently the Class 2 serial data line was shorted-to-ground inside the radio, which took the entire BUS down. When BUS communication is lost, the PCM is not able to energize the fuel pump, ignition, or fuel injectors so the engine would not start. The radio was replaced to solve the no-start condition.

 **FREQUENTLY ASKED QUESTION**

**Which Module Is the Gateway Module?**

The gateway module is responsible for communicating with other modules and acts as the main communications module for scan tool data. Most General Motors vehicles use the body control module (BCM) or the instrument panel control (IPC) module as the gateway. To verify which module is the gateway, check the schematic and look for one that has voltage applied during all of the following conditions:

- Key on, engine off
- Engine cranking
- Engine running

# OBD-II DATA LINK CONNECTOR

All OBD-II vehicles use a 16-pin connector that includes the following:

    Pin 4 = chassis ground

    Pin 5 = signal ground

    Pin 16 = battery power (4 A max)

    ● **SEE FIGURE 12–30**.

## GENERAL MOTORS VEHICLES

- SAE J-1850 (VPW, Class 2, 10.4 Kbs) standard, which uses pins 2, 4, 5, and 16 but not 10

| PIN NO. | ASSIGNMENTS |
|---|---|
| 1. | MANUFACTURER'S DISCRETION |
| 2. | BUS + LINE, SAE J1850 |
| 3. | MANUFACTURER'S DISCRETION |
| 4. | CHASSIS GROUND |
| 5. | SIGNAL GROUND |
| 6. | MANUFACTURER'S DISCRETION |
| 7. | K LINE, ISO 9141 |
| 8. | MANUFACTURER'S DISCRETION |
| 9. | MANUFACTURER'S DISCRETION |
| 10. | BUS – LINE, SAE J1850 |
| 11. | MANUFACTURER'S DISCRETION |
| 12. | MANUFACTURER'S DISCRETION |
| 13. | MANUFACTURER'S DISCRETION |
| 14. | MANUFACTURER'S DISCRETION |
| 15. | L LINE, ISO 9141 |
| 16. | VEHICLE BATTERY POSITIVE (4A MAX) |

**OBD-II DLC**

**FIGURE 12–30** A 16-pin OBD-II DLC with terminals identified. Scan tools use the power pin (16) and ground pin (4) for power so that a separate cigarette lighter plug is not necessary on OBD-II vehicles.

 **TECH TIP**

**Check Computer Data Line Circuit Schematic**

Many General Motors vehicles use more than one type of BUS communications protocol. Check service information (SI) and look at the schematic for computer data line circuits, which should show all of the data BUSes and their connectors to the diagnostic link connector (DLC). ● **SEE FIGURE 12–31**.

- GM Domestic OBD-II

  Pins 1 and 9: CCM (comprehensive component monitor) slow baud rate, 8,192 UART

  Pin 1 (2006 and newer): low speed GMLAN

  Pins 2 and 10: OEM enhanced, fast rate, 40,500 baud rate

  Pins 7 and 15: generic OBD-II, ISO 9141, 10,400 baud rate

  Pins 6 and 14: GMLAN

## ASIAN, CHRYSLER, AND EUROPEAN VEHICLES

- ISO 9141-2 standard, which uses pins 4, 5, 7, 15, and 16
- Chrysler Domestic Group OBD-II

  Pins 2 and 10: CCM

  Pins 3 and 14: OEM enhanced, 60,500 baud rate

  Pins 7 and 15: generic OBD-II, ISO 9141, 10,400 baud rate

## FORD VEHICLES

- SAE J-1850 (PWM, 41.6 Kbs) standard, which uses pins 2, 4, 5, 10, and 16
- Ford Domestic OBD-II

  Pins 2 and 10: CCM

  Pins 6 and 14: OEM enhanced, Class C, 40,500 baud rate

  Pins 7 and 15: generic OBD-II, ISO 9141, 10,400 baud rate

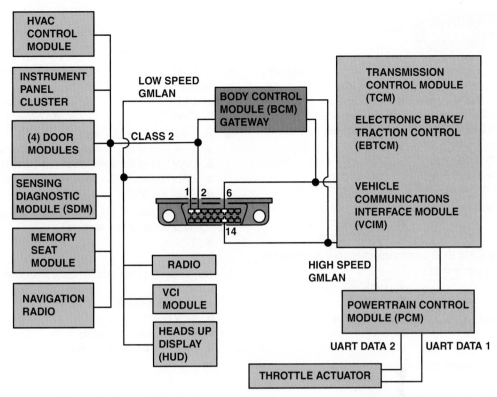

**FIGURE 12–31** This schematic of a Chevrolet Equinox shows that the vehicle uses a GMLAN BUS (DLC pins 6 and 14), plus a Class 2 (pin 2) and UART. Pin 1 connects to the low-speed GMLAN network.

## SUMMARY

1. The use of a network for module communications reduces the number of wires and connections needed.

2. Module communication configurations include ring link, star link, and ring/star hybrid systems.

3. The SAE communication classifications for vehicle communications systems include Class A (low speed), Class B (medium speed), and Class C (high speed).

4. Various module communications used on General Motors vehicles include UART, E & C, Class 2, keyword communications, and GMLAN (CAN).

5. Types of module communications used on Ford vehicles include SCP, UBP, and CAN.

6. Chrysler brand vehicles use SCI, CCD, PCI, and CAN communications protocols.

7. Many European vehicles use an underhood electrical connector that can be used to access electrical components and modules using a breakout box (BOB) or special tester.

8. Diagnosis of network communications includes checking the terminating resistors and checking for changing voltage signals at the DLC.

## REVIEW QUESTIONS

1. Why is a communication network used?

2. Why are the two wires twisted if used for network communications?

3. Why is a gateway module used?

4. What are U codes?

1. Technician A says that module communications networks are used to reduce the number of wires in a vehicle. Technician B says that a communications network is used to share data from sensors, which can be used by many different modules. Which technician is correct?
   a. Technician A only
   b. Technician B only
   c. Both Technicians A and B
   d. Neither Technician A nor B

2. A module is also known as a _____.
   a. BUS
   b. Node
   c. Terminator
   d. Resistor pack

3. A high-speed CAN BUS communicates with a scan tool through which terminal(s)?
   a. 6 and 14
   b. 2
   c. 7 and 15
   d. 4 and 16

4. UART uses a(n) _____ signal that toggles 0 V.
   a. 5-V
   b. 7-V
   c. 8-V
   d. 12-V

5. GM Class 2 communication toggles between _____.
   a. 5 and 7 V
   b. 0 and 12 V
   c. 7 and 12 V
   d. 0 and 7 V

6. Which terminal of the data link connector does General Motors use for Class 2 communication?
   a. 1
   b. 2
   c. 3
   d. 4

7. GMLAN is the General Motors term for which type of module communication?
   a. UART
   b. Class 2
   c. High-speed CAN
   d. Keyword 2000

8. How do CAN H and CAN L operate?
   a. CAN H is at 2.5 volts when not transmitting.
   b. CAN L is at 2.5 volts when not transmitting.
   c. CAN H goes to 3.5 volts when transmitting.
   d. All of the above

9. Which terminal of the OBD-II data link connector is the signal ground for all vehicles?
   a. 1
   b. 3
   c. 4
   d. 5

10. Terminal 16 of the OBD-II data link connector is used for what?
    a. Chassis ground
    b. 12 V positive
    c. Module (signal ground)
    d. Manufacturer's discretion

# chapter 13
# TEMPERATURE SENSORS

**LEARNING OBJECTIVES:** **After studying this chapter, the reader should be able to:** • Discuss the purpose and function of engine coolant temperature sensors. • Explain the procedure for inspecting and testing engine coolant temperature sensors. • Explain the function of intake air temperature sensors and the procedure to test them. • Explain transmission fluid, cylinder head, engine fuel, and exhaust gas recirculation temperature sensors.

**KEY TERMS:** Cylinder head temperature (CHT) 187 • Engine coolant temperature (ECT) 179 • Engine fuel temperature (EFT) 187 • Negative temperature coefficient (NTC) 179 • Throttle-body temperature (TBT) 185 • Transmission fluid temperature (TFT) 186

## ENGINE COOLANT TEMPERATURE SENSORS

**PURPOSE AND FUNCTION** Computer-equipped vehicles use an **engine coolant temperature (ECT)** sensor. When the engine is cold, the fuel mixture must be richer to prevent stalling and engine stumble. When the engine is warm, the fuel mixture can be leaner to provide maximum fuel economy with the lowest possible exhaust emissions. Because the computer controls spark timing and fuel mixture, it will need to know the engine temperature. An engine coolant temperature (ECT) sensor screwed into the engine coolant passage will provide the computer with this information. ● **SEE FIGURE 13–1.** This will be the most important (high-authority) sensor while the engine is cold. The ignition timing can also be tailored to engine (coolant) temperature. A hot engine cannot have the spark timing as far advanced as can a cold engine. The ECT sensor is also used as an important input for the following:

- Idle air control (IAC) position
- Oxygen sensor closed-loop status
- Canister purge on/off times
- Idle speed

**ECT SENSOR CONSTRUCTION** Engine coolant temperature sensors are constructed of a semiconductor material that decreases in resistance as the temperature of the sensor increases. Coolant sensors have very high resistance when the coolant is cold and low resistance when the coolant is hot. This is referred to as having a **negative temperature coefficient (NTC),** which is opposite to the situation with most other electrical components. ● **SEE FIGURE 13–2.** Therefore, if the coolant sensor has a poor connection (high resistance) at the wiring connector,

**FIGURE 13–1** A typical engine coolant temperature (ECT) sensor. ECT sensors are located near the thermostat housing on most engines.

the computer will supply a richer-than-normal fuel mixture based on the resistance of the coolant sensor. Poor fuel economy and a possible-rich code can be caused by a defective sensor or high resistance in the sensor wiring. If the sensor was shorted or defective and had too low a resistance, a leaner-than-normal fuel mixture would be supplied to the engine. A too-lean fuel mixture can cause driveability problems and a possible-lean computer code.

**STEPPED ECT CIRCUITS** Some vehicle manufacturers use a step-up resistor to effectively broaden the range of the ECT sensor. Chrysler and General Motors vehicles use the same sensor as a nonstepped ECT circuit but instead apply the sensor voltage through two different resistors:

- When the temperature is cold, usually below 120°F (50°C), the ECT sensor voltage is applied through a high-value resistor inside the PCM.

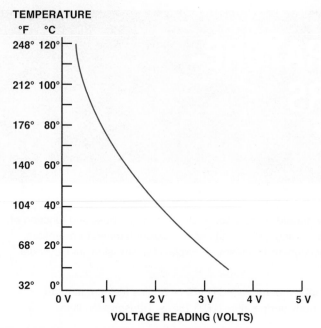

FIGURE 13–2 A typical ECT sensor temperature versus voltage curve.

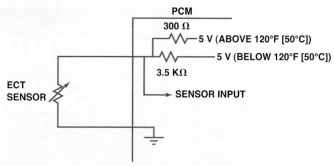

FIGURE 13–3 A typical two-step ECT circuit showing that when the coolant temperature is low, the PCM applies a 5-volt reference voltage to the ECT sensor through a higher resistance compared to when the temperature is higher.

- When the temperature is warm, usually above 120°F (50°C), the ECT sensor voltage is applied through a much lower resistance value inside the PCM. ●SEE FIGURE 13–3.

The purpose of this extra circuit is to give the PCM a more accurate reading of the engine coolant temperature compared to the same sensor with only one circuit. ●SEE FIGURE 13–4.

## TESTING THE ENGINE COOLANT TEMPERATURE SENSOR

**TESTING THE ENGINE COOLANT TEMPERATURE BY VISUAL INSPECTION** The correct functioning of the engine coolant temperature (ECT) sensor depends on the following items that should be checked or inspected:

- **Properly filled cooling system.** Check that the radiator reservoir bottle is full and that the radiator itself is filled to the top.

    CAUTION: Be sure that the radiator is cool before removing the radiator cap to avoid being scalded by hot coolant.

    The ECT sensor must be submerged in coolant to be able to indicate the proper coolant temperature.

- **Proper pressure maintained by the radiator cap.** If the radiator cap is defective and cannot allow the cooling system to become pressurized, air pockets could develop. These air pockets could cause the engine to operate at a hotter-than-normal temperature and prevent proper temperature measurement, especially if the air pockets occur around the sensor.

- **Proper antifreeze–water mixture.** Most vehicle manufacturers recommend a 50/50 mixture of antifreeze and water as the best compromise between freezing protection and heat transfer ability.

- **Proper operation of the cooling fan.** If the cooling fan does not operate correctly, the engine may overheat.

**TESTING THE ECT USING A MULTIMETER** Both the resistance (in ohms) and the voltage drop across the sensor can be measured and compared with specifications. ●SEE FIGURE 13–5. See the following charts showing examples of typical engine coolant temperature sensor specifications. Some vehicles use the PCM to attach another resistor in the ECT circuit to provide a more accurate measure of the engine temperature. ●SEE FIGURE 13–6.

If resistance values match the approximate coolant temperature and there is still a coolant sensor trouble code, the problem is generally in the wiring between the sensor and the computer. Always consult the manufacturer's recommended procedures for checking this wiring. If the resistance values do not match, the sensor may need to be replaced.

| General Motors ECT Sensor with Pull-Up Resistor | | | |
|---|---|---|---|
| °F | °C | Ohms | Voltage Drop Across Sensor |
| −40 | −40 | 100,000+ | 4.95 |
| 18 | −8 | 14,628 | 4.68 |
| 32 | 0 | 9,420 | 4.52 |
| 50 | 10 | 5,670 | 4.25 |
| 68 | 20 | 3,520 | 3.89 |
| 86 | 30 | 2,238 | 3.46 |
| 104 | 40 | 1,459 | 2.97 |
| 122 | 50 | 973 | 2.47 |
| 140 | 60 | 667 | 2.00 |
| 158 | 70 | 467 | 1.59 |
| 176 | 80 | 332 | 1.25 |
| 194 | 90 | 241 | 0.97 |
| 212 | 100 | 177 | 0.75 |

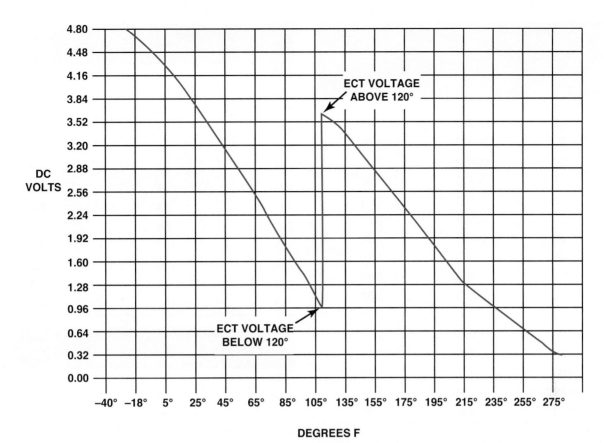

DC VOLTS

4.80
4.48
4.16
3.84
3.52
3.20
2.88
2.56
2.24
1.92
1.60
1.28
0.96
0.64
0.32
0.00

−40° −18° 5° 25° 45° 65° 85° 105° 135° 155° 175° 195° 215° 235° 255° 275°

DEGREES F

ECT VOLTAGE ABOVE 120°

ECT VOLTAGE BELOW 120°

**FIGURE 13–4** The transition between steps usually occurs at a temperature that would not interfere with cold engine starts or the cooling fan operation. In this example, the switch point between the two resistors occurs when the sensor voltage is about 1 volt and rises to about 3.6 volts.

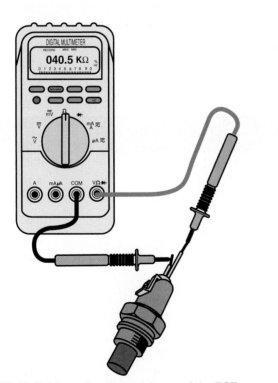

**FIGURE 13–5** Measuring the resistance of the ECT sensor. The resistance measurement can then be compared with specifications.

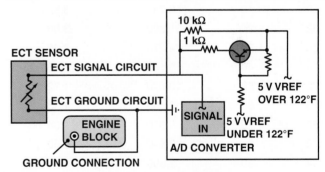

**FIGURE 13–6** When the voltage drop reaches approximately 1.2 volts, the PCM turns on a transistor. The transistor connects a 1-kΩ resistor in parallel with the 10-kΩ resistor. Total circuit resistance now drops to around 909 ohms. This function allows the PCM to have full control at cold temperatures up to approximately 122°F and a second full control at temperatures greater than 122°F.

| General Motors ECT Sensor Without Pull-Up Resistor | | | |
|---|---|---|---|
| °F | °C | Ohms | Voltage Drop Across Sensor |
| −40 | −40 | 100,000 | 5 |
| −22 | −30 | 53,000 | 4.78 |
| −4 | −20 | 29,000 | 4.34 |
| 14 | −10 | 16,000 | 3.89 |
| 32 | 0 | 9,400 | 3.45 |
| 50 | 10 | 5,700 | 3.01 |
| 68 | 20 | 3,500 | 2.56 |
| 86 | 30 | 2,200 | 1.80 |
| 104 | 40 | 1,500 | 1.10 |
| 122 | 50 | 970 | 3.25 |
| 140 | 60 | 670 | 2.88 |
| 158 | 70 | 470 | 2.56 |
| 176 | 80 | 330 | 2.24 |
| 194 | 90 | 240 | 1.70 |
| 212 | 100 | 177 | 1.42 |
| 230 | 110 | 132 | 1.15 |
| 248 | 120 | 100 | .87 |

| Ford ECT Sensor | | | |
|---|---|---|---|
| °F | °C | Resistance (Ω) | Voltage (V) |
| 50 | 10 | 58,750 | 3.52 |
| 68 | 20 | 37,300 | 3.06 |
| 86 | 30 | 24,270 | 2.26 |
| 104 | 40 | 16,150 | 2.16 |
| 122 | 50 | 10,970 | 1.72 |
| 140 | 60 | 7,600 | 1.35 |
| 158 | 70 | 5,370 | 1.04 |
| 176 | 80 | 3,840 | 0.80 |
| 194 | 90 | 2,800 | 0.61 |
| 212 | 100 | 2,070 | 0.47 |
| 230 | 110 | 1,550 | 0.36 |
| 248 | 120 | 1,180 | 0.28 |

| Chrysler ECT Sensor Without Pull-Up Resistor | | |
|---|---|---|
| °F | °C | Voltage (V) |
| 130 | 54 | 3.77 |
| 140 | 60 | 3.60 |
| 150 | 66 | 3.40 |
| 160 | 71 | 3.20 |
| 170 | 77 | 3.02 |
| 180 | 82 | 2.80 |
| 190 | 88 | 2.60 |
| 200 | 93 | 2.40 |
| 210 | 99 | 2.20 |
| 220 | 104 | 2.00 |
| 230 | 110 | 1.80 |
| 240 | 116 | 1.62 |
| 250 | 121 | 1.45 |

| Chrysler ECT Sensor with Pull-Up Resistor | | |
|---|---|---|
| °F | °C | Volts |
| −20 | −29 | 4.70 |
| −10 | −23 | 4.57 |
| 0 | −18 | 4.45 |
| 10 | −12 | 4.30 |
| 20 | −7 | 4.10 |
| 30 | −1 | 3.90 |
| 40 | 4 | 3.60 |
| 50 | 10 | 3.30 |
| 60 | 16 | 3.00 |
| 70 | 21 | 2.75 |
| 80 | 27 | 2.44 |
| 90 | 32 | 2.15 |
| 100 | 38 | 1.83 |

| | | Pull-Up Resistor Switched by PCM |
|---|---|---|
| 110 | 43 | 4.20 |
| 120 | 49 | 4.10 |
| 130 | 54 | 4.00 |
| 140 | 60 | 3.60 |
| 150 | 66 | 3.40 |
| 160 | 71 | 3.20 |
| 170 | 77 | 3.02 |
| 180 | 82 | 2.80 |
| 190 | 88 | 2.60 |
| 200 | 93 | 2.40 |
| 210 | 99 | 2.20 |
| 220 | 104 | 2.00 |
| 230 | 110 | 1.80 |
| 240 | 116 | 1.62 |
| 250 | 121 | 1.45 |

| Nissan ECT Sensor | | |
| --- | --- | --- |
| °F | °C | Resistance (Ω) |
| 14 | −10 | 7,000–11,400 |
| 68 | 20 | 2,100–2,900 |
| 122 | 50 | 680–1,000 |
| 176 | 80 | 260–390 |
| 212 | 100 | 180–200 |

| Mercedes ECT | | |
| --- | --- | --- |
| °F | °C | Voltage (DCV) |
| 60 | 20 | 3.5 |
| 86 | 30 | 3.1 |
| 104 | 40 | 2.7 |
| 122 | 50 | 2.3 |
| 140 | 60 | 1.9 |
| 158 | 70 | 1.5 |
| 176 | 80 | 1.2 |
| 194 | 90 | 1.0 |
| 212 | 100 | 0.8 |

| European Bosch ECT Sensor | | |
| --- | --- | --- |
| °F | °C | Resistance (Ω) |
| 32 | 0 | 6,500 |
| 50 | 10 | 4,000 |
| 68 | 20 | 3,000 |
| 86 | 30 | 2,000 |
| 104 | 40 | 1,500 |
| 122 | 50 | 900 |
| 140 | 60 | 650 |
| 158 | 70 | 500 |
| 176 | 80 | 375 |
| 194 | 90 | 295 |
| 212 | 100 | 230 |

| Honda ECT Sensor (Resistance Chart) | | |
| --- | --- | --- |
| °F | °C | Resistance (Ω) |
| 0 | −18 | 15,000 |
| 32 | 0 | 5,000 |
| 68 | 20 | 3,000 |
| 104 | 40 | 1,000 |
| 140 | 60 | 500 |
| 176 | 80 | 400 |
| 212 | 100 | 250 |

| Honda ECT Sensor (Voltage Chart) | | |
| --- | --- | --- |
| °F | °C | Voltage (V) |
| 0 | −18 | 4.70 |
| 10 | −12 | 4.50 |
| 20 | −7 | 4.29 |
| 30 | −1 | 4.10 |
| 40 | 4 | 3.86 |
| 50 | 10 | 3.61 |
| 60 | 16 | 3.35 |
| 70 | 21 | 3.08 |
| 80 | 27 | 2.81 |
| 90 | 32 | 2.50 |
| 100 | 38 | 2.26 |
| 110 | 43 | 2.00 |
| 120 | 49 | 1.74 |
| 130 | 54 | 1.52 |
| 140 | 60 | 1.33 |
| 150 | 66 | 1.15 |
| 160 | 71 | 1.00 |
| 170 | 77 | 0.88 |
| 180 | 82 | 0.74 |
| 190 | 88 | 0.64 |
| 200 | 93 | 0.55 |
| 210 | 99 | 0.47 |

Normal operating temperature varies with vehicle make and model. Some vehicles are equipped with a thermostat with an opening temperature of 180°F (82°C), whereas other vehicles use a thermostat that is 195°F (90°C) or higher. Before replacing the ECT sensor, be sure that the engine is operating at the temperature specified by the manufacturer. Most manufacturers recommend checking the ECT sensor after the cooling fan has cycled twice, indicating a fully warmed engine. To test for voltage at the ECT sensor, select DC volts on a digital meter and carefully back probe the sensor wire and read the voltage. ● SEE FIGURE 13–7.

NOTE: Many manufacturers install another resistor in parallel inside the computer to change the voltage drop across the ECT sensor. This is done to expand the scale of the ECT sensor and to make the sensor more sensitive. Therefore, if measuring *voltage* at the ECT sensor, check with the service manual for the proper voltage at each temperature.

## TESTING THE ECT SENSOR USING A SCAN TOOL
Follow the scan tool manufacturer's instructions and connect a scan tool to the data link connector (DLC) of the vehicle. Comparing the temperature of the engine coolant as displayed on a scan tool with the actual temperature of the engine is an excellent method to test an engine coolant temperature sensor:

| REMARKS: | ECT Voltage |
| | 2001 Jeep Wrangler Warm-up Cycle |
| | AUTO 202 - Fuel and Emissions Systems |

| FORM SAVED TIME: | 2/18/04 4:11:55 PM |
| UPLOAD TIME: | 2/18/04 4:09:05 PM |
| METER ID: | FLUKE 189 V2.02 0085510089 |

**SHOW DATA: ALL GRAPH VIEW: ALL**

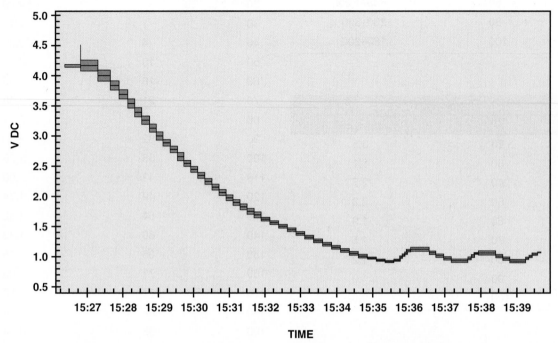

TIME

**FIGURE 13–7** An ECT sensor being tested using a digital meter set to DC volts and in record mode. A chart showing the voltage decrease of the ECT sensor as the temperature increases from a cold start. The bumps at the bottom of the waveform represent temperature decreases when the thermostat opens and is controlling coolant temperature.

1. Record the scan tool temperature of the coolant (ECT).
2. Measure the actual temperature of the coolant using an infrared pyrometer or contact-type temperature probe.

**NOTE: Often the coolant temperature gauge in the dash of the vehicle can be used to compare with the scan tool temperature. Although not necessarily accurate, it may help to diagnose a faulty sensor, especially if the temperature shown on the scan tool varies greatly from the temperature indicated on the dash gauge.**

The maximum difference between the two readings should be 10°F (5°C). If the actual temperature varies by more than 10°F from the temperature indicated on the scan tool, check the ECT sensor wiring and connector for damage or corrosion. If the connector and wiring are okay, check the sensor with a DVOM for resistance and compare to the actual engine temperature chart. If that checks out okay, check the computer.

**NOTE: Some manufacturers use two coolant sensors, one for the dash gauge and another one for the computer.**

## INTAKE AIR TEMPERATURE SENSOR

**PURPOSE AND FUNCTION** The intake air temperature (IAT) sensor is a negative temperature coefficient (NTC) thermistor that decreases in resistance as the temperature of the sensor increases. The IAT sensor can be located in one of the following locations:

- In the air cleaner housing
- In the air duct between the air filter and the throttle body, as shown in ●**FIGURE 13–8**
- Built into the mass airflow (MAF) or airflow sensor
- Screwed into the intake manifold, where it senses the temperature of the air entering the cylinders

**NOTE: An IAT installed in the intake manifold is the most likely to suffer damage because of an engine backfire, which can often destroy the sensor.**

The purpose and function of the intake air temperature sensor is to provide the engine computer (PCM) the temperature

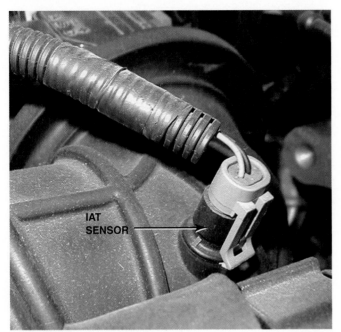

FIGURE 13–8 The IAT sensor on this General Motors 3800 V-6 engine is in the air passage duct between the air cleaner housing and the throttle body.

of the air entering the engine. The IAT sensor information is used for fuel control (adding or subtracting fuel) and spark timing, depending on the temperature of incoming air:

- If the air temperature is cold, the PCM will modify the amount of fuel delivery and add fuel.

- If the air temperature is hot, the PCM will subtract the calculated amount of fuel.

- Spark timing is also changed, depending on the temperature of the air entering the engine. The timing is advanced if the temperature is cold and retarded from the base-programmed timing if the temperature is hot.

- Cold air is more dense, contains more oxygen, and therefore requires a richer mixture to achieve the proper air–fuel mixture. Air at 32°F (0°C) is 14% denser than air at 100°F (38°C).

- Hot air is less dense, contains less oxygen, and therefore requires less fuel to achieve the proper air–fuel mixture.

The IAT sensor is a low-authority sensor and is used by the computer to modify the amount of fuel and ignition timing as determined by the engine coolant temperature sensor.

The IAT sensor is used by the PCM as a backup in the event that the ECT sensor is determined to be inoperative.

NOTE: Some engines use a **throttle-body temperature (TBT) sensor to sense the temperature of the air entering the engine, instead of an intake air temperature sensor.**

Engine temperature is most accurately determined by looking at the engine coolant temperature (ECT) sensor. In certain conditions, the IAT has an effect on performance and drive-

TECH TIP

**Quick and Easy ECT Test**

To check that the wiring and the computer are functioning, regarding the ECT sensor, connect a scan tool and look at the ECT temperature display:

STEP 1    Unplug the connector from the ECT sensor. The temperature displayed on the scan tool should read about −40.

**NOTE: −40° Celsius is also −40° Fahrenheit. This is the point where both temperature scales meet.**

STEP 2    With the connector still removed from the ECT sensor, use a fused jumper lead and connect the two terminals of the connector together. The scan tool should display about 285°F (140°C).

This same test procedure will work for the IAT and most other temperature sensors.

TECH TIP

**Poor Fuel Economy? Black Exhaust Smoke? Look at the IAT**

If the intake air temperature sensor is defective, it may be signaling the computer that the intake air temperature is extremely cold when in fact it is warm. In such a case the computer will supply a mixture that is much richer than normal.

If a sensor is physically damaged or electrically open, the computer will often set a diagnostic trouble code (DTC). This DTC is based on the fact that the sensor temperature did not change for a certain amount of time, usually about eight minutes. If, however, the wiring or the sensor itself has excessive resistance, a DTC will not be set and the result will be lower-than-normal fuel economy and, in serious cases, black exhaust smoke from the tailpipe during acceleration.

ability. One such condition is a warm engine being stopped in very cold weather. In this case, when the engine is restarted, the ECT may be near normal operating temperature such as 200°F (93°C), yet the air temperature could be −20°F (−30°C). In this case, the engine requires a richer mixture because of the cold air than the ECT would seem to indicate.

## TESTING THE INTAKE AIR TEMPERATURE SENSOR

If the intake air temperature sensor circuit is damaged or faulty, a diagnostic trouble code (DTC) is set, and the malfunction indicator lamp (MIL) may or may not turn on, depending on the condition and the type and model of the vehicle. To diagnose the IAT sensor follow these steps:

**STEP 1** After the vehicle has been allowed to cool for several hours, use a scan tool, observe the IAT, and compare it to the engine coolant temperature (ECT). The two temperatures should be within 5°F of each other.

**STEP 2** Perform a thorough visual inspection of the sensor and the wiring. If the IAT is screwed into the intake manifold, remove the sensor and check for damage.

**STEP 3** Check the voltage and compare to the following chart.

| Intake Air Temperature Sensor Temperature Versus Resistance and Voltage Drop (Approximate) | | | |
|---|---|---|---|
| °F | °C | Ohms | Voltage Drop Across the Sensor |
| −40 | −40 | 100,000 | 4.95 |
| +18 | −8 | 15,000 | 4.68 |
| 32 | 0 | 9,400 | 4.52 |
| 50 | 10 | 5,700 | 4.25 |
| 68 | 20 | 3,500 | 3.89 |
| 86 | 30 | 2,200 | 3.46 |
| 104 | 40 | 1,500 | 2.97 |
| 122 | 50 | 1,000 | 2.47 |
| 140 | 60 | 700 | 2.00 |
| 158 | 70 | 500 | 1.59 |
| 176 | 80 | 300 | 1.25 |
| 194 | 90 | 250 | 0.97 |
| 212 | 100 | 200 | 0.75 |

## TRANSMISSION FLUID TEMPERATURE SENSOR

The **transmission fluid temperature (TFT),** also called *transmission oil temperature (TOT),* sensor is an important sensor for the proper operation of the automatic transmission. A TFT sensor is a negative temperature coefficient (NTC) thermistor that decreases in resistance as the temperature of the sensor increases.

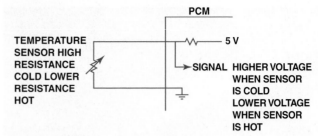

**FIGURE 13–9** A typical temperature sensor circuit.

### ? FREQUENTLY ASKED QUESTION

**What Exactly Is an NTC Sensor?**

A negative temperature coefficient (NTC) thermistor is a semiconductor whose resistance decreases as the temperature increases. In other words, the sensor becomes more electrically conductive as the temperature increases. Therefore, when a voltage is applied, typically 5 volts, the signal voltage is high when the sensor is cold because the sensor has a high resistance and little current flows through to ground. ● **SEE FIGURE 13–9.**

However, when the temperature increases, the sensor becomes more electrically conductive and takes more of the 5 volts to ground, resulting in a lower signal voltage as the sensor warms.

| General Motors Transaxle Sensor—Temperature to Resistance (approximate) | | |
|---|---|---|
| °F | °C | Resistance Ohms |
| 32 | 0 | 7,987–10,859 |
| 50 | 10 | 4,934–6,407 |
| 68 | 20 | 3,106–3,923 |
| 86 | 30 | 1,991–2,483 |
| 104 | 40 | 1,307–1,611 |
| 122 | 50 | 878–1,067 |
| 140 | 60 | 605–728 |
| 158 | 70 | 425–507 |
| 176 | 80 | 304–359 |
| 194 | 90 | 221–259 |
| 212 | 100 | 163–190 |

| Chrysler<br>Sensor Resistance (Ohms)—Transmission<br>Temperature Sensor | | |
|---|---|---|
| °F | °C | Resistance Ohms |
| −40 | −40 | 291,490–381,710 |
| −4 | −20 | 85,850–108,390 |
| 14 | −10 | 49,250–61,430 |
| 32 | 0 | 29,330–35,990 |
| 50 | 10 | 17,990–21,810 |
| 68 | 20 | 11,370–13,610 |
| 77 | 25 | 9,120–10,880 |
| 86 | 30 | 7,370–8,750 |
| 104 | 40 | 4,900–5,750 |
| 122 | 50 | 3,330–3,880 |
| 140 | 60 | 2,310–2,670 |
| 158 | 70 | 1,630–1,870 |
| 176 | 80 | 1,170–1,340 |
| 194 | 90 | 860–970 |
| 212 | 100 | 640–720 |
| 230 | 110 | 480–540 |
| 248 | 120 | 370–410 |

| Ford<br>Transmission Fluid Temperature | | |
|---|---|---|
| °F | °C | Resistance Ohms |
| −40 to −4 | −40 to −20 | 967K–284K |
| −3 to 31 | −19 to −1 | 284K–100K |
| 32 to 68 | 0 to 20 | 100K–37K |
| 69 to 104 | 21 to 40 | 37K–16K |
| 105 to 158 | 41 to 70 | 16K–5K |
| 159 to 194 | 71 to 90 | 5K–2.7K |
| 195 to 230 | 91 to 110 | 2.7K–1.5K |
| 231 to 266 | 111 to 130 | 1.5K–0.8K |
| 267 to 302 | 131 to 150 | 0.8K–0.54K |

The transmission fluid temperature signal is used by the powertrain control module (PCM) to perform certain strategies based on the temperature of the automatic transmission fluid. For example:

- If the temperature of the automatic transmission fluid is low (typically below 32°F [0°C]), the shift points may be delayed and overdrive disabled. The torque converter clutch also may not be applied to assist in the heating of the fluid.
- If the temperature of the automatic transmission fluid is high (typically above 260°F [130°C]), the overdrive is disabled and the torque converter clutch is applied to help reduce the temperature of the fluid.

NOTE: Check service information for the exact shift strategy based on high and low transmission fluid temperatures for the vehicle being serviced.

# CYLINDER HEAD TEMPERATURE SENSOR

Some vehicles are equipped with **cylinder head temperature (CHT)** sensors:

**VW Golf**

$$14°F (−10°C) = 11,600 \ \Omega$$

$$68°F (20°C) = 2,900 \ \Omega$$

$$176°F (80°C) = 390 \ \Omega$$

# ENGINE FUEL TEMPERATURE (EFT) SENSOR

Some vehicles, such as many Ford vehicles that are equipped with an electronic returnless type of fuel injection, use an **engine fuel temperature (EFT)** sensor to give the PCM information regarding the temperature and, therefore, the density of the fuel.

# EXHAUST GAS RECIRCULATION (EGR) TEMPERATURE SENSOR

Some engines, such as Toyota, are equipped with exhaust gas recirculation (EGR) temperature sensors. EGR is a well-established method for reduction of $NO_x$ emissions in internal combustion engines. The exhaust gas contains unburned hydrocarbons, which are recirculated in the combustion process. Recirculation is controlled by valves, which operate as a function of exhaust gas speed, load, and temperature. The gas reaches a temperature of about 850°F (450°C) for which a special heavy-duty glass-encapsulated NTC sensor is available.

The PCM monitors the temperature in the exhaust passage between the EGR valve and the intake manifold. If the temperature increases when the EGR is commanded on, the PCM can determine that the valve or related components are functioning.

## ENGINE OIL TEMPERATURE SENSOR

Engine oil temperature sensors are used on many General Motors vehicles and are used as an input to the oil life monitoring system. The computer program inside the PCM calculates engine oil life based on run time, engine RPM, and oil temperature.

## TEMPERATURE SENSOR DIAGNOSTIC TROUBLE CODES

The OBD-II diagnostic trouble codes that relate to temperature sensors include both high- and low-voltage codes as well as intermittent codes.

| Diagnostic Trouble Code | Description | Possible Causes |
|---|---|---|
| P0112 | IAT sensor low voltage | • IAT sensor internally shorted-to-ground<br>• IAT sensor wiring shorted-to-ground<br>• IAT sensor damaged by backfire (usually associated with IAT sensors that are mounted in the intake manifold)<br>• Possible defective PCM |
| P0113 | IAT sensor high voltage | • IAT sensor internally (electrically) open<br>• IAT sensor signal, circuit, or ground circuit open<br>• Possible defective PCM |
| P0117 | ECT sensor low voltage | • ECT sensor internally shorted-to-ground<br>• The ECT sensor circuit wiring shorted-to-ground<br>• Possible defective PCM |
| P0118 | ECT sensor high voltage | • ECT sensor internally (electrically) open<br>• ECT sensor signal, circuit, or ground circuit open<br>• Engine operating in an overheated condition<br>• Possible defective PCM |

## SUMMARY

1. The ECT sensor is a high-authority sensor at engine start-up and is used for closed-loop control as well as idle speed.
2. All temperature sensors decrease in resistance as the temperature increases. This is called negative temperature coefficient (NTC).
3. The ECT and IAT sensors can be tested visually as well as by using a digital multimeter or a scan tool.
4. Some vehicle manufacturers use a stepped ECT circuit inside the PCM to broaden the accuracy of the sensor.
5. Other temperature sensors include transmission fluid temperature (TFT), engine fuel temperature (EFT), exhaust gas recirculation (EGR) temperature, and engine oil temperature.

## REVIEW QUESTIONS

1. How does a typical NTC temperature sensor work?
2. What is the difference between a stepped and a nonstepped ECT circuit?
3. What temperature should be displayed on a scan tool if the ECT sensor is unplugged with the key on, engine off?
4. What are the three ways that temperature sensors can be tested?
5. If the transmission fluid temperature (TFT) sensor were to fail open (as if it were unplugged), what would the PCM do to the transmission shifting points?

1. The sensor that most determines fuel delivery when a fuel-injected engine is first started is the _____.
   a. O2S
   b. ECT sensor
   c. Engine MAP sensor
   d. IAT sensor

2. What happens to the voltage measured at the ECT sensor when the thermostat opens?
   a. Increases slightly
   b. Increases about 1 volt
   c. Decreases slightly
   d. Decreases about 1 volt

3. Two technicians are discussing a stepped ECT circuit. Technician A says that the sensor used for a stepped circuit is different than one used in a nonstepped circuit. Technician B says that a stepped ECT circuit uses different internal resistance inside the PCM. Which technician is correct?
   a. Technician A only
   b. Technician B only
   c. Both Technicians A and B
   d. Neither Technician A nor B

4. When testing an ECT sensor on a vehicle, a digital multimeter can be used and the signal wire tested with the connector attached the ignition on (engine off). What setting should the technician use to test the sensor?
   a. AC volts
   b. DC volts
   c. Ohms
   d. Hz (hertz)

5. When testing the ECT sensor with the connector disconnected, the technician should select what position on the DMM?
   a. AC volts
   b. DC volts
   c. Ohms
   d. Hz (hertz)

6. When checking the ECT sensor with a scan tool, about what temperature should be displayed if the connector is removed from the sensor with the key on, engine off?
   a. 284°F (140°C)
   b. 230°F (110°C)
   c. 120°F (50°C)
   d. −40°F (−40°C)

7. Two technicians are discussing the IAT sensor. Technician A says that the IAT sensor is more important to the operation of the engine (higher authority) than the ECT sensor. Technician B says that the PCM will add fuel if the IAT indicates that the incoming air temperature is cold. Which technician is correct?
   a. Technician A only
   b. Technician B only
   c. Both Technicians A and B
   d. Neither Technician A nor B

8. A typical IAT or ECT sensor reads about 3,000 ohms when tested using a DMM. This resistance represents a temperature of about _____.
   a. −40°F (−40°C)
   b. 70°F (20°C)
   c. 120°F (50°C)
   d. 284°F (140°C)

9. If the transmission fluid temperature (TFT) sensor indicates cold automatic transmission fluid temperature, what would the PCM do to the shifts?
   a. Normal shifts and normal operation of the torque converter clutch
   b. Disable torque converter clutch; normal shift points
   c. Delayed shift points and torque converter clutch disabled
   d. Normal shifts, but overdrive will be disabled

10. A P0118 DTC is being discussed. Technician A says that the ECT sensor could be shorted internally. Technician B says that the signal wire could be open. Which technician is correct?
    a. Technician A only
    b. Technician B only
    c. Both Technicians A and B
    d. Neither Technician A nor B

# THROTTLE POSITION SENSORS

**LEARNING OBJECTIVES:** **After studying this chapter, the reader should be able to:** • Discuss the purpose and function of throttle position (TP) sensors. • Describe the powertrain control module (PCM) uses for the TP sensor. • Describe how to test the TP sensor and interpret the TP sensor diagnostic trouble codes.

**KEY TERMS:** Potentiometer 190 • Skewed 193 • Throttle position (TP) sensor 190

## THROTTLE POSITION SENSOR CONSTRUCTION

Most computer-equipped engines use a **throttle position (TP) sensor** to signal to the computer the position of the throttle. ● **SEE FIGURE 14–1.** The TP sensor consists of a **potentiometer,** a type of variable resistor.

**POTENTIOMETERS** A potentiometer is a variable-resistance sensor with three terminals. One end of the resistor receives reference voltage, while the other end is grounded. The third terminal is attached to a movable contact that slides across the resistor to vary its resistance. Depending on whether the contact is near the supply end or the ground end of the resistor, return voltage is high or low. ● **SEE FIGURE 14–2.**

Throttle position sensors are among the most common potentiometer-type sensors. The computer uses their input to determine the amount of throttle opening and the rate of change.

A typical sensor has three wires:

- A 5-volt reference feed wire from the computer
- Signal return (a ground wire back to the computer)
- A voltage signal wire back to the computer; as the throttle is opened, the voltage to the computer changes

Normal throttle position voltage on most vehicles is about 0.5 volt at idle (closed throttle) and 4.5 volts at wide-open throttle (WOT).

**NOTE: The TP sensor voltage at idle is usually about 10% of the TP sensor voltage when the throttle is wide open but can vary from as low as 0.3 to 1.2 volts, depending on the make and model of vehicle.**

**FIGURE 14–1** A typical TP sensor mounted on the throttle plate of this port-injected engine.

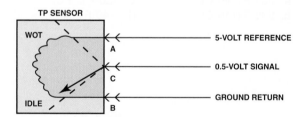

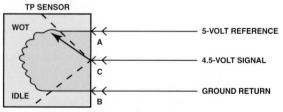

**FIGURE 14–2** The signal voltage from a throttle position increases as the throttle is opened because the wiper arm is closer to the 5-volt reference. At idle, the resistance of the sensor winding effectively reduces the signal voltage output to the computer.

# TP SENSOR COMPUTER INPUT FUNCTIONS

- The computer senses any change in throttle position and changes the fuel mixture and ignition timing. The actual change in fuel mixture and ignition timing is also partly determined by the other sensors, such as the manifold pressure (engine vacuum), engine RPM, the coolant temperature, and oxygen sensor(s). Some throttle position sensors are adjustable and should be set according to the exact engine manufacturer's specifications.

- The throttle position (TP) sensor used on fuel-injected vehicles acts as an "electronic accelerator pump." This means that the computer will pulse additional fuel from the injectors when the throttle is depressed. Because the air can quickly flow into the engine when the throttle is opened, additional fuel must be supplied to prevent the air–fuel mixture from going lean, causing the engine to hesitate when the throttle is depressed. If the TP sensor is unplugged or defective, the engine may still operate satisfactorily but hesitate upon acceleration.

- The PCM supplies the TP sensor with a regulated voltage that ranges from 4.8 to 5.1 volts. This reference voltage is usually referred to as a 5-volt reference or "Vref." The TP output signal is an input to the PCM, and the TP sensor ground also flows through the PCM.

See the Ford throttle position (TP) sensor chart for an example of how sensor voltage changes with throttle angle.

| Ford Throttle Position (TP) Sensor Chart | |
|---|---|
| Throttle Angle (Degrees) | Voltage (V) |
| 0 | 0.50 |
| 10 | 0.97 |
| 20 | 1.44 |
| 30 | 1.90 |
| 40 | 2.37 |
| 50 | 2.84 |
| 60 | 3.31 |
| 70 | 3.78 |
| 80 | 4.24 |

**NOTE: Generally, any reading higher than 80% represents wide-open throttle to the computer.**

# PCM USES FOR THE TP SENSOR

The TP sensor is used by the powertrain control module (PCM) for the following reasons.

**CLEAR FLOOD MODE** If the throttle is depressed to the floor during engine cranking, the PCM will either greatly reduce or entirely eliminate any fuel-injector pulses to aid in cleaning a flooded engine. If the throttle is depressed to the floor and the engine is not flooded with excessive fuel, the engine may not start.

**TORQUE CONVERTER CLUTCH ENGAGEMENT AND RELEASE** The torque converter clutch will be released if the PCM detects rapid acceleration to help the transmission deliver maximum torque to the drive wheels. The torque converter clutch is applied when the vehicle is lightly accelerating and during cruise conditions to improve fuel economy.

**RATIONALITY TESTING FOR MAP AND MAF SENSORS** As part of the rationality tests for the MAP and/or MAF sensor, the TP sensor signal is compared to the reading from other sensors to determine if they match. For example, if the throttle position sensor is showing wide-open throttle (WOT), the MAP and/or MAF reading should also indicate that this engine is under a heavy load. If not, a diagnostic trouble code could be set for the TP, as well as the MAP and/or MAF sensors.

**AUTOMATIC TRANSMISSION SHIFT POINTS** The shift points are delayed if the throttle is opened wide to allow the engine speed to increase, thereby producing more power and aiding in the acceleration of the vehicle. If the throttle is barely open, the shift point occurs at the minimum speed designed for the vehicle.

**TARGET IDLE SPEED (IDLE CONTROL STRATEGY)** When the TP sensor voltage is at idle, the PCM controls idle speed using the idle air control (IAC) and/or spark timing variation to maintain the commanded idle speed. If the TP sensor indicates that the throttle has moved off idle, fuel delivery and spark timing are programmed for acceleration. Therefore, if the throttle linkage is stuck or binding, the idle speed may not be correct.

**AIR-CONDITIONING COMPRESSOR OPERATION** The TP sensor is also used as an input sensor for air-conditioning compressor operation. If the PCM detects that the throttle is at or close to wide open, the air-conditioning compressor is disengaged.

**BACKS UP OTHER SENSORS** The TP sensor is used as a backup to the MAP sensor and/or MAF in the event the PCM detects that one or both are not functioning correctly. The PCM then calculates fuel needs and spark timing based on the engine speed (RPM) and throttle position.

**FIGURE 14–3** A meter lead connected to a T-pin that was gently pushed along the signal wire of the TP sensor until the point of the pin touched the metal terminal inside the plastic connector.

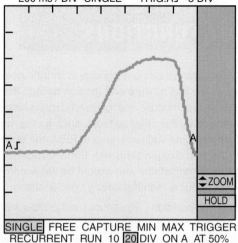

A 1 V DC 1:1 PROBE B 200 mV OFF 1:1 PROBE
200 ms / DIV SINGLE TRIG:A⌐ –3 DIV

SINGLE FREE CAPTURE MIN MAX TRIGGER
RECURRENT RUN 10 20 DIV ON A AT 50%

**FIGURE 14–4** A typical waveform of a TP sensor signal as recorded on a DSO when the accelerator pedal was depressed with the ignition switch on (engine off). Clean transitions and the lack of any glitches in this waveform indicate a good sensor.

# TESTING THE THROTTLE POSITION SENSOR

A TP sensor can be tested using one or more of the following tools:

- A digital voltmeter with three test leads connected in series between the sensor and the wiring harness connector or back probing using T-pins or other recommended tool that will not cause harm to the connector or wiring.
- A scan tool or a specific tool recommended by the vehicle manufacturer.
- A breakout box that is connected in series between the computer and the wiring harness connector(s). A typical breakout box includes test points at which TP voltages can be measured with a digital voltmeter.
- An oscilloscope.

Use jumper wires, T-pins to back-probe the wires, or a breakout box to gain electrical access to the wiring to the TP sensor. ● **SEE FIGURE 14–3.**

**NOTE: The procedure that follows is the method used by many manufacturers. Always refer to service information for the exact recommended procedure and specifications for the vehicle being tested.**

The procedure for testing the sensor using a digital multimeter is as follows:

1. Turn the ignition switch on (engine off).
2. Set the digital meter to read to DC volts and measure the voltage between the signal wire and ground (reference low) wire. The voltage should be about 0.5 volt.

**NOTE: Consult the service information for exact wire colors or locations.**

3. With the engine still not running (but with the ignition still on), slowly increase the throttle opening. The voltage signal from the TP sensor should also increase. Look for any "dead spots" or open circuit readings as the throttle is increased to the wide-open position. ● **SEE FIGURE 14–4** for an example of how a good TP sensor would look when tested with a digital storage oscilloscope (DSO).

**NOTE: Use the accelerator pedal to depress the throttle because this applies the same forces on the TP sensor as the driver does during normal driving. Moving the throttle by hand under the hood may not accurately test the TP sensor.**

4. With the voltmeter still connected, slowly return the throttle down to the idle position. The voltage from the TP sensor should also decrease evenly on the return to idle.

The TP sensor voltage at idle should be within the acceptable range as specified by the manufacturer. Some TP sensors can be adjusted by loosening their retaining screws and moving the sensor in relation to the throttle opening. This movement changes the output voltage of the sensor.

All TP sensors should also provide a smooth transition voltage reading from idle to WOT and back to idle. Replace the TP sensor if erratic voltage readings are obtained or if the correct setting at idle cannot be obtained.

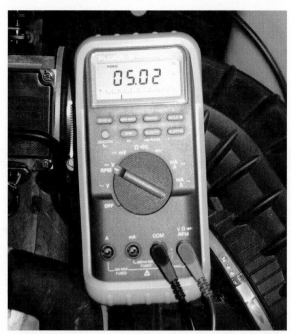

**FIGURE 14–5** Checking the 5-volt reference from the computer being applied to the TP sensor with the ignition switch on (engine off). The reading for this vehicle (5.02 volts DC) is within the normal range for the reference voltage of 4.9 to 5.1 volts.

 **TECH TIP**

**Check Power and Ground Before Condemning a Bad Sensor**

Most engine sensors use a 5-volt reference and a ground. If the 5 volts to the sensor is too high (shorted to voltage) or too low (high resistance), then the sensor output will be **skewed** or out of range. Before replacing the sensor that did not read correctly, measure both the 5-volt reference and ground. To measure the ground, simply turn the ignition on (engine off) and touch one test lead of a DMM set to read DC volts to the sensor ground and the other to the negative terminal of the battery. Any reading higher than 0.2 volt (200 mV) represents a poor ground. ● **SEE FIGURES 14–5 AND 14–6.**

# TESTING A TP SENSOR USING THE MIN/MAX FUNCTION

Many digital multimeters are capable of recording voltage readings over time and then displaying the minimum, maximum, and average readings. To perform a MIN/MAX test of the TP sensor, manually set the meter to read higher than 4 volts:

**FIGURE 14–6** Checking the voltage drop between the TP sensor ground and a good engine ground with the ignition on (engine off). A reading of greater than 0.2 volt (200 mV) represents a bad computer ground.

**STEP 1** Connect the red meter lead to the signal wire and the black meter lead to a good ground or the ground return wire at the TP sensor.

**STEP 2** With the ignition on, engine off, slowly depress and release the accelerator pedal from inside the vehicle.

**STEP 3** Check the minimum and maximum voltage reading on the meter display. Any 0-volt or 5-volt reading would indicate a fault or short in the TP sensor.

# TESTING THE TP SENSOR USING A SCAN TOOL

A scan tool can be used to check for proper operation of the throttle position sensor using the following steps:

**STEP 1** With the key on, engine off, the TP sensor voltage display should be about 0.5 volt but can vary from as low as 0.3 volt to as high as 1.2 volts.

**STEP 2** Check the scan tool display for the percentage of throttle opening. The reading should be zero and gradually increase in percentage as the throttle is depressed.

**STEP 3** The idle air control (IAC) counts should increase as the throttle is opened and decrease as the throttle is closed. Start the engine and observe the IAC counts as the throttle is depressed.

**STEP 4** Start the engine and observe the TP sensor reading. Use a wedge at the throttle stop to increase the throttle opening slightly. The throttle percentage reading should increase. Shut off and restart the engine. If the percentage of throttle opening returns to 0%, the PCM determines that the increased throttle opening is now the new minimum and resets the idle position of the TP sensor. Remove the wedge and cycle the ignition key. The throttle position sensor should again read 0%.

**NOTE:** Some engine computers are not capable of resetting the throttle position sensor.

# TP SENSOR DIAGNOSTIC TROUBLE CODES

The diagnostic trouble codes (DTCs) associated with the throttle position sensor include the following:

| Diagnostic Trouble Code | Description | Possible Causes |
|---|---|---|
| P0122 | TP sensor low voltage | • TP sensor internally shorted-to-ground<br>• TP sensor wiring shorted-to-ground<br>• TP sensor or wiring open |
| P0123 | TP sensor high voltage | • TP sensor internally shorted to 5-volt reference<br>• TP sensor ground open<br>• TP sensor wiring shorted-to-voltage |
| P0121 | TP sensor signal does not agree with MAP | • Defective TP sensor<br>• Incorrect vehicle-speed (VS) sensor signal<br>• MAP sensor out-of-calibration or defective |

## SUMMARY

1. A throttle position (TP) sensor is a three-wire variable resistor called a potentiometer.

2. The three wires on the TP sensor include a 5-volt reference voltage from the PCM, plus the signal wire to the PCM, and a ground, which also goes to the PCM.

3. The TP sensor is used by the PCM for clear flood mode, torque converter engagement and release, and automotive transmission shift points as well as rationality testing for the MAP and MAF sensors.

4. The TP sensor signal voltage should be about 0.5 volt at idle and increase to about 4.5 volts at wide-open throttle (WOT).

5. A TP sensor can be tested using a digital multimeter, a digital storage oscilloscope (DSO), or a scan tool.

## REVIEW QUESTIONS

1. What is the purpose of each of the three wires on a typical TP sensor?

2. What all does the PCM do with the TP sensor signal voltage?

3. What is the procedure to follow when checking the 5-volt reference and TP sensor ground?

4. How can a TP sensor be diagnosed using a scan tool?

## CHAPTER QUIZ

1. Which sensor is generally considered to be the electronic accelerator pump of a fuel-injected engine?
   a. O2S
   b. ECT sensor
   c. Engine MAP sensor
   d. TP sensor

2. Typical TP sensor voltage at idle is about _____.
   a. 2.5 to 2.8 volts
   b. 0.5 volt or 10% of WOT TP sensor voltage
   c. 1.5 to 2.8 volts
   d. 13.5 to 15 volts

3. A TP sensor is what type of sensor?
   a. Rheostat
   b. Voltage generating
   c. Potentiometer
   d. Piezoelectric

4. Most TP sensors have how many wires?
   a. One
   b. Two
   c. Three
   d. Four

5. Which sensor does the TP sensor back up if the PCM determines that a failure has occurred?
   a. Oxygen sensor
   c. MAP sensor
   b. MAF sensor
   d. Either b or c

6. Which wire on a TP sensor should be back-probed to check the voltage signal to the PCM?
   a. 5-volt reference (Vref)
   b. Signal
   c. Ground
   d. Meter should be connected between the 5-volt reference and the ground

7. After a TP sensor has been tested using the MIN/MAX function on a DMM, a reading of 0 volts is displayed. What does this reading indicate?
   a. The TP sensor is open at one point during the test.
   b. The TP sensor is shorted.
   c. The TP sensor signal is shorted to 5-volt reference.
   d. Both b and c are possible.

8. After a TP sensor has been tested using the MIN/MAX function on a DMM, a reading of 5 volts is displayed. What does this reading indicate?
   a. The TP sensor is open at one point during the test.
   b. The TP sensor is shorted.
   c. The TP sensor signal is shorted to 5-volt reference.
   d. Both b and c are possible.

9. A technician attaches one lead of a digital voltmeter to the ground terminal of the TP sensor and the other meter lead to the negative terminal of the battery. The ignition is switched to on, engine off, and the meter displays 37.3 mV. Technician A says that this is the signal voltage and is a little low. Technician B says that the TP sensor ground circuit has excessive resistance. Which technician is correct?
   a. Technician A only
   b. Technician B only
   c. Both Technicians A and B
   d. Neither Technician A nor B

10. A P0122 DTC is retrieved using a scan tool. This DTC means _____.
    a. The TP sensor voltage is low
    b. The TP sensor could be shorted-to-ground
    c. The TP sensor signal circuit could be shorted-to-ground
    d. All of the above

# chapter
# 15

# MAP/BARO SENSORS

**LEARNING OBJECTIVES:** **After studying this chapter, the reader should be able to:** • Discuss purpose and function of manifold absolute pressure (MAP) sensors. • Explain the PCM uses of MAP sensors. • Explain the purpose and function of barometric pressure (BARO) sensors. • List the methods that can be used to test MAP sensors.

**KEY TERMS:** Barometric manifold absolute pressure (BMAP) sensor 201 • Barometric pressure (BARO) sensor 201 • Manifold absolute pressure (MAP) sensor 196 • Piezoresistivity 198 • Pressure differential 196 • Speed density 199 • Vacuum 196

## AIR PRESSURE—HIGH AND LOW

Think of an internal combustion engine as a big air pump. As the pistons move up and down in the cylinders, they pump in air and fuel for combustion and pump out exhaust gases. They do this by creating a difference in air pressure. The air outside an engine has weight and exerts pressure, as does the air inside an engine.

As a piston moves down on an intake stroke with the intake valve open, it creates a larger area inside the cylinder for the air to fill. This lowers the air pressure within the engine. Because the pressure inside the engine is lower than the pressure outside, air flows into the engine to fill the low-pressure area and equalize the pressure.

The low pressure within the engine is called **vacuum.** Vacuum causes the higher-pressure air on the outside to flow into the low-pressure area inside the cylinder. The difference in pressure between the two areas is called a **pressure differential.** ● SEE FIGURE 15–1.

## PRINCIPLES OF PRESSURE SENSORS

Intake manifold pressure changes with changing throttle positions. At wide-open throttle, manifold pressure is almost the same as atmospheric pressure. On deceleration or at idle, manifold pressure is below atmospheric pressure, thus creating a vacuum. In cases where turbo- or supercharging is used, under part- or full-load condition, intake manifold pressure rises above

atmospheric pressure. Also, oxygen content and barometric pressure change with differences in altitude, and the computer must be able to compensate by making changes in the flow of fuel entering the engine. To provide the computer with changing airflow information, a fuel-injection system may use the following:

■ Manifold absolute pressure (MAP) sensor

■ Manifold absolute pressure (MAP) sensor plus barometric absolute pressure (BARO) sensor

■ Barometric and manifold absolute pressure sensors combined (BMAP)

The **manifold absolute pressure (MAP) sensor** may be a ceramic capacitor diaphragm, an aneroid bellows, or a piezoresistive crystal. It has a sealed vacuum reference input on one side; the other side is connected (vented) to the intake manifold. This sensor housing also contains signal conditioning circuitry. ● SEE FIGURE 15–2. Pressure changes in the manifold cause the sensor to deflect, varying its analog or digital return signal to the computer. As the air pressure increases, the MAP sensor generates a higher voltage or frequency return signal to the computer.

## CONSTRUCTION OF MAP SENSORS

The MAP sensor is used by the engine computer to sense engine load. The typical MAP sensor consists of a ceramic or silicon wafer sealed on one side with a perfect vacuum and exposed to intake manifold vacuum on the other side. As the

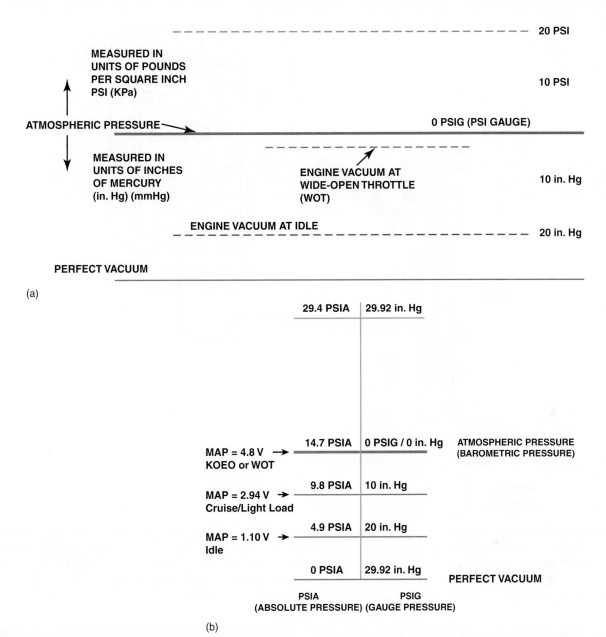

FIGURE 15–1 (a) As an engine is accelerated under a load, the engine vacuum drops. This drop in vacuum is actually an increase in absolute pressure in the intake manifold. A MAP sensor senses all pressures greater than that of a perfect vacuum. (b) The relationship between absolute pressure, vacuum, and gauge pressure.

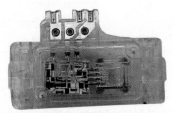

FIGURE 15–2 A clear plastic MAP sensor used for training purposes showing the electronic circuit board and electrical connections.

engine vacuum changes, the pressure difference on the wafer changes the output voltage or frequency of the MAP sensor.

A MAP sensor is used on many engines for the PCM to determine the load on the engine. The relationship among barometer pressure, engine vacuum, and MAP sensor voltage includes the following:

- Absolute pressure is equal to barometric pressure minus intake manifold vacuum.

- A decrease in manifold vacuum means an increase in manifold pressure.

- The MAP sensor compares manifold vacuum to a perfect vacuum.

- Barometric pressure minus MAP sensor reading equals intake manifold vacuum. Normal engine vacuum is 17 to 21 in. Hg.

- Supercharged and turbocharged engines require a MAP sensor that is calibrated for pressures above atmospheric as well as for vacuum.

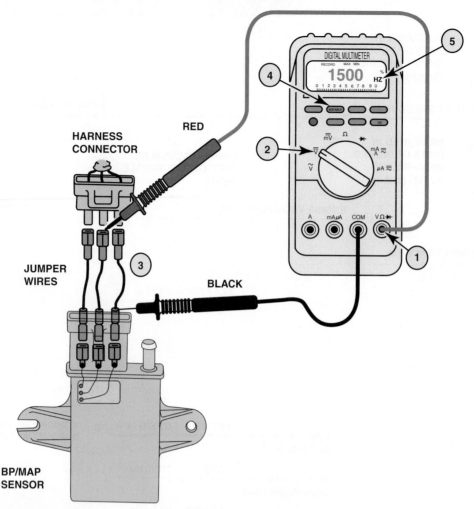

**FIGURE 15–3** MAP sensors use three wires: 1. 5-volt reference from the PCM. 2. Sensor signal (output signal). 3. Ground. A DMM set to test a MAP sensor. (1) Connect the red meter lead to the V meter terminal and the black meter lead to the COM meter terminal. (2) Select DC volts. (3) Connect the test leads to the sensor signal wire and the ground wire. (4) Select hertz (Hz) if testing a MAP sensor whose output is a varying frequency; otherwise, keep it on DC volts. (5) Read the change of voltage (frequency) as the vacuum is applied to the sensor. Compare the vacuum reading and the frequency (or voltage) reading to the specifications.

## SILICON-DIAPHRAGM STRAIN GAUGE MAP SENSOR

This is the most commonly used design for a MAP sensor, and the output is a DC analog (variable) voltage. One side of a silicon wafer is exposed to engine vacuum, and the other side is exposed to a perfect vacuum.

There are four resistors attached to the silicon wafer, which changes in resistance when strain is applied to the wafer. This change in resistance due to strain is called **piezoresistivity.** The resistors are electrically connected to a Wheatstone bridge circuit and then to a differential amplifier, which creates a voltage in proportion to the vacuum applied.

A typical General Motors MAP sensor voltage varies from 0.88 to 1.62 at engine idle:

- 17 in. Hg is equal to about 1.62 volts.
- 21 in. Hg is equal to about 0.88 volts.

Therefore, a good reading should be about 1 volt from the MAP sensor on a sound engine at idle speed. See the following chart that shows engine load, engine vacuum, and MAP.

| Engine Load | Manifold Vacuum | Manifold Absolute Pressure | MAP Sensor Volt Signal |
|---|---|---|---|
| Heavy (WOT) | Low (almost 0 in. Hg) | High (almost atmospheric) | High (4.6–4.8 V) |
| Light (idle) | High (17–21 in. Hg) | Low (lower than atmospheric) | Low (0.8–1.6 V) |

**CAPACITOR-CAPSULE MAP SENSOR** A capacitor-capsule is a type of MAP sensor used by Ford which uses two ceramic (alumina) plates with an insulating washer spacer in the center to create a capacitor. Changes in engine vacuum cause the plates to deflect, which changes the capacitance. The electronics in the sensor then generate a varying digital frequency output signal, which is proportional to the engine vacuum. ● **SEE FIGURE 15–3.** ● **SEE FIGURE 15–4** for a scope waveform of a digital MAP sensor. Also see the Ford MAP sensor chart.

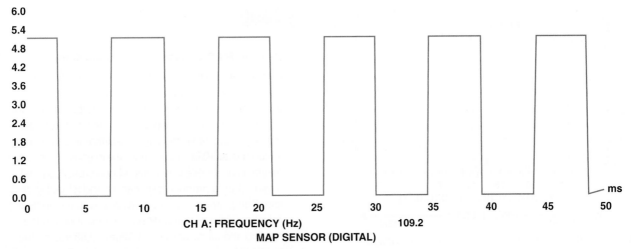

CH A: FREQUENCY (Hz)                                    109.2
MAP SENSOR (DIGITAL)

**FIGURE 15–4** A waveform of a typical digital MAP sensor.

**FIGURE 15–5** Shown is the electronic circuit inside a ceramic disc MAP sensor used on many Chrysler engines. The black areas are carbon resistors that are applied to the ceramic, and lasers are used to cut lines into these resistors during testing to achieve the proper operating calibration.

| Ford MAP Sensor Chart | | |
|---|---|---|
| **MAP Sensor Output** | **Engine Operating Conditions** | **Intake Manifold Vacuum** |
| 156–159 Hz | Key on, engine off | 0 in. Hg |
| 102–109 Hz | Engine at idle (sea level) | 17–21 in. Hg |
| 156–159 Hz | Engine at wide-open throttle (WOT) | About 0 in. Hg |

**CERAMIC DISC MAP SENSOR** The ceramic disc MAP sensor is used by Chrysler and it converts manifold pressure into a capacitance discharge. The discharge controls the amount of voltage delivered by the sensor to the PCM. The output is the same as the previously used strain gauge/Wheatstone bridge design and is interchangeable. ● **SEE FIGURE 15–5.** See the Chrysler MAP sensor chart.

  **TECH TIP**

**If It's Green, It's a Signal Wire**

Ford-built vehicles often use a green wire as the signal wire back to the computer from the sensors. It may not be a solid green, but if there is green somewhere on the wire, then it is the signal wire. The other wires are the power and ground wires to the sensor.

| Chrysler MAP Sensor Chart | |
|---|---|
| **Vacuum (in. Hg)** | **MAP Sensor Signal Voltage (V)** |
| 0.5 | 4.8 |
| 1.0 | 4.6 |
| 3.0 | 4.1 |
| 5.0 | 3.8 |
| 7.0 | 3.5 |
| 10.0 | 2.9 |
| 15.0 | 2.1 |
| 20.0 | 1.2 |
| 25.0 | 0.5 |

# PCM USES OF THE MAP SENSOR

The PCM uses the MAP sensor to determine the following:

■ **Load on the engine.** The MAP sensor is used on a **speed density**-type fuel-injection system to determine engine load, and therefore the amount of fuel needed. On engines equipped with a mass airflow (MAF) sensor, the

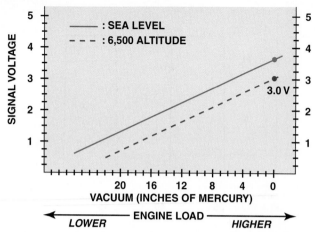

FIGURE 15–6 Altitude affects the MAP sensor voltage.

MAP is used as a backup to the MAF, for diagnosis of other sensors, and systems such as the EGR system.

- **Altitude, fuel, and spark control calculations.** At key on, the MAP sensor determines the altitude (acts as a BARO sensor) and adjusts the fuel delivery and spark timing accordingly:

  - If the altitude is high, generally over 5,000 feet (1,500 m), the PCM will reduce fuel delivery and advance the ignition timing.

  - The altitude is also reset when the engine is accelerated to wide-open throttle and the MAP sensor is used to reset the altitude reading. ● **SEE FIGURE 15–6.**

- **EGR system operation.** As part of the OBD-II standards, the exhaust gas recirculation (EGR) system must be checked for proper operation. One method used by many vehicle manufacturers is to command the EGR valve on and then watch the MAP sensor signal. The opening of the EGR pintle should decrease engine vacuum. If the MAP sensor does not react with the specified drop in manifold vacuum (increase in manifold pressure), an EGR flow rate problem diagnostic trouble code is set.

- **Detect deceleration (vacuum increases).** The engine vacuum rises when the accelerator is released, which changes the MAP sensor voltage. When deceleration is detected by the PCM, fuel is either stopped or greatly reduced to improve exhaust emissions.

- **Monitor engine condition.** As an engine wears, the intake manifold vacuum usually decreases. The PCM is programmed to detect the gradual change in vacuum and is able to keep the air–fuel mixture in the correct range. If the PCM were not capable of making adjustments for engine wear, the lower vacuum could be interpreted as increased load on the engine, resulting in too much

**TECH TIP**

### Use the MAP Sensor as a Vacuum Gauge

A MAP sensor measures the pressure inside the intake manifold compared with absolute zero (perfect vacuum). For example, an idling engine that has 20 in. Hg of vacuum has a lower pressure inside the intake manifold than when the engine is under a load and the vacuum is at 10 in. Hg. A decrease in engine vacuum results in an increase in manifold pressure. A normal engine should produce between 17 and 21 in. Hg at idle. Comparing the vacuum reading with the voltage reading output of the MAP sensor indicates that the reading should be between 1.62 and 0.88 volt or 109 to 102 Hz or lower on Ford MAP sensors. Therefore, a digital multimeter (DMM), scan tool, or scope can be used to measure the MAP sensor voltage and be used instead of a vacuum gauge.

**NOTE: This chart was developed by testing a MAP sensor at a location about 600 feet above sea level. For best results, a chart based on your altitude should be made by applying a known vacuum, and reading the voltage of a known-good MAP sensor. Vacuum usually drops about 1 inch per 1,000 feet of altitude.**

| Vacuum (in. Hg) | GM (DC volts) | Ford (Hz) |
|---|---|---|
| 0 | 4.80 | 156–159 |
| 1 | 4.52 | |
| 2 | 4.46 | |
| 3 | 4.26 | |
| 4 | 4.06 | |
| 5 | 3.88 | 141–143 |
| 6 | 3.66 | |
| 7 | 3.50 | |
| 8 | 3.30 | |
| 9 | 3.10 | |
| 10 | 2.94 | 127–130 |
| 11 | 2.76 | |
| 12 | 2.54 | |
| 13 | 2.36 | |
| 14 | 2.20 | |
| 15 | 2.00 | 114–117 |
| 16 | 1.80 | |
| 17 | 1.62 | |
| 18 | 1.42 | 108–109 |
| 19 | 1.20 | |
| 20 | 1.10 | 102–104 |
| 21 | 0.88 | |
| 22 | 0.66 | |

fuel being injected, thereby reducing fuel economy and increasing exhaust emissions.

- **Load detection for returnless-type fuel injection.** On fuel delivery systems that do not use a return line back to the fuel tank, the engine load calculation for the fuel needed is determined by the signals from the MAP sensor.

- **Altitude and MAP sensor values.** On an engine equipped with a speed density-type fuel injection, the MAP sensor is the most important sensor needed to determine injection pulse width. Changes in altitude change the air density as well as weather conditions. Barometric pressure and altitude are inversely related:

  - As altitude increases, barometric pressure decreases.
  - As altitude decreases, barometric pressure increases.

As the ignition switch is turned from off to the start position, the PCM reads the MAP sensor value to determine atmospheric and air pressure conditions. This barometric pressure reading is updated every time the engine is started and whenever wide-open throttle is detected. The barometric pressure reading at that time is updated. See the chart that compares altitude to MAP sensor voltage.

| Altitude and MAP Sensor Voltage | |
|---|---|
| Altitude | MAP Sensor Voltage (key on, engine off) |
| Sea level | 4.6–4.8 volts |
| 2,500 ft (760 m) | 4.0 volts |
| 5,000 ft (1,520 m) | 3.7 volts |
| 7,500 ft (2,300 m) | 3.35 volts |
| 10,000 ft (3,050 m) | 3.05 volts |
| 12,500 ft (3,800 m) | 2.80 volts |
| 15,000 ft (4,600 m) | 2.45 volts |

# BAROMETRIC PRESSURE SENSOR

A **barometric pressure (BARO) sensor** is similar in design, but senses more subtle changes in barometric absolute pressure (atmospheric air pressure). It is vented directly to the atmosphere. The **barometric manifold absolute pressure (BMAP) sensor** is actually a combination of a BARO and MAP sensor in the same housing. The BMAP sensor has individual circuits to measure barometric and manifold pressure. This input not

**The Cavalier Convertible Story**

The owner of a Cavalier convertible stated to a service technician that the "check engine" (MIL) was on. The technician found a diagnostic trouble code (DTC) for a MAP sensor. The technician removed the hose at the MAP sensor and discovered that gasoline had accumulated in the sensor and dripped out of the hose as it was being removed. The technician replaced the MAP sensor and test-drove the vehicle to confirm the repair. Almost at once the check engine light came on with the same MAP sensor code. After several hours of troubleshooting without success in determining the cause, the technician decided to start over again. Almost at once, the technician discovered that no vacuum was getting to the MAP sensor where a vacuum gauge was connected with a T-fitting in the vacuum line to the MAP sensor. The vacuum port in the base of the throttle body was clogged with carbon. After a thorough cleaning and clearing the DTC, the Cavalier again performed properly, and the check engine light did not come on again. The technician had assumed that if gasoline was able to reach the sensor through the vacuum hose, surely vacuum could reach the sensor. The technician learned to stop assuming when diagnosing a vehicle and concentrate more on testing the simple things first.

only allows the computer to adjust for changes in atmospheric pressure due to weather but also is the primary sensor used to determine altitude.

**NOTE: A MAP sensor and a BARO sensor are usually the same sensor, but the MAP sensor is connected to the manifold and a BARO sensor is open to the atmosphere. The MAP sensor is capable of reading barometric pressure just as the ignition switch is turned to the on position before the engine starts. Therefore, altitude and weather changes are available to the computer. During mountainous driving, it may be an advantage to stop and then restart the engine so that the engine computer can take another barometric pressure reading and recalibrate fuel delivery based on the new altitude. See the Ford/BARO altitude chart for an example of how altitude affects intake manifold pressure. The computer on some vehicles will monitor the throttle position sensor and use the MAP sensor reading at wide-open throttle (WOT) to update the BARO sensor if it has changed during driving.**

| Ford MAP/BARO Altitude Chart | |
|---|---|
| **Altitude (ft)** | **Volts (V)** |
| 0 | 1.59 |
| 1,000 | 1.56 |
| 2,000 | 1.53 |
| 3,000 | 1.50 |
| 4,000 | 1.47 |
| 5,000 | 1.44 |
| 6,000 | 1.41 |
| 7,000 | 1.39 |

**NOTE: Some older Chrysler brand vehicles were equipped with a combination BARO and IAT sensor. The sensor was mounted on the bulkhead (firewall) and sensed the underhood air temperature.**

# TESTING THE MAP SENSOR

Most pressure sensors operate on 5 volts from the computer and return a signal (voltage or frequency) based on the pressure (vacuum) applied to the sensor. If a MAP sensor is being tested, make certain that the vacuum hose and hose fittings are sound and making a good, tight connection to a manifold vacuum source on the engine.

Four different types of test instruments can be used to test a pressure sensor:

1. A digital voltmeter with three test leads connected in series between the sensor and the wiring harness connector or back-probe the terminals
2. A scope connected to the sensor output, power, and ground
3. A scan tool or a specific tool recommended by the vehicle manufacturer
4. A breakout box connected in series between the computer and the wiring harness connection(s) (A typical breakout box includes test points at which pressure sensor values can be measured with a digital voltmeter set on DC volts— or frequency counter, if a frequency-type MAP sensor is being tested.)

**NOTE: Always check service information for the exact testing procedures and specifications for the vehicle being tested.**

## TESTING THE MAP SENSOR USING A DMM OR SCOPE
Use jumper wires, T-pins to back-probe the connector, or a breakout box to gain electrical access to the wiring to the pressure sensor. Most pressure sensors use three wires:

1. A 5-volt wire from the computer
2. A variable-signal wire back to the computer
3. A ground or reference low wire

## TECH TIP

### Visual Check of the MAP Sensor
A defective vacuum hose to a MAP sensor can cause a variety of driveability problems including poor fuel economy, hesitation, stalling, and rough idle. A small air leak (vacuum leak) around the hose can cause these symptoms and often set a trouble code in the vehicle computer. When working on a vehicle that uses a MAP sensor, make certain that the vacuum hose travels consistently *downward* on its route from the sensor to the source of manifold vacuum. Inspect the hose, especially if another technician has previously replaced the factory-original hose. It should not be so long that it sags down at any point. Condensed fuel and/or moisture can become trapped in this low spot in the hose and cause all types of driveability problems and MAP sensor codes.

When checking the MAP sensor, if anything comes out of the sensor itself, it should be replaced. This includes water, gasoline, or any other substance.

The procedure for testing the sensor is as follows:

1. Turn the ignition on (engine off).
2. Measure the voltage (or frequency) of the sensor output.
3. Using a hand-operated vacuum pump (or other variable vacuum source), apply vacuum to the sensor.

A good pressure sensor should change voltage (or frequency) in relation to the applied vacuum. If the signal does not change or the values are out of range according to the manufacturer's specifications, the sensor must be replaced.

**TESTING THE MAP SENSOR USING A SCAN TOOL** A scan tool can be used to test a MAP sensor by monitoring the injector pulse width (in milliseconds) when vacuum is being applied to the MAP sensor using a hand-operated vacuum pump. ● SEE FIGURE 15–7.

**STEP 1** Apply about 20 in. Hg of vacuum to the MAP sensor and start the engine.

**STEP 2** Observe the injector pulse width. On a warm engine, the injector pulse width will normally be 1.5 to 3.5 ms.

**STEP 3** Slowly reduce the vacuum to the MAP sensor and observe the pulse width. A lower vacuum to the MAP sensor indicates a heavier load on the engine, and the injector pulse width should increase.

**NOTE: If 23 in. Hg or more vacuum is applied to the MAP sensor with the engine running, this high vacuum will often stall the engine. The engine stalls because the high vacuum is interpreted by the PCM to indicate that the engine is being decelerated, which shuts off the fuel. During engine deceleration, the PCM shuts off the fuel injectors to reduce exhaust emissions and increase fuel economy.**

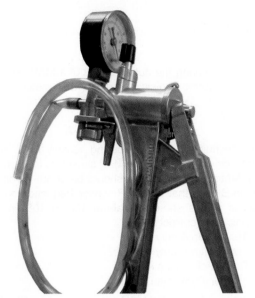

**FIGURE 15–7** A typical hand-operated vacuum pump.

## MAP/BARO DIAGNOSTIC TROUBLE CODES

The diagnostic trouble codes (DTCs) associated with the MAP and BARO sensors include the following:

| Diagnostic Trouble Code | Description | Possible Causes |
|---|---|---|
| P0106 | BARO sensor out-of-range at key on | • MAP sensor fault<br>• MAP sensor O-ring damaged or missing |
| P0107 | MAP sensor low voltage | • MAP sensor fault<br>• MAP sensor signal circuit shorted-to-ground<br>• MAP sensor 5-volt supply circuit open |
| P0108 | Map sensor high voltage | • MAP sensor fault<br>• MAP sensor O-ring damaged or missing<br>• MAP sensor signal circuit shorted-to-voltage |

## FUEL-RAIL PRESSURE SENSOR

A fuel-rail pressure (FRP) sensor is used on some vehicles such as Fords that are equipped with electronic returnless fuel injection. This sensor provides fuel pressure information to the PCM for fuel-injection pulse-width calculations.

## SUMMARY

1. Pressure below atmospheric pressure is called vacuum and is measured in inches of mercury.
2. A manifold absolute pressure sensor uses a perfect vacuum (zero absolute pressure) in the sensor to determine the pressure.
3. Three types of MAP sensors include the following:
   • Silicon-diaphragm strain gauge
   • Capacitor-capsule design
   • Ceramic disc design

4. A heavy engine load results in low intake manifold vacuum and a high MAP sensor signal voltage.
5. A light engine load results in high intake manifold vacuum and a low MAP sensor signal voltage.
6. A MAP sensor is used to detect changes in altitude as well as check other sensors and engine systems.
7. A MAP sensor can be tested by visual inspection, testing the output using a digital meter or scan tool.

## REVIEW QUESTIONS

1. What is the relationship among atmospheric pressure, vacuum, and boost pressure in PSI?
2. What are two types (construction) of MAP sensors?

3. What is the MAP sensor signal voltage or frequency at idle on a typical General Motors, Chrysler, and Ford engine?
4. What are three uses of a MAP sensor by the PCM?

1. As the load on an engine increases, the manifold vacuum decreases and the manifold absolute pressure _____.
   a. Increases
   b. Decreases
   c. Changes with barometric pressure only (altitude or weather)
   d. Remains constant (absolute)

2. A typical MAP sensor compares the vacuum in the intake manifold to _____.
   a. Atmospheric pressure
   b. A perfect vacuum
   c. Barometric pressure
   d. The value of the IAT sensor

3. Which statement is *false*?
   a. Absolute pressure is equal to barometric pressure plus intake manifold vacuum.
   b. A decrease in manifold vacuum means an increase in manifold pressure.
   c. The MAP sensor compares manifold vacuum to a perfect vacuum.
   d. Barometric pressure minus the MAP sensor reading equals intake manifold vacuum.

4. Which design of MAP sensor produces a frequency (digital) output signal?
   a. Silicon-diaphragm strain gauge
   b. Piezoresistivity design
   c. Capacitor-capsule
   d. Ceramic disc

5. The frequency output of a digital MAP sensor is reading 114 Hz. What is the approximate engine vacuum?
   a. Zero
   b. 5 in. Hg
   c. 10 in. Hg
   d. 15 in. Hg

6. Which is *not* a purpose or function of the MAP sensor?
   a. Measures the load on the engine
   b. Measures engine speed
   c. Calculates fuel delivery based on altitude
   d. Helps diagnose the EGR system

7. When measuring the output signal of a MAP sensor on a General Motors vehicle, the digital multimeter should be set to read _____.
   a. DC V
   b. AC V
   c. Hz
   d. DC A

8. Two technicians are discussing testing MAP sensors. Technician A says that the MAP sensor voltage on a General Motors vehicle at idle should be about 1 volt. Technician B says that the MAP sensor frequency on a Ford vehicle at idle should be about 105 to 108 Hz. Which technician is correct?
   a. Technician A only
   b. Technician B only
   c. Both Technicians A and B
   d. Neither Technician A nor B

9. Technician A says that MAP sensors use a 5-volt reference voltage from the PCM. Technician B says that the MAP sensor voltage will be higher at idle at high altitudes compared to when the engine is operating at near sea level. Which technician is correct?
   a. Technician A only
   b. Technician B only
   c. Both Technicians A and B
   d. Neither Technician A nor B

10. A P0107 DTC is being discussed. Technician A says that a defective MAP sensor could be the cause. Technician B says that a MAP sensor signal wire shorted-to-ground could be the cause. Which technician is correct?
    a. Technician A only
    b. Technician B only
    c. Both Technicians A and B
    d. Neither Technician A nor B

# chapter 16

# MASS AIRFLOW SENSORS

**LEARNING OBJECTIVES:** **After studying this chapter, the reader should be able to:** • Describe the purpose and function of mass airflow (MAF) sensors. • List the methods that can be used to test MAF sensors. • Discuss the symptoms of a failed MAF sensor.

**KEY TERMS:** False air 209 • Mass airflow (MAF) sensor 206 • Speed density 205 • Tap test 208 • Vane airflow (VAF) sensor 205

## AIRFLOW SENSORS

Electronic fuel injection systems that do not use the "speed density" system for fuel calculation measure the airflow volume delivered to the engine. Older systems use a movable vane in the intake stream called a vane airflow (VAF) sensor. The vane is part of the **vane airflow (VAF) sensor.** The vane is deflected by intake airflow. ● **SEE FIGURE 16–1.**

The VAF sensor used in Bosch L-Jetronic, Ford, and most Japanese electronic port fuel-injection systems is a movable vane connected to a laser-calibrated potentiometer. The vane is mounted on a pivot pin and is deflected by intake airflow proportionate to air velocity. As the vane moves, it also moves the potentiometer. This causes a change in the signal voltage supplied to the computer. ● **SEE FIGURE 16–2.** For example, if the reference voltage is 5 volts, the potentiometer's signal to the computer will vary from a zero voltage signal (no airflow) to almost a 5-volt signal (maximum airflow). In this way, the potentiometer provides the information the computer needs to vary the injector pulse width proportionate to airflow. There is a special "dampening chamber" built into the VAF to smooth out vane pulsations that would be created by intake manifold air-pressure fluctuations caused by the valve opening and closing. Many VAF sensors include a switch to energize the electric fuel pump. This is a safety feature that prevents the operation of the fuel pump if the engine stalls.

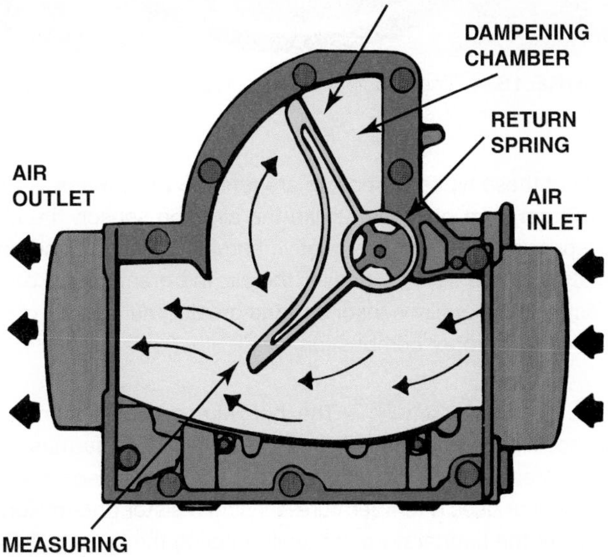

FIGURE 16–1 A vane airflow (VAF) sensor.

## MASS AIRFLOW SENSOR TYPES

Most newer fuel injection systems use a **mass airflow (MAF)** sensor to calculate the amount of air volume delivered to the engine.

There are several types of mass airflow sensors.

**HOT FILM SENSOR** The hot film sensor uses a temperature-sensing resistor (thermistor) to measure the temperature of the incoming air. Through the electronics within the sensor, a

FIGURE 16–2 A typical air vane sensor with the cover removed. The movable arm contacts a carbon resistance path as the vane opens. Many air vane sensors also have contacts that close to supply voltage to the electric fuel pump as the air vane starts to open when the engine is being cranked and air is being drawn into the engine.

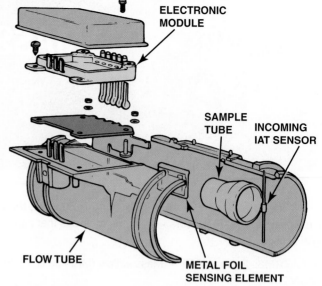

FIGURE 16–3 This five-wire mass airflow sensor consists of a metal foil sensing unit, an intake air temperature (IAT) sensor, and the electronic module.

FIGURE 16–4 The sensing wire in a typical hot wire mass airflow sensor.

conductive film is kept at a temperature 70°C above the temperature of the incoming air. ● SEE FIGURE 16–3.

Because both the amount and the density of the air tend to contribute to the cooling effect as the air passes through the sensor, this type of sensor can actually produce an output based on the mass of the airflow. Mass equals volume times density. For example, cold air is denser than warm air, so a small amount of cold air may have the same mass as a larger amount of warm air. Therefore, a mass airflow sensor is designed to measure the mass, not the volume, of the air entering the engine.

The output of this type of sensor is usually a frequency based on the amount of air entering the sensor. The more air that enters the sensor, the more the hot film is cooled. The electronics inside the sensor, therefore, increase the current flow through the hot film to maintain the 70°C temperature differential between the air temperature and the temperature of the hot film. This change in current flow is converted to a frequency output that the computer can use as a measurement of airflow.

Most of these types of sensors are referred to as **mass airflow (MAF) sensors** because, unlike the air vane sensor, the MAF sensor takes into account relative humidity, altitude, and temperature of the air. The denser the air, the greater the cooling effect on the hot film sensor and the greater the amount of fuel required for proper combustion.

**HOT WIRE SENSOR**  The hot wire sensor is similar to the hot film type but uses a hot wire to sense the mass airflow instead of the hot film. Like the hot film sensor, the hot wire sensor uses a temperature-sensing resistor (thermistor) to measure the temperature of the air entering the sensor. ● SEE FIGURE 16–4. The electronic circuitry within the sensor keeps the temperature of the wire at 70°C above the temperature of the incoming air.

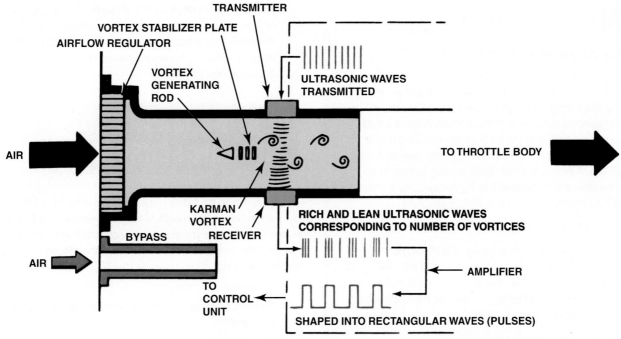

**FIGURE 16–5** A Karman vortex airflow sensor uses a triangle-shaped rod to create vortexes as the air flows through the sensor. The electronics in the sensor itself converts these vortexes to a digital square wave signal.

Both designs operate in essentially the same way. A resistor wire or screen installed in the path of intake airflow is heated to a constant temperature by electric current provided by the computer. Air flowing past the screen or wire cools it. The degree of cooling varies with air velocity, temperature, density, and humidity. These factors combine to indicate the mass of air entering the engine. As the screen or wire cools, more current is required to maintain the specified temperature. As the screen or wire heats up, less current is required. The operating principle can be summarized as follows:

- More intake air volume = cooler sensor, more current
- Less intake air volume = warmer sensor, less current

The computer constantly monitors the change in current and translates it into a voltage signal that is used to determine injector pulse width.

**BURN-OFF CIRCUIT.** Some hot wire-type MAF sensors use a burn-off circuit to keep the sensing wire clean of dust and dirt. A high current is passed through the sensing wire for a short time but long enough to cause the wire to glow because of the heat. The burn-off circuit is turned on when the ignition switch is switched off after the engine has been operating long enough to achieve normal operating temperature.

## KARMAN VORTEX SENSORS

In 1912, a Hungarian scientist named Theodore Van Karman observed that vortexes were created when air passed over a pointed surface. This type of sensor sends a sound wave through the turbulence created by incoming air passing through the sensor. Air mass is calculated based on the time required for the sound waves to cross the turbulent air passage.

There are two basic designs of Karman vortex airflow sensors:

- **Ultrasonic.** This type of sensor uses ultrasonic waves to detect the vortexes that are produced and produces a digital (on-and-off) signal where frequency is proportional to the amount of air passing through the sensor. ● **SEE FIGURE 16–5.**
- **Pressure type.** Chrysler uses a pressure-type Karman vortex sensor that uses a pressure sensor to detect the vortexes. As the airflow through the sensor increases, so do the number of pressure variations. The electronics in the sensor convert these pressure variations to a square wave (digital DC voltage) signal, whose frequency is in proportion to the airflow through the sensor.

## PCM USES FOR AIRFLOW SENSORS

The PCM uses the information from the airflow sensor for the following purposes:

- Airflow sensors are used mostly to determine the amount of fuel needed and base pulse-width numbers. The greater the mass of the incoming air, the longer the injectors are pulsed on.

**The Dirty MAF Sensor Story**

The owner of a Buick Park Avenue equipped with a 3800 V-6 engine complained that the engine would hesitate during acceleration, showed lack of power, and seemed to surge or miss at times. A visual inspection found everything to be like new, including a new air filter. There were no stored diagnostic trouble codes (DTCs). A look at the scan data showed airflow to be within the recommended 3 to 7 g per second. A check of the frequency output showed the problem:

**Idle frequency = 2.177 kHz (2,177 Hz)**

Normal frequency at idle speed should be 2.37 to 2.52 kHz. Cleaning the hot wire of the MAF sensor restored proper operation. The sensor wire was covered with what looked like fine fibers, possibly from the replacement air filter.

**NOTE: Older GM MAF sensors operated at a lower frequency of 32 to 150 Hz, with 32 Hz being the average reading at idle and 150 Hz for wide-open throttle.**

**? FREQUENTLY ASKED QUESTION**

**What Is Meant by a "High-Authority Sensor"?**

A high-authority sensor is a sensor that has a major influence over the amount of fuel being delivered to the engine. For example, at engine start-up, the engine coolant temperature (ECT) sensor is a high-authority sensor, and the oxygen sensor (O2S) is a low-authority sensor. However, as the engine reaches operating temperature, the oxygen sensor becomes a high-authority sensor and can greatly affect the amount of fuel being supplied to the engine. See the following chart.

| High-Authority Sensors | Low-Authority Sensors |
|---|---|
| ECT (especially when the engine starts and is warming up) | IAT (intake air temperature) sensors modify and back up the ECT |
| O2S (after the engine reaches closed-loop operation) | TFT (transmission fluid temperature) |
| MAP | PRNDL (shift position sensor) |
| MAF | KS (knock sensor) |
| TP (high authority during acceleration and deceleration) | EFT (engine fuel temperature) |

- Airflow sensors back up the TP sensor in the event of a loss of signal or an inaccurate throttle position sensor signal. If the MAF sensor fails, then the PCM will calculate the fuel delivery needs of the engine based on throttle position and engine speed (RPM).

# TESTING MASS AIRFLOW SENSORS

**VISUAL INSPECTION** Start the testing of a MAF sensor by performing a thorough visual inspection. Look at all the hoses that direct and send air, especially between the MAF sensor and the throttle body. Also check the electrical connector for the following:

- Corrosion
- Terminals that are bent or pushed out of the plastic connector
- Frayed wiring

**MAF SENSOR OUTPUT TEST** MAF sensors calculate air mass by weight in a given amount of time usually in grams per second (gm/sec). A digital multimeter, set to read DC volts on the signal wire circuit, can be used to check the MAF sensor. See the chart that shows the voltage output compared with the grams per second of airflow through the sensor. Normal airflow is 3 to 7 g per second.

| Analog MAF Sensor Grams per Second/Voltage Chart | |
|---|---|
| **Grams per Second** | **Sensor Voltage** |
| 0 | 0.2 |
| 2 | 0.7 |
| 4 | 1.0 (typical idle value) |
| 8 | 1.5 |
| 15 | 2.0 |
| 30 | 2.5 |
| 50 | 3.0 |
| 80 | 3.5 |
| 110 | 4.0 |
| 150 | 4.5 |
| 175 | 4.8 |

**TAP TEST** With the engine running at idle speed, gently tap the MAF sensor with the fingers of an open hand. If the engine stumbles or stalls, the MAF sensor is defective. This test is commonly called the **tap test.**

**DIGITAL METER TEST OF A MAF SENSOR** A digital multimeter can be used to measure the frequency (Hz) output of the sensor and compare the reading with specifications.

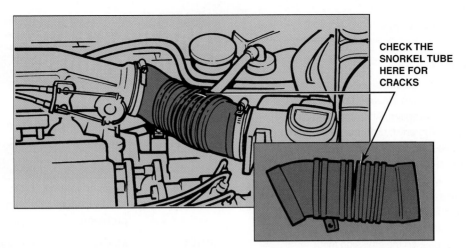

CHECK THE
SNORKEL TUBE
HERE FOR
CRACKS

**FIGURE 16–6** Carefully check the hose between the MAF sensor and the throttle body assembly for cracks or splits that could create extra (false) air into the engine that is not measured by the MAF sensor.

 **FREQUENTLY ASKED QUESTION**

### What Is False Air?

Airflow sensors and mass airflow (MAF) sensors are designed to measure *all* the air entering the engine. If an air hose between the MAF sensor and the throttle body was loose or had a hole, extra air could enter the engine without being measured. This extra air is often called **false air**. ●**SEE FIGURE 16–6.** Because this extra air is unmeasured, the computer does not provide enough fuel delivery, and the engine operates too lean, especially at idle. A small hole in the air inlet hose would represent a fairly large percentage of false air at idle but would represent a very small percentage of extra air at highway speeds.

To diagnose for false air, look at long-term fuel trim numbers at idle and at 3000 RPM.

NOTE: If the engine runs well in reverse yet runs terrible in any forward gear, carefully look at the inlet hose for air leaks that would open when the engine torque moves the engine slightly on its mounts.

 **TECH TIP**

### The Unplug-It Test

If a sensor is defective yet still produces a signal to the computer, the computer will often accept the reading and make the required changes in fuel delivery and spark advance. If, however, the sensor is not reading correctly, the computer will process this wrong information and perform an action assuming that information being supplied is accurate. For example, if a mass airflow (MAF) sensor is telling the computer that 12 g of air per second is going into the engine, then the computer will pulse the injector for 6.4 ms or whatever figure it is programmed to provide. However, if the air going into the engine is actually 14 g per second, the amount of fuel supplied by the injectors will not be enough to provide proper engine operation. If the MAF sensor is unplugged, the computer knows that the sensor is not capable of supplying airflow information, so it defaults to a fixed amount of fuel based on the values of other sensors, such as the TP and MAP sensors. "If in doubt, take it out."

If the engine operates better with a sensor unplugged, then suspect that the sensor is defective. A sensor that is not supplying the correct information is said to be skewed. The computer will not set a diagnostic trouble code for this condition because the computer can often not detect that the sensor is supplying wrong information.

The frequency output and engine speed in RPM can also be plotted on a graph to check to see if the frequency and RPM are proportional, resulting in a straight line on the graph.

## MAF SENSOR CONTAMINATION

Dirt, oil, silicon, or even spiderwebs can coat the sensing wire. Because it tends to insulate the sensing wire at low airflow rates, a contaminated sensor often overestimates the amount of air entering the engine at idle and therefore causes the fuel system to go rich. At higher engine speeds near wide-open throttle (WOT), the contamination can cause the sensor to underestimate the amount of air entering the engine. As a result, the fuel system will go lean, causing spark knock and lack of power concerns. To check for contamination, check the fuel trim numbers.

**The Rich-Running Toyota**

A Toyota failed an enhanced emission test for excessive carbon monoxide, which is caused by a rich (too much fuel) air–fuel ratio problem. After checking all of the basics and not finding any fault in the fuel system, the technician checked the archives of the International Automotive Technicians Network *(www.iatn.net)* and discovered that a broken spring inside the airflow sensor was a possible cause. The sensor was checked, and a broken vane return spring was discovered. Replacing the airflow sensor restored the engine to proper operating conditions, and it passed the emission test.

If the fuel trim is negative (removing fuel) at idle yet is positive (adding fuel) at higher engine speeds, a contaminated MAF sensor is a likely cause. Other tests for a contaminated MAF sensor include the following:

- At WOT, the grams per second, as read on a scan tool, should exceed 100 g.
- At WOT, the voltage, as read on a digital voltmeter, should exceed 4 volts for an analog sensor.
- At WOT, the frequency, as read on a meter or scan tool, should exceed 7 kHz for a digital sensor.

If the readings do not exceed these values, then the MAF sensor is contaminated.

# MAF-RELATED DIAGNOSTIC TROUBLE CODES

The diagnostic trouble codes (DTCs) associated with the mass airflow and air vane sensors include the following:

| Diagnostic Trouble Code | Description | Possible Causes |
|---|---|---|
| P0100 | Mass or volume airflow circuit problems | • Open or short in mass airflow circuit<br>• Defective MAF sensor |
| P0101 | Mass airflow circuit range problems | • Defective MAF sensor (check for false air) |
| P0102 | Mass airflow circuit low output | • Defective MAF sensor<br>• MAF sensor circuit open or shorted-to-ground<br>• Open 12-volt supply voltage circuit |
| P0103 | Mass airflow circuit high output | • Defective MAF sensor<br>• MAF sensor circuit shorted-to-voltage |

# SUMMARY

1. A mass airflow sensor actually measures the density and amount of air flowing into the engine, which results in accurate engine control.
2. An air vane sensor measures the volume of the air, and the intake air temperature sensor is used by the PCM to calculate the mass of the air entering the engine.
3. A hot wire MAF sensor uses the electronics in the sensor itself to heat a wire 70°C above the temperature of the air entering the engine.

# REVIEW QUESTIONS

1. How does a hot film MAF sensor work?
2. What type of voltage signal is produced by a MAF?
3. What change in the signal will occur if engine speed is increased?
4. How is a MAF sensor tested?
5. What is the purpose of a MAF sensor?
6. What are the types of airflow sensors?

1. A fuel-injection system that does not use a sensor to measure the amount (or mass) of air entering the engine is usually called a(n) _____ type of system.
   a. Air vane-controlled
   b. Speed density
   c. Mass airflow
   d. Hot wire

2. Which type of sensor uses a burn-off circuit?
   a. Hot wire MAF sensor
   b. Hot film MAF sensor
   c. Vane-type airflow sensor
   d. Both a and b

3. Which sensor has a switch that controls the electric fuel pump?
   a. VAF
   b. Hot wire MAF
   c. Hot filter MAF
   d. Karman vortex sensor

4. Two technicians are discussing Karman vortex sensors. Technician A says that they contain a burn-off circuit to keep them clean. Technician B says that they contain a movable vane. Which technician is correct?
   a. Technician A only
   b. Technicians B only
   c. Both Technicians A and B
   d. Neither Technician A nor B

5. The typical MAF reading on a scan tool with the engine at idle speed and normal operating temperature is _____.
   a. 1 to 3 g per second
   b. 3 to 7 g per second
   c. 8 to 12 g per second
   d. 14 to 24 g per second

6. Two technicians are diagnosing a poorly running engine. There are no diagnostic trouble codes. When the MAF sensor is unplugged, the engine runs better. Technician A says that this means that the MAF is supplying incorrect airflow information to the PCM. Technician B says that this indicates that the PCM is defective. Which technician is correct?
   a. Technician A only
   b. Technician B only
   c. Both Technicians A and B
   d. Neither Technician A nor B

7. A MAF sensor on a General Motors 3800 V-6 is being tested for contamination. Technician A says that the sensor should show over 100 g per second on a scan tool display when the accelerator is depressed to WOT on a running engine. Technician B says that the output frequency should exceed 7,000 Hz when the accelerator pedal is depressed to WOT on a running engine. Which technician is correct?
   a. Technician A only
   b. Technician B only
   c. Both Technicians A and B
   d. Neither Technician A nor B

8. Which airflow sensor has a dampening chamber?
   a. Vane airflow
   b. Hot film MAF
   c. Hot wire MAF
   d. Karman vortex

9. Air that enters the engine without passing through the airflow sensor is called _____.
   a. Bypass air
   b. Dirty air
   c. False air
   d. Measured air

10. A P0102 DTC is being discussed. Technician A says that a sensor circuit shorted-to-ground can be the cause. Technician B says that an open sensor voltage supply circuit could be the cause. Which technician is correct?
    a. Technician A only
    b. Technician B only
    c. Both Technicians A and B
    d. Neither Technician A nor B

# chapter 17

# OXYGEN SENSORS

**LEARNING OBJECTIVES:** **After studying this chapter, the reader should be able to:** • Discuss the purpose and function of oxygen sensors (O2S). • Understand the PCM uses of O2S. • Explain the ways of diagnosing O2S. • Describe the waveform analysis of O2S. • Understand the voltage readings of O2S.

**KEY TERMS:** Bias voltage 215 • Closed-loop operation 214 • Cross counts 217 • False lean indication 225 • False rich indication 225 • Open-loop operation 214 • Oxygen sensor (O2S) 212

## OXYGEN SENSORS

**PURPOSE AND FUNCTION** Automotive computer systems use a sensor in the exhaust system to measure the oxygen content of the exhaust. These sensors are called **oxygen sensors (O2S).** The oxygen sensor is installed in the exhaust manifold or located downstream from the manifold in the exhaust pipe. ● **SEE FIGURE 17–1.** The oxygen sensor is directly in the path of the exhaust gas stream, where it monitors oxygen level in both the exhaust stream and the ambient air. In a zirconia oxygen sensor, the tip contains a thimble made of zirconium dioxide ($ZrO_2$), an electrically conductive material capable of generating a small voltage in the presence of oxygen. The oxygen sensor is used by the PCM to control fuel delivery.

**CONSTRUCTION AND OPERATION** Exhaust from the engine passes through the end of the sensor, where the gases contact the outer side of the thimble. Atmospheric air enters through the other end of the sensor or through the wire of the sensor and contacts the inner side of the thimble. The inner and outer surfaces of the thimble are plated with platinum. The inner surface becomes a negative electrode; the outer surface is a positive electrode. The atmosphere contains a relatively constant 21% of oxygen. Rich exhaust gases contain little oxygen. Exhaust from a lean mixture contains more oxygen.

Negatively charged oxygen ions are drawn to the thimble, where they collect on both the inner and outer surfaces. ● **SEE FIGURE 17–2.** Because the percentage of oxygen present in the atmosphere exceeds that in the exhaust gases, the atmosphere side of the thimble draws more negative oxygen ions than the exhaust side. The difference between the two sides creates an electrical potential, or voltage. When the concentration

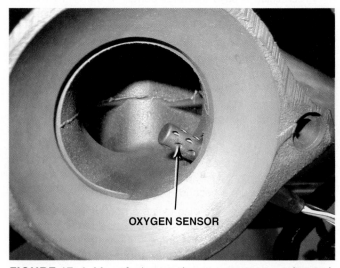

**OXYGEN SENSOR**

**FIGURE 17–1** Many fuel-control oxygen sensors are located in the exhaust manifold near its outlet so that the sensor can detect the air-fuel mixture in the exhaust stream for all cylinders that feed into the manifold.

of oxygen on the exhaust side of the thimble is low (rich exhaust), a high voltage (0.6 to 1 volt) is generated between the electrodes. As the oxygen concentration on the exhaust side increases (lean exhaust), the voltage generated drops low (0.0 to 0.3 volt). ● **SEE FIGURE 17–3.**

This voltage signal is sent to the computer, where it passes through the input conditioner for amplification. The computer interprets a high-voltage signal (low-oxygen content) as a rich air–fuel ratio and a low-voltage signal (high-oxygen content) as a lean air–fuel ratio. Based on the O2S signal (above or below 0.45 volt), the computer compensates by making the mixture either leaner or richer as required to continually vary close to a 14.7:1 air–fuel ratio to satisfy the needs of the three-way catalytic converter.

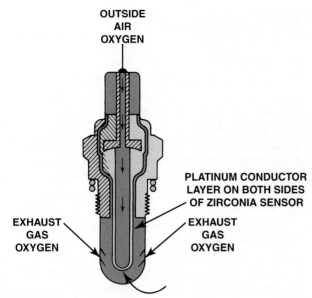

**FIGURE 17–2** A cross-sectional view of a typical zirconia oxygen sensor.

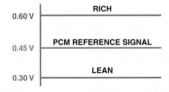

**FIGURE 17–3** A difference in oxygen content between the atmosphere and the exhaust gases enables an O2S to generate voltage.

The O2S is the key sensor of an electronically controlled fuel metering system for emission control.

An O2S does not send a voltage signal until its tip reaches a temperature of about 572°F (300°C). Also, oxygen sensors provide their fastest response to mixture changes at about 1,472°F (800°C). When the engine starts and the O2S is cold, the computer runs the engine in the open-loop mode, drawing on prerecorded data in the PROM for fuel control on a cold engine or when O2S output is not within certain limits.

If the exhaust contains very little oxygen (O2S), the computer assumes that the intake charge is rich (too much fuel) and reduces fuel delivery. ● **SEE FIGURE 17–4.** However, when the oxygen level is high, the computer assumes that the intake

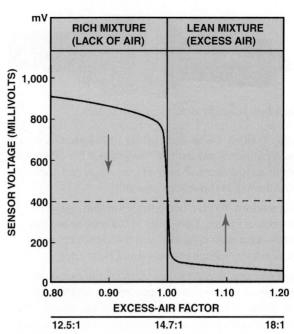

**FIGURE 17–4** The oxygen sensor provides a quick response at the stoichiometric air–fuel ratio of 14.7:1.

charge is lean (not enough fuel) and increases fuel delivery. There are several different designs of oxygen sensors, including the following:

- **One-wire oxygen sensor.** The single wire of a one-wire oxygen sensor is the O2S signal wire. The ground for the O2S is through the shell and threads of the sensor and through the exhaust manifold.

- **Two-wire oxygen sensor.** The two-wire sensor has a signal wire and a ground wire for the O2S.

- **Three-wire oxygen sensor.** The three-wire sensor design uses an electric resistance heater to help get the O2S up to temperature more quickly and to help keep the sensor at operating temperature even at idle speeds. The three wires include the O2S signal, the power, and ground for the heater.

- **Four-wire oxygen sensor.** The four-wire sensor is a heated O2S (HO2S) that uses an O2S signal wire and signal ground. The other two wires are the power and ground for the heater.

## ZIRCONIA OXYGEN SENSORS

The most common type of oxygen sensor is made from zirconia (zirconium dioxide). It is usually constructed using powder that is pressed into a thimble shape and coated with porous platinum material that acts as electrodes. All zirconia sensors use 18-mm-diameter threads with a washer. ● **SEE FIGURE 17–5.**

Zirconia oxygen sensors are constructed so that oxygen ions flow through the sensor when there is a difference between

**FIGURE 17–5** A typical zirconia oxygen sensor.

the oxygen content inside and outside of the sensor. An ion is an electrically charged particle. The greater the differences between the oxygen content between the inside and outside of the sensor the higher the voltage created:

- **Rich mixture.** A rich mixture results in little oxygen in the exhaust stream. Compared to the outside air, this represents a large difference and the sensor creates a relatively high voltage of about 1 volt (1,000 mV).

- **Lean mixture.** A lean mixture leaves some oxygen in the exhaust stream that did not combine with the fuel. This leftover oxygen reduces the difference between the oxygen content of the exhaust compared to the oxygen content of the outside air. As a result, the sensor voltage is low or almost zero volts.

- **O2S voltage above 450 mV.** This is produced by the sensor when the oxygen content in the exhaust is low. This is interpreted by the engine computer (PCM) as being a rich exhaust.

- **O2S voltage below 450 mV.** This is produced by the sensor when the oxygen content is high. This is interpreted by the engine computer (PCM) as being a lean exhaust.

## TITANIA OXYGEN SENSOR

The titania (titanium dioxide) oxygen sensor does not produce a voltage but rather changes in resistance with the presence of oxygen in the exhaust. All titania oxygen sensors use a four-terminal variable-resistance unit with a heating element. A titania sensor samples exhaust air only and uses a reference voltage from the PCM. Titania oxide oxygen sensors use a 14-mm thread and are not interchangeable with zirconia oxygen sensors. One volt is applied to the sensor, and the changing resistance of the titania oxygen sensor changes the voltage of the sensor circuit. As with a zirconia oxygen sensor, the voltage signal is above 450 mV when the exhaust is rich and low (below 450 mV) when the exhaust is lean.

## CLOSED LOOP AND OPEN LOOP

The amount of fuel delivered to an engine is determined by the powertrain control module (PCM) based on inputs from the engine coolant temperature (ECT), throttle position (TP)

sensor, and others until the oxygen sensor is capable of supplying a usable signal. When the PCM alone (without feedback) is determining the amount of fuel needed, it is called **open-loop operation.** As soon as the oxygen sensor is capable of supplying rich and lean signals, adjustments by the computer can be made to fine-tune the correct air–fuel mixture. This checking and adjusting by the computer is called **closed-loop operation.**

## PCM USES OF THE OXYGEN SENSOR

**FUEL CONTROL** The upstream oxygen sensors are among the high-authority sensors used for fuel control while operating in closed loop. Before the oxygen sensors are hot enough to give accurate exhaust oxygen information to the computer, fuel control is determined by other sensors and the anticipated injector pulse width determined by those sensors. After the control system achieves closed-loop status, the oxygen sensor provides feedback with actual exhaust gas oxygen content.

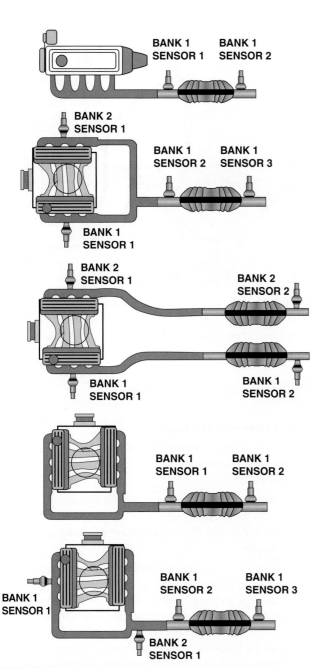

**FIGURE 17–6** Number and label designations for oxygen sensors. Bank 1 is the bank where cylinder number 1 is located.

**FUEL TRIM** Fuel trim is a computer program that is used to compensate for a too rich or a too lean air–fuel exhaust as detected by the oxygen sensor(s). Fuel trim is necessary to keep the air–fuel mixture within limits to allow the catalytic converter to operate efficiently. If the exhaust is too lean or too rich for a long time, the catalytic converter can be damaged. The fuel trim numbers are determined from the signals from the oxygen sensor(s). If the engine has been operating too lean, short-term and long-term fuel time programming inside the PCM can cause an increase in the commanded injector pulse width to bring the air–fuel mixture back into the proper range. Fuel trim can be negative (subtracting fuel) or positive (adding fuel).

 **FREQUENTLY ASKED QUESTION**

### What Happens to the Bias Voltage?

Some vehicle manufacturers such as General Motors Corporation have the computer apply 450 mV (0.450 V) to the O2S signal wire. This voltage is called the **bias voltage** and represents the threshold voltage for the transition from rich to lean.

This bias voltage is displayed on a scan tool when the ignition switch is turned on with the engine off. When the engine is started, the O2S becomes warm enough to produce a usable voltage, and bias voltage "disappears" as the O2S responds to a rich and lean mixture. What happened to the bias voltage that the computer applied to the O2S? The voltage from the O2S simply overcame the very weak voltage signal from the computer. This bias voltage is so weak that even a 20-megohm impedance DMM will affect the strength enough to cause the voltage to drop to 426 mV. Other meters with only 10 megohms of impedance will cause the bias voltage to read less than 400 mV.

Therefore, even though the O2S voltage is relatively low powered, it is more than strong enough to override the very weak bias voltage the computer sends to the O2S.

## OXYGEN SENSOR DIAGNOSIS

The oxygen sensors are used for diagnosis of other systems and components. For example, the exhaust gas recirculation (EGR) system is tested by the PCM by commanding the valve to open during the test. Some PCMs determine whether enough exhaust gas flows into the engine by looking at the oxygen sensor response (fuel trim numbers). The upstream and downstream oxygen sensors are also used to determine the efficiency of the catalytic converter. ● **SEE FIGURE 17–7.**

**TESTING AN OXYGEN SENSOR USING A DIGITAL VOLTMETER** The oxygen sensor can be checked for proper operation using a digital high-impedance voltmeter:

1. With the engine off, connect the red lead of the meter to the oxygen sensor signal wire and the black meter lead to a good engine ground. ● **SEE FIGURE 17–8.**
2. Start the engine and allow it to reach closed-loop operation.
3. In closed-loop operation, the oxygen sensor voltage should be constantly changing as the fuel mixture is being controlled.

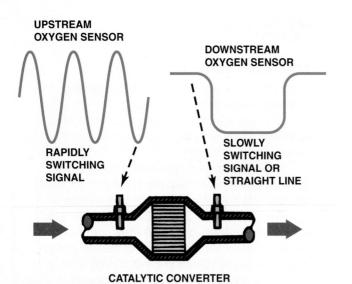

**FIGURE 17–7** The OBD-II catalytic converter monitor compares the signals of the upstream and downstream oxygen sensor to determine converter efficiency.

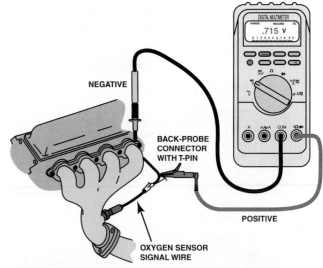

**FIGURE 17–8** Testing an oxygen sensor using a DMM set on DC volts. With the engine operating in closed loop, the oxygen voltage should read over 800 mV and lower than 200 mV and be constantly fluctuating.

---

 **REAL WORLD FIX**

### The Oxygen Sensor Is Lying to You

A technician was trying to solve a driveability problem with an older V-6 passenger car. The car idled roughly, hesitated, and accelerated poorly. A thorough visual inspection did not indicate any possible problems, and there were no diagnostic trouble codes stored.

A check was made on the oxygen sensor activity using a DMM. The voltage stayed above 600 mV most of the time. If a large vacuum hose was removed, the oxygen sensor voltage would temporarily drop to below 450 mV and then return to a reading of over 600 mV. Remember the following:

- High O2S readings = rich exhaust (low $O_2$ content in the exhaust)
- Low O2S readings = lean exhaust (high $O_2$ content in the exhaust)

As part of a thorough visual inspection, the technician removed and inspected the spark plugs. All the spark plugs were white, indicating a lean mixture, not the rich mixture that the oxygen sensor was indicating. The high O2S reading signaled the computer to reduce the amount of fuel, resulting in an excessively lean operation.

After replacing the oxygen sensor, the engine ran great. But what killed the oxygen sensor? The technician finally learned from the owner that the head gasket had been replaced over a year ago. The phosphate and silicate additives in the antifreeze coolant had coated the oxygen sensor. Because the oxygen sensor was coated, the oxygen content of the exhaust could not be detected—the result: a false rich signal from the oxygen sensor.

---

 **REAL WORLD FIX**

### The Missing Ford Escort

A Ford Escort was being analyzed for poor engine operation. The engine ran perfectly during the following conditions:

1. With the engine cold or operating in open loop
2. With the engine at idle
3. With the engine operating at or near wide-open throttle

After hours of troubleshooting, the cause was found to be a poor ground connection for the oxygen sensor. The engine ran okay during times when the computer ignored the oxygen sensor. Unfortunately, the service technician did not have a definite plan during the diagnostic process and as a result checked and replaced many unnecessary parts. An oxygen sensor test early in the diagnostic procedure would have indicated that the oxygen (O2S) signal was not correct. The poor ground caused the oxygen sensor voltage level to be too high, indicating to the computer that the mixture was too rich. The computer then subtracted fuel, which caused the engine to miss and run rough as the result of the now too lean air–fuel mixture.

The results should be interpreted as follows:

- If the oxygen sensor fails to respond and its voltage remains at about 450 mV, the sensor may be defective and require replacement. Before replacing the oxygen sensor, check the manufacturer's recommended procedures.
- If the oxygen sensor reads high all the time (above 550 mV), the fuel system could be supplying too rich a

---

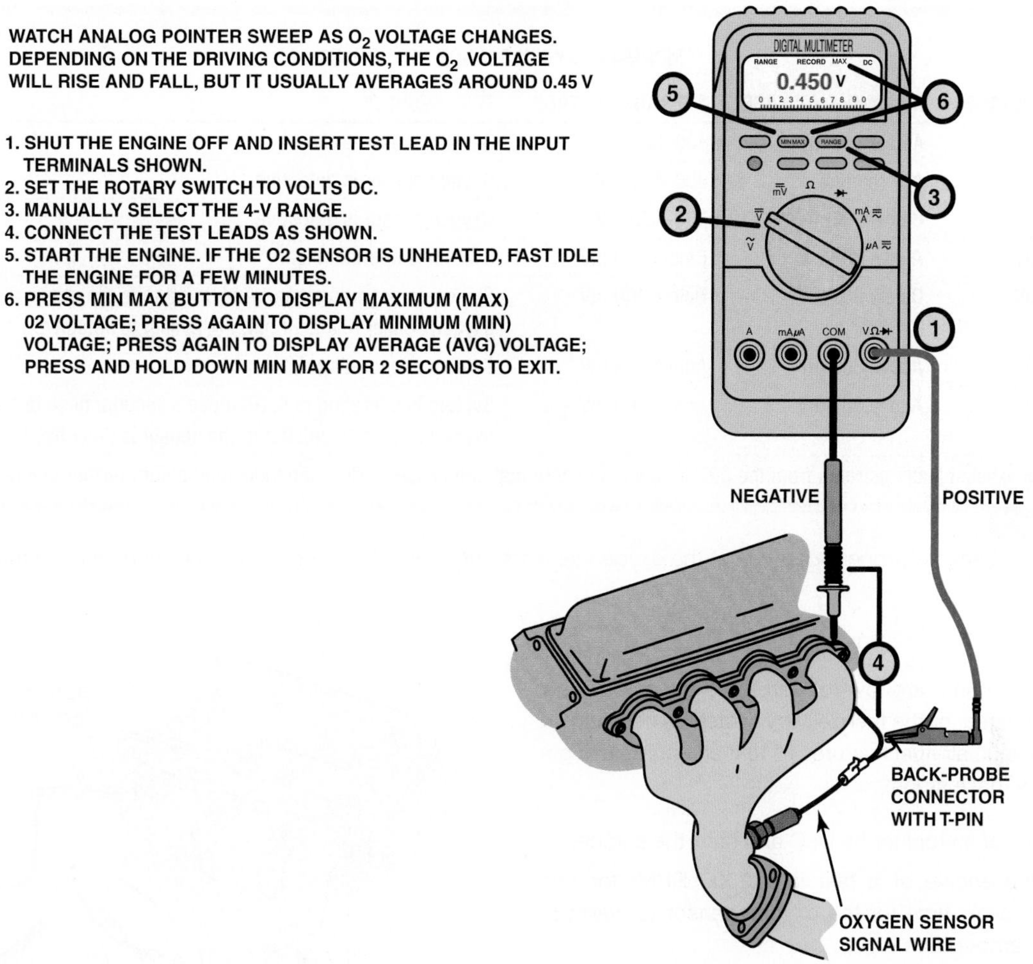

WATCH ANALOG POINTER SWEEP AS O$_2$ VOLTAGE CHANGES.
DEPENDING ON THE DRIVING CONDITIONS, THE O$_2$ VOLTAGE
WILL RISE AND FALL, BUT IT USUALLY AVERAGES AROUND 0.45 V

1. SHUT THE ENGINE OFF AND INSERT TEST LEAD IN THE INPUT
   TERMINALS SHOWN.
2. SET THE ROTARY SWITCH TO VOLTS DC.
3. MANUALLY SELECT THE 4-V RANGE.
4. CONNECT THE TEST LEADS AS SHOWN.
5. START THE ENGINE. IF THE O2 SENSOR IS UNHEATED, FAST IDLE
   THE ENGINE FOR A FEW MINUTES.
6. PRESS MIN MAX BUTTON TO DISPLAY MAXIMUM (MAX)
   02 VOLTAGE; PRESS AGAIN TO DISPLAY MINIMUM (MIN)
   VOLTAGE; PRESS AGAIN TO DISPLAY AVERAGE (AVG) VOLTAGE;
   PRESS AND HOLD DOWN MIN MAX FOR 2 SECONDS TO EXIT.

NEGATIVE        POSITIVE

BACK-PROBE
CONNECTOR
WITH T-PIN

OXYGEN SENSOR
SIGNAL WIRE

**FIGURE 17–9** Using a digital multimeter to test an oxygen sensor using the MIN/MAX record function of the meter.

**Why Does the Oxygen Sensor Voltage Read 5 Volts on Many Chrysler Vehicles?**

Many Chrysler vehicles apply a 5-volt reference to the signal wire of the oxygen sensor. The purpose of this voltage is to allow the computer to detect if the oxygen sensor signal circuit is open or grounded:

- If the voltage on the signal wire is 4.5 volts or more, the computer assumes that the sensor is open.
- If the voltage on the signal wire is zero, the computer assumes that the sensor is shorted-to-ground.

   If either condition exists, the computer can set a diagnostic trouble code (DTC).

fuel mixture, or the oxygen sensor may be contaminated.

- If the oxygen sensor voltage remains low (below 350 mV), the fuel system could be supplying too lean a fuel mixture. Check for a vacuum leak or partially clogged fuel injector(s). Before replacing the oxygen sensor, check the manufacturer's recommended procedures.

## TESTING THE OXYGEN SENSOR USING THE MIN/MAX METHOD
A digital meter set on DC volts can be used to record the minimum and maximum voltage with the engine running. A good oxygen sensor should be able to produce a value of less than 300 mV and a maximum voltage above 800 mV. Replace any oxygen sensor that fails to go above 700 mV or lower than 300 mV. ● **SEE FIGURE 17–9 AND CHART 17–1.**

## TESTING AN OXYGEN SENSOR USING A SCAN TOOL
A good oxygen sensor should be able to sense the oxygen content and change voltage outputs rapidly. How fast an oxygen sensor switches from high (above 450 mV) to low (below 350 mV) is measured in oxygen sensor **cross counts.** Cross counts are the number of times an oxygen sensor changes voltage from high to low (from low to high voltage is not counted) in 1 second (or 1.25 seconds, depending on scan tool and computer speed).

**NOTE: On a fuel-injected engine at 2,000 engine RPM, 8 to 10 cross counts is normal.**

   Oxygen sensor cross counts can be determined only by using a scan tool or other suitable tester that reads computer data.

| MIN/MAX OXYGEN SENSOR TEST CHART | | | |
|---|---|---|---|
| MINIMUM VOLTAGE | MAXIMUM VOLTAGE | AVERAGE VOLTAGE | TEST RESULTS |
| Below 200 mV | Above 800 mV | 400–500 mV | Oxygen sensor is okay. |
| Above 200 mV | Any reading | 400–500 mV | Oxygen sensor is defective. |
| Any reading | Below 800 mV | 400–500 mV | Oxygen sensor is defective. |
| Below 200 mV | Above 800 mV | Below 400 mV | System is operating lean.* |
| Below 200 mV | Below 800 mV | Below 400 mV | System is operating lean. (Add propane to the intake air to see if the oxygen sensor reacts. If not, the sensor is defective.) |
| Below 200 mV | Above 800 mV | Above 500 mV | System is operating rich. |
| Above 200 mV | Above 800 mV | Above 500 mV | System is operating rich. (Remove a vacuum hose to see if the oxygen sensor reacts. If not, the sensor is defective.) |
| *Check for an exhaust leak upstream from the O2S or ignition misfire that can cause a false lean indication before further diagnosis. | | | |

**CHART 17–1**

Use this chart to check for proper operation of the oxygen sensors and fuel system after checking them using a multimeter set to read MIN/MAX.

If the cross counts are low (or zero), the oxygen sensor may be contaminated, or the fuel delivery system is delivering a constant rich or lean air–fuel mixture. To test an engine using a scan tool, follow these steps:

1. Connect the scan tool to the DLC and start the engine.

2. Operate the engine at a fast idle (2500 RPM) for two minutes to allow time for the oxygen sensor to warm to operating temperature.

3. Observe the oxygen sensor activity on the scan tool to verify closed-loop operation. Select "snapshot" mode and hold the engine speed steady and start recording.

4. Play back snapshot and place a mark beside each range of oxygen sensor voltage for each frame of the snapshot.

A good oxygen sensor and computer system should result in most snapshot values at both ends (0 to 300 mV and 600 to 1,000 mV). If most of the readings are in the middle, the oxygen sensor is not working correctly.

## TESTING AN OXYGEN SENSOR USING A SCOPE
An oscilloscope (scope) can also be used to test an oxygen sensor. Connect the scope to the signal wire and ground for the sensor (if it is so equipped). ● SEE FIGURE 17–10. With the engine operating in closed loop, the voltage signal of the sensor should be constantly changing. ● SEE FIGURE 17–11. Check for rapid switching from rich to lean and lean to rich and change between once every two seconds and five times per second (0.5 to 5 Hz). ● SEE FIGURES 17–12, 17–13, AND 17–14.

**NOTE: General Motors warns not to base the diagnosis of an oxygen sensor problem solely on its scope pattern. The varying voltage output of an oxygen sensor can easily be mistaken for a fault in the sensor itself rather than a fault in the fuel delivery system.**

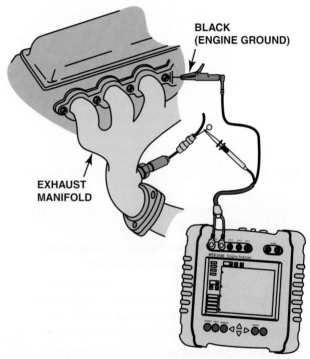

**FIGURE 17–10** Connecting a handheld digital storage oscilloscope to an oxygen sensor signal wire. Check the instructions for the scope as some require the use of a filter to be installed in the test lead to reduce electromagnetic interference that can affect the oxygen sensor waveform.

# OXYGEN SENSOR WAVEFORM ANALYSIS

As the O2S sensor warms up, the sensor voltage begins to rise. When the sensor voltage rises above 450 mV, the PCM determines that the sensor is up to operating temperature, takes

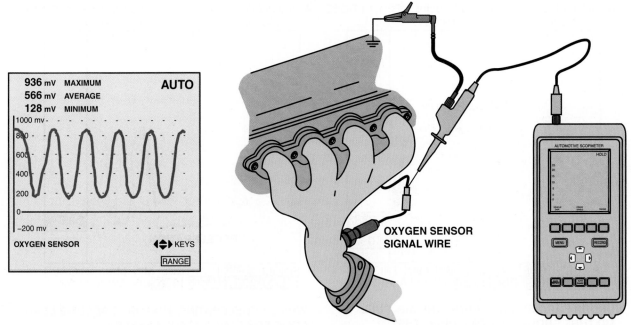

**FIGURE 17–11** The waveform of a good oxygen sensor as displayed on a digital storage oscilloscope (DSO). Note that the maximum reading is above 800 mV and that the minimum reading is less than 200 mV.

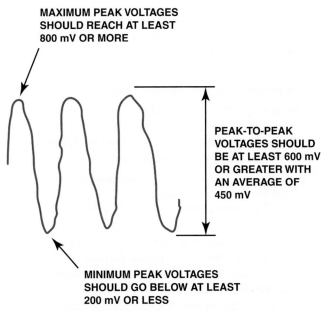

MAXIMUM PEAK VOLTAGES SHOULD REACH AT LEAST 800 mV OR MORE

PEAK-TO-PEAK VOLTAGES SHOULD BE AT LEAST 600 mV OR GREATER WITH AN AVERAGE OF 450 mV

MINIMUM PEAK VOLTAGES SHOULD GO BELOW AT LEAST 200 mV OR LESS

**FIGURE 17–12** A typical good oxygen sensor waveform as displayed on a digital storage oscilloscope. Look for transitions that occur between once every two seconds at idle and five times per second at higher engine speeds (0.5 and 5 Hz).

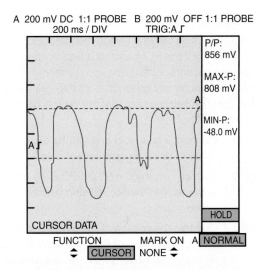

ONCE YOU'VE ACTIVATED "PEAK-TO-PEAK," "MAX-PEAK," AND "MIN-PEAK," FRAME THE WAVEFORM WITH CURSORS. LOOK FOR THE MINIMUM AND MAXIMUM VOLTAGES AND THE DIFFERENCE BETWEEN THEM IN THE RIGHT DISPLAY.

**FIGURE 17–13** Using the cursors on the oscilloscope, the high- and low-oxygen sensor values can be displayed on the screen.

A  200 mV DC  1:1 PROBE   B  200 mV  OFF 1:1 PROBE
200 ms /                              TRIG:A ⌐

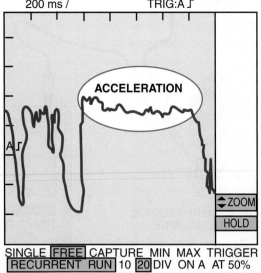

ACCELERATION

A ⌐

⬍ZOOM

HOLD

SINGLE FREE CAPTURE MIN MAX TRIGGER
RECURRENT RUN 10 20 DIV ON A AT 50%

**UNDER HARD ACCELERATION, THE AIR–FUEL
MIXTURE SHOULD BECOME RICH - THE
VOLTAGE SHOULD STAY FAIRLY HIGH**

A  200 mV DC  1:1 PROBE   B  200 mV  OFF 1:1 PROBE
200 ms /                              TRIG:A ⌐-1 DIV

A ⌐

DECELERATION

⬍ZOOM

HOLD

SINGLE FREE CAPTURE MIN MAX TRIGGER
RECURRENT RUN 10 20 DIV ON A AT 50%

**WHILE DECELERATING, MIXTURES BECOME LEAN.
LOOK FOR LOW VOLTAGE LEVELS.**

**FIGURE 17–14** When the air–fuel mixture rapidly changes such as during a rapid acceleration, look for a rapid response. The transition from low to high should be less than 100 ms.

---

🔧 **TECH TIP**

**The Key On, Engine Off Oxygen Sensor Test**

This test works on General Motors vehicles and may work on others if the PCM applies a bias voltage to the oxygen sensors. Zirconia oxygen sensors become more electrically conductive as they get hot. To perform this test, be sure that the vehicle has not run for several hours:

**STEP 1**   Connect a scan tool and get the display ready to show oxygen sensor data.

**STEP 2**   Turn key on, engine off (KOEO). The heater in the oxygen sensor will start heating the sensor.

**STEP 3**   Observe the voltage of the oxygen sensor. The applied bias voltage of 450 mV should slowly decrease for all oxygen sensors as they become more electrically conductive and other bias voltage is flowing to ground.

**STEP 4**   A good oxygen sensor should indicate a voltage of less than 100 mV after three minutes. Any sensor that displays a higher-than-usual voltage or seems to stay higher longer than the others could be defective or skewed high.

🔧 **TECH TIP**

**The Propane Oxygen Sensor Test**

Adding propane to the air inlet of a running engine is an excellent way to check if the oxygen sensor is able to react to changes in air–fuel mixture. Follow these steps in performing the propane trick:

1. Connect a digital storage oscilloscope to the oxygen sensor signal wire.
2. Start and operate the engine until up to operating temperature and in closed-loop fuel control.
3. While watching the scope display, add some propane to the air inlet. The scope display should read full rich (over 800 mV), as shown in ● **FIGURE 17–15.**
4. Shut off the propane. The waveform should drop to less than 200 mV (0.200 V), as shown in ● **FIGURE 17–16.**
5. Quickly add some propane while the oxygen sensor is reading low and watch for a rapid transition to rich. The transition should occur in less than 100 milliseconds (ms).

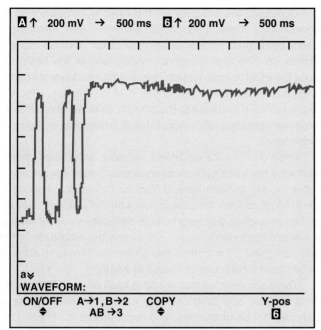

FIGURE 17–15 Adding propane to the air inlet of an engine operating in closed loop with a working oxygen sensor causes the oxygen sensor voltage to read high.

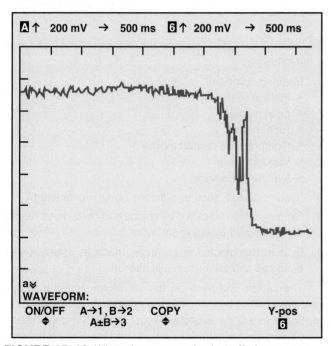

FIGURE 17–16 When the propane is shut off, the oxygen sensor should read below 200 mV.

control of the fuel mixture, and begins to cycle rich and lean. At this point, the system is considered to be in closed loop. ● SEE FIGURE 17–17.

**FREQUENCY** The frequency of the O2S is important in determining the condition of the fuel control system. The higher the frequency, the better, but the frequency must not exceed

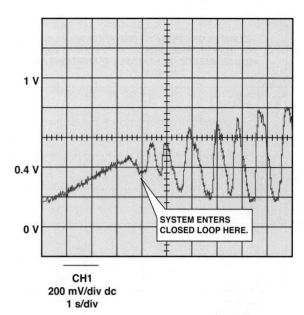

FIGURE 17–17 When the O2S voltage rises above 450 mV, the PCM starts to control the fuel mixture based on oxygen sensor activity.

6 Hz. For its OBD-II standards, the government has stated that a frequency greater than 6 Hz represents a misfire.

**THROTTLE-BODY FUEL-INJECTION SYSTEMS.** Normal TBI system rich/lean switching frequencies are from about 0.5 Hz at idle to about 3 Hz at 2500 RPM. Additionally, because of the TBI design limitations, fuel distribution to individual cylinders may not always be equal (due to unequal intake runner length, etc.). This may be normal unless certain other conditions are present at the same time.

**PORT FUEL-INJECTION SYSTEMS.** Specification for port fuel-injection systems is 0.5 Hz at idle to 5 Hz at 2500 RPM. ● SEE FIGURE 17–18. Port fuel-injection systems have more rich/lean O2S voltage transitions (cross counts) for a given amount of time than any other type of system because of the greatly improved system design compared to TBI units.

Port fuel-injection systems take the least amount of time to react to the fuel adaptive command (for example, changing injector pulse width).

## HASH

**BACKGROUND INFORMATION** Hash on the O2S waveform is defined as a series of high-frequency spikes, or the fuzz (or noise) viewed on some O2S waveforms, or, more specifically, oscillation frequencies higher than those created by the PCM normal feedback operation (normal rich/lean oscillations).

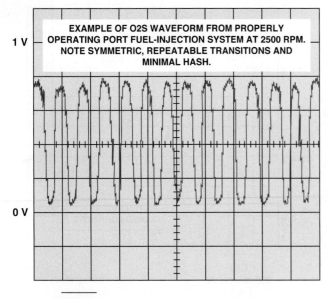

EXAMPLE OF O2S WAVEFORM FROM PROPERLY OPERATING PORT FUEL-INJECTION SYSTEM AT 2500 RPM. NOTE SYMMETRIC, REPEATABLE TRANSITIONS AND MINIMAL HASH.

1 V

0 V

CH1
200 mV/div dc
1 s/div

**FIGURE 17–18** Normal oxygen sensor frequency is from about one to five times per second.

Hash is the critical indicator of reduced combustion efficiency. Hash on the O2S waveform can warn of reduced performance in individual engine cylinders. Hash also impedes proper operation of the PCM feedback fuel control program. The feedback program is the active software program that interprets the O2S voltage and calculates a corrective mixture control command.

Generally, the program for the PCM is not designed to process O2S signal frequencies efficiently that result from events other than normal system operation and fuel control commands. The high-frequency oscillations of the hash can cause the PCM to lose control. This, in turn, has several effects. When the operating strategy of the PCM is adversely affected, the air–fuel ratio drifts out of the catalyst window, which affects converter operating efficiency, exhaust emissions, and engine performance.

Hash on the O2S waveform indicates an exhaust charge imbalance from one cylinder to another, or, more specifically, a higher oxygen content sensed from an individual combustion event. Most oxygen sensors, when working properly, can react fast enough to generate voltage deflections corresponding to a single combustion event. The bigger the amplitude of the deflection (hash), the greater the differential in oxygen content sensed from a particular combustion event.

There are vehicles that will have hash on their O2S waveforms and are operating perfectly normal. Small amounts of hash may not be of concern, and larger amounts of hash may be all important. A good rule concerning hash is, if engine performance is good, there are no vacuum leaks, and if exhaust (HC) hydrocarbon and oxygen levels are okay while hash is present on the O2S waveform, then the hash is nothing to worry about.

**CAUSES OF HASH**  Hash on the O2S signal can be caused by the following:

1. Misfiring cylinders
   - Ignition misfire
   - Lean misfire
   - Rich misfire
   - Compression-related misfire
   - Vacuum leaks
   - Injector imbalance
2. System design, such as different intake runner length
3. System design amplified by engine and component degradation caused by aging and wear
4. System manufacturing variances, such as intake runner blockage and valve stem mismachining

The spikes and hash on the waveform during a misfire event are created by incomplete combustion, which results in only partial use of the available oxygen in the cylinder. The leftover oxygen goes out the exhaust port and travels past the oxygen sensor. When the oxygen sensor "sees" the oxygen-filled exhaust charge, it quickly generates a low voltage, or spike. A series of these high-frequency spikes make up what we are calling "hash."

## CLASSIFICATIONS OF HASH

**CLASS 1: AMPLIFIED AND SIGNIFICANT HASH.**  Amplified hash is the somewhat unimportant hash that is often present between 300 and 600 mV on the O2S waveform. This type of hash is usually not important for diagnosis. That is because amplified hash is created largely as a result of the electrochemical

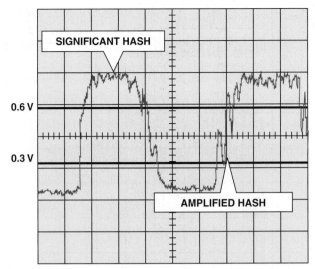

FIGURE 17–19 Significant hash can be caused by faults in one or more cylinders, whereas amplified hash is not as important for diagnosis.

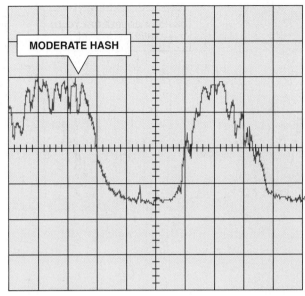

200 mV/div
1 s/div

FIGURE 17–20 Moderate hash may or may not be significant for diagnosis.

properties of the O2S itself and many times not an engine or other unrelated problem. Hash between 300 and 600 mV is not particularly conclusive, so for all practical purposes it is insignificant. ● SEE FIGURE 17–19.

Significant hash is defined as the hash that occurs above 600 mV and below 300 mV on the O2S waveform. This is the area of the waveform that the PCM is watching to determine the fuel mixture. Significant hash is important for diagnosis because it is caused by a combustion event. If the waveform exhibits class 1 hash, the combustion event problem is probably occurring in only one of the cylinders. If the event happens in a greater number of the cylinders, the waveform will become class 3 or be fixed lean or rich the majority of the time.

CLASS 2: MODERATE HASH. Moderate hash is defined as spikes shooting downward from the top arc of the waveform as the waveform carves its arc through the rich phase. Moderate hash spikes are not greater than 150 mV in amplitude. They may get as large as 200 mV in amplitude as the O2S waveform goes through 450 mV. Moderate hash may or may not be significant to a particular diagnosis. ● SEE FIGURE 17–20. For instance, most vehicles will exhibit more hash on the O2S waveform at idle. Additionally, the engine family or type of O2S could be important factors when considering the significance of moderate hash on the O2S waveform.

CLASS 3: SEVERE HASH. Severe hash is defined as hash whose amplitude is greater than 200 mV. Severe hash may even cover the entire voltage range of the sensor for an extended period of operation. Severe hash on the DSO display appears as spikes that shoot downward, over 200 mV from the top of the operating range of the sensor, or as far as to the bottom of the sensor's operating range. ● SEE FIGURE 17–21. If severe hash is present for several seconds during a steady-state engine operating mode, say 2500 RPM, it is almost always significant to the diagnosis of any vehicle. Severe hash of this nature is almost never caused by a normal system design. It is caused by cylinder misfire or mixture imbalance.

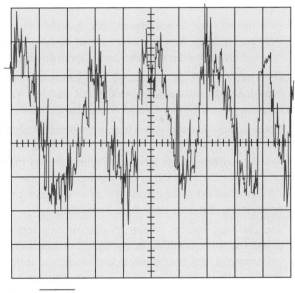

CH1
200 mV/div dc
500 ms/div

FIGURE 17–21 Severe hash is almost always caused by cylinder misfire conditions.

## HASH INTERPRETATION

### TYPES OF MISFIRES THAT CAN CAUSE HASH

1. Ignition misfire caused by a bad spark plug, spark plug wire, distributor cap, rotor, ignition coil, or ignition primary problem. Usually an engine analyzer is used to eliminate these possibilities or confirm these problems. ● SEE FIGURE 17–22.

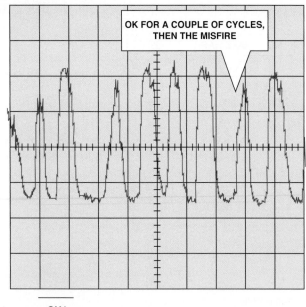

CH1
200 mV/div dc
500 ms/div

**FIGURE 17–22** An ignition- or mixture-related misfire can cause hash on the oxygen sensor waveform.

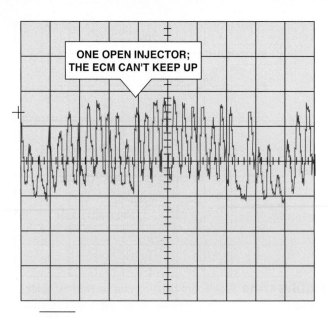

Memory 4
200 mV/div
200 ms/div

**FIGURE 17–23** An injector imbalance can cause a lean or a rich misfire.

2. Rich misfire from an excessively rich fuel delivery to an individual cylinder (various potential root causes). Air–fuel ratio in a given cylinder ventured below approximately 13:1.

3. Lean misfire from an excessively lean fuel delivery to an individual cylinder (various potential root causes). Air–fuel ratio in a given cylinder ventured above approximately 17:1.

4. Compression-related misfire from a mechanical problem that reduces compression to the point that not enough heat is generated from compressing the air–fuel mixture prior to ignition, preventing combustion. This raises O2S content in the exhaust (for example, a burned valve, broken or worn ring, flat cam lobe, or sticking valve).

5. Vacuum leak misfire unique to one or two individual cylinders. This possibility is eliminated or confirmed by inducing propane around any potential vacuum leak area (intake runners, intake manifold gaskets, vacuum hoses, etc.) while watching the DSO to see when the signal goes rich and the hash changes from ingesting the propane. Vacuum leak misfires are caused when a vacuum leak unique to one cylinder or a few individual cylinders causes the air–fuel ratio in the affected cylinder(s) to venture above approximately 17:1, causing a lean misfire.

6. Injector imbalance misfire (on port fuel-injected engines only); one cylinder has a rich or lean misfire because of an individual injector(s) delivering the wrong quantity of fuel. Injector imbalance misfires are caused when an injector on one cylinder or a few individual cylinders causes the air–fuel ratio in its cylinder(s) to venture above approximately 17:1, causing a lean misfire, or

below approximately 13.7:1, causing a rich misfire. ● SEE FIGURE 17–23.

**OTHER RULES CONCERNING HASH ON THE O2S WAVEFORM** If there is significant hash on the O2S signal that is not normal for that type of system, it will usually be accompanied by a repeatable and generally detectable engine miss at idle (for example, a thump, thump, thump every time the cylinder fires). Generally, if the hash is significant, the engine miss will correlate in time with individual spikes seen on the waveform.

Hash that may be difficult to get rid of (and is normal in some cases) will not be accompanied by a significant engine miss that corresponds with the hash. When the individual spikes that make up the hash on the waveform do not correlate in time with an engine miss, less success can usually be found in getting rid of them by performing repairs.

A fair rule of thumb is that if you are sure there are no intake vacuum leaks, the exhaust gas HC (hydrocarbon) and oxygen levels are normal, and the engine does not run or idle rough, the hash is probably acceptable or normal.

## NEGATIVE O2S VOLTAGE

When testing O2S waveforms, some oxygen sensors will exhibit some negative voltage. The acceptable amount of negative O2S voltage is −0.75 mV, provided that the maximum

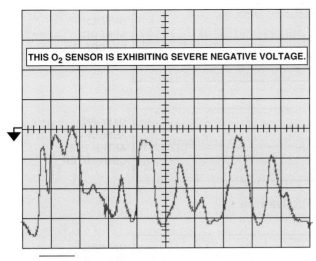

THIS O₂ SENSOR IS EXHIBITING SEVERE NEGATIVE VOLTAGE.

**WAVE 1**
**200 mV/div**
**1 s/div**

**FIGURE 17–24** Negative reading oxygen sensor voltage can be caused by several problems.

---

● SEE FIGURE 17–24. voltage peak exceeds 850 mV. Testing has shown that negative voltage signals from an oxygen sensor have usually been caused by the following:

1. Chemical poisoning of sensing element (silicon, oil, etc.)
2. Overheated engines
3. Mishandling of new oxygen sensors (dropped and banged around, resulting in a cracked insulator)
4. Poor oxygen sensor ground

## LOW O2S READINGS

An oxygen sensor reading that is low could be due to other things besides a lean air–fuel mixture. Remember, an O2S senses oxygen, not unburned gas, even though a high reading generally indicates a rich exhaust (lack of oxygen) and a low reading indicates a lean mixture (excess oxygen).

**FALSE LEAN**   If an oxygen sensor reads low as a result of a factor besides a lean mixture, it is often called a **false lean indication.**

False lean indications (low O2S readings) can be attributed to the following:

1. **Ignition misfire.** An ignition misfire due to a defective spark plug wire, fouled spark plug, and so forth causes no burned air and fuel to be exhausted past the O2S. The O2S "sees" the oxygen (not the unburned gasoline), and the O2S voltage is low.
2. **Exhaust leak in front of the O2S.** An exhaust leak between the engine and the oxygen sensor causes outside oxygen to be drawn into the exhaust and past the O2S. This oxygen is "read" by the O2S and produces a lower-than-normal voltage. The computer interrupts the lower-than-normal voltage signal from the O2S as meaning that the air–fuel mixture is lean. The computer will cause the fuel system to deliver a richer air–fuel mixture.
3. **A spark plug misfire represents a false lean signal to the oxygen sensor.** The computer does not know that the extra oxygen going past the oxygen sensor is not due to a lean air–fuel mixture. The computer commands a richer mixture, which could cause the spark plugs to foul, increasing the rate of misfirings.

## HIGH O2S READINGS

An oxygen sensor reading that is high could be due to other things beside a rich air–fuel mixture. When the O2S reads high as a result of other factors besides a rich mixture, it is often called a **false rich indication.**

False rich indication (high O2S readings) can be attributed to the following:

1. Contaminated O2S due to additives in the engine coolant or due to silicon poisoning
2. A stuck-open EGR valve (especially at idle)
3. A spark plug wire too close to the oxygen sensor signal wire, which can induce a higher-than-normal voltage in the signal wire, thereby indicating to the computer a false rich condition
4. A loose oxygen sensor ground connection, which can cause a higher-than-normal voltage and a false rich signal
5. A break or contamination of the wiring and its connectors, which could prevent reference oxygen from reaching the oxygen sensor, resulting in a false rich indication (All oxygen sensors require an oxygen supply inside the sensor itself for reference to be able to sense exhaust gas oxygen.)

---

### TECH TIP

**Look for Missing Shield**

In rare (very rare) instances, the metal shield on the exhaust side of the oxygen sensor (the shield over the zirconia thimble) may be damaged (or missing) and may create hash on the O2S waveform that could be mistaken for bad injectors or other misfires, vacuum leaks, or compression problems. After you have checked everything and possibly replaced the injectors, pull the O2S to check for rare situations.

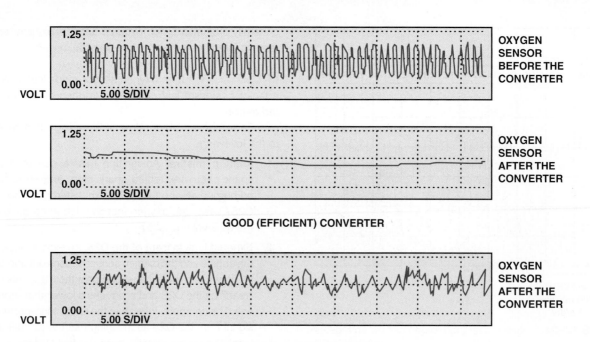

OXYGEN
SENSOR
BEFORE THE
CONVERTER

OXYGEN
SENSOR
AFTER THE
CONVERTER

**GOOD (EFFICIENT) CONVERTER**

OXYGEN
SENSOR
AFTER THE
CONVERTER

**BAD (INEFFICIENT) CONVERTER**

**FIGURE 17–25** The post-catalytic converter oxygen sensor should display very little activity if the catalytic converter is efficient.

## POST-CATALYTIC CONVERTER OXYGEN SENSOR TESTING

The oxygen sensor located behind the catalytic converter is used on OBD-II vehicles to monitor converter efficiency. A changing air–fuel mixture is required for the most efficient operation of the converter. If the converter is working correctly, the oxygen content after the converter should be fairly constant. ●**SEE FIGURE 17–25.**

## OXYGEN SENSOR VISUAL INSPECTION

Whenever an oxygen sensor is replaced, the old sensor should be carefully inspected to help determine the cause of the failure. This is an important step because if the cause of the failure is not discovered, it could lead to another sensor failure.

Inspection may reveal the following:

1. **Black sooty deposits** usually indicate a rich air–fuel mixture.

 **FREQUENTLY ASKED QUESTION**

**What Is Lambda?**

An oxygen sensor is also called a lambda sensor because the voltage changes at the air–fuel ratio of 14.7:1, which is the stoichiometric rate for gasoline. If this mixture of gasoline and air is burned, all of the gasoline is burned and uses all of the oxygen in the mixture. This exact ratio represents a lambda of 1.0. If the mixture is richer (more fuel or less air), the number is less than 1.0, such as 0.95 (5% rich). If the mixture is leaner than 14.7:1 (less fuel or more air), the lambda number is higher than 1.0, such as 1.05 (5% lean). Often, the target lambda is displayed on a scan tool. ●**SEE FIGURE 17–26.**

2. **White chalky deposits** are characteristic of silica contamination. Usual causes for this type of sensor failure include silica deposits in the fuel or a technician having used the wrong type of silicone sealant during the servicing of the engine.

3. **White sandy or gritty deposits** are characteristic of antifreeze (ethylene glycol) contamination. A defective cylinder head or intake manifold gasket could be the cause, as could a cracked cylinder head or engine block. Antifreeze

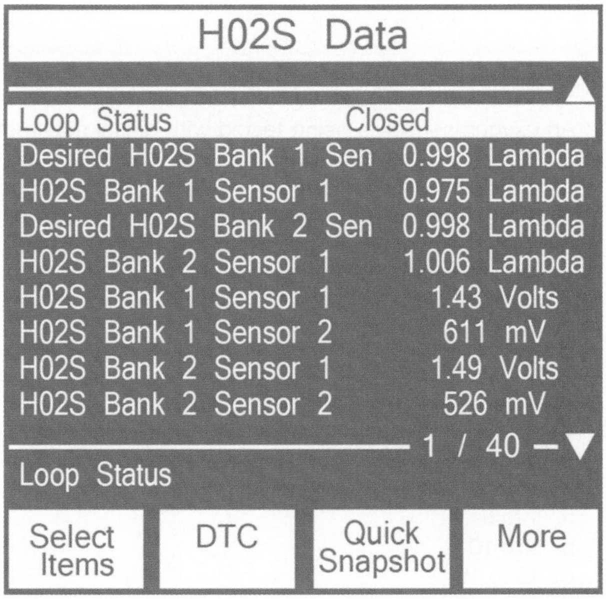

```
┌─────────────────────────────────────┐
│          H02S  Data                  │
├─────────────────────────────────────┤
│                              ▲        │
│ Loop  Status          Closed         │
│ Desired  H02S  Bank  1  Sen   0.998 Lambda │
│ H02S  Bank  1  Sensor  1      0.975 Lambda │
│ Desired  H02S  Bank  2  Sen   0.998 Lambda │
│ H02S  Bank  2  Sensor  1      1.006 Lambda │
│ H02S  Bank  1  Sensor  1       1.43 Volts  │
│ H02S  Bank  1  Sensor  2        611 mV     │
│ H02S  Bank  2  Sensor  1       1.49 Volts  │
│ H02S  Bank  2  Sensor  2        526 mV     │
│                        — 1 / 40 — ▼        │
│ Loop  Status                          │
├────────┬────────┬──────────┬──────────┤
│ Select │  DTC   │  Quick   │   More   │
│ Items  │        │ Snapshot │          │
└────────┴────────┴──────────┴──────────┘
```

**FIGURE 17–26** The target lambda on this vehicle is slightly lower than 1.0, indicating that the PCM is attempting to supply the engine with an air–fuel mixture that is slightly richer than stoichiometric. Multiply the lambda number by 14.7 to find the actual air–fuel ratio.

may also cause the oxygen sensor to become green as a result of the dye used in antifreeze.

4. **Dark brown deposits** are an indication of excessive oil consumption. Possible causes include a defective positive crankcase ventilation (PCV) system or a mechanical engine problem, such as defective valve stem seals or piston rings.

**CAUTION: Do not spray any silicone spray near the engine where the engine vacuum could draw the fumes into the engine. This can also cause silica damage to the oxygen sensor. Also be sure that the silicone sealer used for gaskets is rated oxygen-sensor safe.**

# O2S-RELATED DIAGNOSTIC TROUBLE CODES

Diagnostic trouble codes (DTCs) associated with the oxygen sensor include the following:

| Diagnostic Trouble Code | Description | Possible Causes |
|---|---|---|
| P0131 | Upstream HO2S grounded | • Exhaust leak upstream of HO2S (bank 1)<br>• Extremely lean air–fuel mixture<br>• HO2S defective or contaminated<br>• HO2S signal wire shorted-to-ground |
| P0132 | Upstream HO2S shorted | • Upstream HO2S (bank 1) shorted<br>• Defective HO2S<br>• Fuel-contaminated HO2S |
| P0133 | Upstream HO2S slow response | • Open or short in heater circuit<br>• Defective or fuel-contaminated HO2S<br>• EGR or fuel-system fault |

# SUMMARY

1. An oxygen sensor produces a voltage output signal based on the oxygen content of the exhaust stream.
2. If the exhaust has little oxygen, the voltage of the oxygen sensor will be close to 1 volt (1,000 mV) and close to zero if there is high oxygen content in the exhaust.
3. Oxygen sensors can have one, two, three, four, or more wires, depending on the style and design.
4. The oxygen sensor signal determines fuel trim, which is used to tailor the air–fuel mixture for the catalytic converter.
5. Conditions can occur that cause the oxygen sensor to be fooled and give a false lean or false rich signals to the PCM.
6. Oxygen sensors can be tested using a digital meter, a scope, or a scan tool.

# REVIEW QUESTIONS

1. How does an oxygen sensor detect oxygen levels in the exhaust?
2. What are four basic designs of oxygen sensors, and how many wires may be used for each?
3. What is the difference between open-loop and closed-loop engine operation?
4. What are three ways oxygen sensors can be tested?
5. How can the oxygen sensor be fooled and provide the wrong information to the PCM?

1. The sensor that must be warmed and functioning before the engine management computer will go to closed loop is the _____.
   a. O2S
   b. ECT sensor
   c. Engine MAP sensor
   d. BARO sensor

2. The voltage output of a zirconia oxygen sensor when the exhaust stream is lean (excess oxygen) is _____.
   a. Relatively high (close to 1 volt)
   b. About in the middle of the voltage range
   c. Relatively low (close to zero volt)
   d. Either a or b, depending on atmospheric pressure

3. Where is sensor 1, bank 1 located on a V-type engine?
   a. On the same bank where number 1 cylinder is located
   b. In the exhaust system upstream of the catalytic converter
   c. On the bank opposite cylinder number 1
   d. Both a and b

4. A heated zirconia oxygen sensor will have how many wires?
   a. Two
   b. Three
   c. Four
   d. Either b or c

5. A high O2S voltage could be due to a _____.
   a. Rich exhaust
   b. Lean exhaust
   c. Defective spark plug wire
   d. Both a and c

6. A low O2S voltage could be due to a _____.
   a. Rich exhaust
   b. Lean exhaust
   c. Defective spark plug wire
   d. Both b and c

7. An oxygen sensor is being tested with digital multimeter (DMM), using the MIN/MAX function. The readings are minimum = 78 mV, maximum = 932 mV, and average = 442 mV. Technician A says that the engine is operating correctly. Technician B says that the oxygen sensor is skewed too rich. Which technician is correct?
   a. Technician A only
   b. Technician B only
   c. Both Technicians A and B
   d. Neither Technician A nor B

8. An oxygen sensor is being tested using a digital storage oscilloscope (DSO). A good oxygen sensor should display how many switches per second?
   a. 1 to 5
   b. 5 to 10
   c. 10 to 15
   d. 15 to 20

9. When testing an oxygen sensor using a digital storage oscilloscope (DSO), how quickly should the voltage change when either propane is added to the intake stream or a vacuum leak is created?
   a. Less than 50 ms
   b. 1 to 3 seconds
   c. Less than 100 ms
   d. 450 to 550 ms

10. A P0133 DTC is being discussed. Technician A says that a defective heater circuit could be the cause. Technician B says that a contaminated sensor could be the cause. Which technician is correct?
    a. Technician A only
    b. Technician B only
    c. Both Technicians A and B
    d. Neither Technician A nor B

# chapter 18

# WIDE-BAND OXYGEN SENSORS

**LEARNING OBJECTIVES:** **After studying this chapter, the reader should be able to:** • Discuss the need for wide-band oxygen sensors compared to a narrow-band O2S. • Explain the working of dual-cell planar wide-band sensors and their diagnosis. • Explain the working of single-cell wide-band oxygen sensors and their diagnosis.

**KEY TERMS:** Air–fuel ratio sensor 234 • Air reference chamber 232 • Ambient air electrode 231 • Ambient side electrode 231 • Cup design 231 • Diffusion chamber 232 • Dual cell 232 • Exhaust side electrode 231 • Finger design 231 • Linear air–fuel (LAF) sensor 229 • Light-off time (LOT) 231 • Nernst cell 232 • Planar design 231 • Pump cell 232 • Reference electrode 231 • Reference voltage 232 • Signal electrode 231 • Single-cell 234 • Thimble design 231

## TERMINOLOGY

Honda was the first manufacturer to use wide-band oxygen sensors beginning in 1992. Wide-band oxygen sensors are used by most vehicle manufacturers to ensure that the exhaust emissions can meet the current standard. Wide-band oxygen sensors are also called by various names, depending on the vehicle and/or oxygen sensor manufacturer. The terms used include following:

- **Wide-band oxygen sensor**
- **Broadband oxygen sensor**
- **Wide-range oxygen sensor**
- **Air–fuel ratio (AFR) sensor**
- **Wide-range air–fuel (WRAF) sensor**
- **Linear air–fuel (LAF) sensor**
- **Air–fuel (AF) sensor**

Wide-band oxygen sensors are also manufactured in dual-cell and single-cell designs.

## NEED FOR WIDE-BAND SENSORS

**INTRODUCTION** A conventional zirconia oxygen sensor reacts to an air–fuel mixture that is either richer or leaner than 14.7:1. This means that the sensor cannot be used to detect the exact air–fuel mixture. ● **SEE FIGURE 18–1.**

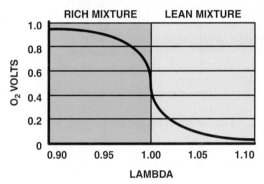

**FIGURE 18–1** A conventional zirconia oxygen sensor can only reset to exhaust mixtures that are richer or leaner than 14.7:1 (lambda 1.00).

The need for more stringent exhaust emission standards such as the national low-emission vehicle (NLEV) plus the ultra low-emission vehicle (ULEV) and the super ultra low-emission vehicle (SULEV) require more accurate fuel control than can be provided by a traditional oxygen sensor.

**PURPOSE AND FUNCTION** A wide-band oxygen sensor is capable of supplying air–fuel ratio information to the PCM over a much broader range. The use of a wide-band oxygen sensor compared with a conventional zirconia oxygen sensor differs as follows:

1. Able to detect exhaust air–fuel ratio from as rich as 10:1 and as lean as 23:1 in some cases

2. Cold-start activity within as little as 10 seconds

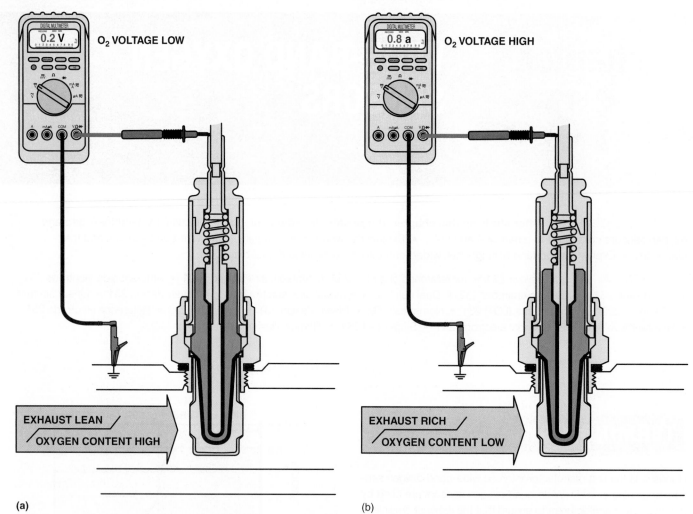

**O₂ VOLTAGE LOW**

**0.2 V**

**O₂ VOLTAGE HIGH**

**0.8 a**

EXHAUST LEAN / OXYGEN CONTENT HIGH

EXHAUST RICH / OXYGEN CONTENT LOW

(a)　　　　　　　　　　　　　　　　　　　　(b)

**FIGURE 18–2** (a) When the exhaust is lean, the output of a zirconia oxygen sensor is below 450 mV. (b) When the exhaust is rich, the output of a zirconia oxygen sensor is above 450 mV.

**? FREQUENTLY ASKED QUESTION**

**How Quickly Can a Wide-Band Oxygen Sensor Achieve Closed Loop?**

In a Toyota Highlander hybrid electric vehicle, the operation of the gasoline engine is delayed for a short time when the vehicle is first driven. During this time of electric operation, the oxygen sensor heaters are turned on in readiness for the gasoline engine starting. The gasoline engine often achieves closed-loop operation during *cranking* because the oxygen sensors are fully warm and ready to go at the same time the engine is started. Having the gasoline engine achieve closed loop quickly allows it to meet the stringent SULEV standards.

# CONVENTIONAL O2S REVIEW

**NARROW BAND**　A conventional zirconia oxygen sensor (O2S) is only able to detect if the exhaust is richer or leaner than 14.7:1. A conventional oxygen sensor is therefore referred to as follows:

- **2-step sensor**—either rich or lean
- **Narrow band sensor**—informs the PCM whether the exhaust is rich or lean only

The voltage value where a zirconia oxygen sensor switches from rich to lean or from lean to rich is 0.450 V (450 mV):

- Above 0.450 V = rich
- Below 0.450 V = lean
- **SEE FIGURE 18–2.**

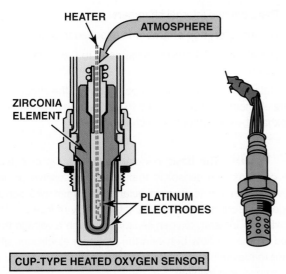

**FIGURE 18–3** Most conventional zirconia oxygen sensors and some wide-band oxygen sensors use the cup-type design.

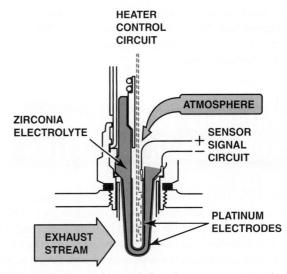

**FIGURE 18–4** A typical heated zirconia oxygen sensor, showing the sensor signal circuit that uses the outer (exhaust) electrode as negative and the ambient air side electrode as the positive.

## CONSTRUCTION
A typical zirconia oxygen sensor has the sensing element in the shape of a thimble and is often referred to as follows:

- **Thimble design**
- **Cup design**
- **Finger design**
  ● **SEE FIGURE 18–3.**

A typical zirconia oxygen sensor has a heater inside the thimble and does not touch the inside of the sensor. The sensor is similar to a battery that has two electrodes and an electrolyte. The electrolyte is solid and is the zirconia (zirconium dioxide). There are also two porous platinum electrodes, which have the following functions:

- **Exhaust side electrode**—This electrode is exposed to the exhaust stream.
- **Ambient side electrode**—This electrode is exposed to outside (ambient) air and is the **signal electrode,** also called the **reference electrode** or **ambient air electrode.**
  ● **SEE FIGURE 18–4.**

The electrolyte (zirconia) is able to conduct electrons as follows:

- If the exhaust is rich, $O_2$ from the reference (inner) electrode wants to flow to the exhaust side electrode, which results in the generation of a voltage.
- If the exhaust is lean, $O_2$ flow is not needed, and, as a result, there is little, if any, electron movement, and therefore no voltage is being produced.

## HEATER CIRCUITS
The heater circuit on conventional oxygen sensors requires 0.8 to 2 amperes, and it keeps the sensor at about 600°F (315°C).

A wide-band oxygen sensor operates at a higher temperature than a conventional HO2S from 1,200°F to 1,400°F (650°C to 760°C). The amount of electrical current needed for a wide-band oxygen sensor is about 8 to 10 amperes.

## PLANAR DESIGN
In 1998, Bosch introduced a wide-band oxygen sensor that is flat and thin (1.5 mm, or 0.006 inches), known as a planar design and not in the shape of a thimble, as previously constructed. Now several manufacturers produce a similar planar design wide-band oxygen sensor. Because it is thin, it is easier to heat than older styles of oxygen sensors and as a result can achieve closed loop in less than 10 seconds. This fast heating, called **light-off time (LOT),** helps improve fuel economy and reduces cold-start exhaust emissions. The type of construction is not noticed by the technician, nor does it affect the testing procedures.

A conventional oxygen sensor can be constructed using a **planar design** instead of the thimble-type design. A planar design has the following features:

- The elements including the zirconia electrolyte and the two electrodes and heater are stacked together in a flat-type design.
- The planar design allows faster warm-up because the heater is in direct contact with the other elements.
- Planar oxygen sensors are the most commonly used. Some planar designs are used as a conventional narrow-band oxygen sensor.

The sandwich-type design of the planar style of oxygen sensor has the same elements and operates the same but is stacked in the following way from the exhaust side to the ambient air side:

Exhaust stream

Outer electrode

Zirconia ($ZiO_2$) (electrolyte)

Inner electrode (reference or signal)

Outside (ambient) air

Heater

● **SEE FIGURE 18–5**

**NOTE:** Another name for a conventional oxygen sensor is a **Nernst cell.** The Nernst cell is named for Walther Nernst, 1864–1941, a German physicist known for his work in electrochemistry.

# DUAL-CELL PLANAR WIDE-BAND SENSOR OPERATION

**CONSTRUCTION** In a conventional zirconia oxygen sensor, a bias or **reference voltage** can be applied to the two platinum electrodes, and then oxygen ions can be forced (pumped) from the ambient reference air side to the exhaust side of the sensor. If the polarity is reversed, the oxygen ion can be forced to travel in the opposite direction.

A **dual cell** planar-type wide-band oxygen sensor is made like a conventional planar O2S and is labeled Nernst cell. Above the Nernst cell is another zirconia layer with two electrodes, which is called the **pump cell.** The two cells share a common ground, which is called the reference.

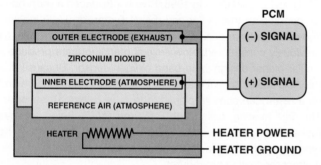

**FIGURE 18–5** A planar design zirconia oxygen sensor places all of the elements together, which allows the sensor to reach operating temperature quickly.

There are two internal chambers:

- The **air reference chamber** is exposed to ambient air.
- The **diffusion chamber** is exposed to the exhaust gases.

Platinum electrodes are on both sides of the zirconia electrolyte elements, which separate the air reference chamber and the exhaust-exposed diffusion chamber.

**OPERATION** The basic principle of operation of a typical wide-band oxygen sensor is that it uses a positive or negative voltage signal to keep a balance between two sensors. Oxygen sensors do not measure the quantity of free oxygen in the exhaust. Instead, oxygen sensors produce a voltage that is based on the ion flow between the platinum electrodes of the sensor to maintain a stoichiometric balance.

An example follows:

- If there is a lean exhaust, there is oxygen in the exhaust, and the ion flow from the ambient side to the exhaust side is low.
- If there were rich exhaust, the ion flow is increased to help maintain balance between the ambient air side and the exhaust side of the sensor.

The PCM can apply a small current to the pump cell electrodes, which causes oxygen ions through the zirconia into or out of the diffusion chamber. The PCM pumps $O_2$ ions in and out of the diffusion chamber to bring the voltage back to 0.45, using the pump cell.

The operation of a wide-band oxygen sensor is best described by looking at what occurs when the exhaust is stoichiometric, rich, and lean. ● **SEE FIGURE 18–6.**

## STOICHIOMETRIC

- When the exhaust is at stoichiometric (14.7:1 air–fuel ratio), the voltage of the Nernst cell is 450 mV (0.450 V).

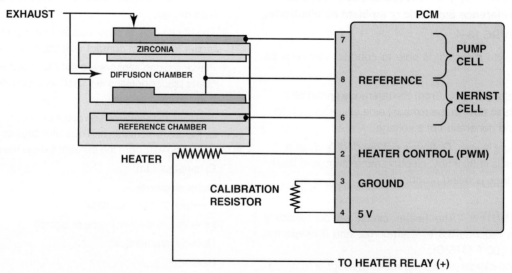

**FIGURE 18–6** The reference electrodes are shared by the Nernst cell and the pump cell.

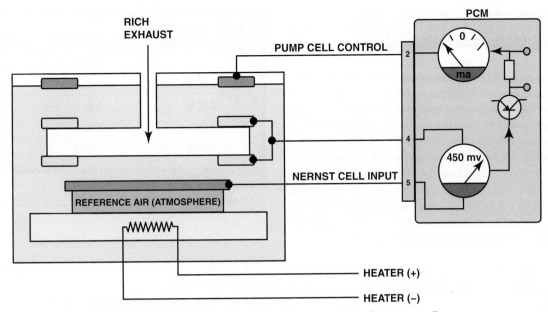

**FIGURE 18–7** When the exhaust is rich, the PCM applies a negative current into the pump cell.

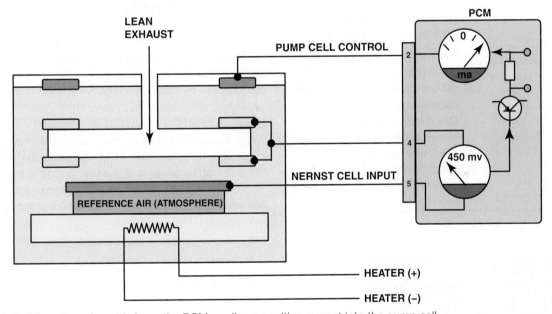

**FIGURE 18–8** When the exhaust is lean, the PCM applies a positive current into the pump cell.

- The voltage between the diffusion chamber and the air reference chamber changes from 0.45 V. This voltage will be as follows:

  - Higher if the exhaust is rich
  - Lower if the exhaust is lean

The reference voltage remains constant, usually at 2.5 volts, but can vary depending on the year, make, and model of vehicle and the type of sensor. Typical reference voltages include the following:

- 2.2
- 2.5
- 2.7
- 3.3
- 3.6

**RICH EXHAUST.** When the exhaust is rich, the voltage between the common (reference) electrode and the Nernst cell electrode that is exposed to ambient air is higher than 0.45 V. The PCM applies a negative current in milliamperes to the pump cell electrode to bring the circuit back into balance. ● **SEE FIGURE 18–7.**

**LEAN EXHAUST.** When the exhaust is lean, the voltage between the common (reference) electrode and the Nernst cell electrode is lower than 0.45 V. The PCM applies a positive current in milliamperes to the pump cell to bring the circuit back into balance. ● **SEE FIGURE 18–8.**

# DUAL-CELL DIAGNOSIS

## SCAN TOOL DIAGNOSIS

Most service information specifies that a scan tool be used to check the wide-band oxygen sensor. This is because the PCM performs tests of the unit and can identify faults. However, even wide-band oxygen sensors can be fooled if there is an exhaust manifold leak or other fault that could lead to false or inaccurate readings. If the oxygen sensor reading is false, the PCM will command an incorrect amount of fuel. The scan data shown on a generic (global) OBD-II scan tool will often be different than the reading on the factory scan tool. ● **SEE CHART 18–1** for an example of a Toyota wide-band oxygen sensor being tested using a factory scan tool and a generic OBD-II scan tool.

## SCAN TOOL DATA (PID)

The following information will be displayed on a scan tool when looking at data for a wide-band oxygen sensor:

| | |
|---|---|
| HO2S1 = _____ mA | If the current is positive, it means that the PCM is pumping current in the diffusion gap because of a rich exhaust. |
| | If the current is negative, the PCM is pumping current out of the diffusion gap because of a lean exhaust. |
| Air–fuel ratio = _____ | Usually expressed in lambda. One means that the exhaust is at stoichiometric (14.7:1 air–fuel ratio), and numbers higher than 1 indicate a lean exhaust, and numbers lower than 1 indicate a rich exhaust. |

# DIGITAL MULTIMETER TESTING

When testing a wide-band oxygen sensor for proper operation, perform the following steps:

**STEP 1** Check service information and determine the circuit and connector terminal identification.

**STEP 2** Measure the calibration resistor. While the value of this resistor can vary widely, depending on the type of sensor, the calibrating resistor should still be checked for opens and shorts.

NOTE: **The calibration resistor is usually located within the connector itself.**

- If open, the ohmmeter will read OL (infinity ohms).
- If shorted, the ohmmeter will read zero or close to zero.

| MASTER TECH TOYOTA (FACTORY SCAN TOOL) | OBD-II SCAN TOOL | AIR–FUEL RATIO |
|---|---|---|
| 2.50 V | 0.50 V | 12.5:1 |
| 3.00 V | 0.60 V | 14.0:1 |
| 3.30 V | 0.66 V | 14.7:1 |
| 3.50 V | 0.70 V | 15.5:1 |
| 4.00 V | 0.80 V | 18.5:1 |

**CHART 18–1**

A comparison showing what a factory scan tool and a generic OBD-II scan tool might display at various air–fuel ratios.

**STEP 3** Measure the heater circuit for proper resistance or current flow.

**STEP 4** Measure the reference voltage relative to ground. This can vary but is generally 2.4 to 2.6 volts.

**STEP 5** Using jumper wires, connect an ammeter and measure the current in the pump cell control wire.

## RICH EXHAUST (LAMBDA LESS THAN 1.00)

When the exhaust is rich, the Nernst cell voltage will move higher than 0.45 volt. The PCM will pump oxygen from the exhaust into the diffusion gap by applying a negative voltage to the pump cell.

## LEAN EXHAUST (LAMBDA HIGHER THAN 1.00)

When the exhaust is lean, the Nernst cell voltage will move lower than 0.45 volt. The PCM will pump oxygen out of the diffusion gap by applying a positive voltage to the pump cell.

The pump cell is used to pump oxygen into the diffusion gap when the exhaust is rich. The pump cell applies a negative voltage to do this:

- Positive current = lean exhaust
- Negative current = rich exhaust
- ● **SEE FIGURE 18–9.**

# SINGLE-CELL WIDE-BAND OXYGEN SENSORS

## CONSTRUCTION

A typical **single-cell** wide-band oxygen sensor looks similar to a conventional four-wire zirconia oxygen sensor. The typical single-cell wide-band oxygen sensor, usually called an **air–fuel ratio sensor,** has the following construction features:

- It can be made using the cup or planar design.
- Oxygen ($O_2$) is pumped into the diffusion layer similar to the operation of a dual-cell wide-band oxygen sensor. ● **SEE FIGURE 18–10.**

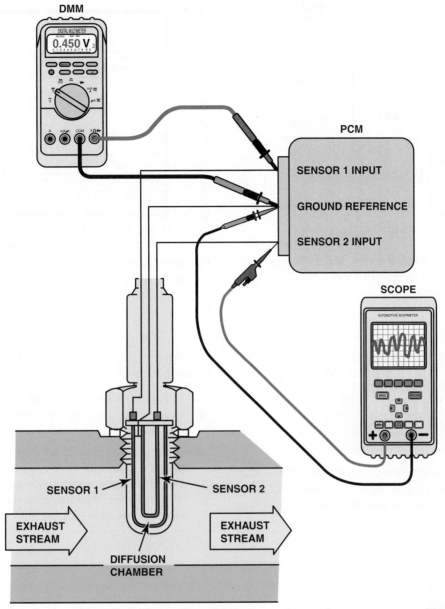

**FIGURE 18–9** Testing a dual-cell wide-band oxygen sensor can be done using a voltmeter or a scope. The meter reading is attached to the Nernst cell (sensor 1) and should read stoichiometric (450 mV) at all times. The scope is showing activity to the pump cell (sensor 2) with commands from the PCM to keep the Nernst cell at 14.7:1 air–fuel ratio.

- Current flow reverses positive and negative.
- It consists of two cell wires and two heater wires (power and ground).
- The heater usually requires 6 amperes and the ground side is pulse-width modulated.

**TESTING WITH A MILLIAMMETER**   The PCM controls the single-cell wide-band oxygen sensor by maintaining a voltage difference of 300 mV (0.3 V) between the two sensor leads. The PCM keeps the voltage difference constant under all operating conditions by increasing or decreasing current between the element of the cell:

- Zero (0 mA) represents lambda or stoichiometric air–fuel ratio of 14.7:1
- +10 mA indicates a lean condition
- −10 mA indicates a rich condition

**TESTING USING A SCAN TOOL**   A scan tool will display a voltage reading but can vary depending on the type and maker of scan tool. ● **SEE FIGURE 18–11.**

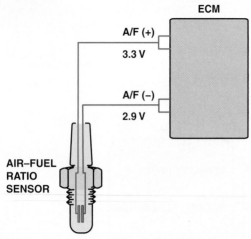

FIGURE 18–10 A single-cell wide-band oxygen sensor has four wires with two for the heater and two for the sensor itself. The voltage applied to the sensor is 0.4 volt (3.3 − 2.9 = 0.4) across the two leads of the sensor.

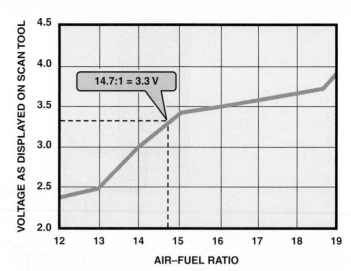

FIGURE 18–11 The scan tool can display various voltages but will often show 3.3 volts because the PCM is controlling the sensor by applying a low current to the sensor to achieve balance.

# WIDE-BAND OXYGEN SENSOR PATTERN FAILURES

Wide-band oxygen sensors have a long life but can fail. Most of the failures will cause a diagnostic trouble code (DTC) to set, usually causing the malfunction indicator (check engine) lamp to light.

However, one type of failure may not set a DTC when the following occurs:

1. Voltage from the heater circuit bleeds into the Nernst cell.
2. This voltage will cause the engine to operate extremely lean and may or may not set a diagnostic trouble code.
3. When testing indicates an extremely lean condition, unplug the connector to the oxygen sensor. If the engine starts to operate correctly with the sensor unplugged, this is confirmation that the wide-band oxygen sensor has failed and requires replacement.

## SUMMARY

1. Wide-band oxygen sensors are known by many different terms, including the following:
   - Broadband oxygen sensor
   - Wide-range oxygen sensor
   - Air–fuel ratio (AFR) sensor
   - Wide-range air–fuel (WRAF) sensor
   - Linear air–fuel (LAF) sensor
   - Air–fuel (AF) sensor
2. Wide-band oxygen sensors are manufactured using a cup or planar design and are dual-cell or single-cell design.
3. A wide-band oxygen sensor is capable of furnishing the PCM with exhaust air–fuel ratios as rich as 10:1 and as lean as 23:1.
4. The use of a wide-band oxygen sensor allows the engine to achieve more stringent exhaust emission standards.
5. A conventional zirconia oxygen sensor can be made in a cup shape or planar design and is sometimes called a narrow band or two-step sensor.

6. The heater used on a conventional zirconia oxygen sensor uses up to 2 amperes and heats the sensor to about 600°F (315°C). A broadband sensor heater has to heat the sensor to 1,200°F to 1,400°F (650°C to 760°C) and requires up to 8 to 10 amperes.
7. A typical dual-cell wide-band oxygen sensor uses the PCM to apply a current to the pump cell to keep the Nernst cell at 14.7:1:
   - When the exhaust is rich, the PCM applies a negative current to the pump cell.
   - When the exhaust is lean, the PCM applies a positive current to the pump cell.
8. Wide-band oxygen sensors can also be made using a single-cell design.
9. Wide-band oxygen sensors can be best tested using a scan tool, but dual-cell sensors can be checked with a voltmeter or scope. Single-cell sensors can be checked using a milliammeter.

1. What type of construction is used to make wide-band oxygen sensors?

2. Why are wide-band oxygen sensors used instead of conventional zirconia sensors?

3. How is the heater different for a wide-band oxygen sensor compared with a conventional zirconia oxygen sensor?

4. How does a wide-range oxygen sensor work?

5. How can a wide-band oxygen sensor be tested?

## CHAPTER QUIZ

1. A wide-band oxygen sensor was first used on a Honda in what model year?
   a. 1992
   b. 1996
   c. 2000
   d. 2006

2. A wide-band oxygen sensor is capable of detecting the air–fuel mixture in the exhaust from _____ (rich) to _____ (lean).
   a. 12:1 to 15:1
   b. 13:1 to 16.7:1
   c. 10:1 to 23:1
   d. 8:1 to 18:1

3. A conventional zirconia oxygen sensor can be made with what designs?
   a. Cup and thimble
   b. Cup and planar
   c. Finger and thimble
   d. Dual cell and single cell

4. A wide-band oxygen sensor can be made using what design?
   a. Cup and thimble
   b. Cup and planar
   c. Finger and thimble
   d. Dual cell and single cell

5. A wide-band oxygen sensor heater could draw how much current (amperes)?
   a. 0.8 to 2 A
   b. 2 to 4 A
   c. 6 to 8 A
   d. 8 to 10 A

6. A wide-band oxygen sensor needs to be heated to what operating temperature?
   a. 600°F (315°C)
   b. 800°F (427°C)
   c. 1,400°F (760°C)
   d. 2,000°F (1,093°C)

7. The two internal chambers of a dual-cell wide-band oxygen sensor include _____.
   a. Single and dual
   b. Nernst and pump
   c. Air reference and diffusion
   d. Inside and outside

8. When the exhaust is rich, the PCM applies a _____ current into the pump cell.
   a. Positive
   b. Negative

9. When the exhaust is lean, the PCM applies a _____ current into the pump cell.
   a. Positive
   b. Negative

10. A dual-cell wide-band oxygen sensor can be tested using a _____.
    a. Scan tool
    b. Voltmeter
    c. Scope
    d. All of the above

# FUEL PUMPS, LINES, AND FILTERS

**LEARNING OBJECTIVES:** After studying this chapter, the reader should be able to: • Discuss the purpose and function of the fuel delivery system. • Explain the types of fuel lines. • Discuss the different types of electric fuel pumps. • Describe the purpose and function of fuel filters. • Describe how to test and replace fuel pumps.

**KEY TERMS:** Accumulator 247 • Baffle 238 • Check valve 240 • Delivery system 238 • Filter basket 249 • Gerotor 244 • Hydrokinetic pump 244 • Inertia switch 240 • Onboard refueling vapor recovery (ORVR) 239 • Peripheral pump 244 • Residual or rest pressure 243 • Roller cell 243 • Rotary vane pump 243 • Side-channel pump 244 • Turbine pump 244 • Vacuum lock 240 • Vapor lock 240 • Volatile organic compound (VOC) 243

## FUEL DELIVERY SYSTEM

Creating and maintaining a correct air–fuel mixture requires a properly functioning fuel and air **delivery system**. Fuel delivery (and return) systems use many if not all of the following components to make certain that fuel is available under the right conditions to the fuel-injection system:

- Fuel storage tank, filler neck, and gas cap
- Fuel tank pressure sensor
- Fuel pump
- Fuel filter(s)
- Fuel delivery lines and fuel rail
- Fuel-pressure regulator
- Fuel return line (if equipped with a return-type fuel delivery system)

## FUEL TANKS

A vehicle fuel tank is made of corrosion-resistant steel or polyethylene plastic. Some models, such as sport utility vehicles (SUVs) and light trucks, may have an auxiliary fuel tank.

Tank design and capacity are a compromise between available space, filler location, fuel expansion room, and fuel movement. Some later-model tanks deliberately limit tank capacity by extending the filler tube neck into the tank low enough to prevent complete filling or by providing for expansion room. ● **SEE FIGURE 19–1.** A vertical **baffle** in fuel tanks limits fuel sloshing as the vehicle moves.

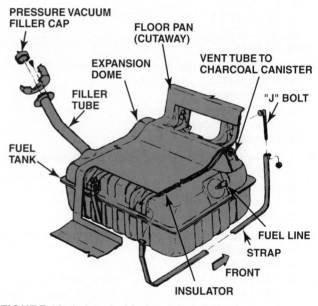

**FIGURE 19–1** A typical fuel tank installation.

Regardless of size and shape, all fuel tanks incorporate most if not all of the following features:

- Inlet or filler tube through which fuel enters the tank
- Filler cap with pressure holding and relief features
- An outlet to the fuel line leading to the fuel pump or fuel injector
- Fuel pump mounted within the tank
- Tank vent system
- Fuel pickup tube and fuel level sending unit

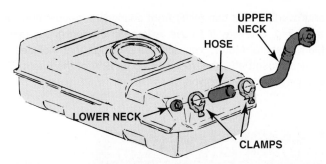

FIGURE 19–2 A three-piece filler tube assembly. The main three parts include the upper neck, hose, and lower neck.

FIGURE 19–3 A view of a typical filler tube with the fuel tank removed. Notice the ground strap used to help prevent the buildup of static electricity as the fuel flows into the plastic tank. The check ball looks exactly like a ping-pong ball.

FIGURE 19–4 Vehicles equipped with onboard refueling vapor recovery usually have a reduced-size fill tube.

## TANK LOCATION AND MOUNTING
Most vehicles use a horizontally suspended fuel tank, usually mounted below the rear of the floor pan, just ahead of or behind the rear axle. Fuel tanks are located there so that frame rails and body components protect the tank in the event of a crash. To prevent squeaks, some models have insulated strips cemented on the top or sides of the tank wherever it contacts the underbody.

Fuel inlet location depends on the tank design and filler tube placement. It is located behind a filler cap and is often a hinged door in the outer side of either rear fender panel.

Generally, a pair of metal retaining straps holds a fuel tank in place. Underbody brackets or support panels hold the strap ends using bolts. The free ends are drawn underneath the tank to hold it in place, then bolted to other support brackets or to a frame member on the opposite side of the tank.

## FILLER TUBES
Fuel enters the tank through a large tube extending from the tank to an opening on the outside of the vehicle. ● SEE FIGURE 19–2.

Effective in 1993, federal regulations require manufacturers to install a device to prevent fuel from being siphoned through the filler neck. Federal authorities recognized methanol as a poison, and methanol used in gasoline is a definite health hazard. Additionally, gasoline is a suspected carcinogen (cancer-causing agent). To prevent siphoning, manufacturers welded a filler-neck check-ball tube in fuel tanks. To drain check-ball-equipped fuel tanks, a technician must disconnect the check-ball tube at the tank and attach a siphon directly to the tank. ● SEE FIGURE 19–3.

**Onboard refueling vapor recovery (ORVR)** systems have been developed to reduce evaporative emissions during refueling. ● SEE FIGURE 19–4. These systems add components to the filler neck and the tank. One ORVR system utilizes a tapered filler neck with a smaller diameter tube and a check valve. When fuel flows down the neck, it opens the normally closed check valve. The vapor passage to the charcoal canister is opened. The decreased size neck and the opened air passage allow fuel and vapor to flow rapidly into the tank and the canister respectively. When the fuel has reached a pre-determined level, the check valve closes, and the fuel tank

pressure increases. This forces the nozzle to shut off, thereby preventing the tank from being overfilled.

## PRESSURE-VACUUM FILLER CAP
Fuel and vapors are sealed in the tank by the safety filler cap. The safety cap must release excess pressure or excess vacuum. Either condition could cause fuel tank damage, fuel spills, and vapor escape. Typically, the cap will release if the pressure is over 1.5 to 2 PSI (10 to 14 kPa) or if the vacuum is 0.15 to 0.3 PSI (1 to 2 kPa).

## FUEL PICKUP TUBE
The fuel pickup tube is usually a part of the fuel sender assembly or the electric fuel pump assembly. Since dirt and sediment eventually gather on the bottom of a fuel tank, the fuel pickup tube is fitted with a filter sock or strainer to prevent contamination from entering the fuel lines. The woven plastic strainer also acts as a water separator by preventing water from being drawn up with the fuel. The filter sock usually is designed to filter out particles that are larger than 70 to 100 microns, or 30 microns if a gerotor-type fuel pump is used. One micron is 0.000039 inches. ● SEE FIGURE 19–5.

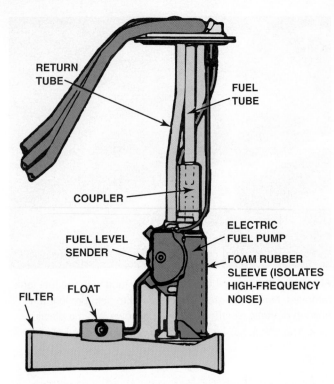

**FIGURE 19–5** The fuel pickup tube is part of the fuel sender and pump assembly.

RETURN TUBE

FUEL TUBE

COUPLER

FUEL LEVEL SENDER

ELECTRIC FUEL PUMP

FOAM RUBBER SLEEVE (ISOLATES HIGH-FREQUENCY NOISE)

FILTER

FLOAT

**NOTE: The human eye cannot see anything smaller than about 40 microns.**

The filter is made from woven Saran resin (copolymer of vinylidene chloride and vinyl chloride). The filter blocks any water that may be in the fuel tank, unless it is completely submerged in water. In that case, it will allow water through the filter. This filter should be replaced whenever the fuel pump is replaced.

**TANK VENTING REQUIREMENTS** Fuel tanks must be vented to prevent a **vacuum lock** as fuel is drawn from the tank. As fuel is used and its level drops in the tank, the space above the fuel increases. As the air in the tank expands to fill this greater space, its pressure drops. Without a vent, the air pressure inside the tank would drop below atmospheric pressure, developing a vacuum which prevents the flow of fuel. Under extreme pressure variance, the tank could collapse. Venting the tank allows outside air to enter as the fuel level drops, preventing a vacuum from developing.

An EVAP system vents gasoline vapors from the fuel tank directly to a charcoal-filled vapor storage canister and uses an unvented filler cap. Many filler caps contain valves that open to relieve pressure or vacuum above specified safety levels. Systems that use completely sealed caps have separate pressure and vacuum relief valves for venting.

Because fuel tanks are not vented directly to the atmosphere, the tank must allow for fuel expansion, contraction, and overflow that can result from changes in temperature or overfilling. One way is to use a dome in the top of the tank. Many General Motors vehicles use a design that includes a vertical slosh baffle that reserves up to 12% of the total tank capacity for fuel expansion.

## ROLLOVER LEAKAGE PROTECTION

All vehicles have one or more devices to prevent fuel leaks in case of vehicle rollover or a collision in which fuel may spill.

Variations of the basic one-way **check valve** may be installed in any number of places between the fuel tank and the engine. The valve may be installed in the fuel return line, vapor vent line, or fuel tank filler cap.

In addition to the rollover protection devices, some vehicles use devices to ensure that the fuel pump shuts off when an accident occurs. On some air vane sensors, a microswitch is built into the sensor to switch on the fuel pump as soon as intake airflow causes the vane to lift from its rest position. ● **SEE FIGURE 19–6.**

Ford vehicles use an **inertia switch.** ● **SEE FIGURE 19–7.** The inertia switch is installed in the rear of the vehicle between the electric fuel pump and its power supply. With any sudden impact, such as a jolt from another vehicle in a parking lot, the inertia switch opens and shuts off power to the fuel pump. The switch must be reset manually by pushing a button to restore current to the pump.

## FUEL LINES

Fuel and vapor lines made of steel, nylon tubing, or fuel-resistant rubber hoses connect the parts of the fuel system. Fuel lines supply fuel to the throttle body or fuel rail. They also return excess fuel and vapors to the tank. Depending on their function, fuel and vapor lines may be either rigid or flexible.

Fuel lines must remain as cool as possible. If any part of the line is located near too much heat, the gasoline passing through it vaporizes, and **vapor lock** occurs. When this happens, the fuel pump supplies only vapor that passes into the injectors. Without liquid gasoline, the engine stalls, and a hot restart problem develops.

The fuel delivery system supplies 10 to 15 PSI (69 to 103 kPa) or up to 35 PSI (241 kPa) to many throttle-body injection units and up to 60 PSI (414 kPa) for multiport fuel-injection systems. Fuel-injection systems retain residual or rest pressure in the lines for a half hour or longer when the engine is turned off to prevent hot engine restart problems. Higher-pressure systems such as these require special fuel lines.

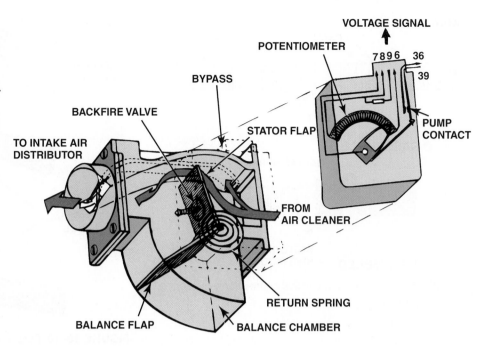

**FIGURE 19–6** On some vehicles equipped with an airflow sensor, a switch is used to energize the fuel pump. In the event of a collision, the switch opens, and the fuel flow stops.

VOLTAGE SIGNAL

POTENTIOMETER

7 8 9 6   36

39

BYPASS

BACKFIRE VALVE

PUMP CONTACT

STATOR FLAP

TO INTAKE AIR DISTRIBUTOR

FROM AIR CLEANER

RETURN SPRING

BALANCE FLAP

BALANCE CHAMBER

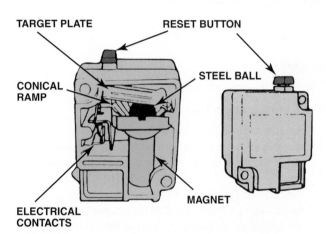

TARGET PLATE

RESET BUTTON

CONICAL RAMP

STEEL BALL

ELECTRICAL CONTACTS

MAGNET

**FIGURE 19–7** Ford uses an inertia switch to turn off the electric fuel pump in an accident.

**RIGID LINES**    All fuel lines fastened to the body, frame, or engine are made from nylon reinforced plastic or seamless steel tubing. Steel springs may be wound around the tubing at certain points to protect against impact damage.

Only steel tubing, or that recommended by the manufacturer, should be used when replacing rigid fuel lines. *Never substitute copper or aluminum tubing for steel tubing.* These materials do not withstand normal vehicle vibration and could combine with the fuel to cause a chemical reaction.

**FLEXIBLE LINES**    Most fuel systems use synthetic rubber hose sections where flexibility is needed. Short hose sections often connect steel fuel lines to other system components. The fuel delivery hose inside diameter (ID), is usually 3/16" or 3/8" (8 or 10 millimeters) and the return line ID is normally 1/4" (6 millimeters).

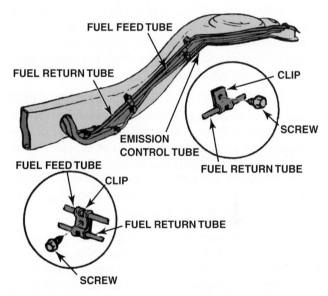

FUEL FEED TUBE

FUEL RETURN TUBE

CLIP

SCREW

EMISSION CONTROL TUBE

FUEL RETURN TUBE

FUEL FEED TUBE

CLIP

FUEL RETURN TUBE

SCREW

**FIGURE 19–8** Fuel lines are routed along the frame or body and secured with clips.

Fuel-injection systems require special-composition reinforced hoses specifically made for these higher-pressure systems. Similarly, vapor vent lines must be made of materials that resist fuel vapors. Replacement vent hoses are usually marked with the designation "EVAP" to indicate their intended use.

**FUEL LINE MOUNTING**    Fuel supply lines from the tank to a throttle body or fuel rail are routed to follow the frame along the underbody of the vehicle. Vapor and return lines may be routed with the fuel supply line. All rigid lines are fastened to the frame rail or underbody with screws and clamps or with clips.
● **SEE FIGURE 19–8.**

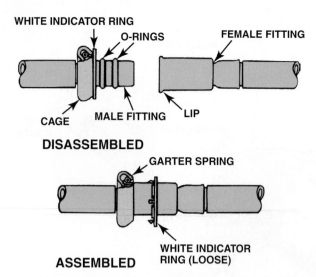

**FIGURE 19–9** Some Ford metal line connections use spring locks and O-rings.

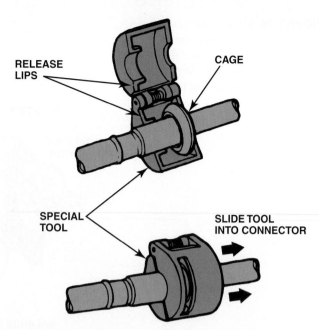

**FIGURE 19–10** Ford spring-lock connectors require a special tool for disassembly.

**FUEL-INJECTION LINES AND CLAMPS** Hoses used for fuel-injection systems are made of materials with high resistance to oxidation and deterioration. Replacement hoses for injection systems should always be equivalent to original equipment manufacturer (OEM) hoses.

**FUEL-INJECTION FITTINGS AND NYLON LINES**
Because of their operating pressures, fuel-injection systems often use special kinds of fittings to ensure leakproof connections. Some high-pressure fittings on GM vehicles with port fuel-injection systems use O-ring seals instead of the traditional flare connections. When disconnecting such a fitting, inspect the O-ring for damage and replace it if necessary. *Always* tighten O-ring fittings to the specified torque value to prevent damage.

Other manufacturers also use O-ring seals on fuel line connections. In all cases, the O-rings are made of special materials that withstand contact with gasoline and oxygenated fuel blends. Some manufacturers specify that the O-rings be replaced every time the fuel system connection is opened. When replacing one of these O-rings, a new part specifically designed for fuel system service must be used.

Ford also uses spring-lock connectors to join male and female ends of steel tubing. ● **SEE FIGURE 19–9.** The coupling is held together by a garter spring inside a circular cage. The flared end of the female fitting slips behind the spring to lock the coupling together.

General Motors has used nylon fuel lines with quick-connect fittings at the fuel tank and fuel filter since the early 1990s. Like the GM threaded couplings used with steel lines, nylon line couplings use internal O-ring seals. Unlocking the metal connectors requires a special quick-connector separator tool; plastic connectors can be released without the tool. ● **SEE FIGURES 19–10 AND 19–11.**

**?  FREQUENTLY ASKED QUESTION**

**Just How Much Fuel Is Recirculated?**

Approximately 80% of the available fuel pump volume is released to the fuel tank through the fuel–pressure regulator at idle speed. As an example, a passenger vehicle cruising down the road at 60 miles per hour gets 30 miles per gallon. With a typical return-style fuel system pumping about 30 gallons per hour from the tank, it would burn 2 gallons per hour and return about 28 gallons per hour to the tank!

**FUEL LINE LAYOUT** All fuel lines are routed to avoid engine and exhaust heat from heating the fuel. Fuel pressures have tended to become higher to prevent vapor lock, and a major portion of the fuel routed to the fuel-injection system returns to the tank by way of a fuel return line or return-type systems. This allows better control, within limits, of heat absorbed by the gasoline as it is routed through the engine compartment. Throttle-body and multiport injection systems have typically used a pressure regulator to control fuel pressure in the throttle body or fuel rail and also allow excess fuel not used by the injectors to return to the tank. However, the warmer fuel in the tank may create problems, such as an excessive rise in fuel vapor pressures in the tank.

**FIGURE 19–11** Typical quick-connect steps.

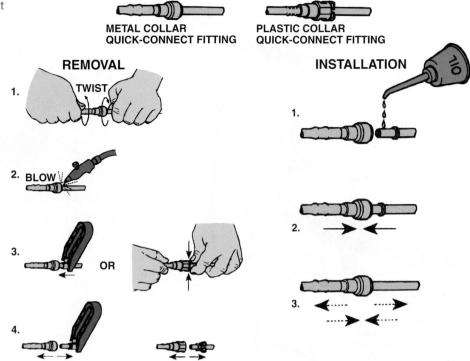

METAL COLLAR
QUICK-CONNECT FITTING

PLASTIC COLLAR
QUICK-CONNECT FITTING

REMOVAL

1. TWIST

2. BLOW

3. OR

4.

INSTALLATION

1.

2.

3.

With late-model vehicles, there has been some concern about too much heat being sent back to the fuel tank, causing rising in-tank temperatures and increases in fuel vaporization and **volatile organic compound (VOC)** (hydrocarbon) emissions. To combat this problem, manufacturers have placed the pressure regulator back by the tank instead of under the hood on mechanical returnless systems. In this way, returned fuel is not subjected to the heat generated by the engine and the underhood environment. To prevent vapor lock in these systems, pressures have been raised in the fuel rail, and injectors tend to have smaller openings to maintain control of the fuel spray under pressure.

Not only must the fuel be filtered and supplied under adequate pressure, but there must also be a consistent *volume* of fuel to assure smooth engine performance even under the heaviest of loads.

## ELECTRIC FUEL PUMPS

The electric fuel pump is a pusher unit. When the pump is mounted in the tank, the entire fuel supply line to the engine can be pressurized. Because the fuel, when pressurized, has a higher boiling point, it is unlikely that vapor will form to interfere with fuel flow.

Most vehicles use the impeller or turbine pumps. ● **SEE FIGURE 19–12.** All electrical pumps are driven by a small electric motor, but the turbine pump turns at higher speeds and is quieter than the others.

> **? FREQUENTLY ASKED QUESTION**
>
> **How Can an Electric Pump Work Inside a Gas Tank and Not Cause a Fire?**
>
> Even though fuel fills the entire pump, no burnable mixture exists inside the pump because there is no air and no danger of commutator brush arcing, igniting the fuel.

**POSITIVE DISPLACEMENT PUMP** A positive displacement pump is a design that forces everything that enters the pump to leave the pump.

In the **roller cell** or vane pump, the impeller draws fuel into the pump and then pushes it out through the fuel line to the injection system. All designs of pumps use a variable-sized chamber to draw in fuel. When the maximum volume has been reached, the supply port closes, and the discharge opens. Fuel is then forced out the discharge as this volume decreases. The chambers are formed by rollers or gears in a rotor plate. Because this type of pump uses no valves to move the fuel, the fuel flows steadily through the entire pump housing, including the electrical portion which keeps the pump cool. Usually, only when a vehicle runs out of fuel is there a risk of pump damage.

Most electric fuel pumps are equipped with a fuel outlet check valve that closes to maintain fuel pressure when the pump shuts off. **Residual** or **rest pressure** prevents vapor lock and hot-start problems on these systems.

● **FIGURE 19–13** shows the pumping action of a **rotary vane pump.** The pump consists of a central impeller disk,

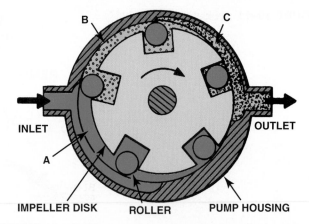

**FIGURE 19–13** The pumping action of an impeller or rotary vane pump.

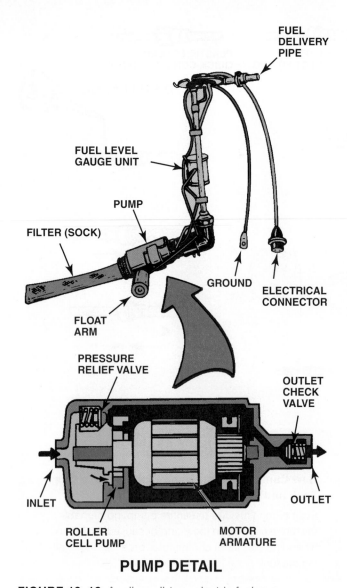

**PUMP DETAIL**

**FIGURE 19–12** A roller cell-type electric fuel pump.

several rollers or vanes that ride in notches in the impeller, and a pump housing that is offset from the impeller centerline. The impeller is mounted on the end of the motor armature and spins whenever the motor is running. The rollers are free to slide in and out within the notches in the impeller to maintain sealing contact. Unpressurized fuel enters the pump, fills the spaces between the rollers, and is trapped between the impeller, the housing, and two rollers. An internal gear pump, called a **gerotor**, is another type of positive displacement pump that is often used in engine oil pumps. It uses the meshing of internal and external gear teeth to pressurize the fuel. ● **SEE FIGURE 19–14** for an example of a gerotor-type fuel pump that uses an impeller as the first stage and is used to move the fuel gerotor section where it is pressurized.

**HYDROKINETIC FLOW PUMP DESIGN** The word *hydro* means liquid, and the term *kinetic* refers to motion, so the term **hydrokinetic pump** means that this design of pump rapidly

moves the fuel to create pressure. This design of pump is a nonpositive displacement pump design.

A **turbine pump** is the most common because it tends to be less noisy. Also known as **peripheral** and **side-channel pumps,** these units use an impeller that accelerates the fuel particles before actually discharging them into a tract where they generate pressure via pulse exchange. Actual pump volume is controlled by using a different number of impeller blades, and in some cases a higher number of impellers, or different shapes along the side discharge channels. These units are fitted more toward lower operating pressures of less than 60 PSI. ● **SEE FIGURE 19–15** for an example of a two-stage turbine pump. The turbine impeller has a staggered blade design to minimize pump harmonic noise and to separate vapor from the liquid fuel. The end cap assembly contains a pressure relief valve and a radio-frequency interference (RFI) suppression module. The check valve is usually located in the upper fuel pipe connector assembly.

After fuel passes through the strainer, it is drawn into the lower housing inlet port by the impellers. It is pressurized and delivered to the convoluted fuel tube for transfer through a check valve into the fuel feed pipe. A typical electric fuel pump used on a fuel-injection system delivers about 40 to 50 gallons per hour or 0.6 to 0.8 gallon per minute at a pressure of 70 to 90 PSI.

**MODULAR FUEL SENDER ASSEMBLY** The modular fuel sender consists of a fuel level sensor, a turbine pump, and a jet pump. The reservoir housing is attached to the cover containing fuel pipes and the electrical connector. Fuel is transferred from the pump to the fuel pipe through a convoluted (flexible) fuel pipe. The convoluted fuel pipe eliminates the need for rubber hoses, nylon pipes, and clamps. The reservoir dampens fuel slosh to maintain a constant fuel level available to the roller vane pump; it also reduces noise.

Some of the flow, however, is returned to the jet pump for recirculation. Excess fuel is returned to the reservoir through one of the three hollow support pipes. The hot fuel quickly

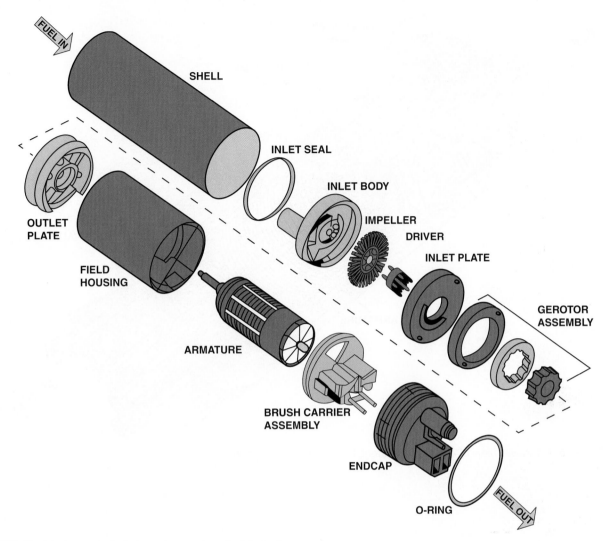

**FIGURE 19–14** An exploded view of a gerotor electric fuel pump.

mixes with the cooler fuel in the reservoir; this minimizes the possibility of vapor lock. In these modules, the reservoir is filled by the jet pump. Some of the fuel from the pump is sent through the jet pump to lift fuel from the tank into the reservoir.

**ELECTRIC PUMP CONTROL CIRCUITS** Fuel pump circuits are controlled by the fuel pump relay. Fuel pump relays are activated initially by turning the ignition key to on, which allows the pump to pressurize the fuel system. As a safety precaution, the relay deenergizes after a few seconds until the key is moved to the crank position. Once an ignition coil signal, or "tach" signal, is received by the engine control computer, indicating the engine is rotating, the relay remains energized even with the key released to the run position.

**CHRYSLER.** On older Chrysler vehicles, the PCM must receive an engine speed (RPM) signal during cranking before it can energize a circuit driver inside the power module to activate an

 **FREQUENTLY ASKED QUESTION**

**Why Are Many Fuel Pump Modules Spring Loaded?**

Fuel modules that contain the fuel pickup sock, fuel pump, and fuel level sensor are often spring loaded when fitted to a plastic fuel tank. The plastic material shrinks when cold and expands when hot, so having the fuel module spring loaded ensures that the fuel pickup sock will always be the same distance from the bottom of the tank. ● **SEE FIGURE 19–16**.

automatic shutdown (ASD) relay to power the fuel pump, ignition coil, and injectors. As a safety precaution, if the RPM signal to the logic module is interrupted, the logic module signals the power module to deactivate the ASD, turning off the pump, coil,

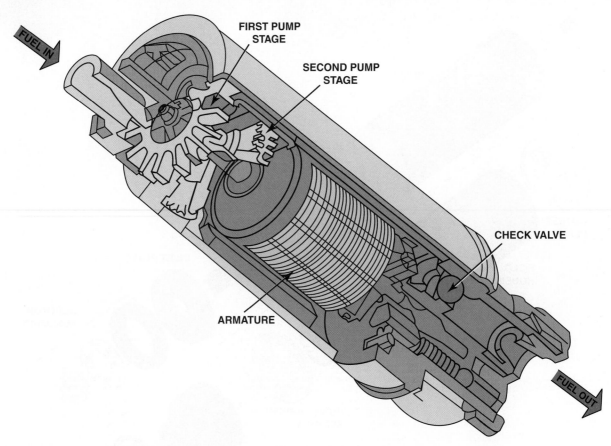

**FIGURE 19–15** A cutaway view of a typical two-stage turbine electric fuel pump.

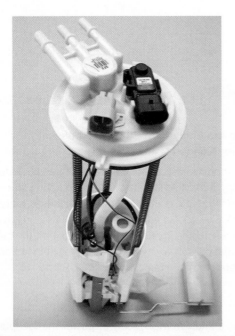

**FIGURE 19–16** A typical fuel pump module assembly, which includes the pickup strainer and fuel pump, as well as the fuel-pressure sensor and fuel level sensing unit.

and injectors. In some vehicles, the oil pressure switch circuit may be used as a safety circuit to activate the pump in the ignition switch run position.

**GENERAL MOTORS.** General Motors systems energize the pump with the ignition switch to initially pressurize the fuel lines but then deactivate the pump if an RPM signal is not received within one or two seconds. The pump is reactivated as soon as engine cranking is detected. The oil pressure sending unit serves as a backup to the fuel pump relay on some vehicles. In case of pump relay failure, the oil pressure switch will operate the fuel pump once oil pressure reaches about 4 PSI (28 kPa).

**FORD.** Older fuel-injected Fords used an inertia switch between the fuel pump relay and fuel pump.

The inertia switch opens under a specified impact, such as a collision. When the switch opens, current to the pump shuts off because the fuel pump relay will not energize. The switch must be reset manually by opening the trunk and depressing the reset button before current flow to the pump can

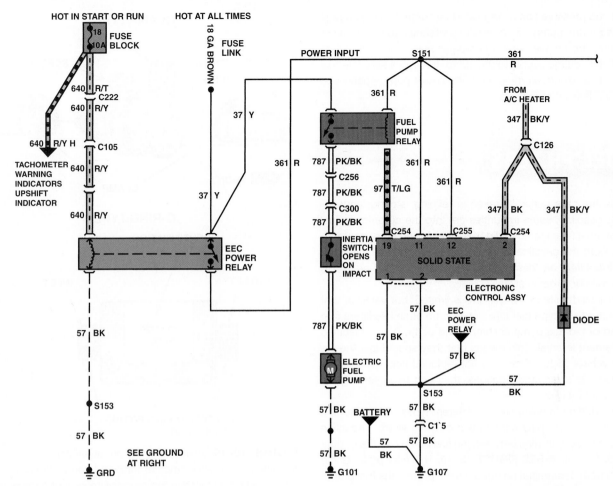

**FIGURE 19–17** A schematic showing that an inertia switch is connected in series between the fuel pump relay and the fuel pump.

be restored. ● **SEE FIGURE 19–17** for a schematic of a typical fuel system that uses an inertia switch in the power feed circuit to the electric fuel pump.

Since about 2008, the inertial switch has been replaced with a signal input from the airbag module. If the airbag is deployed, the circuit to the fuel pump is opened and the fuel pump stops.

**PUMP PULSATION DAMPENING** Some manufacturers use an **accumulator** in the system to reduce pressure pulses and noise. Others use a pulsator located at the outlet of the fuel pump to absorb pressure pulsations that are created by the pump. These pulsators are usually used on roller vane pumps and are a source of many internal fuel leaks. ● **SEE FIGURE 19–18.**

**NOTE: Some experts suggest that the pulsator be removed and replaced with a standard section of fuel line to prevent the loss of fuel pressure that results when the connections on the pulsator loosen and leak fuel back into the tank.**

**VARIABLE SPEED PUMPS** Another way to help reduce noise, current draw, and pump wear is to reduce the speed of the pump when less than maximum output is required. Pump

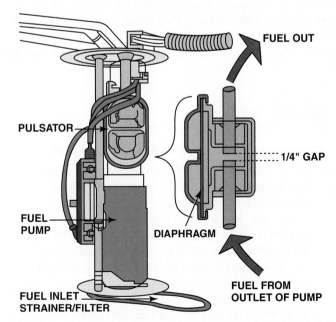

**FIGURE 19–18** A typical fuel pulsator used mostly with roller vane-type pumps to help even out the pulsation in pressure that can cause noise.

speed and pressure can be regulated by controlling the voltage supplied to the pump with a resistor switched into the circuit or by letting the engine-control computer pulse-width modulate (PWM) the voltage supply to the pump, through a separate fuel pump driver electronic module. With slower pump speed and pressure, less noise is produced.

## FUEL FILTERS

Despite the care generally taken in refining, storing, and delivering gasoline, some impurities get into the automotive fuel system. Fuel filters remove dirt, rust, water, and other contamination from the gasoline before it can reach the fuel injectors. Most fuel filters are designed to filter particles that are 10 to 20 microns or larger in size.

The useful life of many filters is limited, but vehicles that use a returnless-type fuel-injection system often use filters that are part of the fuel pump assembly and do have any specified replacement interval. This means that they should last the life of the vehicle. If fuel filters are not replaced according to the manufacturer's recommendations, they can become clogged and restrict fuel flow.

In addition to using several different types of fuel filters, a single fuel system may contain two or more filters. The inline filter is located in the line between the fuel pump and the throttle body or fuel rail. ● SEE FIGURE 19–19. This filter protects the system from contamination but does not protect the fuel pump. The inline filter usually is a metal or plastic container with a pleated paper element sealed inside.

Fuel filters may be mounted on a bracket on the fender panel, a shock tower, or another convenient place in the engine compartment. They may also be installed under the vehicle near the fuel tank. Fuel filters should be replaced according to the vehicle manufacturer's recommendations, which range from every 30,000 miles (48,000 km) to

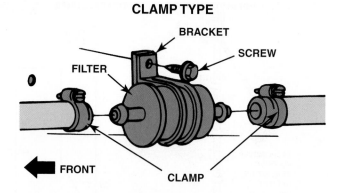

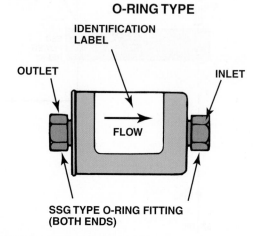

**FIGURE 19–19** Inline fuel filters are usually attached to the fuel line with screw clamps or threaded connections. The fuel filter must be installed in the proper direction or a restricted fuel flow can result.

100,000 miles (160,000 km) or longer. Fuel filters that are part of the fuel pump module assemblies usually do not have any specified service interval.

## FUEL PUMP TESTING

Fuel pump testing includes many different tests and procedures. Even though a fuel pump can pass one test, it does not mean that there is not a fuel pump problem. For example, if the pump motor is rotating slower than normal, it may be able to produce the specified pressure but not enough volume to meet the needs of the engine while operating under a heavy load.

**TESTING FUEL PUMP PRESSURE** Fuel pump–regulated pressure has become more important than ever with a more exact fuel control. Although an increase in fuel pressure does increase fuel volume to the engine, this is *not* the preferred method to add additional fuel as some units will not open correctly at the increased fuel pressure. On the other side of the discussion, many newer engines will not start when fuel pressure is just a few

**TECH TIP**

**Use a Headlight to Test for Power and Ground**

When replacing a fuel pump, always check for proper power and ground. If the supply voltage is low due to resistance in the circuit or the ground connection is poor, the lower available voltage to the pump will result in lower pump output and could also reduce the life of the pump. While a voltage drop test can be preformed, a quick and easy test is to use a headlight connected to the circuit. If the headlight is bright, then both the power side and the ground side of the pump circuit are normal. If the headlight is dim, then more testing will be needed to find the source of the resistance in the circuit(s). ● SEE FIGURE 19–20.

**FIGURE 19–20** A dim headlight indicates excessive resistance in fuel pump circuit.

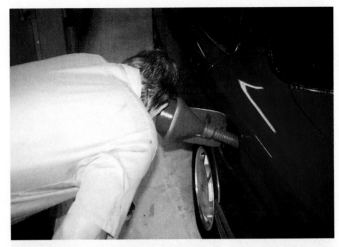

(a)

(b)

PSI low. Correct fuel pressure is very important for proper engine operation. Check fuel pressure and then compare with factory specifications.

| Fuel Pump Pressure | Possible Causes |
|---|---|
| Zero pressure | Defective fuel pump relay, fuel pressure regulator, fuel pump, or wiring. |
| Within factory specifications | If drivability problem, continue diagnosis |
| Lower than specifications | Possible defective fuel pressure regulator or pump |
| Higher than specifications | Possible defective fuel pressure regulator or restricted return line (on return-type systems) |

**FIGURE 19–21** (a) A funnel helps in hearing if the electric fuel pump inside the gas tank is working. (b) If the pump is not running, check the wiring and current flow before going through the process of dropping the fuel tank to remove the pump.

In both types of systems, maximum fuel pump pressure is about double the normal operating pressure to ensure that a continuous flow of cool fuel is being supplied to the injector(s) to help prevent vapor from forming in the fuel system. Although

 **TECH TIP**

**The Ear Test**

No, this is not a test of your hearing but rather using your ear to check that the electric fuel pump is operating. The electric fuel pump inside the fuel tank is often difficult to hear running, especially in a noisy shop environment. A commonly used trick to better hear the pump is to use a funnel in the fuel filter neck. ● **SEE FIGURE 19–21.**

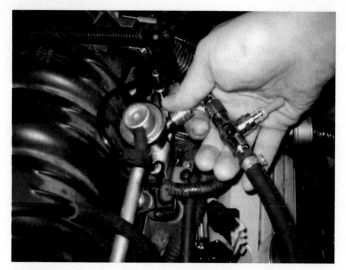

**FIGURE 19–22** The Schrader valve on this General Motors 3800 V-6 is located next to the fuel-pressure regulator.

 **TECH TIP**

**The Rubber Mallet Trick**

Often a no-start condition is due to an inoperative electric fuel pump. A common trick is to tap on the bottom of the fuel tank with a rubber mallet in an attempt to jar the pump motor enough to work. Instead of pushing a vehicle into the shop, simply tap on the fuel tank and attempt to start the engine. This is not a repair but rather a confirmation that the fuel pump does indeed require replacement.

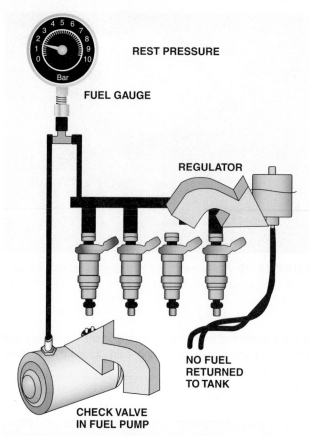

**FIGURE 19–23** The fuel system should hold pressure if the system is leak free.

vapor or foaming in a fuel system can greatly affect engine operation, the cooling and lubricating flow of the fuel must be maintained to ensure the durability of injector nozzles.

To measure fuel pump pressure, locate the Schrader valve and attach a fuel-pressure gauge. ● **SEE FIGURE 19–22.**

**REST PRESSURE TEST** If the fuel pressure is acceptable, then check the system for leakdown. Observe the pressure gauge after five minutes. ● **SEE FIGURE 19–23.** The pressure should be the same as the initial reading. If not, then the pressure regulator, fuel pump check valve, or the injectors are leaking.

**DYNAMIC PRESSURE TEST** To test the pressure dynamically, start the engine. If the pressure is vacuum referenced, then the pressure should change when the throttle is cycled. If

it does not, then check the vacuum supply circuit. Remove the vacuum line from the regulator and inspect for any presence of fuel. ● **SEE FIGURE 19–24.** There should never be any fuel present on the vacuum side of the regulator diaphragm. When the engine speed is increased, the pressure reading should remain within the specifications.

Some engines do not use a vacuum-referenced regulator. The running pressure remains constant, which is typical for a mechanical returnless-type fuel system. On these systems, the pressure is higher than on return-type systems to help reduce the formation of fuel vapors in the system.

 **TECH TIP**

**The Fuel-Pressure Stethoscope Test**

When the fuel pump is energized and the engine is not running, fuel should be heard flowing back to the fuel tank at the outlet of the fuel-pressure regulator. ● **SEE FIGURE 19–25.** If fuel is heard flowing through the return line, the fuel pump pressure is higher than the regulator pressure. If no sound of fuel is heard, either the fuel pump or the fuel-pressure regulator is at fault.

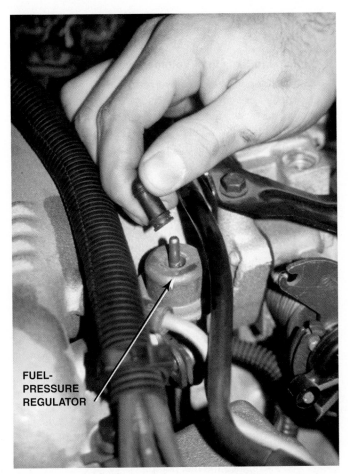

FIGURE 19–24 If the vacuum hose is removed from the fuel-pressure regulator when the engine is running, the fuel pressure should increase. If it does not increase, then the fuel pump is not capable of supplying adequate pressure or the fuel-pressure regulator is defective. If gasoline is visible in the vacuum hose, the regulator is leaking and should be replaced.

**TESTING FUEL PUMP VOLUME** Fuel pressure alone is not enough for proper engine operation. ●**SEE FIGURE 19–26.** Sufficient fuel capacity (flow) should be at least 2 pints (1 liter) every 30 seconds or 1 pint in 15 seconds. Fuel flow specifications are usually expressed in gallons per minute. A typical specification would be 0.5 gallon per minute or more. Volume testing is shown in ●**SEE FIGURES 19–27 AND 19–28.**

All fuel must be filtered to prevent dirt and impurities from damaging the fuel system components and/or engine. The first filter is inside the gas tank and is usually not replaceable separately but is attached to the fuel pump (if the pump is electric) and/or fuel gauge sending unit. The replaceable fuel filter is usually located between the fuel tank and the fuel rail or inlet to the fuel-injection system. Most vehicle manufacturers state in service information when to replace the fuel filter. Most newer vehicles that use returnless-type fuel-injection systems do not have replaceable filters as they are built into the fuel pump module assembly. (Check the

FIGURE 19–25 Fuel should be heard returning to the fuel tank at the fuel return line if the fuel pump and fuel-pressure regulator are functioning correctly.

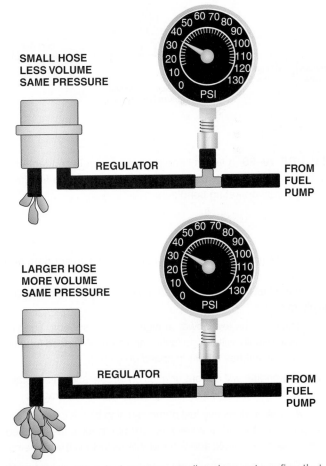

FIGURE 19–26 A fuel-pressure reading does not confirm that there is enough fuel volume for the engine to operate correctly.

**FIGURE 19–27** A fuel system tester connected in series in the fuel system so all of the fuel used flows through the meter, which displays the rate flow and the fuel pressure.

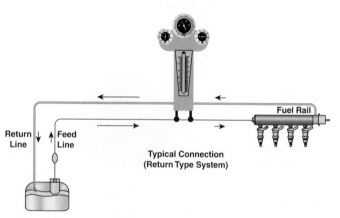

Return Line    Feed Line

**Fuel Rail**

**Typical Connection
(Return Type System)**

**FIGURE 19–28** A fuel volume/pressure tester is connected in series in the fuel delivery line so it can measure the fuel flow under actual operating conditions.

vehicle manufacturer's recommendations for exact time and mileage intervals.)

If the fuel filter becomes partially clogged, the following are likely to occur:

1. There will be low power at higher engine speeds. The vehicle usually will not go faster than a certain speed (engine acts as if it has a built-in speed governor).
2. The engine will cut out or miss on acceleration, especially when climbing hills or during heavy-load acceleration.

A weak or defective fuel pump can also be the cause of the symptoms just listed. If an electric fuel pump for a fuel-injected engine becomes weak, additional problems include the following:

1. The engine may be hard to start.
2. There may be a rough idle and stalling.

**TECH TIP**

**Quick and Easy Fuel Volume Test**

Testing for pump volume involves using a specialized tester or a fuel-pressure gauge equipped with a hose to allow the fuel to be drawn from the system into a container with volume markings to allow for a volume measurement. This test can be hazardous because of flammable gasoline vapors. An alternative test involves connecting a fuel-pressure gauge to the system with the following steps:

**STEP 1**    Start the engine and observe the fuel-pressure gauge. The reading should be within factory specifications (typically between 35 PSI and 45 PSI).

**STEP 2**    Remove the hose from the fuel-pressure regulator. The pressure should increase if the system uses a demand-type regulator.

**STEP 3**    Rapidly accelerate the engine while watching the fuel-pressure gauge. If the fuel volume is okay, the fuel pressure should not drop more than 2 PSI. If the fuel pressure drops more than 2 PSI, replace the fuel filter and retest.

**STEP 4**    After replacing the fuel filter, accelerate the engine and observe the pressure gauge. If the pressure drops more than 2 PSI, replace the fuel pump.

**NOTE: The fuel pump could still be delivering less than the specified volume of fuel, but as long as the volume needed by the engine is met, the pressure will not drop. If, however, the vehicle is pulling a heavy load, the demand for fuel volume may exceed the capacity of the pump.**

**TECH TIP**

**Remove the Bed to Save Time?**

The electric fuel pump is easier to replace on many General Motors pickup trucks if the bed is removed. Access to the top of the fuel tank, where the access hole is located, for the removal of the fuel tank sender unit and pump is restricted by the bottom of the pickup truck bed. Rather than drop the tank, it is often much easier to use an engine hoist or a couple of other technicians to lift the bed from the frame after removing only a few fasteners.

**CAUTION: Be sure to clean around the fuel pump opening so that dirt or debris does not enter the tank when the fuel pump is removed.**

3. There may be erratic shifting of the automatic transmission as a result of engine missing due to lack of fuel pump pressure and/or volume.

**CAUTION:** Be certain to consult the vehicle manufacturer's recommended service and testing procedures before attempting to test or replace any component of a high-pressure electronic fuel-injection system.

## FUEL PUMP CURRENT DRAW TEST

Another test that can and should be performed on a fuel pump is to measure the current draw in amperes. This test is most often performed by connecting a digital multimeter set to read DC amperes and test the current draw. ● **SEE FIGURE 19–29** for the hookup for vehicles equipped with a fuel pump relay. Compare the reading to factory specifications. See the following chart for an example of typical fuel pump current draw readings.

**NOTE: Testing the current draw of an electric fuel pump may not indicate whether the pump is good. A pump that is not rotating may draw normal current.**

Using a mini clamp on ammeter is a quick and easy way to measure fuel pump current. Clamp the inductive probe around a wire to the fuel pump or add a fused jumper wire to replace the fuel pump fuse. Start the engine and read the meter display.

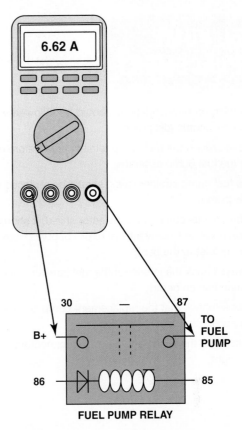

**FIGURE 19–29** Hookup for testing fuel pump current draw on any vehicle equipped with a fuel pump relay.

**Fuel Pump Current Draw Table**

| Amperage Reading | Expected Value | Amperage Too High | Amperage Too Low |
|---|---|---|---|
| Throttle-body fuel-injection engines | 2–5 amps | • Check the fuel filter.<br>• Check for restrictions in other fuel line areas.<br>• Replace the fuel pump. | • Check for a high-resistance connection.<br>• Check for a high-resistance ground fault.<br>• Replace the fuel pump. |
| Port fuel-injection engines | 4–8 amps | • Check the fuel filter.<br>• Check for restrictions in other fuel line areas.<br>• Replace the fuel pump. | • Check for a high-resistance connection.<br>• Check for a high-resistance ground fault.<br>• Replace the fuel pump. |
| Turbo engines | 6–10 amps | • Check the fuel filter.<br>• Check for restrictions in other fuel line areas.<br>• Replace the fuel pump. | • Check for a high-resistance connection.<br>• Check for a high-resistance ground fault.<br>• Replace the fuel pump. |
| GM CPI truck engines | 8–12 amps | • Check the fuel filter.<br>• Check for restrictions in other fuel line areas.<br>• Replace the fuel pump. | • Check for a high-resistance connection.<br>• Check for a high-resistance ground fault.<br>• Replace the fuel pump. |

# FUEL PUMP REPLACEMENT

The following recommendations should be followed whenever replacing an electric fuel pump.

- Clean around the fuel pump retainer area before removing the fuel pump assembly.
- The fuel pump strainer (sock) should be replaced with the new pump.
- If the original pump had a defector shield, it should always be used to prevent fuel return bubbles from blocking the inlet to the pump.
- Always check the interior of the fuel tank for evidence of contamination or dirt.
- Double-check that the replacement pump is correct for the application.
- Check that the wiring and electrical connectors are clean and tight.

| Fuel Supply-Related Symptom Guide | |
|---|---|
| **Problem** | **Possible Causes** |
| **Pressure too high after engine start-up.** | 1. Defective fuel-pressure regulator<br>2. Restricted fuel return line<br>3. Excessive system voltage<br>4. Wrong fuel pump |
| **Pressure too low after engine start-up.** | 1. Stuck-open pressure regulator<br>2. Low voltage<br>3. Poor ground<br>4. Plugged fuel filter<br>5. Faulty inline fuel pump<br>6. Faulty in-tank fuel pump<br>7. Partially clogged filter sock<br>8. Faulty hose coupling<br>9. Leaking fuel line<br>10. Wrong fuel pump<br>11. Leaking pulsator<br>12. Restricted accumulator<br>13. Faulty pump check valves<br>14. Faulty pump installation |
| **Pressure drops off with key on/engine off. With key off, the pressure does not hold.** | 1. Leaky pulsator<br>2. Leaking fuel pump coupling hose<br>3. Faulty fuel pump (check valves)<br>4. Faulty pressure regulator<br>5. Leaking fuel injector<br>6. Faulty installation<br>7. Lines leaking |

# SUMMARY

1. The fuel delivery system includes the following items:
   - Fuel tank
   - Fuel pump
   - Fuel filter(s)
   - Fuel lines
2. A fuel tank is either constructed of steel with a tin plating for corrosion resistance or polyethylene plastic.
3. Fuel tank filler tubes contain an antisiphoning device.
4. Accident and rollover protection devices include check valves and inertia switches.
5. Most fuel lines are made of nylon plastic.
6. Electric fuel pump types include: roller cell, gerotor, and turbine.
7. Fuel filters remove particles that are 10 to 20 microns or larger in size and should be replaced regularly.
8. Fuel pumps can be tested by checking the following:
   - Pressure
   - Volume
   - Specified current draw

# REVIEW QUESTIONS

1. What are the two materials used to construct fuel tanks?
2. What are the three most commonly used pump designs?
3. What is the proper way to disconnect and connect plastic fuel line connections?
4. Where are the fuel filters located in the fuel system?
5. What accident and rollover devices are installed in a fuel delivery system?
6. What three methods can be used to test a fuel pump?

1. The first fuel filter in the sock inside the fuel tank normally filters particles larger than _____.
   a. 0.001 to 0.003 inches
   c. 10 to 20 microns
   b. 0.01 to 0.03 inches
   d. 70 to 100 microns

2. If it is tripped, which type of safety device will keep the electric fuel pump from operating?
   a. Rollover valve
   c. Antisiphoning valve
   b. Inertia switch
   d. Check valve

3. Fuel lines are constructed from _____.
   a. Seamless steel tubing
   b. Nylon plastic
   c. Copper and/or aluminum tubing
   d. Both a and b

4. What prevents the fuel pump inside the fuel tank from catching the gasoline on fire?
   a. Electricity is not used to power the pump.
   b. No air is around the motor brushes.
   c. Gasoline is hard to ignite in a closed space.
   d. All of the above

5. A good fuel pump should be able to supply how much fuel per minute?
   a. 1/4 pint
   c. 1 pint
   b. 1/2 pint
   d. 0.5 to 0.8 gallon

6. Technician A says that fuel pump modules are spring loaded so that they can be compressed to fit into the opening. Technician B says that they are spring loaded to allow for expansion and contraction of plastic fuel tanks. Which technician is correct?
   a. Technician A only
   b. Technician B only
   c. Both Technicians A and B
   d. Neither Technician A nor B

7. Most fuel filters are designed to remove particles larger than _____.
   a. 10 microns
   c. 70 microns
   b. 20 microns
   d. 100 microns

8. The amperage draw of an electric fuel pump is higher than specified. All of the following are possible causes *except* _____.
   a. Corroded electrical connections at the pump motor
   b. Clogged fuel filter
   c. Restriction in the fuel line
   d. Defective fuel pump

9. A fuel pump is being replaced for the third time. Technician A says that the gasoline could be contaminated. Technician B says that wiring to the pump could be corroded. Which technician is correct?
   a. Technician A only
   b. Technician B only
   c. Both Technicians A and B
   d. Neither Technician A nor B

10. A fuel pump was measured for current draw and amperage exceeded factory specifications. This can be caused by a_____.
    a. Clogged fuel filter
    b. Restricted fuel line
    c. Defective fuel pump
    d. Any of the above

# FUEL-INJECTION COMPONENTS AND OPERATION

**LEARNING OBJECTIVES:** **After studying this chapter, the reader should be able to:** • Describe the operation of electronic fuel-injection systems and compare speed-density and mass airflow fuel-injection-type systems. • Explain the operation of throttle-body injection and port fuel-injection systems. • Understand the purpose and function of a fuel-pressure regulator. • Differentiate between electronic and mechanical returnless fuel systems and discuss demand delivery systems. • List the types of fuel-injection systems and explain their modes of operation. • Understand the use of idle control and stepper motors in fuel-injection systems.

**KEY TERMS:** Demand delivery system (DDS) 263 • Electronic air control (EAC) 267 • Electronic returnless fuel system (ERFS) 262 • Flare 267 • Fuel rail 264 • Gang fired 259 • Idle speed control (ISC) motor 268 • Mechanical returnless fuel system (MRFS) 263 • Nonchecking 262 • Port fuel injection 256 • Pressure control valve (PCV) 263 • Pressure vent valve (PVV) 263 • Sequential fuel injection (SFI) 259 • Throttle-body injection (TBI) 256

## ELECTRONIC FUEL-INJECTION OPERATION

Electronic fuel-injection systems use the powertrain control module (PCM) to control the operation of fuel injectors and other functions based on information sent to the PCM from the various sensors. Most electronic fuel-injection systems share the following:

1. Electric fuel pump (usually located inside the fuel tank)
2. Fuel-pump relay (usually controlled by the computer)
3. Fuel-pressure regulator (mechanically operated spring-loaded rubber diaphragm maintains proper fuel pressure)
4. Fuel-injector nozzle or nozzles

●**SEE FIGURE 20–1.** Most electronic fuel-injection systems use the computer to control the following aspects of their operation:

1. **Pulsing the fuel injectors on and off.** The longer the injectors are held open, the greater the amount of fuel injected into the cylinder.
2. **Operating the fuel-pump relay circuit.** The computer usually controls the operation of the electric fuel pump located inside (or near) the fuel tank. The computer uses signals from the ignition switch and RPM signals from the ignition module or system to energize the fuel-pump relay circuit.

**NOTE: This is a safety feature because if the engine stalls and the tachometer (engine speed) signal is lost, the computer will shut off (deenergize) the fuel-pump relay and stop the fuel pump.**

Computer-controlled fuel-injection systems are normally reliable systems if the proper service procedures are followed. Fuel-injection systems use the gasoline flowing through the injectors to lubricate and cool the injector electrical windings and pintle valves.

**NOTE: The fuel does not actually make contact with the electrical windings because the injectors have O-rings at the top and bottom of the winding spool to keep fuel out.**

There are two types of electronic fuel-injection systems:

■ **Throttle-body-injection (TBI)** type. A TBI system delivers fuel from a nozzle(s) into the air above the throttle plate. ●**SEE FIGURE 20–2.**
■ **Port fuel-injection** type. A port fuel-injection design uses a nozzle for each cylinder, and the fuel is squirted into the intake manifold about 2 to 3 inches (70 to 100 mm) from the intake valve. ●**SEE FIGURE 20–3.**

## SPEED-DENSITY FUEL-INJECTION SYSTEMS

Fuel-injection computer systems require a method for measuring the amount of air the engine is breathing in, in order to match the correct fuel delivery. There are two basic methods used:

1. Speed density
2. Mass airflow

The speed-density method does not require an air quantity sensor but rather calculates the amount of fuel required by the

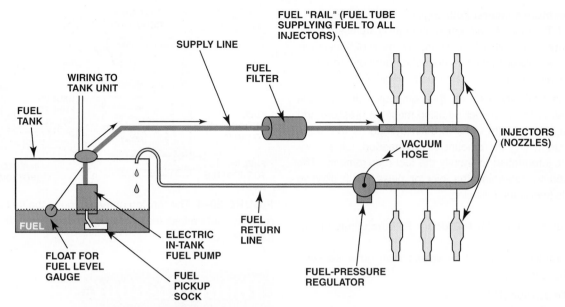

FIGURE 20–1 Typical port fuel-injection system, indicating the location of various components. Notice that the fuel-pressure regulator is located on the fuel return side of the system. The computer does not control fuel pressure but does control the operation of the electric fuel pump (on most systems) and the pulsing on and off of the injectors.

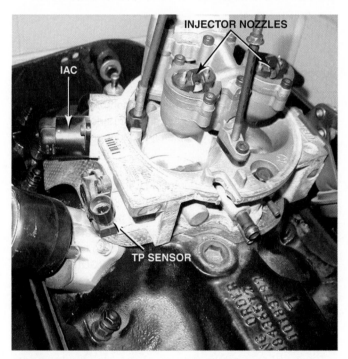

FIGURE 20–2 A dual-nozzle TBI unit on a Chevrolet 5.0 L V-8 engine. The fuel is squirted above the throttle plate, where the fuel mixes with air before entering the intake manifold.

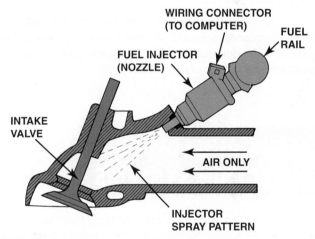

FIGURE 20–3 A typical port fuel-injection system squirts fuel into the low pressure (vacuum) of the intake manifold, about 2 to 3 inches (70 to 100 mm) from the intake valve.

**TECH TIP**

**Two Must-Dos**

For long service life of the fuel system, always do the following:

1. Avoid operating the vehicle on a near-empty tank of fuel. The water or alcohol that may be in the tank becomes more concentrated when the fuel level is low. Dirt that settles near the bottom of the fuel tank can be drawn through the fuel system and cause damage to the pump and injector nozzles.
2. Replace the fuel filter at regular service intervals.

engine. The computer uses information from sensors such as the MAP and TP to calculate the needed amount of fuel:

- **MAP sensor.** The value of the intake (inlet) manifold pressure (vacuum) is a direct indication of engine load.
- **TP sensor.** The position of the throttle plate and its rate of change are used as part of the equation to calculate the proper amount of fuel to inject.

- **Temperature sensors.** Both engine coolant temperature (ECT) and intake air temperature (IAT) are used to calculate the density of the air and the need of the engine for fuel. A cold engine (low-coolant temperature) requires a richer air–fuel mixture than a warm engine.

On speed-density systems, the computer calculates the amount of air in each cylinder by using manifold pressure and engine rpm. The amount of air in each cylinder is the major factor in determining the amount of fuel needed. Other sensors provide information to modify the fuel requirements. The formula used to determine the injector pulse width (PW) in milliseconds (ms) is the following:

**Injector pulse width = MAP/BARO × RPM/maximum rpm**

The formula is modified by values from other sensors, including the following:

- Throttle position (TP)
- Engine coolant temperature (ECT)
- Intake air temperature (IAT)
- Oxygen sensor (O2S) voltage
- Adaptive memory

A fuel injector delivers atomized fuel into the airstream, where it is instantly vaporized. All throttle-body (TB) fuel-injection systems and many multipoint (port) injection systems use the speed-density method of fuel calculation.

# MASS AIRFLOW FUEL-INJECTION SYSTEMS

The formula used by fuel-injection systems that use a mass airflow (MAF) sensor to calculate the injection base pulse width is the following:

**Injector pulse width = airflow/rpm**

The formula is modified by other sensor values, such as the following:

- Throttle position
- Engine coolant temperature
- Barometric pressure
- Adaptive memory

**NOTE: Many 4-cylinder engines do not use a MAF sensor because, due to the time interval between intake events, some reverse airflow can occur in the intake manifold. The MAF sensor would "read" this flow of air as being additional air entering the engine, giving the PCM incorrect airflow information. Therefore, most 4-cylinder engines use the speed-density method of fuel control.**

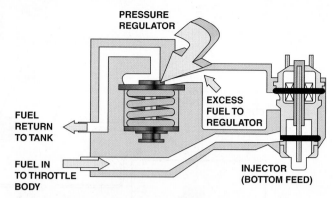

**FIGURE 20–4** The tension of the spring in the fuel-pressure regulator determines the operating pressure on a throttle-body fuel-injection unit.

# THROTTLE-BODY INJECTION

The computer controls injector pulses in one of two ways:

- Synchronized
- Nonsynchronized

If the system uses a synchronized mode, the injector pulses once for each distributor reference pulse. In some vehicles, when dual injectors are used in a synchronized system, the injectors pulse alternately. In a nonsynchronized system, the injectors are pulsed once during a given period (which varies according to calibration) completely independent of distributor reference pulses.

The injector always opens the same distance, and the fuel pressure is maintained at a controlled value by the pressure regulator. The regulators used on throttle-body injection systems are not connected to a vacuum like many port fuel-injection systems. The strength of the spring inside the regulator determines at what pressure the valve is unseated, sending the fuel back to the tank and lowering the pressure. ● **SEE FIGURE 20–4.** The amount of fuel delivered by the injector depends on the amount of time (on-time) that the nozzle is open. This is the injector pulse width—the on-time in milliseconds that the nozzle is open.

The PCM commands a variety of pulse widths to supply the amount of fuel that an engine needs at any specific moment:

- A long pulse width delivers more fuel.
- A short pulse width delivers less fuel.

# PORT FUEL INJECTION

The advantages of port fuel-injection design also are related to characteristics of intake manifolds:

- Fuel distribution is equal to all cylinders because each cylinder has its own injector. ● **SEE FIGURE 20–5.**

### How Do the Sensors Affect the Pulse Width?

The base pulse width of a fuel-injection system is primarily determined by the value of the MAF or MAP sensor and engine speed (RPM). However, the PCM relies on the input from many other sensors to modify the base pulse width as needed:

- **TP Sensor.** This sensor causes the PCM to command up to 500% (five times) the base pulse width if the accelerator pedal is depressed rapidly to the floor. It can also reduce the pulse width by about 70% if the throttle is rapidly closed.
- **ECT.** The value of this sensor determines the temperature of the engine coolant, helps determine the base pulse width, and can account for up to 60% of the determining factors.
- **BARO.** The BARO sensor compensates for altitude and adds up to about 10% under high-pressure conditions and subtracts as much as 50% from the base pulse width at high altitudes.
- **IAT.** The intake air temperature is used to modify the base pulse width based on the temperature of the air entering the engine. It is usually capable of adding as much as 20% if very cold air is entering the engine or reducing the pulse width by up to 20% if very hot air is entering the engine.
- **O2S.** This is one of the main modifiers to the base pulse width and can add or subtract up to about 20% to 25% or more, depending on the oxygen sensor activity.

- The fuel is injected almost directly into the combustion chamber, so there is no chance for it to condense on the walls of a cold intake manifold.
- Because the manifold does not have to carry fuel to properly position a TBI unit, it can be shaped and sized to tune the intake airflow to achieve specific engine performance characteristics.

An EFI injector is simply a specialized solenoid. ● **SEE FIGURE 20–6.** It has an armature winding to create a magnetic field, and a needle (pintle), a disc, or a ball valve. A spring holds the needle, disc, or ball closed against the valve seat, and when energized, the armature winding pulls open the valve when it receives a current pulse from the powertrain control module (PCM). When the solenoid is energized, it unseats the valve to inject fuel.

Electronic fuel-injection systems use a solenoid-operated injector to spray atomized fuel in timed pulses into the manifold or near the intake valve. ● **SEE FIGURE 20–7.** Injectors may be sequenced and fired in one of several ways, but their pulse width is determined and controlled by the engine computer.

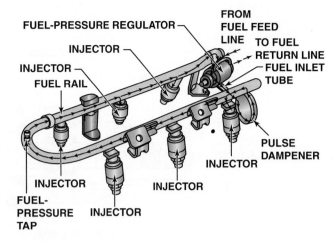

**FIGURE 20–5** The injectors receive fuel and are supported by the fuel rail.

Port systems have an injector for each cylinder, but they do not all fire the injectors in the same way. Domestic systems use one of three ways to trigger the injectors:

- Grouped double-fire
- Simultaneous double-fire
- Sequential

**GROUPED DOUBLE-FIRE** This system divides the injectors into two equalized groups. The groups fire alternately; each group fires once each crankshaft revolution, or twice per four-stroke cycle. The fuel injected remains near the intake valve and enters the engine when the valve opens. This method of pulsing injectors in groups is sometimes called **gang fired.**

**SIMULTANEOUS DOUBLE-FIRE** This design fires all of the injectors at the same time once every engine revolution: two pulses per four-stroke cycle. Many port fuel-injection systems on 4-cylinder engines use this pattern of injector firing. It is easier for engineers to program this system and it can make relatively quick adjustments in the air–fuel ratio, but it still requires the intake charge to wait in the manifold for varying lengths of time.

**SEQUENTIAL** Sequential firing of the injectors according to engine firing order is the most accurate and desirable method of regulating port fuel injection. However, it is also the most complex and expensive to design and manufacture. In this system, the injectors are timed and pulsed individually, much like the spark plugs are sequentially operated in firing order of the engine. This system is often called **sequential fuel injection,** or **SFI.** Each cylinder receives one charge every two crankshaft revolutions, just before the intake valve opens. This means that the mixture is never static in the intake manifold, and mixture adjustments can be made almost instantaneously between the firing of one injector and the next. A camshaft position sensor (CMP) signal or a special distributor reference pulse informs the PCM when the number 1 cylinder is on its compression stroke. If the sensor fails or the reference pulse is interrupted, some injection systems shut down, while others revert to pulsing the injectors simultaneously.

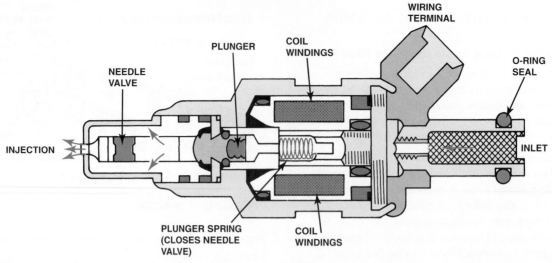

FIGURE 20–6 Cross section of a typical port fuel-injection nozzle assembly. These injectors are serviced as an assembly only; no part replacement or service is possible except for replacement of external O-ring seals.

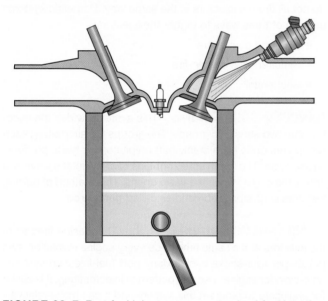

FIGURE 20–7 Port fuel injectors spray atomized fuel into the intake manifold about 3 inches (75 mm) from the intake valve.

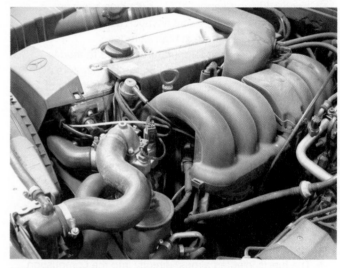

FIGURE 20–8 A port fuel-injected engine that is equipped with long, tuned intake manifold runners.

 **FREQUENTLY ASKED QUESTION**

**How Can It Be Determined If the Injection System Is Sequential?**

Look at the color of the wires at the injectors. If a sequentially fired injector is used, then one wire color (the pulse wire) will be a different color for each injector. The other wire is usually the same color because all injectors receive voltage from some source. If a group- or batch-fired injection system is being used, then the wire colors will be the same for the injectors that are group fired. For example, a V-6 group-fired engine will have three injectors with a pink and blue wire (power and pulse), and the other three will have pink and green wires.

The major advantage of using port injection instead of the simpler throttle-body injection is that the intake manifolds on port fuel-injected engines contain only air, not a mixture of air and fuel. This allows the engine design engineer the opportunity to design long, "tuned" intake-manifold runners that help the engine produce increased torque at low engine speeds. ● SEE FIGURE 20–8.

NOTE: Some port fuel-injection systems used on engines with four or more valves per cylinder may use two injectors per cylinder. One injector is used all the time, and the second injector is operated by the computer when high engine speed and high-load conditions are detected by the computer. Typically, the second injector injects fuel into the high-speed intake ports of the manifold. This system permits good low-speed power and throttle responses as well as superior high-speed power.

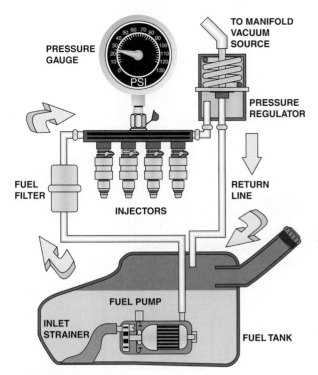

FIGURE 20–9 A typical port fuel-injected system showing a vacuum-controlled fuel-pressure regulator.

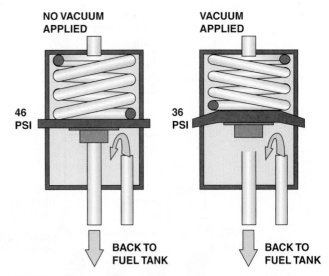

FIGURE 20–10 A typical fuel-pressure regulator that has a spring that exerts 46 pounds of force against the fuel. If 20 inches of vacuum are applied above the spring, the vacuum reduces the force exerted by the spring on the fuel, allowing the fuel to return to the tank at a lower pressure.

# FUEL-PRESSURE REGULATOR

The pressure regulator and fuel pump work together to maintain the required pressure drop at the injector tips. The fuel-pressure regulator typically consists of a spring-loaded, diaphragm-operated valve in a metal housing.

Fuel-pressure regulators on fuel-return-type fuel-injection systems are installed on the return (downstream) side of the injectors at the end of the fuel rail or are built into or mounted upon the throttle-body housing. Downstream regulation minimizes fuel-pressure pulsations caused by pressure drop across the injectors as the nozzles open. It also ensures positive fuel pressure at the injectors at all times and holds residual pressure in the lines when the engine is off. On mechanical returnless systems, the regulator is located back at the tank with the fuel filter.

In order for excess fuel (about 80% to 90% of the fuel delivered) to return to the tank, fuel pressure must overcome spring pressure on the spring-loaded diaphragm to uncover the return line to the tank. This happens when system pressure exceeds operating requirements. With TBI, the regulator is close to the injector tip, so the regulator senses essentially the same air pressure as the injector.

The pressure regulator used in a port fuel-injection system has an intake manifold vacuum line connection on the regulator vacuum chamber. This allows fuel pressure to be modulated by a combination of spring pressure and manifold vacuum acting on the diaphragm. ● SEE FIGURES 20–9 AND 20–10.

FIGURE 20–11 A lack of fuel flow could be due to a restricted fuel-pressure regulator. Notice the fine screen filter. If this filter were to become clogged, higher-than-normal fuel pressure would occur.

TECH TIP

**Don't Forget the Regulator**

Some fuel-pressure regulators contain a 10-micron filter. If this filter becomes clogged, a lack of fuel flow would result. ● SEE FIGURE 20–11.

In both TBI and port fuel-injection systems, the regulator shuts off the return line when the fuel pump is not running. This maintains pressure at the injectors for easy restarting after hot soak as well as reducing vapor lock.

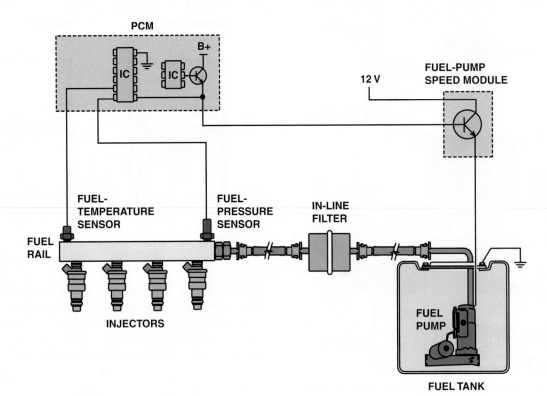

**FIGURE 20–12** The fuel-pressure sensor and fuel-temperature sensor are often constructed together in one assembly to help give the PCM the needed data to control the fuel-pump speed.

**NOTE: Some General Motors throttle-body units do not hold pressure and are called nonchecking.**

Port fuel-injection systems generally operate with pressures at the injector of about 30 to 55 PSI (207 to 379 kPa), while TBI systems work with injector pressures of about 10 to 20 PSI (69 to 138 kPa). The difference in system pressures results from the difference in how the systems operate. Since injectors in a TBI system inject the fuel into the airflow at the manifold inlet (above the throttle), there is more time for atomization in the manifold before the air–fuel charge reaches the intake valve. This allows TBI injectors to work at lower pressures than injectors used in a port system.

| Engine Operating Condition | Intake Manifold Vacuum | Fuel Pressure |
|---|---|---|
| Idle or cruise | High | Lower |
| Heavy load | Low | Higher |

The computer can best calculate injector pulse width based on all sensors if the pressure drop across the injector is the same under all operating conditions. A vacuum-controlled fuel-pressure regulator allows the equal pressure drop by reducing the force exerted by the regulator spring at high vacuum (low-load condition) yet allowing the full force of the regulator spring to be exerted when the vacuum is low (high-engine-load condition).

# VACUUM-BIASED FUEL-PRESSURE REGULATOR

The primary reason why many port fuel-injected systems use a vacuum-controlled fuel-pressure regulator is to ensure that there is a constant pressure drop across the injectors. In a throttle-body fuel-injection system, the injector squirts into the atmospheric pressure regardless of the load on the engine. In a port fuel-injected engine, however, the pressure inside the intake manifold changes as the load on the engine increases.

# ELECTRONIC RETURNLESS FUEL SYSTEM

This system is unique because it does not use a mechanical valve to regulate rail pressure. Fuel pressure at the rail is sensed by a pressure transducer, which sends a low-level signal to a controller. The controller contains logic to calculate a signal to the pump power driver. The power driver contains a high-current transistor that controls the pump speed using pulse width modulation (PWM). This system is called the **electronic returnless fuel system (ERFS).** ● SEE FIGURE 20–12. This

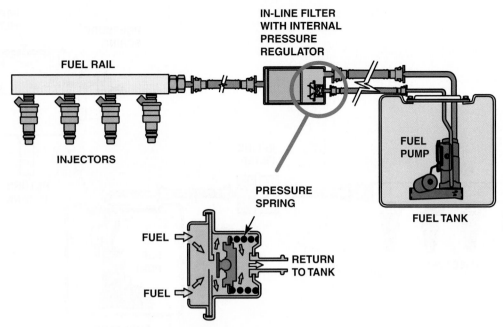

**FIGURE 20–13** A mechanical returnless fuel system. The bypass regulator in the fuel filter controls fuel line pressure.

transducer can be differentially referenced to manifold pressure for closed-loop feedback, correcting and maintaining the output of the pump to a desired rail setting. This system is capable of continuously varying rail pressure as a result of engine vacuum, engine fuel demand, and fuel temperature (as sensed by an external temperature transducer, if necessary). A **pressure vent valve (PVV)** is employed at the tank to relieve overpressure due to thermal expansion of fuel. In addition, a supply-side bleed, by means of an in-tank reservoir using a supply-side jet pump, is necessary for proper pump operation.

## MECHANICAL RETURNLESS FUEL SYSTEM

The first production returnless systems employed the **mechanical returnless fuel system (MRFS)** approach. This system has a bypass regulator to control rail pressure that is located in close proximity to the fuel tank. Fuel is sent by the in-tank pump to a chassis-mounted in-line filter with excess fuel returning to the tank through a short return line. ● **SEE FIGURE 20–13.** The in-line filter may be mounted directly to the tank, thereby eliminating the shortened return line. Supply pressure is regulated on the downstream side of the in-line filter to accommodate changing restrictions throughout the filter's service life. This system is limited to constant rail pressure (*CRP) system calibrations, whereas with ERFS, the pressure transducer can be referenced to atmospheric pressure for CRP systems or differentially referenced to intake manifold pressure for constant differential injector pressure (**CIP) systems.

**NOTE:** *CRP is referenced to atmospheric pressure, has lower operating pressure, and is desirable for calibrations using speed/air density sensing. **CIP is referenced to manifold pressure, varies rail pressure, and is desirable in engines that use mass airflow sensing.

## DEMAND DELIVERY SYSTEM

Given the experience with both ERFS and MRFS, a need was recognized to develop new returnless technologies that could combine the speed control and constant injector pressure attributes of ERFS together with the cost savings, simplicity, and reliability of MRFS. This new technology also needed to address pulsation dampening/hammering and fuel transient response. Therefore, the **demand delivery system (DDS)** technology was developed. A different form of demand pressure regulator has been applied to the fuel rail. It mounts at the head or port entry and regulates the pressure downstream at the injectors by admitting the precise quantity of fuel into the rail as consumed by the engine. Having demand regulation at the rail improves pressure response to flow transients and provides rail pulsation dampening. A fuel pump and a low-cost, high-performance bypass regulator are used within the appropriate fuel sender. ● **SEE FIGURE 20–14.** They supply a pressure somewhat higher than the required rail set pressure to accommodate dynamic line and filter pressure losses. Electronic pump speed control is accomplished using a smart regulator as an integral flow sensor. A **pressure control valve (PCV)** may also be used and can readily reconfigure an existing design fuel sender into a returnless sender.

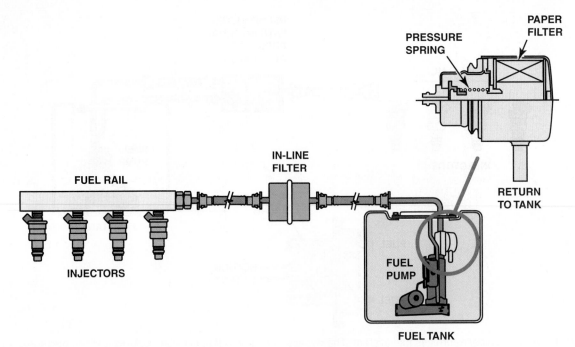

**FIGURE 20–14** A demand delivery system uses a fuel pressure regulator attached to the fuel pump assembly.

**? FREQUENTLY ASKED QUESTION**

**Why Are Some Fuel Rails Rectangle Shaped?**

A port fuel-injection system uses a pipe or tubes to deliver fuel from the fuel line to the intended fuel injectors. This pipe or tube is called the **fuel rail**. Some vehicle manufacturers construct the fuel rail in a rectangular cross section. ● **SEE FIGURE 20–15.** The sides of the fuel rail are able to move in and out slightly, thereby acting as a fuel pulsator evening out the pressure pulses created by the opening and closing of the injectors to reduce underhood noise. A round cross-section fuel rail is not able to deform, and as a result, some manufacturers have had to use a separate dampener.

**FIGURE 20–15** A rectangle shaped fuel rail is used to help dampen fuel system pulsations and noise caused by the injectors opening and closing.

# FUEL INJECTORS

EFI systems use a 12 volt solenoid-operated injectors. ● **SEE FIGURE 20–16.** This electromagnetic device contains an armature and a spring-loaded needle valve or ball valve assembly. When the computer energizes the solenoid, voltage is applied to the solenoid coil until the current reaches a specified level. This permits a quick pull-in of the armature during turn-on. The armature is pulled off of its seat against spring force, allowing fuel to flow through the inlet filter screen to the spray nozzle, where it is sprayed in a pattern that varies with application. ● **SEE FIGURE 20–17.** The injector opens the same amount each time it is energized, so the amount of fuel injected depends on the length of time the injector remains open. By angling the director

hole plates, the injector sprays fuel more directly at the intake valves, which further atomizes and vaporizes the fuel before it enters the combustion chamber. PFI injectors typically are a top-feed design in which fuel enters the top of the injector and passes through its entire length to keep it cool before being injected.

Ford introduced two basic designs of deposit-resistant injectors on some engines. The design, manufactured by Bosch, uses a four-hole director/metering plate similar to that used by the Rochester Multec injectors. The design manufactured by Nippondenso uses an internal upstream orifice in the adjusting tube. It also has a redesigned pintle/seat containing a wider tip opening that tolerates deposit buildup without affecting injector performance.

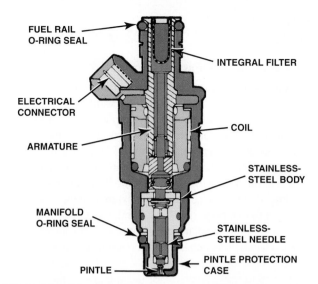

FIGURE 20–16 A multiport fuel injector. Notice that the fuel flows straight through and does not come in contact with the coil windings.

FIGURE 20–17 Each of the eight injectors shown are producing a correct spray pattern for the applications. While all throttle-body injectors spray a conical pattern, most port fuel injections do not.

# CENTRAL PORT INJECTION

A cross between port fuel injection and throttle-body injection, CPI was introduced in the early 1990s by General Motors. The CPI assembly consists of a single fuel injector, a pressure regulator, and six poppet nozzle assemblies with nozzle tubes. ●SEE FIGURE 20–18. The central sequential fuel injection (CSFI) system has six injectors in place of just one used on the CPI unit.

When the injector is energized, its armature lifts off of the six fuel tube seats, and pressurized fuel flows through the nozzle tubes to each poppet nozzle. The increased pressure causes each poppet nozzle ball to also lift from its seat, allowing fuel to flow from the nozzle. This hybrid injection system combines the single injector of a TBI system with the equalized fuel

**?** FREQUENTLY ASKED QUESTION

### How Can the Proper Injector Size Be Determined?

Most people want to increase the output of fuel to increase engine performance. Injector sizing can sometimes be a challenge, especially if the size of injector is not known. In most cases, manufacturers publish the rating of injectors in pounds of fuel per hour (lb/hr). The rate is figured with the injector held open at 3 bars (43.5 PSI). An important consideration is that larger flow injectors have a higher minimum flow rating. Here is a formula to calculate injector sizing when changing the mechanical characteristics of an engine:

**Flow rate = hp × BSFC/# of cylinders × maximum duty cycle (% of on-time of the injectors)**

- **hp** is the projected horsepower. Be realistic!
- **BSFC** is brake-specific fuel consumption in pounds per horsepower-hour. Calculated values are used for this, 0.4 to 0.8 pound. In most cases, start on the low side for naturally aspirated engines and the high side for engines with forced induction.
- **# of cylinders** is actually the number of injectors being used.
- **Maximum duty cycle** is considered at 0.8 (80%). Above this, the injector may overheat, lose consistency, or not work at all.

An example follows:

**5.7 liter V-8 = 240 hp × 0.65/8 cylinders × 8
= 24.37 lb/hr injectors required**

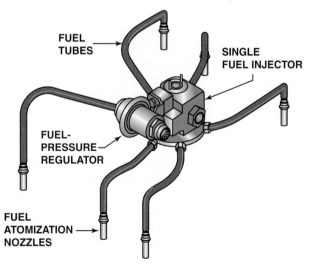

FIGURE 20–18 A central port fuel-injection system.

distribution of a PFI system. It eliminates the individual fuel rail while allowing more efficient manifold tuning than is otherwise possible with a TBI system. Newer versions use six individual solenoids to fire one for each cylinder. ●SEE FIGURE 20–19.

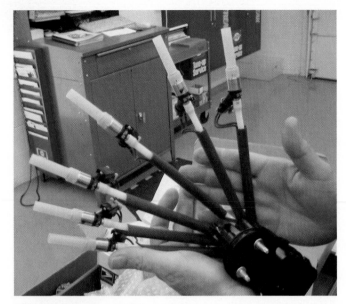

**FIGURE 20–19** A factory replacement unit for a CSFI unit that has individual injectors at the ends that go into the intake manifold instead of poppet valves.

# FUEL-INJECTION MODES OF OPERATION

All fuel-injection systems are designed to supply the correct amount of fuel under a wide range of engine operating conditions. These modes of operation include the following:

| | |
|---|---|
| Starting (cranking) | Acceleration enrichment |
| Clear flood | Deceleration enleanment |
| Idle (run) | Fuel shutoff |

**STARTING MODE**   When the ignition is turned to the start position, the engine cranks, and the PCM energizes the fuel pump relay. The PCM also pulses the injectors on, basing the pulse width on engine speed and engine coolant temperature. The colder the engine is, the greater the pulse width. Cranking mode air–fuel ratio varies from about 1.5:1 at −40°F (−40°C) to 14.7:1 at 200°F (93°C).

**CLEAR FLOOD MODE**   If the engine becomes flooded with too much fuel, the driver can depress the accelerator pedal to greater than 80% to enter the clear flood mode. When the PCM detects that the engine speed is low (usually below 600 RPM) and the throttle-position (TP) sensor voltage is high (WOT), the injector pulse width is greatly reduced or even shut off entirely, depending on the vehicle.

**OPEN-LOOP MODE**   Open-loop operation occurs during warm-up before the oxygen sensor can supply accurate information to the PCM. The PCM determines injector pulse width based on values from the MAF, MAP, TP, ECT, and IAT sensors.

**?** **FREQUENTLY ASKED QUESTION**

**What Is Battery Voltage Correction?**

Battery voltage correction is a program built into the PCM that causes the injector pulse width to increase if there is a drop in electrical system voltage. Lower battery voltage would cause the fuel injectors to open slower than normal and the fuel pump to run slower. Both of these conditions can cause the engine to run leaner than normal if the battery voltage is low. Because a lean air–fuel mixture can cause the engine to overheat, the PCM compensates for the lower voltage by adding a percentage to the injector pulse width. This richer condition will help prevent serious engine damage. The idle speed is also increased to turn the alternator faster if low battery voltage is detected.

**CLOSED-LOOP MODE**   Closed-loop operation is used to modify the base injector pulse width as determined by feedback from the oxygen sensor to achieve proper fuel control.

**ACCELERATION ENRICHMENT MODE**   During acceleration, the throttle-position (TP) voltage increases, indicating that a richer air–fuel mixture is required. The PCM then supplies a longer injector pulse width and may even supply extra pulses to supply the needed fuel for acceleration.

**DECELERATION ENLEANMENT MODE**   When the engine decelerates, a leaner air–fuel mixture is required to help reduce emissions and to prevent deceleration backfire. If the deceleration is rapid, the injector may be shut off entirely for a short time and then pulsed on enough to keep the engine running.

**FUEL SHUTOFF MODE**   Besides shutting off fuel entirely during periods of rapid deceleration, PCM also shuts off the injector when the ignition is turned off to prevent the engine from continuing to run.

# IDLE CONTROL

Port fuel-injection systems generally use an auxiliary air bypass to control idle speed. **● SEE FIGURE 20–20.** This air bypass or regulator provides needed additional airflow and thus more fuel. The engine needs more power when cold to maintain its normal idle speed to overcome the increased friction from cold lubricating oil. It does this by opening an intake air passage to let more air into the engine just as depressing the accelerator pedal would open the throttle valve, allowing more air into the engine. The system is calibrated to maintain engine idle speed at a specified value regardless of engine temperature.

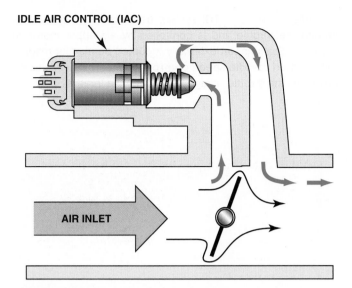

FIGURE 20–20 The small arrows indicate the air bypassing the throttle plate in the closed throttle position. This air is called minimum air. The air flowing through the IAC (blue arrows) is the airflow that determines the idle speed.

Most PFI systems use an idle air control (IAC) motor to regulate idle bypass air. The IAC is computer controlled and is either a solenoid-operated valve or a stepper motor that regulates the airflow around the throttle. The idle air control valve is also called an **electronic air control (EAC)** valve.

When the engine stops, most IAC units will retract outward to get ready for the next engine start. When the engine starts, the engine speed is high to provide for proper operation when the engine is cold. Then, as the engine gets warmer, the computer reduces engine idle speed gradually by reducing the number of counts or steps commanded by the IAC.

When the engine is warm and restarted, the idle speed should momentarily increase, then decrease to normal idle speed. This increase and then decrease in engine speed is often called an engine **flare.** If the engine speed does not flare, then the IAC may not be working (it may be stuck in one position).

# STEPPER MOTOR OPERATION

A digital output is used to control stepper motors. Stepper motors are direct-current motors that move in fixed steps or increments from deenergized (no voltage) to fully energized (full voltage). A stepper motor often has as many as 120 steps of motion.

A common use for stepper motors is as an idle air control (IAC) valve, which controls engine idle speeds and prevents stalls due to changes in engine load. When used as an IAC, the stepper motor is usually a reversible DC motor that moves in increments, or steps. The motor moves a shaft back and forth to operate a conical valve. When the conical valve is moved back,

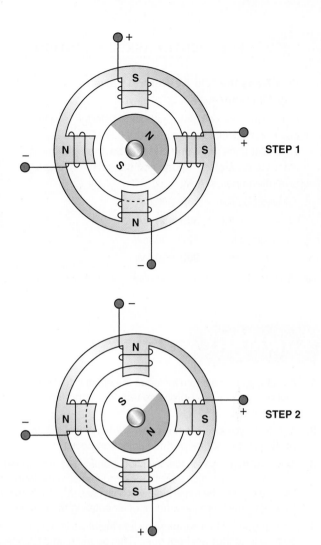

FIGURE 20–21 Most stepper motors use four wires, which are pulsed by the computer to rotate the armature in steps.

more air bypasses the throttle plates and enters the engine, increasing idle speed. As the conical valve moves inward, the idle speed decreases.

When using a stepper motor that is controlled by the PCM, it is very easy for the PCM to keep track of the position of the stepper motor. By counting the number of steps that have been sent to the stepper motor, the PCM can determine the relative position of the stepper motor. While the PCM does not actually receive a feedback signal from the stepper motor, it does know how many steps forward or backward the motor should have moved.

A typical stepper motor uses a permanent magnet and two electromagnets. Each of the two electromagnetic windings is controlled by the computer. The computer pulses the windings and changes the polarity of the windings to cause the armature of the stepper motor to rotate 90 degrees at a time. Each 90-degree pulse is recorded by the computer as a "count" or "step," thus the name given to this type of motor. ● **SEE FIGURE 20–21.**

**Why Does the Idle Air Control Valve Use Milliamperes?**

Some Chrysler vehicles, such as the Dodge minivan, use linear solenoid idle air control (LSIAC) valves. The PCM uses regulated current flow through the solenoid to control idle speed and the scan tool display is in milliamperes (mA):

Closed position = 180–200 mA

Idle = 300–450 mA

Light cruise = 500–700 mA

Fully open = 900–950 mA

Idle airflow in a TBI system travels through a passage around the throttle and is controlled by a stepper motor. In some applications, an externally mounted permanent magnet motor called the **idle speed control (ISC) motor** mechanically advances the throttle linkage to advance the throttle opening.

## SUMMARY

1. A fuel-injection system includes the electric fuel pump and fuel pump relay, fuel-pressure regulator, and fuel injectors (nozzles).

2. The two types of fuel-injection systems are the throttle-body design and the port fuel-injection design.

3. The two methods of fuel-injection control are the speed-density system, which uses the MAP to measure the load on the engine, and the mass airflow, which uses the MAF sensor to directly measure the amount of air entering the engine.

4. The amount of fuel supplied by fuel injectors is determined by how long they are kept open. This opening time is called the pulse width and is measured in milliseconds.

5. The fuel-pressure regulator is usually located on the fuel return on return-type fuel-injection systems.

6. TBI-type fuel-injection systems do not use a vacuum-controlled fuel-pressure regulator, whereas many port fuel-injection systems use a vacuum-controlled regulator to monitor equal pressure drop across the injectors.

7. Other fuel designs include the electronic returnless, the mechanical returnless, and the demand delivery systems.

## REVIEW QUESTIONS

1. What are the two basic types of fuel-injection systems?

2. What is the purpose of the vacuum-controlled (biased) fuel-pressure regulator?

3. How many sensors are used to determine the base pulse width on a speed-density system?

4. How many sensors are used to determine the base pulse width on a mass airflow system?

5. What are the three types of returnless fuel-injection systems?

## CHAPTER QUIZ

1. Technician A says that the fuel-pump relay is usually controlled by the PCM. Technician B says that a TBI injector squirts fuel above the throttle plate. Which technician is correct?
   a. Technician A only
   b. Technician B only
   c. Both Technicians A and B
   d. Neither Technician A nor B

2. Why are some fuel rails rectangular in shape?
   a. Increases fuel pressure
   b. Helps keep air out of the injectors
   c. Reduces noise
   d. Increases the speed of the fuel through the fuel rail

3. Which fuel-injection system uses the MAP sensor as the primary sensor to determine the base pulse width?
   a. Speed density
   b. Mass airflow
   c. Demand delivery
   d. Mechanical returnless

4. Why is a vacuum line attached to a fuel-pressure regulator on many port fuel-injected engines?
   a. To draw fuel back into the intake manifold through the vacuum hose
   b. To create an equal pressure drop across the injectors
   c. To raise the fuel pressure at idle
   d. To lower the fuel pressure under heavy engine load conditions to help improve fuel economy

5. Which sensor has the greatest influence on injector pulse width besides the MAF sensor?
   a. IAT          c. ECT
   b. BARO         d. TP

6. Technician A says that the port fuel-injection injectors operate using 5 volts from the computer. Technician B says that sequential fuel injectors all use a different wire color on the injectors. Which technician is correct?
   a. Technician A only
   b. Technician B only
   c. Both Technicians A and B
   d. Neither Technician A nor B

7. Which type of port fuel-injection system uses a fuel-temperature and/or fuel-pressure sensor?
   a. All port fuel-injected engines
   b. TBI units only
   c. Electronic returnless systems
   d. Demand delivery systems

8. Dampeners are used on some fuel rails to _____.
   a. Increase the fuel pressure in the rail
   b. Reduce (decrease) the fuel pressure in the rail
   c. Reduce noise
   d. Trap dirt and keep it away from the injectors

9. Where is the fuel-pressure regulator located on a vacuum-biased port fuel-injection system?
   a. In the tank
   b. At the inlet of the fuel rail
   c. At the outlet of the fuel rail
   d. Near or on the fuel filter

10. What type of device is used in a typical idle air control?
    a. DC motor
    b. Stepper motor
    c. Pulsator-type actuator
    d. Solenoid

# chapter 21
# GASOLINE DIRECT-INJECTION SYSTEMS

**LEARNING OBJECTIVES:** **After studying this chapter, the reader should be able to:** • Explain the operation of a direct-injection fuel delivery system. • Understand a gasoline direct-injection fuel injector, the modes of operation. • Describe the port- and direct-injection systems used in Lexus vehicles. • Describe how to diagnose a gasoline direct-injection system.

**KEY TERMS:** Gasoline direct injection (GDI) 270 • Homogeneous mode 272 • Spark ignition direct injection (SIDI) 270 • Stratified mode 272

## DIRECT FUEL INJECTION

Several vehicle manufacturers such as Audi, Mitsubishi, Mercedes, BMW, Toyota/Lexus, Mazda, Ford, and General Motors are using **gasoline direct-injection (GDI)** systems, which General Motors refers to as a **spark ignition direct injection (SIDI)** system. A direct-injection system sprays high-pressure fuel, up to 2,900 PSI, into the combustion chamber as the piston approaches the top of the compression stroke. With the combination of high-pressure swirl injectors and modified combustion chamber, almost instantaneous vaporization of the fuel occurs. This, combined with a higher compression ratio, allows a direct-injected engine to operate using a leaner-than-normal air–fuel ratio, which results in improved fuel economy with higher power output and reduced exhaust emissions.
● **SEE FIGURES 21–1.**

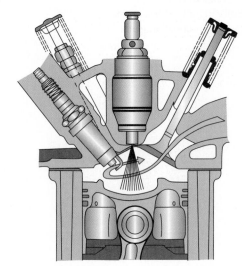

**FIGURE 21–1** A gasoline direct-injection (GDI) system injects fuel under high pressure directly into the combustion chamber.

**ADVANTAGES OF GDI** The use of direct injection compared with port fuel injection has many advantages, including the following:

- Improved fuel economy due to reduced pumping losses and heat loss
- Allows a higher compression ratio for higher engine efficiency
- Allows the use of lower-octane gasoline
- The volumetric efficiency is higher
- Less need for extra fuel for acceleration
- Improved cold starting and throttle response
- Allows the use of higher percentage of EGR to reduce exhaust emissions

- Up to 25% improvement in fuel economy
- 12% to 15% reduction in exhaust emissions

### DISADVANTAGES OF GDI

- Higher cost due to high-pressure pump and injectors
- More components compared with port fuel injection
- Because of the high compression, a $NO_x$ storage catalyst is sometimes required to meet emission standards, especially in Europe.
- Uses up to six operating modes, depending on engine load and speed, which requires more calculations to be performed by the powertrain control module (PCM).

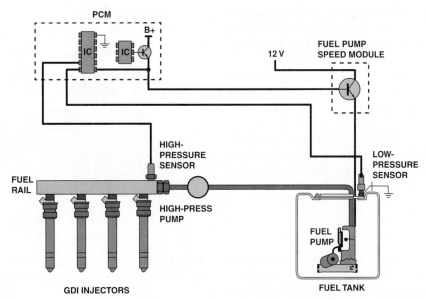

**FIGURE 21–2** A GDI system uses a low-pressure pump in the gas tank similar to other types of fuel-injection systems. The PCM controls the pressure of the high-pressure pump using sensor inputs.

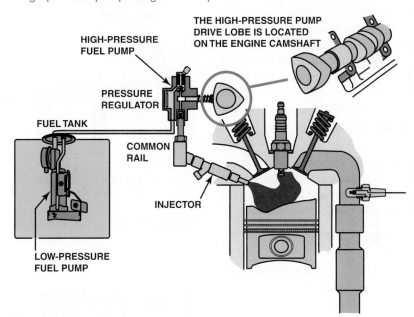

**FIGURE 21–3** A typical direct-injection system uses two pumps—one low-pressure electric pump in the fuel tank and a high-pressure pump driven by the camshaft. The high pressure fuel system operates at a pressure as low as 500 PSI during light-load conditions and as high as 2,900 PSI under heavy loads.

## DIRECT-INJECTION FUEL DELIVERY SYSTEM

**LOW-PRESSURE SUPPLY PUMP** The fuel pump in the fuel tank supplies fuel to the high-pressure fuel pump at a pressure of approximately 60 PSI. The fuel filter is located in the fuel tank and is part of the fuel-pump assembly. It is not usually serviceable as a separate component. The fuel pump is pulse-width-modulated (PWM) and pump speed is changed by varying the duty cycle (on-off time). A fuel pressure sensor is used to provide fuel pressure feedback to the fuel pump control module. It uses a 5-volt feed

plus a signal pump flow control module. The fuel pressure sensor is a serviceable unit that has a 5-volt power feed, plus a ground and signal wire.

**HIGH-PRESSURE PUMP** The powertrain control module (PCM) controls the output of the high-pressure pump, which has a range between 500 PSI (3,440 kPa) and 2,900 PSI (15,200 kPa) during engine operation. ● **SEE FIGURES 21–2 AND 21–3.** In a General Motors system, the engine control module (ECM) controls the output of the high-pressure pump, which has a range between 500 PSI (3,440 kPa) and 2,900 PSI (15,200 kPa) during engine operation. The high-pressure fuel pump connects to the pump in the fuel tank through the low-pressure fuel line. The pump consists of a single-barrel piston pump, which is driven by the engine camshaft. The pump plunger rides on a three-lobed cam on the camshaft. The high-pressure pump is cooled and lubricated by the fuel itself. ● SEE FIGURE 21–4.

**FUEL RAIL** The fuel rail stores fuel from the high pressure pump for use by each injector. All injectors receive the same pressure as what is present in the fuel rail.

**FUEL-PRESSURE REGULATOR** An electric pressure-control valve is installed between the pump inlet and outlet valves. The fuel rail pressure sensor connects to the PCM with three wires:

- 5-volt reference
- Ground
- Signal

The sensor signal provides an analog signal to the PCM that varies in voltage as fuel rail pressure changes. Low pressure results in a low-voltage signal, and high pressure results in a high-voltage signal.

The PCM uses internal drivers to control the power feed and ground for the pressure control valve. When both PCM drivers are deactivated, the inlet valve is held open by spring pressure. This causes the high-pressure fuel pump to default to low-pressure mode. The fuel from the high-pressure fuel pump flows through a line to the fuel rail and injectors. The actual operating pressure can vary from as low as 500 PSI (6,200 kPa) at idle to over 2,000 PSI (13,800 kPa) during high-speed or heavy-load conditions. ● SEE FIGURE 21–5.

NOTE: Unlike a port fuel-injection system, a gasoline direct injection system varies the fuel pressure to achieves greater fuel delivery using a very short pulse time, which is usually less than one millisecond. To summarize:

- **Port Fuel Injection = constant fuel pressure but variable injector pulse-width.**
- **GDI = almost constant injector pulse-width with varying fuel pressure.**

## GDI FUEL INJECTORS

Each high-pressure fuel injector assembly is an electrically magnetic injector mounted in the cylinder head. In the GDI system, the PCM controls each fuel injector with 50 to 90 volts (usually 60 to 70 volts), depending on the system, which is created by a boost capacitor in the PCM. During the high-voltage boost phase, the capacitor is discharged through an injector, allowing for initial injector opening. The injector is then held open with 12 volts. The high-pressure fuel injector has a small slit or six precision-machined holes that generate the desired spray pattern. The injector also has an extended tip to allow for cooling from a water jacket in the cylinder head.
● **SEE CHART 21–1** for an overview of the differences between a port fuel-injection system and a GDI system.

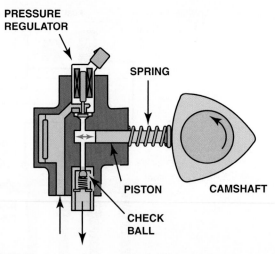

**FIGURE 21–4** A typical camshaft-driven high-pressure pump used to increase fuel pressure to 2,000 PSI or higher.

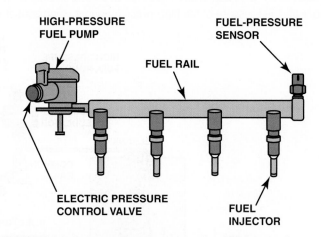

**FIGURE 21–5** A gasoline direct-injection (GDI) fuel rail and pump assembly with the electric pressure control valve.

## MODES OF OPERATION

The two basic modes of operation include the following:

1. **Stratified mode.** In this mode of operation, the air–fuel mixture is richer around the spark plug than it is in the rest of the cylinder.

2. **Homogeneous mode.** In this mode of operation, the air–fuel mixture is the same throughout the cylinder.

There are variations of these modes that can be used to fine-tune the air–fuel mixture inside the cylinder. For example, Bosch, a supplier to many vehicle manufacturers, uses six modes of operation, including the following:

- **Homogeneous mode.** In this mode, the injector is pulsed one time to create an even air–fuel mixture in the cylinder. The injection occurs during the intake stroke. This mode is used during high-speed and/or high-torque conditions.

## PORT FUEL-INJECTION SYSTEM COMPARED WITH GDI SYSTEM

| | PORT FUEL INJECTION | GDI |
|---|---|---|
| Fuel pressure | 35–60 PSI | Lift pump—50–60 PSI<br>High-pressure pump—500–2,900 PSI |
| Injection pulse width at idle | 1.5–3.5 ms | About 0.4 ms (400 μs) |
| Injector resistance | 12–16 ohms | 1–3 ohms |
| Injector voltage | 6 V for low-resistance injectors, 12 V for most injectors | 50–90 V |
| Number of injections per event | One | 1–3 |
| Engine compression ratio | 8:1–11:1 | 11:1–13:1 |

**CHART 21–1**

A comparison chart showing the major differences between a port fuel-injection system and a gasoline direct-injection (GDI) system.

- **Homogeneous lean mode.** Similar to the homogeneous mode except that the overall air–fuel mixture is slightly lean for better fuel economy. The injection occurs during the intake stroke. This mode is used under steady, light-load conditions.

- **Stratified mode.** In this mode of operation, the injection occurs just before the spark occurs, resulting in lean combustion and reducing fuel consumption.

- **Homogeneous stratified mode.** In this mode, there are two injections of fuel:
  - The first injection is during the intake stroke.
  - The second injection is during the compression stroke. As a result of these double injections, the rich air–fuel mixture around the spark plug is ignited first. Then the rich mixture ignites the leaner mixture. The advantages of this mode include lower exhaust emissions than the stratified mode and less fuel consumption than the homogeneous lean mode.

- **Homogeneous knock protection mode.** The purpose of this mode is to reduce the possibility of spark knock from occurring under heavy loads at low engine speeds. There are two injections of fuel:
  - The first injection occurs on the intake stroke.
  - The second injection occurs during the compression stroke with the overall mixture being stoichiometric.

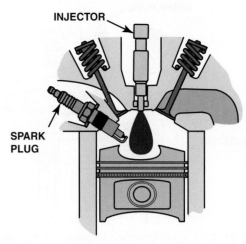

**SPRAY-GUIDED COMBUSTION**

INJECTOR

SPARK PLUG

**FIGURE 21–6** In this design, the fuel injector is at the top of the cylinder and sprays fuel into the cavity of the piston.

As a result of this mode, the PCM does not need to retard ignition timing as much to operate knock free.

- **Stratified catalyst heating mode.** In this mode, there are two injections:
  - The first injection is on the compression stroke just before combustion.
  - The second injection occurs after combustion occurs to heat the exhaust. This mode is used to quickly warm the catalytic converter and to burn the sulfur from the $NO_X$ catalyst.

## PISTON TOP DESIGNS

GDI systems use a variety of shapes of piston and injector locations depending on make and model of engine. Three of the most commonly used designs include the following:

- **Spray-guided combustion.** In this design, the injector is placed in the center of the combustion chamber and injects fuel into the dished-out portion of the piston. The shape of the piston helps guide and direct the mist of fuel in the combustion chamber. ● SEE **FIGURE 21–6.**

- **Swirl combustion.** This design uses the shape of the piston and the position of the injector at the side of the combustion chamber to create turbulence and swirl of the air–fuel mixture. ● SEE FIGURE 21–7.

- **Tumble combustion.** Depending on when the fuel is injected into the combustion chamber, this helps determine how the air–fuel mixture is moved or tumbled. ● SEE FIGURE 21–8.

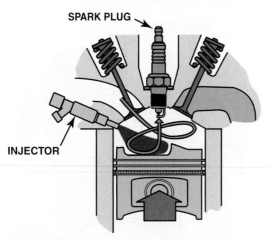

FIGURE 21–7 The side injector combines with the shape of the piston to create a swirl as the piston moves up on the compression stroke.

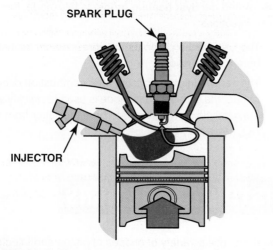

FIGURE 21–8 The piston creates a tumbling force as the piston moves upward.

## LEXUS PORT- AND DIRECT-INJECTION SYSTEMS

**OVERVIEW** Many Lexus vehicles use GDI, and in some engines, they also use a conventional port fuel-injection system. The Lexus D-4S system combines direct-injection injectors located in the combustion chamber with port fuel injectors in the intake manifold near the intake valve. The two injection systems work together to supply the fuel needed by the

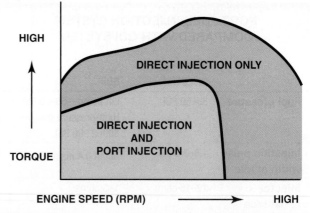

FIGURE 21–9 Notice that there are conditions when both the port fuel injector located in the intake manifold and the gasoline direct injector located in the cylinder operate to provide the proper air–fuel mixture.

engine. ● SEE FIGURE 21–9 for how the two systems are used throughout the various stages of engine operation.

**COLD-START WARM-UP** To help reduce exhaust emissions after a cold start, the fuel system uses a stratified change mode. This results in a richer air–fuel mixture near the spark plug and allows for the spark to be retarded to increase the temperature of the exhaust. As a result of the increased exhaust temperature, the catalytic converter rapidly reaches operating temperature, which reduces exhaust emissions.

## ENGINE START SYSTEM

An engine equipped with GDI could use the system to start the engine. This is most useful during idle stop mode when the engine is stopped while the vehicle is at a traffic light to save fuel. The steps used in the Mazda start-stop system, called the *smart idle stop system (SISS)*, allow the engine to be started without a starter motor and include the following steps:

STEP 1    The engine is stopped. The normal stopping position of an engine when it stops is 70 degrees before top dead center, plus or minus 20 degrees. This is because the engine stops with one cylinder on the compression stroke, and the PCM can determine the cylinder position, using the crankshaft and camshaft position sensors.

STEP 2    When a command is made to start the engine by the PCM, fuel is injected into the cylinder that is on the compression stroke and ignited by the spark plug.

STEP 3    The piston on the compression stroke is forced downward, forcing the crankshaft to rotate counterclockwise or in the opposite direction to normal operation.

**STEP 4** The rotation of the crankshaft then forces the companion cylinder toward the top of the cylinder.

**STEP 5** Fuel is injected, and the spark plug is fired, forcing the piston down, causing the crankshaft to rotate in the normal (clockwise) direction. Normal combustion events continue, allowing the engine to keep running.

## GDI SERVICE

**NOISE ISSUES** GDI systems operate at high pressure, and the injectors can often be heard with the engine running and the hood open. This noise can be a customer concern because the clicking sound is similar to noisy valves. If a noise issue is the customer concern, check the following:

- Check a similar vehicle to determine if the sound is louder or more noticeable than normal.
- Check that nothing under the hood is touching the fuel rail. If another line or hose is in contact with the fuel rail, the sound of the injectors clicking can be transmitted throughout the engine, making the sound more noticeable.
- Check for any technical service bulletins (TSBs) that may include new clips or sound insulators to help reduce the noise.

**CARBON ISSUES** Carbon is often an issue in engines equipped with GDI systems. Carbon can affect engine operation by accumulating in two places:

- **On the injector itself.** Because the injector tip is in the combustion chamber, fuel residue can accumulate on the injector, reducing its ability to provide the proper spray pattern and amount of fuel. Some injector designs are more likely to be affected by carbon than others. For example, if the injector uses small holes, these tend to become clogged more often than an injector that uses a single slit opening, where the fuel being sprayed out tends to blast away any carbon. ● **SEE FIGURE 21–10.**

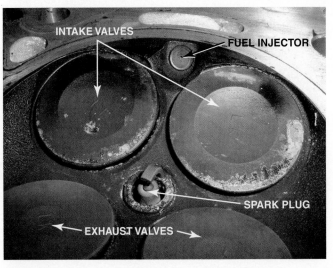

**FIGURE 21–10** There may become a driveability issue because the gasoline direct-injection (GDI) injector is exposed to combustion carbon and fuel residue.

- **The backside of the intake valve.** This is a common place for fuel residue and carbon to accumulate on engines equipped with GDI. The accumulation of carbon on the intake valve can become so severe that the engine will start and idle but lack power to accelerate the vehicle. The carbon deposits restrict the airflow into the cylinder enough to decrease engine power.

NOTE: Lexus engines that use both port and GDI injectors do not show intake valve deposits. It is thought that the fuel being sprayed onto the intake valve from the port injector helps keep the intake valve clean.

**CARBON CLEANING.** Most experts recommend the use of Techron®, a fuel system dispersant, to help keep carbon from accumulating. The use of a dispersant every six months or every 6,000 miles has proven to help prevent injector and intake valve deposits.

If the lack of power is discovered and there are no stored diagnostic trouble codes, a conventional carbon cleaning procedure will likely restore power if the intake valves are coated.

## SUMMARY

1. A GDI system uses a fuel injector that delivers a short squirt of fuel directly into the combustion chamber rather than in the intake manifold, near the intake valve on a port fuel injection system.

2. The advantages of using GDI instead of port fuel injection include the following:
   - Improved fuel economy
   - Reduced exhaust emissions
   - Greater engine power

3. Some of the disadvantages of GDI systems compared with a port fuel-injection system include the following:
   - Higher cost
   - The need for $NO_x$ storage catalyst in some applications
   - More components

4. The operating pressure can vary from as low as 500 PSI during some low-demand conditions to as high as 2,900 PSI.

5. The fuel injectors are open for a very short period of time and are pulsed using a 50- to 90-V pulse from a capacitor circuit.

6. GDI systems can operate in many modes, which are separated into the two basic modes:
   • Stratified mode
   • Homogeneous mode
7. GDI can be used to start an engine without the use of a starter motor for idle-stop functions.

8. GDI does create a louder clicking noise from the fuel injectors than port fuel-injection injectors.
9. Carbon deposits on the injector and the backside of the intake valve are a common problem with engines equipped with GDI systems.

## REVIEW QUESTIONS

1. What are two advantages of GDI compared with port fuel injection?
2. What are two disadvantages of GDI compared with port fuel injection?
3. How is the fuel delivery system different from a port fuel-injection system?
4. What are the basic modes of operation of a GDI system?

## CHAPTER QUIZ

1. Where is the fuel injected in an engine equipped with GDI?
   a. Into the intake manifold near the intake valve
   b. Directly into the combustion chamber
   c. Above the intake port
   d. In the exhaust port
2. The fuel pump inside the fuel tank on a vehicle equipped with GDI produces about what fuel pressure?
   a. 5 to 10 PSI
   b. 10 to 20 PSI
   c. 20 to 40 PSI
   d. 50 to 60 PSI
3. The high-pressure fuel pumps used in GDI systems are powered by _____.
   a. Electricity (DC motor)
   b. Electricity (AC motor)
   c. The camshaft
   d. The crankshaft
4. The high-pressure fuel pump pressure is regulated by using _____.
   a. An electric pressure-control valve
   b. A vacuum-biased regulator
   c. A mechanical regulator at the inlet to the fuel rail
   d. A non-vacuum-biased regulator
5. The fuel injectors operate under a fuel pressure of about _____.
   a. 35 to 45 PSI
   b. 90 to 150 PSI
   c. 500 to 2,900 PSI
   d. 2,000 to 5,000 PSI

6. The fuel injectors used on a GDI system are pulsed on using what voltage?
   a. 12 to 14 V
   b. 50 to 90 V
   c. 100 to 110 V
   d. 200 to 220 V
7. Which mode of operation results in a richer air–fuel mixture near the spark plug?
   a. Stoichiometric
   b. Homogeneous
   c. Stratified
   d. Knock protection
8. Some engines that use a GDI system also have port injection.
   a. True
   b. False
9. A GDI system can be used to start an engine without the need for a starter.
   a. True
   b. False
10. A lack of power from an engine equipped with GDI could be due to _____.
    a. Noisy injectors
    b. Carbon on the injectors
    c. Carbon on the intake valves
    d. Both b and c

# chapter 22
# ELECTRONIC THROTTLE CONTROL SYSTEM

**LEARNING OBJECTIVES:** **After studying this chapter, the reader should be able to:** • Describe the purpose and function of electronic throttle control (ETC) systems. • Describe the operation of the throttle body assembly, accelerator pedal position (APP), and throttle position sensors. • Explain how to diagnose an electronic throttle control system. • Describe how to service an electronic throttle control (ETC) system.

**KEY TERMS:** Accelerator pedal position (APP) sensor  277 • Coast-down stall  283 • Default position  279 • Drive-by-wire  277 • Electronic throttle control (ETC)  277 • Fail-safe position  279 • Neutral position  279 • Servomotor  279 • Throttle position (TP) sensor  277

## ELECTRONIC THROTTLE CONTROL (ETC) SYSTEM

**ADVANTAGES OF ETC** The absence of any mechanical linkage between the throttle pedal and the throttle body requires the use of an electric actuator motor. The electronic throttle system has the following advantages over the conventional cable:

- Eliminates the mechanical throttle cable, thereby reducing the number of moving parts.

- Eliminates the need for cruise control actuators and controllers.

- Helps reduce engine power for traction control (TC) and electronic stability control (ESC) systems.

- Used to delay rapid applications of torque to the transmission/transaxle to help improve driveability and to smooth shifts.

- Helps reduce pumping losses by using the electronic throttle to open at highway speeds with greater fuel economy. The ETC opens the throttle to maintain engine and vehicle speed as the powertrain control module (PCM) leans the air–fuel ratio, retards ignition timing, and introduces additional exhaust gas recirculation (EGR) to reducing pumping losses.

- Used to provide smooth engine operation, especially during rapid acceleration.

- Eliminates the need for an idle air control valve.

The electronic throttle can be called **drive-by-wire,** but most vehicle manufacturers use the term **electronic throttle** **control (ETC)** to describe the system that opens the throttle valve electrically.

**PARTS INVOLVED** The typical ETC system includes the following components:

1. **Accelerator pedal position (APP)** sensor, also called accelerator pedal sensor (APS)

2. **Electronic throttle actuator (servomotor), which is part of the electronic throttle body**

3. **Throttle position (TP) sensor**

4. Electronic control unit, which is usually the powertrain control module (PCM)
   ● **SEE FIGURE 22–1.**

## NORMAL OPERATION OF THE ETC SYSTEM

Driving a vehicle equipped with an ETC system is about the same as driving a vehicle with a conventional mechanical throttle cable and throttle valve. However, the driver may notice some differences that are to be considered normal. These normal conditions include the following:

- The engine may not increase above idle speed when depressing the accelerator pedal when the gear selector is in Park.

- If the engine speed does increase when the accelerator is depressed with the transmission in Park or Neutral, the engine speed will likely be limited to less than 2000 RPM.

- While accelerating rapidly, there is often a slight delay before the engine responds. ● **SEE FIGURE 22–2.**

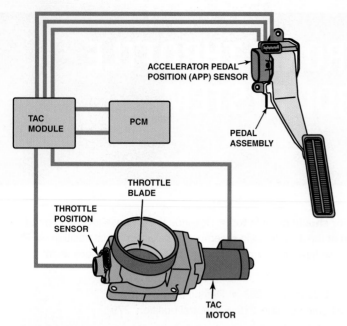

**FIGURE 22–1** The throttle pedal is connected to the accelerator pedal position (APP) sensor. The electronic throttle body includes a throttle position (TP) sensor to provide throttle angle feedback to the vehicle computer. Some systems use a throttle actuator control (TAC) module to operate the throttle blade (plate).

- While at cruise speed, the accelerator pedal may or may not cause the engine speed to increase if the accelerator pedal is moved slightly.

## ACCELERATOR PEDAL POSITION SENSOR

**CABLE-OPERATED SYSTEM** Honda Accords until 2008 model year used a cable attached to the accelerator pedal to operate the APP sensor located under the hood. A similar arrangement was used in Dodge RAM trucks in 2003. In both of these applications, the throttle cable was simply moving the APP sensor and not moving the throttle plate. The throttle plate is controlled by the PCM and moved by the ETC motor.

**TWO SENSORS** The accelerator pedal position sensor uses two and sometimes three separate sensors, which act together to give accurate accelerator pedal position information to the controller but also are used to check that the sensor is working properly. They function just like a throttle position (TP) sensor, and two are needed for proper system function. One APP sensor output signal increases as the pedal is depressed, and the other signal decreases. The controller compares the signals with a lookup table to determine the pedal position. Using two or three signals improves redundancy should one sensor fail and allows the PCM to quickly detect a malfunction. When three sensors

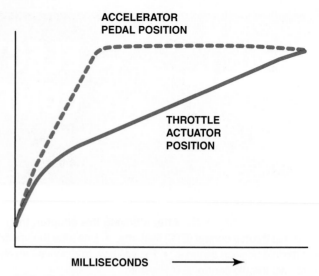

**FIGURE 22–2** The opening of the throttle plate can be delayed as long as 30 milliseconds (0.03 sec.) to allow time for the amount of fuel needed to catch up to the opening of the throttle plate.

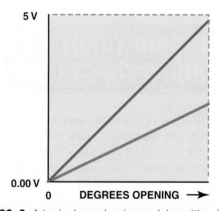

**FIGURE 22–3** A typical accelerator pedal position (APP) sensor, showing two different output voltage signals that are used by the PCM to determine accelerator pedal position. Two (or three in some applications) are used as a double check because this is a safety-related sensor.

are used, the third signal can either decrease or increase with pedal position, but its voltage range will still be different from the other two. ● **SEE FIGURE 22–3.**

## THROTTLE BODY ASSEMBLY

The throttle body assembly contains the following components:

- Throttle plate
- Electric actuator DC motor
- Dual TP sensors
- Gears used to multiply the torque of the DC motor
- Springs used to hold the throttle plate in the default location

### What Is the "Spring Test"?

The spring test is a self-test performed by the PCM whenever the engine is started. The PCM operates the throttle to check if it can react to the command and return to the default (home) position. This self-test is used by the PCM to determine that the spring and motor are working correctly and may be noticed by some vehicle owners by the following factors:

- A slight delay in the operation of the starter motor. The PCM performs this test when the ignition switch is turned to the "on" position. While it takes just a short time to perform the test, it can be sensed by the driver that there could be a fault in the ignition switch or starter motor circuits.
- A slight "clicking" sound may also be heard coming from under the hood when the ignition is turned on. This is normal and is related to the self-test on the throttle as it opens and closes.

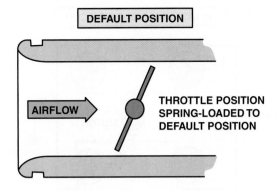

**THROTTLE PLATE AND SPRING** The throttle plate is held slightly open by a concentric clock spring. The spring applies a force that will close the throttle plate if power is lost to the actuator motor. The spring is also used to open the throttle plate slightly from the fully closed position.

**ELECTRONIC THROTTLE BODY MOTOR** The actuator is a DC electric motor and is often called a **servomotor**. The throttle plate is held in a **default position** by a spring inside the throttle body assembly. This partially open position, also called the **neutral position** or the **fail-safe position,** is about 16% to 20% open. This default position varies depending on the vehicle and usually results in an engine speed of 1200 to 1500 RPM:

- The throttle plate is driven closed to achieve speeds lower than the default position, such as idle speed.
- The throttle plate is driven open to achieve speeds higher than the default position, such as during acceleration.
  ● **SEE FIGURE 22–4.**

The throttle plate motor is driven by a bidirectional pulse-width-modulated (PWM) signal from the PCM or ETC module using an H-bridge circuit. ● **SEE FIGURE 22–5.**

The H-bridge circuit is controlled by the PCM by the following:

- Reversing the polarity of power and ground brushes to the DC motor
- Pulse-width modulating the current through the motor

The PCM monitors the position of the throttle from the two TP sensors. The PCM then commands the throttle plate to the desired position. ● **SEE FIGURE 22–6.**

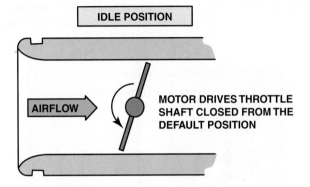

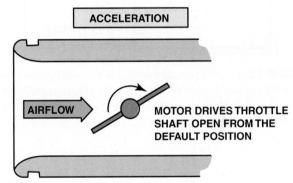

**FIGURE 22–4** The default position for the throttle plate is in slightly open position. The servomotor then is used to close it for idle and open it during acceleration.

### Why Not Use a Stepper Motor for ETC?

A stepper motor is a type of motor that has multiple windings and is pulsed by a computer to rotate a certain number of degrees when pulsed. The disadvantage is that a stepper motor is too slow to react compared with a conventional DC electric motor and is the reason a stepper motor is not used in ETC systems.

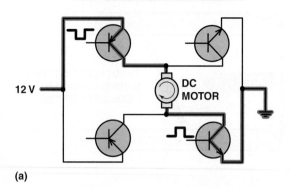

(a)

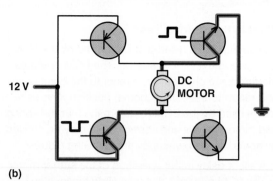

(b)

**FIGURE 22–5** (a) An H-bridge circuit is used to control the direction of the DC electric motor of the electronic throttle control (ETC) unit. (b) To reverse the direction of operation, the polarity of the current through the motor is reversed.

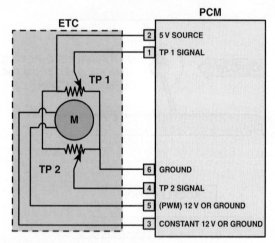

**FIGURE 22–6** Schematic of a typical electronic throttle control (ETC) system. Note that terminal #5 is always pulse-width modulated and that terminal #3 is always constant, but both power and ground are switched to change the direction of the motor.

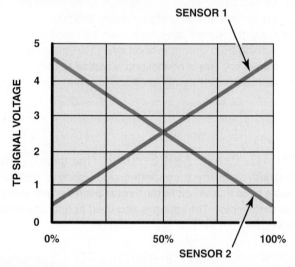

**FIGURE 22–7** The two TP sensors used on the throttle body of an electronic throttle body assembly produce opposite voltage signals as the throttle is opened. The total voltage of both combined at any throttle plate position is 5 volts.

## TP SENSOR

Two TP sensors are used in the throttle body assembly to provide TP signals to the PCM. Two sensors are used as a fail-safe measure and for diagnosis. There are two types of TP sensors used in ETC systems: potentiometers and Hall effect.

**THREE-WIRE POTENTIOMETER SENSORS** These sensors use a 5-volt reference from the PCM and produce an analog (variable) voltage signal that is proportional to the throttle plate position. The two sensors produce opposite signals as the throttle plate opens:

- One sensor starts at low voltage (about 0.5 V) and increases as the throttle plate is opened.
- The second sensor starts at a higher voltage (about 4.5 V) and produces a lower voltage as the throttle plate is opened. ● **SEE FIGURE 22–7.**

(a)

(b)

**FIGURE 22–8** (a) A "reduced power" warning light indicates a fault with the electronic throttle control (ETC) system on some General Motors vehicles. (b) A symbol showing an engine with an arrow pointing down is used on some General Motors vehicles to indicate a fault with the ETC system.

**HALL-EFFECT TP SENSORS** Some vehicle manufacturers, Honda, for example, use a noncontact Hall-effect TP sensor. Because there is not physical contact, this type of sensor is less likely to fail due to wear.

**FIGURE 22–9** A wrench symbol warning lamp on a Ford vehicle. The symbol can also be green.

## DIAGNOSIS OF ETC SYSTEMS

**FAULT MODE** ETC systems can have faults like any other automatic system. Because of the redundant sensors in accelerator pedal position (APP) sensors and TP sensor, many faults result in a *limp home* situation instead of a total failure. The limp home mode is also called the *fail-safe mode* and indicates the following actions performed by the PCM:

- Engine speed is limited to the default speed (about 1200 to 1600 RPM).
- There is slow or no response when the accelerator pedal is depressed.
- The cruise control system is disabled.
- A diagnostic trouble code (DTC) is set.
- An ETC warning lamp on the dash will light. The warning lamp may be labeled differently, depending on the vehicle manufacturer. Examples include the following:
  - General Motors vehicle—Reduced power lamp (● **SEE FIGURE 22–8**)
  - Ford—Wrench symbol (amber or green) (● **SEE FIGURE 22–9**)

  - Chrysler—Red lightning bolt symbol (● **SEE FIGURE 22–10**)
- The engine will run and can be driven slowly. This limp-in mode operation allows the vehicle to be driven off of the road and to a safe location.

The ETC may enter the limp-in mode if any of the following has occurred:

- Low battery voltage has been detected
- PCM failure
- One TP and the MAP sensor have failed
- Both TP sensors have failed
- The ETC actuator motor has failed
- The ETC throttle spring has failed

FIGURE 22–10 A symbol used on a Chrysler vehicle indicating a fault with the electronic throttle control (ETC).

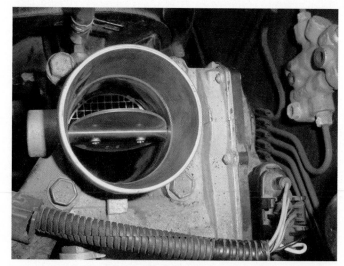

FIGURE 22–11 The throttle plate stayed where it was moved by the technician, which indicates that there is a problem with the electronic throttle body control assembly.

 **REAL WORLD FIX**

### The High-Idle Toyota

The owner of a Toyota Camry complained that the engine would idle at over 1200 RPM compared with a normal 600 to 700 RPM. The vehicle would also not accelerate. Using a scan tool, a check for DTCs showed one code: P2101—"TAC motor circuit low."

Checking service information led to the inspection of the ETC throttle body assembly. With the ignition key out of the ignition and the inlet air duct off the throttle body, the technician used a screwdriver to gently push to see if the throttle plate worked:

**Normal operation**—The throttle plate should move and then spring back quickly to the default position.

**Abnormal operation**—If the throttle plate stays where it is moved or does not return to the default position, there is a fault with the throttle body assembly. ● **SEE FIGURE 22–11.**

**Solution:** The technician replaced the throttle body assembly with an updated version, and proper engine operation was restored. The technician disassembled the old throttle body and found it was corroded inside because of moisture entering the unit through the vent hose. ● **SEE FIGURE 22–12.**

**VACUUM LEAKS** The ETC system is able to compensate for many vacuum leaks. A vacuum leak at the intake manifold for example will allow air into the engine that is not measured by the mass airflow sensor. The ETC system will simply move the throttle as needed to achieve the proper idle speed to compensate for the leak.

**DIAGNOSTIC PROCEDURE** If a fault occurs in the ETC system, check service information for the specified procedure to follow for the vehicle being checked. Most vehicle service information includes the following steps:

**STEP 1** Verify the customer concern.

**STEP 2** Use a factory scan tool or an aftermarket scan tool with original equipment capability and check for DTCs.

**STEP 3** If there are stored DTCs, follow service information instructions for diagnosing the system.

**STEP 4** If there are no stored DTCs, check scan tool data for possible fault areas in the system.

**SCAN TOOL DATA** Scan data related to the ETC system can be confusing. Typical data and the meaning include the following:

- **APP indicated angle.** The scan tool will display a percentage ranging from 0% to 100%. When the throttle is released, the indicated angle should be 0%. When the throttle is depressed to wide open, the reading should indicate 100%.

- **TP desired angle.** The scan tool will display a percentage ranging from 0% to 100%. This represents the desired throttle angle as commanded by the driver of the vehicle.

- **TP indicated angle.** The TP indicated angle is the angle of the measured throttle opening, and it should agree with the TP desired angle.

- **TP sensors 1 and 2.** The scan tool will display "agree" or "disagree." If the PCM or throttle actuator control (TAC) module receives a voltage signal from one of the TP sensors that is not in the proper relationship to the other TP sensor, the scan tool will display "disagree."

**FIGURE 22–12** A corroded electronic throttle control (ETC) assembly shown with the cover removed.

**FIGURE 22–13** Notice the small motor gear on the left drives a larger plastic gear (black), which then drives the small gear in mesh with the section of a gear attached to the throttle plate. This results in a huge torque increase from the small motor and helps explain why it could be dangerous to insert a finger into the throttle body assembly.

# ETC THROTTLE FOLLOWER TEST

On some vehicles, such as many Chrysler vehicles, the operation of the ETC can be tested using a factory or factory-level scan tool. To perform this test, use the "throttle follower test" procedure as shown on the scan tool. An assistant is needed to check that the throttle plate is moving as the accelerator pedal is depressed. This test cannot be done normally because the PCM does not normally allow the throttle plate to be moved unless the engine is running.

# SERVICING ELECTRONIC THROTTLE SYSTEMS

**ETC-RELATED PERFORMANCE ISSUES** The only service that an ETC system may require is a cleaning of the throttle body. Throttle body cleaning is a routine service procedure on port fuel-injected engines and is still needed when the throttle is being opened by an electric motor rather than a throttle cable tied to a mechanical accelerator pedal. The throttle body may need cleaning if one or more of the following symptoms are present:

- Lower-than-normal idle speed
- Rough idle
- Engine stalls when coming to a stop (called a **coast-down stall**)

If any of the above conditions exists, a throttle body cleaning will often correct these faults.

**CAUTION: Some vehicle manufacturers add a nonstick coating to the throttle assembly and warn that cleaning could remove this protective coating. Always follow the vehicle manufacturer's recommended procedures.**

**THROTTLE BODY CLEANING PROCEDURE** Before attempting to clean a throttle body on an engine equipped with an ETC system, be sure that the ignition key is out of the vehicle and the ready light is off if working on a Toyota/Lexus hybrid electric vehicle to avoid the possibility of personal injury.

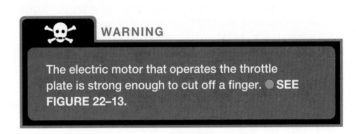

**WARNING**

The electric motor that operates the throttle plate is strong enough to cut off a finger. ● SEE FIGURE 22–13.

To clean the throttle, perform the following steps:

**STEP 1** With the ignition off and the key removed from the ignition, remove the air inlet hose from the throttle body.

**STEP 2** Spray throttle body cleaner onto a shop cloth.

**STEP 3** Open the throttle body and use the shop cloth to remove the varnish and carbon deposits from the throttle body housing and throttle plate.

**CAUTION: Do not spray cleaner into the throttle body assembly. The liquid cleaner could flow into and damage the throttle position (TP) sensors.**

**STEP 4** Reinstall the inlet hose being sure that there are no air leaks between the hose and the throttle body assembly.

**STEP 5** Start the engine and allow the PCM to learn the correct idle. If the idle is not correct, check service information for the specified procedures to follow to perform a throttle relearn.

## THROTTLE BODY RELEARN PROCEDURE

When installing a new throttle body or PCM or sometimes after cleaning the throttle body, the throttle position has to be learned by the PCM. After the following conditions have been met, a typical throttle body relearn procedure for a General Motors vehicle includes the following:

- Accelerator pedal released
- Battery voltage higher than 8 volts
- Vehicle speed must be zero
- Engine coolant temperature (ECT) higher than 40°F (5°C) and lower than 212°F (100°C)
- Intake air temperature (IAT) higher than 40°F (5°C)
- No throttle DTCs set

If all of the above conditions are met, perform the following steps:

**STEP 1** Turn the ignition on (engine off) for 30 seconds.

**STEP 2** Turn the ignition off and wait 30 seconds.

Start the engine and the idle learn procedure should cause the engine to idle at the correct speed.

## SUMMARY

1. Using an electronic throttle control (ETC) system on an engine has many advantages over a conventional method that uses a mechanical cable between the accelerator pedal and the throttle valve.

2. The major components of an ETC system include the following:
   - Accelerator pedal position (APP) sensor
   - ETC actuator motor and spring
   - Throttle position (TP) sensor
   - Electronic control unit

3. The TP sensor is actually two sensors that share the 5-volt reference from the PCM and produce opposite signals as a redundant check.

4. Limp-in mode is commanded if there is a major fault in the system, which can allow the vehicle to be driven enough to be pulled off the road to safety.

5. The diagnostic procedure for the ETC system includes verifying the customer concern, using a scan tool to check for DTCs, and checking the value of the TP and APP sensors.

6. Servicing the ETC system includes cleaning the throttle body and throttle plate.

## REVIEW QUESTIONS

1. What parts can be deleted if an engine uses an electronic throttle control (ETC) system instead of a conventional accelerator pedal and cable to operate the throttle valve?

2. How can the use of an ETC system improve fuel economy?

3. How is the operation of the throttle different on a system that uses an ETC system compared with a conventional mechanical system?

4. What component parts are included in an ETC system?

5. What is the default or limp-in position of the throttle plate?

6. What dash warning light indicates a fault with the ETC system?

## CHAPTER QUIZ

1. The use of an electronic throttle control (ETC) system allows the elimination of all except _____.
   a. Accelerator pedal
   b. Mechanical throttle cable (most systems)
   c. Cruise control actuator
   d. Idle air control

2. The throttle plate is spring loaded to hold the throttle slightly open how far?
   a. 3% to 5%
   b. 8% to 10%
   c. 16% to 20%
   d. 22% to 28%

3. The throttle plate actuator motor is what type of electric motor?
   a. Stepper motor
   b. DC motor
   c. AC motor
   d. Brushless motor

4. The actuator motor is controlled by the PCM through what type of circuit?
   a. Series
   b. Parallel
   c. H-bridge
   d. Series-parallel

5. When does the PCM perform a self-test of the ETC system?
   a. During cruise speed when the throttle is steady
   b. During deceleration
   c. During acceleration
   d. When the ignition switch is first rotated to the on position before the engine starts

6. The throttle position sensor used in the throttle body assembly of an ETC system is what type?
   a. Single potentiometer
   b. Two potentiometers that read in the opposite direction
   c. Hall-effect sensor
   d. Either b or c

7. A green wrench symbol is displayed on the dash. What does this mean?
   a. A fault in the ETC in a Ford has been detected.
   b. A fault in the ETC in a Honda has been detected.
   c. A fault in the ETC in a Chrysler has been detected.
   d. A fault in the ETC in a General Motors vehicle has been detected.

8. A technician is checking the operation of the ETC system by depressing the accelerator pedal with the ignition in the on (run) position (engine off). What is the most likely result if the system is functioning correctly?
   a. The throttle goes to wide open when the accelerator pedal is depressed all the way.
   b. There is no throttle movement.
   c. The throttle will open partially but not all of the way.
   d. The throttle will perform a self-test by closing and then opening to the default position.

9. With the ignition off and the key out of the ignition, what should happen if a technician uses a screwdriver and pushes gently on the throttle plate in an attempt to open the valve?
   a. Nothing. The throttle should be kept from moving by the motor, which is not energized with the key off.
   b. The throttle should move and stay where it is moved and not go back unless moved back.
   c. The throttle should move and then spring back to the home position when released.
   d. The throttle should move closed but not open further than the default position.

10. The throttle body may be cleaned (if recommended by the vehicle manufacturer) if what conditions are occurring?
    a. Coast-down stall
    b. Rough idle
    c. Lower-than-normal idle speed
    d. Any of the above

# FUEL-INJECTION SYSTEM DIAGNOSIS AND SERVICE

**LEARNING OBJECTIVES:** **After studying this chapter, the reader should be able to:** • Explain the diagnosis of electronic fuel-injection systems. • Describe how to test for an injector pulse. • Understand the process of checking fuel-injector resistance. • Explain how to scope-test fuel injectors and conduct pressure-drop balance and injector voltage-drop tests.

**KEY TERMS:** Graphing multimeter (GMM) 287 • Idle air control counts 288 • Idle air control (IAC) 295 • Noid light 289 • Peak-and-hold injector 294 • Pressure transducer 287 • Saturation 294

## PORT FUEL-INJECTION PRESSURE REGULATOR DIAGNOSIS

Most port fuel-injected engines use a vacuum hose connected to the fuel-pressure regulator. At idle, the pressure inside the intake manifold is low (high vacuum). Manifold vacuum is applied above the diaphragm inside the fuel-pressure regulator. This reduces the pressure exerted on the diaphragm and results in a lower, about 10 PSI (69 kPa), fuel pressure applied to the injectors. To test a vacuum-controlled fuel-pressure regulator, follow these steps:

1. Connect a fuel-pressure gauge to monitor the fuel pressure.

2. Locate the fuel-pressure regulator and disconnect the vacuum hose from the regulator.

   **NOTE: If gasoline drips out of the vacuum hose when removed from the fuel-pressure regulator, the regulator is defective and will require replacement.**

3. With the engine running at idle speed, reconnect the vacuum hose to the fuel-pressure regulator while watching the fuel-pressure gauge. The fuel pressure should drop (about 10 PSI, or 69 kPa) when the hose is reattached to the regulator.

4. Using a hand-operated vacuum pump, apply vacuum (20 in. Hg) to the regulator. The regulator should hold

**FIGURE 23–1** If the vacuum hose is removed from the fuel-pressure regulator when the engine is running, the fuel pressure should increase. If it does not increase, then the fuel pump is not capable of supplying adequate pressure or the fuel-pressure regulator is defective. If gasoline is visible in the vacuum hose, the regulator is leaking and should be replaced.

vacuum. If the vacuum drops, replace the fuel-pressure regulator. ● **SEE FIGURE 23–1.**

**NOTE: Some vehicles do not use a vacuum-regulated fuel-pressure regulator. Many of these vehicles use a regulator located inside the fuel tank that supplies a constant fuel pressure to the fuel injectors.**

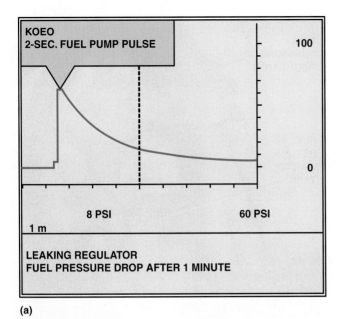

KOEO
2-SEC. FUEL PUMP PULSE

100

0

8 PSI          60 PSI
1 m

LEAKING REGULATOR
FUEL PRESSURE DROP AFTER 1 MINUTE

(a)

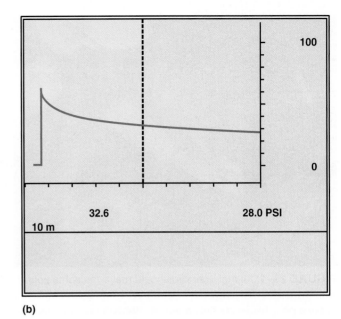

100

0

32.6          28.0 PSI
10 m

(b)

**FIGURE 23–2** (a) A fuel-pressure graph after key on, engine off (KOEO) on a TBI system. (b) Pressure drop after 10 minutes on a normal port fuel-injection system.

🔧 TECH TIP

**Pressure Transducer Fuel-Pressure Test**

Using a **pressure transducer** and a **graphing multimeter (GMM)** or digital storage oscilloscope (DSO) allows the service technician to view the fuel pressure over time. ● **SEE FIGURE 23–2(a).** Note that the fuel pressure dropped from 15 PSI to 6 PSI on a TBI-equipped vehicle after just one minute. A normal pressure holding capability is shown in ● **FIGURE 23–2(b)** when the pressure dropped only about 10% after 10 minutes on a port fuel-injection system.

**FIGURE 23–3** A clogged PCV system caused the engine oil fumes to be drawn into the air cleaner assembly. This is what the technician discovered during a visual inspection.

# DIAGNOSING ELECTRONIC FUEL-INJECTION PROBLEMS USING VISUAL INSPECTION

All fuel-injection systems require the proper amount of clean fuel delivered to the system at the proper pressure and the correct amount of filtered air. The following items should be carefully inspected before proceeding to more detailed tests:

- Check the air filter and replace as needed.
- Check the air induction system for obstructions.
- Check the conditions of all vacuum hoses. Replace any hose that is split, soft (mushy), or brittle.

- Check the positive crankcase ventilation (PCV) valve for proper operation or replacement as needed. ● **SEE FIGURE 23–3.**

  NOTE: The use of an incorrect PCV valve can cause a rough idle or stalling.

- Check all fuel-injection electrical connections for corrosion or damage.
- Check for gasoline at the vacuum port of the fuel-pressure regulator if the vehicle is so equipped. Gasoline in the vacuum hose at the fuel-pressure regulator indicates that the regulator is defective and requires replacement.

**FIGURE 23–4** All fuel injectors should make the same sound with the engine running at idle speed. A lack of sound indicates a possible electrically open injector or a break in the wiring. A defective computer could also be the cause of a lack of clicking (pulsing) of the injectors.

FUEL-PRESSURE REGULATOR

FUEL RETURN LINE TO TANK

**FIGURE 23–5** Fuel should be heard returning to the fuel tank at the fuel return line if the fuel-pump and fuel-pressure regulator are functioning correctly.

---

**TECH TIP**

**Stethoscope Fuel-Injection Test**

A commonly used test for injector operation is to listen to the injector using a stethoscope with the engine operating at idle speed. ● **SEE FIGURE 23–4.** All injectors should produce the same clicking sound. If any injector makes a clunking or rattling sound, it should be tested further or replaced. With the engine still running, place the end of the stethoscope probe to the return line from the fuel-pressure regulator. ● **SEE FIGURE 23–5.** Fuel should be heard flowing back to the fuel tank if the fuel-pump pressure is higher than the fuel-regulator pressure. If no sound of fuel is heard, then either the fuel pump or the fuel-pressure regulator is at fault.

**TECH TIP**

**Quick and Easy Leaking Injector Test**

Leaking injectors may be found by disabling the ignition, unhooking all injectors, and checking exhaust for hydrocarbons (HC) using a gas analyzer while cranking the engine (maximum HC = 300 PPM).

- The computer increases the injector pulse width slightly longer because of the signal from the MAP sensor.
- The air–fuel mixture remains unchanged.
- The idle air control (IAC) counts will decrease, thereby attempting to reduce the engine speed to the target idle speed stored in the computer memory. ● **SEE FIGURE 23–6.**

Therefore, one of the best indicators of a vacuum leak on a speed-density fuel-injection system is to look at the IAC counts or percentage. Normal **IAC counts** or percentage is usually 15 to 25. A reading of less than 5 indicates a vacuum leak.

If a vacuum leak occurs on an engine equipped with a mass airflow type of fuel-injection system, the extra air causes the following to occur:

- The engine will operate leaner than normal because the extra air has not been measured by the MAF sensor.
- The idle speed will likely be lower because of the leaner-than-normal air–fuel mixture.
- The IAC counts or percentage will often increase in an attempt to return the engine speed to the target speed stored in the computer.

# SCAN TOOL VACUUM LEAK DIAGNOSIS

If a vacuum (air) leak occurs on an engine equipped with a speed-density type of fuel injection, the extra air would cause the following to occur:

- The idle speed increases because of the extra air just as if the throttle pedal were depressed.
- The MAP sensor reacts to the increased air from the vacuum leak as an additional load on the engine.

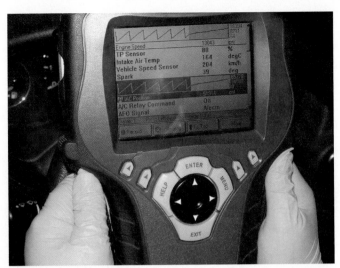

**FIGURE 23–6** Using a scan tool to check for idle air control (IAC) counts or percentage as part of a diagnostic routine.

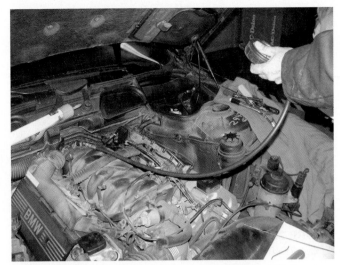

**FIGURE 23–7** Checking the fuel pressure using a fuel-pressure gauge connected to the Schrader valve.

## TECH TIP

### No Spark, No Squirt

Most electronic fuel-injection computer systems use the ignition primary (pickup coil or crank sensor) pulse as the trigger for when to inject (squirt) fuel from the injectors (nozzles). If this signal were not present, no fuel would be injected. Because this pulse is also necessary to trigger the module to create a spark from the coil, it can be said that "no spark" could also mean "no squirt." Therefore, if the cause of a no-start condition is observed to be a lack of fuel injection, do not start testing or replacing fuel-system components until the ignition system is checked for proper operation.

# PORT FUEL-INJECTION SYSTEM DIAGNOSIS

To determine if a port fuel-injection system—including the fuel pump, injectors, and fuel-pressure regulator—is operating correctly, take the following steps:

1. Attach a fuel-pressure gauge to the Schrader valve on the fuel rail. ● **SEE FIGURE 23–7.**

2. Turn the ignition key on or start the engine to build up the fuel-pump pressure (to about 35 to 45 PSI).

3. Wait 20 minutes and observe the fuel pressure retained in the fuel rail and note the PSI reading. The fuel pressure should not drop more than 20 PSI (140 kPa) in 20 minutes. If the drop is less than 20 PSI in 20 minutes, everything is

okay; if the drop is *greater,* then there is a possible problem with the following:

- The check valve in the fuel pump
- Leaking injectors, lines, or fittings
- A defective (leaking) fuel-pressure regulator

To determine which unit is defective, perform the following:

- Reenergize the electric fuel pump.

- Clamp the fuel *supply* line and wait 10 minutes (see Caution box). If the pressure drop does not occur, replace the fuel pump. If the pressure drop still occurs, continue with the next step.

- Repeat the pressure buildup of the electric pump and clamp the fuel return line. If the pressure drop time is now okay, replace the fuel-pressure regulator.

- If the pressure drop still occurs, one or more of the injectors is leaking. Remove the injectors with the fuel rail and hold over paper. Replace those injectors that drip one or more drops after 10 minutes with pressurized fuel.

**CAUTION: Do not clamp plastic fuel lines. Connect shut-off valves to the fuel system to shut off supply and return lines. ● SEE FIGURE 23–8.**

# TESTING FOR AN INJECTOR PULSE

One of the first checks that should be performed when diagnosing a no-start condition is whether the fuel injectors are being pulsed by the computer. Checking for proper pulsing of the injector is also important in diagnosing a weak or dead cylinder.

A **noid light** is designed to electrically replace the injector in the circuit and to flash if the injector circuit is working correctly. ● **SEE FIGURE 23–9.** To use a noid light, disconnect the

FIGURE 23–8 Shutoff valves must be used on vehicles equipped with plastic fuel lines to isolate the cause of a pressure drop in the fuel system.

(a)

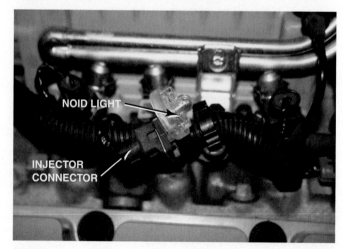

(b)

FIGURE 23–9 (a) Noid lights are usually purchased as an assortment so that one is available for any type or size of injector wiring connector. (b) The connector is unplugged from the injector, and a noid light is plugged into the injector connector. The noid light should flash when the engine is being cranked if the power circuit and the pulsing to ground by the computer are functioning okay.

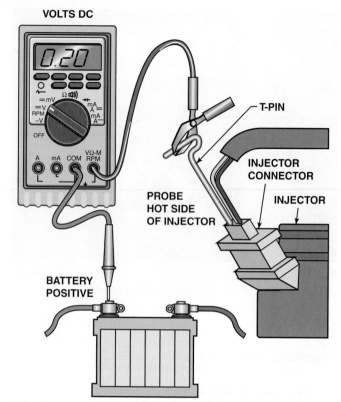

FIGURE 23–10 Use a DMM set to read DC volts to check the voltage drop of the positive circuit to the fuel injector. A reading of 0.5 volt or less is generally considered to be acceptable.

electrical connector at the fuel injector and plug the noid light into the injector harness connections. Crank or start the engine. The noid light should flash regularly.

NOTE: The term *noid* is simply an abbreviation of the word *solenoid*. Injectors use a movable iron core and are therefore solenoids. Therefore, a noid light is a replacement for the solenoid (injector).

Possible noid light problems and causes include the following:

1. **The light is off and does not flash.** The problem could be a defective noid light or an open circuit in either the power side or the ground side (or both) of the injector circuit.

2. **The noid light flashes dimly.** A dim noid light indicates excessive resistance or low voltage available to the injector. Both the power and the ground side must be checked.

3. **The noid light is on and does not flash.** If the noid light is on, then both a power and a ground are present. Because the light does not flash (blink) when the engine is being cranked or started, a short-to-ground fault exists either in the computer itself or in the wiring between the injector and the computer.

CAUTION: A noid lamp must be used with caution. The computer may show a good noid light operation and have low supply voltage. ● SEE FIGURE 23–10.

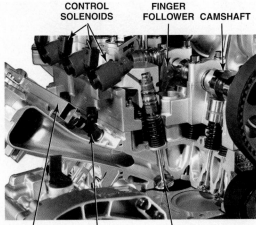

FIGURE 23-11 In a multiair engine design, the exhaust valves are opened by the exhaust camshaft lobes. Intake valves are opened by the high-pressure engine oil, the high pressure being produced by a lobe-actuated piston and controlled by a PCM-controlled solenoid.

 FREQUENTLY ASKED QUESTION

**How Does the Fiat Chrysler Multiair System Work?**

Some Chrysler and Fiat brand vehicles use a type of fuel-injection system that includes the following unique features:

- The engine has one overhead camshaft but only the exhaust cam lobes actually open the exhaust valves.
- The intake camshaft lobes are used to pressurize engine oil, which is then directed to a solenoid that is pulse-width modulated.
- The oil from the solenoid is sent to a piston on top of the intake valves, which are opened by the piston.
- The timing and valve lift are determined by the PCM that pulses the control solenoid to allow oil to open the valve. ● SEE FIGURE 23-11.

Because the intake values are opened using pressured engine oil, it is critical that the specified oil be used and changed at the specified interval. Some customers complain of a "clatter" from the engine, especially at idle, which is normal for this engine and is due to the operation of the solenoid.

# CHECKING FUEL-INJECTOR RESISTANCE

Each port fuel injector must deliver an equal amount of fuel, or the engine will idle roughly or perform poorly.

The electrical balance test involves measuring the injector coil-winding resistance. For best engine operation, all injectors should have the same electrical resistance. To measure the re-

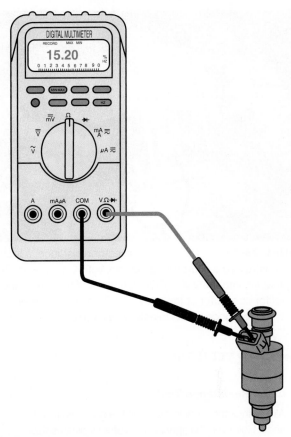

FIGURE 23-12 Connections and settings necessary to measure fuel-injector resistance.

sistance, carefully release the locking feature of the connector and remove the connector from the injector.

**NOTE: Some engines require specific procedures to gain access to the injectors. Always follow the manufacturer's recommended procedures.**

With an ohmmeter, measure the resistance across the injector terminals. Be sure to use the low-ohms feature of the digital ohmmeter to read in tenths (0.1) of an ohm. ● SEE FIGURES 23-12 AND 23-13. Check service information for the resistance specification of the injectors. Measure the resistance of all of the injectors. Replace any injector that does not fall within the resistance range of the specification. The resistance of the injectors should be measured twice—once when the engine (and injectors) are cold and once after the engine has reached normal operating temperature. If any injector measures close to specification, make certain that the terminals of the injector are electrically sound and perform other tests to confirm an injector problem before replacement.

# MEASURING RESISTANCE OF GROUPED INJECTORS

Many vehicles are equipped with a port fuel-injection system that "fires" two or more injectors at a time. For example, a V-6 may group all three injectors on one bank to pulse on at the

FIGURE 23–13 To measure fuel-injector resistance, a technician constructed a short wiring harness with a double banana plug that fits into the V and COM terminals of the meter and an injector connector at the other end. This setup makes checking resistance of fuel injectors quick and easy.

🔧 TECH TIP

**Equal Resistance Test**

All fuel injectors should measure the specified resistance. However, the specification often indicates that the temperature of the injectors be at room temperature and of course will vary according to the temperature. Rather than waiting for all of the injectors to achieve room temperature, measure the resistance and check that they are all within 0.4 ohm of each other. To determine the difference, record the resistance of each injector and then subtract the lowest resistance reading from the highest resistance reading to get the difference. If the difference is more than 0.4 ohm, then further testing will be needed to verify defective injector(s).

same time. Then the other three injectors will be pulsed on. This sequence alternates. To measure the resistance of these injectors, it is often easiest to measure each group of three that is wired in parallel. The resistance of three injectors wired in parallel is one-third of the resistance of each individual injector. An example follows:

**Injector resistance = 12 ohms (Ω)**

**Three injectors in parallel = 4 ohms (Ω)**

A V-6 has two groups of three injectors. Therefore, both groups should measure the same resistance. If both groups measure 4 ohms, then it is likely that all six injectors are okay. However, if one group measures only 2.9 ohms and the other group measures 4 ohms, then it is likely that one or more fuel injectors are defective (shorted). This means that the technician now has reasonable cause to remove the intake manifold to get access to each injector for further testing. ● SEE FIGURE 23–14.

(a)

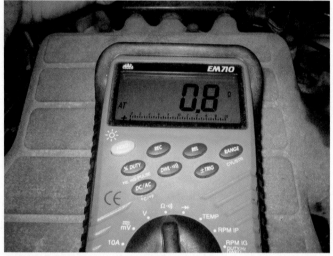

(b)

FIGURE 23–14 (a) The meter is connected to read one group of three 12-ohm injectors. The result should be 4 ohms, and this reading is a little low, indicating that at least one injector is shorted (low resistance). (b) This meter is connected to the other group of three injectors and indicates that most, if not all three, injectors are shorted. The technician replaced all six injectors, and the engine ran great.

# MEASURING RESISTANCE OF INDIVIDUAL INJECTORS

While there are many ways to check injectors, the first test is to measure the resistance of the coil inside and compare it to factory specifications. ● SEE FIGURE 23–15. If the injectors are not accessible, check service information for the location of the electrical connector for the injectors. Unplug the connector and measure the resistance of each injector at the injector side of the connector. Use service information to determine the wire colors for the power side and the pulse side of each injector.

FIGURE 23–15 If an injector has the specified resistance, this does not mean that it is okay. This injector had the specified resistance, yet it did not deliver the correct amount of fuel because it was clogged.

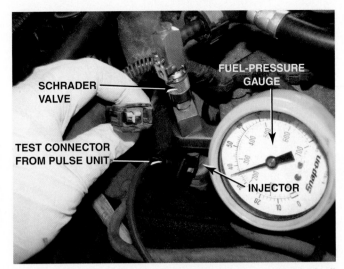

FIGURE 23–16 Connect a fuel-pressure gauge to the fuel rail at the Schrader valve.

## PRESSURE-DROP BALANCE TEST

The pressure balance test involves using an electrical timing device to pulse the fuel injectors on for a given amount of time, usually 500 ms, or 0.5 second, and observing the drop in pressure that accompanies the pulse. If the *fuel flow* through each injector is equal, the drop in pressure in the system will be equal. Most manufacturers recommend that the pressures be within about 1.5 PSI (10 kPa) of each other for satisfactory engine performance. This test method not only tests the electrical functioning of the injector (for definite time and current pulse) but also tests for mechanical defects that could affect fuel flow amounts.

The purpose of running this injector balance test is to determine which injector is restricted, inoperative, or delivering fuel differently than the other injectors. Replacing a complete set of injectors can be expensive. The basic tools needed are the following:

- Accurate pressure gauge with pressure relief
- Injector pulser with time control
- Necessary injector connection adapters
- Safe receptacle for catching and disposing of any fuel released

STEP 1   Attach the pressure gauge to the fuel delivery rail on the supply side. Make sure the connections are safe and leakproof.

STEP 2   Attach the injector pulser to the first injector to be tested.

STEP 3   Turn the ignition key to the on position to prime the fuel rail. Note the static fuel-pressure reading. ● SEE FIGURE 23–16.

STEP 4   Activate the pulser for the timed firing pulses.

STEP 5   Note and record the new static rail pressure after the injector has been pulsed.

STEP 6   Reenergize the fuel pump and repeat this procedure for all of the engine injectors.

STEP 7   Compare the two pressure readings and compute the pressure drop for each injector. Compare the pressure drops of the injectors to each other. Any variation in pressure drops will indicate an uneven fuel delivery rate between the injectors.

An example follows:

| Injector | 1 | 2 | 3 | | 4 | 5 | 6 |
|---|---|---|---|---|---|---|---|
| Initial pressure | 40 | 40 | 40 | | 40 | 40 | 40 |
| Second pressure | 30 | 30 | 35 | | 30 | 20 | 30 |
| Pressure drop | 10 | 10 | 5 | | 10 | 20 | 10 |
| Possible problem | OK | OK | Restriction | | OK | Leak | OK |

## INJECTOR VOLTAGE-DROP TESTS

Another test of injectors involves pulsing the injector and measuring the voltage drop across the windings as current is flowing. A typical voltage-drop tester is shown in ● FIGURE 23–17. The tester, which is recommended for use by General Motors Corporation, pulses the injector while a digital multimeter is connected to the unit, which will display the voltage drop as the current flows through the winding.

CAUTION: Do not test an injector using a pulse-type tester more than once without starting the engine to help avoid a hydrostatic lock caused by the flow of fuel into the cylinder during the pulse test.

FIGURE 23–17 An injector tester being used to check the voltage drop through the injector while the tester is sending current through the injectors. This test is used to check the coil inside the injector. This same tester can be used to check for equal pressure drop of each injector by pulsing the injector on for 500 ms.

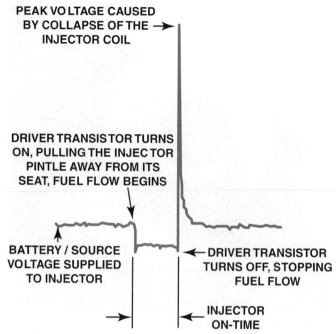

FIGURE 23–19 The injector on-time is called the pulse width.

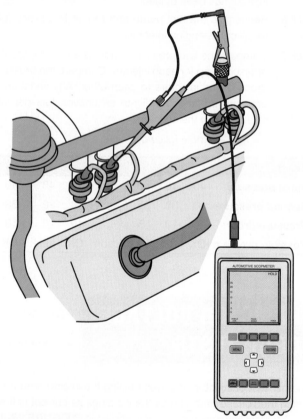

FIGURE 23–18 A digital storage oscilloscope can be easily connected to an injector by carefully back probing the electrical connector.

Record the highest voltage drop observed on the meter display during the test. Repeat the voltage-drop test for all of the injectors. The voltage drop across each injector should be within 0.1 volt of each other. If an injector has a higher-than-normal voltage drop, the injector windings have higher-than-normal resistance.

## SCOPE-TESTING FUEL INJECTORS

A scope (analog or digital storage) can be connected into each injector circuit. There are three types of injector drive circuits, and each type of circuit has its own characteristic pattern. ● SEE FIGURE 23–18 for an example of how to connect a scope to read a fuel-injector waveform.

### SATURATED SWITCH TYPE
In a saturated switch-type injector-driven circuit, voltage (usually a full 12 volts) is applied to the injector. The ground for the injector is provided by the vehicle computer. When the ground connection is completed, current flows through the injector windings. Because of the resistance and inductive reactance of the coil itself, it requires a fraction of a second (about 3 ms, or 0.003 second) for the coil to reach **saturation,** or maximum current flow. Most saturated switch-type fuel injectors have 12 to 16 ohms of resistance. This resistance, as well as the computer switching circuit, control and limit the current flow through the injector. A voltage spike occurs when the computer shuts off (opens the injector ground-side circuit) the injectors. ● SEE FIGURE 23–19.

### PEAK-AND-HOLD TYPE
A **peak-and-hold** type is typically used for TBI and some port low-resistance injectors. Full battery voltage is applied to the injector, and the ground side is controlled through the computer. The computer provides a high initial current flow (about 4 amperes) to flow through the injector windings to open the injector core. Then the computer reduces the current to a lower level (about 1 ampere). The hold current

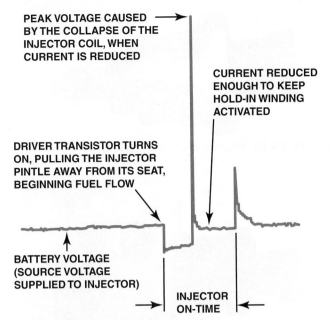

PEAK VOLTAGE CAUSED BY THE COLLAPSE OF THE INJECTOR COIL, WHEN CURRENT IS REDUCED

CURRENT REDUCED ENOUGH TO KEEP HOLD-IN WINDING ACTIVATED

DRIVER TRANSISTOR TURNS ON, PULLING THE INJECTOR PINTLE AWAY FROM ITS SEAT, BEGINNING FUEL FLOW

BATTERY VOLTAGE (SOURCE VOLTAGE SUPPLIED TO INJECTOR)

INJECTOR ON-TIME

**FIGURE 23–20** A typical peak-and-hold fuel-injector waveform. Most fuel injectors that measure less than 6 ohms will usually display a similar waveform.

**FIGURE 23–21** A set of six reconditioned injectors. The sixth injector is barely visible at the far right.

is enough to keep the injector open yet conserves energy and reduces the heat buildup that would occur if the full current flow remains on as long as the injector is commanded on. Typical peak-and-hold-type injector resistance ranges from 2 to 4 ohms.

The scope pattern of a typical peak-and-hold-type injector shows the initial closing of the ground circuit, then a voltage spike as the current flow is reduced. Another voltage spike occurs when the lower level current is turned off (opened) by the computer. ● SEE FIGURE 23–20.

**PULSE-WIDTH MODULATED TYPE** A pulse-width modulated type of injector drive circuit uses lower-resistance coil injectors. Battery voltage is available at the positive terminal of the injector and the computer provides a variable-duration connection to ground on the negative side of the injector.

**If Three of Six Injectors Are Defective, Should I Also Replace the Other Three?**

This is a good question. Many service technicians "recommend" that the three good injectors also be replaced along with the other three that tested as being defective. The reasons given by these technicians include the following:

- All six injectors have been operating under the same fuel, engine, and weather conditions.
- The labor required to replace all six is just about the same as replacing only the three defective injectors.
- Replacing all six at the same time helps ensure that all of the injectors are flowing the same amount of fuel so that the engine is operating most efficiently.

With these ideas in mind, the customer should be informed and offered the choice. Complete sets of injectors such as those in ● **FIGURE 23–21** can be purchased at a reasonable cost.

The computer can vary the time intervals that the injector is grounded for very precise fuel control.

Each time the injector circuit is turned off (ground circuit opened), a small voltage spike occurs. It is normal to see multiple voltage spikes on a scope connected to a pulse-width modulated type of fuel injector.

# IDLE AIR SPEED CONTROL DIAGNOSIS

On an engine equipped with fuel injection (TBI or port injection), the idle speed is controlled by increasing or decreasing the amount of air bypassing the throttle plate. Again, an electronic stepper motor or pulse-width modulated solenoid is used to maintain the correct idle speed. This control is often called the **idle air control (IAC)**. ● SEE FIGURES 23–22 THROUGH 23–24.

When the engine stops, most IAC units will retract outward to get ready for the next engine start. When the engine starts, the engine speed is high to provide for proper operation when the engine is cold. Then, as the engine gets warmer, the computer reduces engine idle speed gradually by reducing the number of counts or steps commanded by the IAC.

When the engine is warm and restarted, the idle speed should momentarily increase, then decrease to normal idle speed. This increase and then decrease in engine speed is often called an engine flare. If the engine speed does not flare, then the IAC may not be working (it may be stuck in one position).

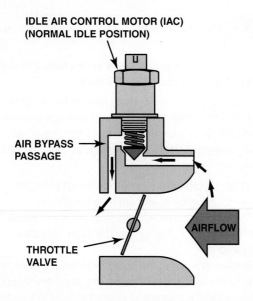

IDLE AIR CONTROL MOTOR (IAC)
(NORMAL IDLE POSITION)

AIR BYPASS
PASSAGE

AIRFLOW

THROTTLE
VALVE

(FULLY EXTENDED POSITION)

AIR BYPASS
PASSAGE

**FIGURE 23–22** An idle air control (IAC) controls idle speed by controlling the amount of air that passes around the throttle plate. More airflow results in a higher idle speed.

**FIGURE 23–23** A typical idle air control (IAC).

**FIGURE 23–24** Some idle air control (IAC) units are purchased with the housing as shown. Carbon buildup in these passages can cause a rough or unstable idling or stalling.

## REAL WORLD FIX

### There Is No Substitute for a Thorough Visual Inspection

An intermittent "check engine" light and a random-misfire diagnostic trouble code (DTC) P0300 was being diagnosed. A scan tool did not provide any help because all systems seemed to be functioning normally. Finally, the technician removed the engine cover and discovered a mouse nest. ● **SEE FIGURE 23–25.**

# FUEL-INJECTION SYSTEM DIAGNOSIS

**FIRST THINGS FIRST** Fuel-injection system diagnosis should include the diagnosis of all engine systems until the fault is narrowed to a fuel system-related fault. The diagnostic steps include:

- Verify the customer concern (complaint).
- Check for any stored diagnostic trouble codes (DTCs).
- Check for any fuel system-related technical service bulletins (TSBs).
- Check scan tool data for injector pulse width and idle speed control/ETC data and compare it to factory specifications.

**FUEL SYSTEM-RELATED SERVICE** Fuel-injection system service should include the following operations:

1. **Check fuel-pump operating pressure and volume.** The missing link here is volume. Most working technicians assume that if the pressure is correct, the volume is also okay. Hook up a fuel-pressure tester to the fuel rail inlet to quickly test the fuel pressure with the engine running. At the same time, test the volume of the pump by sending fuel into the holding tank. (One ounce per second is the usual specification.) ● **SEE FIGURE 23–26.** A two-line system tester is the recommended procedure to use and is attached to the fuel inlet and the return on the fuel rail. The vehicle onboard system is looped and returns fuel to the tank.

FIGURE 23–25 When the cover is removed from the top of the engine, a mouse or some other animal nest is visible. The animal had already eaten through a couple of injector wires. At least the cause of the intermittent misfire was discovered.

2. **Test the fuel-pressure regulator for operation and leakage.** At this time, the fuel-pressure regulator would be tested for operational pressure and proper regulation, including leakage. (This works well, as the operator has total control of rail pressure with a unit control valve.) Below are some points to ponder:

   - Good pressure does not mean proper volume. For example, a clogged filter may test okay on pressure, but the restriction may not allow proper volume under load. ● **SEE FIGURE 23–27.**
   - It is a good idea to use the vehicle's own gasoline to service the system versus a can of shop gasoline that has been sitting around for some time.
   - Pressure regulators do fail, and a lot more do not properly shut off fuel, causing higher-than-normal pump wear and shorter service life.

3. **Flush the entire fuel rail and upper fuel-injector screens, including the fuel-pressure regulator.** Raise the input pressure to a point above regulator setting to allow a constant flow of fuel through the inlet pressure side of the system, through the fuel rail, and out the open fuel-pressure regulator. In most cases, the applied pressure is 75 to 90 PSI (517 to 620 kPa) but will be maintained by the presence of a regulator. At this point, cleaning chemical is added to the fuel at a 5:1 mixture and allowed to flow through the system for 15 to 30 minutes. ● **SEE FIGURE 23–28.** Results are best on a hot engine with the fuel supply looped and the engine not running. Below are some points to ponder:

   - This flush is the fix most vehicles need first. The difference is that the deposits are removed to a remote tank and filter versus attempting to soften the deposits and blow them through the upper screens.
   - Most injectors use a 10-micron final filter screen. A 25% restriction in the upper screen would increase the injector on-time approximately 25%.

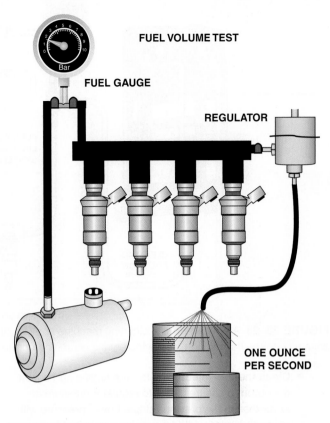

FUEL VOLUME TEST

FUEL GAUGE

REGULATOR

ONE OUNCE PER SECOND

FIGURE 23–26 Checking fuel-pump volume using a hose from the outlet of the fuel-pressure regulator into a calibrated container.

FIGURE 23–27 Testing fuel-pump volume using a fuel-pressure gauge with a bleed hose inserted into a suitable container. The engine is running during this test.

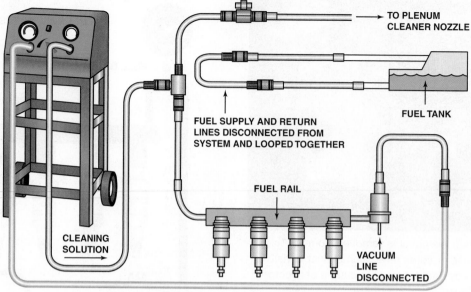

FIGURE 23–28 A typical two-line cleaning machine hookup, showing an extension hose that can be used to squirt a cleaning solution into the throttle body while the engine is running on the cleaning solution and gasoline mixture.

■ **Clean the fuel injectors.** Start the engine and adjust the output pressure closer to regulator pressure or lower than in the previous steps. Lower pressure will cause the pulse width to open up somewhat longer and allow the injectors to be cleaned. Slow speed (idle) position will take a longer time frame, and operating temperature will be reached. Clean injectors are the objective, but the chemical should also decarbon the engine valves, pistons, and oxygen sensor.

4. **Decarbon the engine assembly.** On most vehicles, the injector spray will help the decarboning process. On others, you may need to enhance the operation with external addition of a mixture through the PCV hose, throttle plates, or IACs.

5. **Clean the throttle plate and IAC passages.** Doing this service alone on most late-model engines will show a manifold vacuum increase of up to 2 in. Hg. Stop the engine and clean the areas as needed, then use a handheld fuel injector connected in parallel with the pressure hose, along with a pulser to allow cleaning of the throttle plates with the same chemical as injectors are running on. ●**SEE FIGURE 23–29.** This works well, as air is drawn into IAC passages on a running engine and will clean the passages without IAC removal.

6. **Relearn the PCM.** Some vehicles may have been running in such a poor state of operation that the PCM may need to be relearned. Consult service information for the suggested relearn procedures for each particular vehicle.

This service usually takes approximately one hour for the vehicle to run out of fuel and the entire service to be performed. The good thing is that the technician may do other services while this is being performed. Some technicians may install a set of

FIGURE 23–29 To thoroughly clean a throttle body, it is sometimes best to remove it from the vehicle.

🔧 **TECH TIP**

**Use an Injector Tester**

The best way to check injectors is to remove them all from the engine and test them using an injector tester. A typical injector tester uses a special nonflammable test fluid that has the same viscosity as gasoline. The tester pulses the injectors, and the amount of fuel delivered as well as the spray pattern can be seen. Many testers are capable of varying the frequency of the pulse as well as the duration that helps find intermittent injector faults. ●**SEE FIGURE 23–30.**

plugs or change the fuel filter while the engine is flushing. This service should restore the fuel system to original operations.

All of the previously listed steps may be performed using a *two-line* fuel-injector service unit, such as Carbon Clean, Auto Care, Injector Test, DeCarbon, or Motor-Vac.

**FIGURE 23–30** The amount each injector is able to flow is displayed in glass cylinders are each injector for a quick visual check.

 **TECH TIP**

**Be Sure to Clean the Fuel Rail**

Whenever you service the fuel injectors or if you suspect that there may be a fuel-injector problem, remove the entire fuel rail assembly and check the passages for contamination. Always thoroughly clean the rail when replacing fuel injectors.

 **WARNING**

Before opening any part of the high-pressure section of a gasoline direct injection (GDI) system, the pressure must be bled off. The high pressures of this fuel system can cause injury or death. If any of the high-pressure lines are removed, even temporarily, they MUST be replaced because the ends use a ball-fitting that deforms to create the high-pressure seal. Once this seal has been opened, then a new ball end must be used to insure a proper seal. ● SEE FIGURE 23–31. Always check service information for the exact procedures to follow for the vehicle being serviced.

# FUEL-SYSTEM SCAN TOOL DIAGNOSTICS

Diagnosing a faulty fuel system can be a difficult task. However, it can be made easier by utilizing the information available via the serial data stream. By observing the long-term fuel trim and the short-term fuel trim, we can determine how the fuel system is performing. Short-term fuel trim and long-term fuel trim can help us to zero in on specific areas of trouble. Readings should be taken at idle and at 3000 RPM. Use the following chart as a guide:

HIGH PRESSURE LINE

HIGH PRESSURE PUMP

**FIGURE 23–31** The line that has the yellow tag is a high-pressure line and this line must be replaced with a new part if removed even for a few minutes to gain access to another part.

| Fuel-Injection Symptom Chart | |
|---|---|
| **Symptom** | **Possible Causes** |
| **Hard cold starts** | • Low fuel pressure |
| | • Leaking fuel injectors |
| | • Contaminated fuel |
| | • Low-volatility fuel |
| | • Dirty throttle plate |
| **Garage stalls** | • Low fuel pressure |
| | • Insufficient fuel volume |
| | • Restricted fuel injector |
| | • Contaminated fuel |
| | • Low-volatility fuel |
| **Poor cold performance** | • Low fuel pressure |
| | • Insufficient fuel volume |
| | • Contaminated fuel |
| | • Low-volatility fuel |
| **Tip-in hesitation** (hesitation just as the accelerator pedal is depressed) | • Low fuel pressure |
| | • Insufficient fuel volume |
| | • Intake valve deposits |
| | • Contaminated fuel |
| | • Low-volatility fuel |

| Condition | Long-Term Fuel Trim at Idle | Long-Term Fuel Trim at 3000 RPM |
|---|---|---|
| **System normal** | 0% ± 10% | 0% ± 10% |
| **Vacuum leak** | HIGH | OK |
| **Fuel flow problem** | OK | HIGH |
| **Low fuel pressure** | HIGH | HIGH |
| **High fuel pressure** | *OK or LOW | *OK or LOW |

*High fuel pressure will affect trim at idle, at 3000 RPM, or both.

# FUEL-PUMP RELAY CIRCUIT DIAGNOSIS

**1** The tools needed to diagnose a circuit containing a relay include a digital multimeter (DMM), a fused jumper wire, and an assortment of wiring terminals.

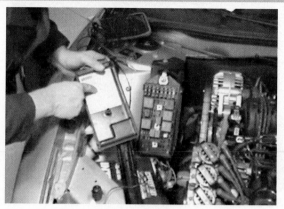

**2** Start the diagnosis by locating the relay center. It is under the hood on this General Motors vehicle, so access is easy. Not all vehicles are this easy.

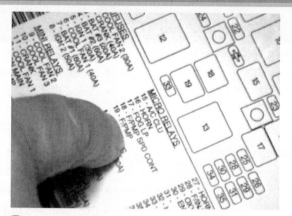

**3** The chart under the cover for the relay center indicates the location of the relay that controls the electric fuel pump.

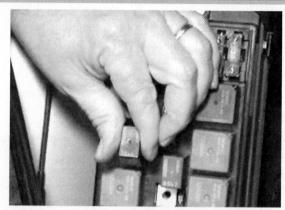

**4** Locate the fuel-pump relay and remove by using a puller if necessary. Do not twist or rock the relay when removing it to avoid damage to the fuse block assembly.

**5** Terminals 85 and 86 represent the coil inside the relay. Terminal 30 is the power terminal, 87a is the normally closed contact, and 87 is the normally open contact.

**6** The terminals are also labeled on most relays.

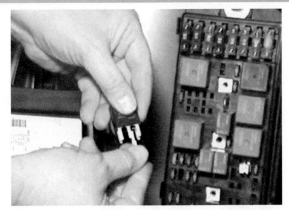

**7** To help make good electrical contact with the terminals without doing any harm, select the proper-size terminal from the terminal assortment.

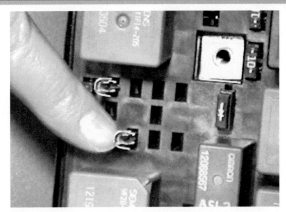

**8** Insert the terminals into the relay socket in 30 and 87.

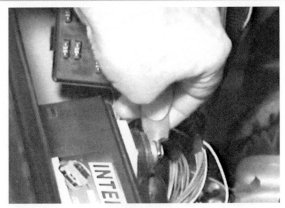

**9** To check for voltage at terminal 30, use a test light or a voltmeter. Start by connecting the alligator clip of the test light to the positive (+) terminal of the battery.

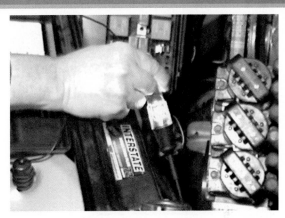

**10** Touch the test light to the negative (−) terminal of the battery or a good engine ground to check the test light.

**11** Use the test light to check for voltage at terminal 30 of the relay. The ignition may have to be in the on (run) position.

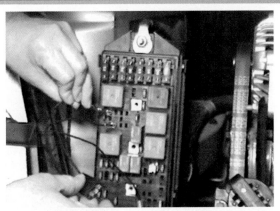

**12** To check to see if the electric fuel pump can be operated from the relay contacts, use a fused jumper wire and touch the relay contacts that correspond to terminals 30 and 87 of the relay.

CONTINUED ▶

**13** Connect the leads of the meter to contacts 30 and 87 of the relay socket. The reading of 4.7 amperes is okay because the specification is 4 to 8 amperes.

**14** Set the meter to read ohms (Ω) and measure the resistance of the relay coil. The usual reading for most relays is between 60 and 100 ohms.

**15** Measure between terminal 30 and 87a. Terminal 87a is the normally closed contact, and there should be little, if any, resistance between these two terminals, as shown.

**16** To test the normally open contacts, connect one meter lead to terminal 30 and the other lead to terminal 87. The ohmmeter should show an open circuit by displaying OL.

**17** Connect a fused jumper wire to supply 12 volts to terminal 86 and a ground to terminal 85 of the relay. If the relay clicks, then the relay coil is able to move the armature (movable arm) of the relay.

**18** After testing, be sure to reinstall the relay and the relay cover.

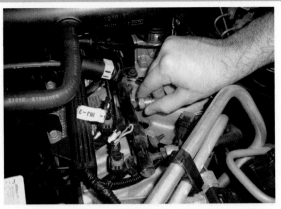

**1** Start the fuel-injector cleaning process by bringing the vehicle's engine up to operating temperature. Shut off the engine, remove the cap from the fuel rail test port, and install the appropriate adapter.

**2** The vehicle's fuel pump is disabled by removing its relay or fuse. In some cases, it may be necessary to disconnect the fuel pump at the tank if the relay or fuse powers more than just the pump.

**3** Turn the outlet valve of the canister to the OFF or CLOSED position.

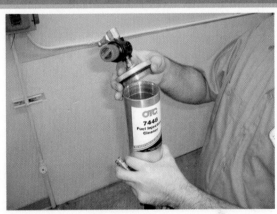

**4** Remove the fuel-injector cleaning canister's top and regulator assembly. Note that there is an O-ring seal located here that must be in place for the canister's top to seal properly.

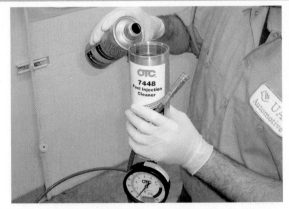

**5** Pour the injection system cleaning fluid into the open canister. Rubber gloves are highly recommended for this step as the fluid is toxic.

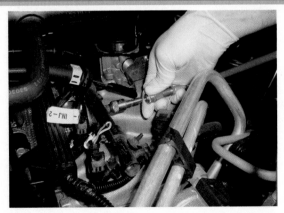

**6** Replace the canister's top (making sure it is tight) and connect its hose to the fuel rail adapter. Be sure that the hose is routed away from exhaust manifolds and other hazards.

CONTINUED ▶

**7** Hang the canister from the vehicle's hood and adjust the air pressure regulator to full OPEN position (CCW).

**8** Connect shop air to the canister and adjust the air pressure regulator to the desired setting. Canister pressure can be read directly from the gauge.

**9** Canister pressure should be adjusted to 5 PSI below system fuel pressure. An alternative for return-type systems is to block the fuel return line to the tank.

**10** Open the outlet valve on the canister.

**11** Start the vehicle's engine and let run at 1000 to 1500 RPM. The engine is now running on fuel-injector cleaning fluid provided by the canister.

**12** Continue the process until the canister is empty and the engine stalls. Remove the cleaning equipment, enable the vehicle's fuel pump, and run the engine to check for leaks.

## SUMMARY

1. A typical throttle-body fuel injector uses a computer-controlled injector solenoid to spray fuel into the throttle-body unit above the throttle plates.

2. A typical port fuel-injection system uses an individual fuel injector for each cylinder and squirts fuel directly into the intake manifold about 3 inches (80 mm) from the intake valve.

3. A typical fuel-injection system fuel pressure should not drop more than 20 PSI in 20 minutes.

4. A noid light can be used to check for the presence of an injector pulse.

5. Injectors can be tested for resistance and should be within 0.3 to 0.4 ohm of each other.

6. Different designs of injectors have a different scope waveform depending on how the computer pulses the injector on and off.

7. An idle air control unit controls idle speed and can be tested for proper operation using a scan tool or scope.

## REVIEW QUESTIONS

1. List the ways fuel injectors can be tested.
2. List the steps necessary to test a fuel-pressure regulator.
3. Describe why it may be necessary to clean the throttle plate of a port fuel-injected engine.

## CHAPTER QUIZ

1. Most port fuel-injected engines operate on how much fuel pressure?
   a. 3 to 5 PSI (21 to 35 kPa)
   b. 9 to 13 PSI (62 to 90 kPa)
   c. 35 to 45 PSI (240 to 310 kPa)
   d. 55 to 65 PSI (380 to 450 kPa)

2. Fuel injectors can be tested using _____.
   a. An ohmmeter
   b. A stethoscope
   c. A scope
   d. All of the above

3. Throttle-body fuel-injection systems use what type of injector driver?
   a. Peak and hold
   b. Saturated switch
   c. Pulse-width modulated
   d. Pulsed

4. Port fuel-injection systems generally use what type of injector driver?
   a. Peak and hold
   b. Saturated switch
   c. Pulse-width modulated
   d. Pulsed

5. The vacuum hose from the fuel-pressure regulator was removed from the regulator and gasoline dripped out of the hose. Technician A says that is normal and that everything is okay. Technician B says that one or more of the injectors may be defective, causing the fuel to get into the hose. Which technician is correct?
   a. Technician A only
   b. Technician B only
   c. Both Technicians A and B
   d. Neither Technician A nor B

6. The fuel pressure drops rapidly when the engine is turned off. Technician A says that one or more injectors could be leaking. Technician B says that a defective check valve in the fuel pump could be the cause. Which technician is correct?
   a. Technician A only
   b. Technician B only
   c. Both Technicians A and B
   d. Neither Technician A nor B

7. In a typical port fuel-injection system, which injectors are most subject to becoming restricted?
   a. Any of them equally
   b. The injectors at the end of the rail on a returnless system
   c. The injectors at the bends in the rail
   d. Either b or c

8. What component pulses the fuel injector on most vehicles?
   a. Electronic control unit (computer)
   b. Ignition module
   c. Crankshaft sensor
   d. Both b and c

9. Fuel-injection service is being discussed. Technician A says that the throttle plate(s) should be cleaned. Technician B says that the fuel rail should be cleaned. Which technician is correct?
   a. Technician A only
   b. Technician B only
   c. Both Technicians A and B
   d. Neither Technician A nor B

10. If the throttle plate needs to be cleaned, what symptoms will be present regarding the operation of the engine?
    a. Stalls
    b. Rough idle
    c. Hesitation on acceleration
    d. All of the above

# chapter 24
# VEHICLE EMISSION STANDARDS AND TESTING

**LEARNING OBJECTIVES:** **After studying this chapter, the reader should be able to:** • Discuss the emissions standards for vehicles. • Discuss exhaust analysis testing procedures. • Identify the reasons for excessive HC, CO, and $NO_x$ emissions.

**KEY TERMS:** Acceleration simulation mode (ASM) 309 • ASM 25/25 test 309 • ASM 50/15 test 309 • Clean Air Act Amendments (CAAA) 306 • Federal Test Procedure (FTP) 308 • I/M 240 test 310 • Lean indicator 313 • Non-methane hydrocarbon (NMHC) 311 • Ozone 313 • Rich indicator 312 • Sealed Housing for Evaporative Determination (SHED) test 308 • Smog 313 • State Implementation Plan (SIP) 308

## EMISSION STANDARDS IN THE UNITED STATES

In the United States, emissions standards are managed by the Environmental Protection Agency (EPA) as well as some U.S. state governments. Some of the strictest standards in the world are formulated in California by the California Air Resources Board (CARB).

**TIER 1 AND TIER 2** Federal emission standards are set by the **Clean Air Act Amendments (CAAA)** of 1990 grouped by tier. All vehicles sold in the United States must meet Tier 1 standards that went into effect in 1994 and are the least stringent. Additional Tier 2 standards have been optional since 2001 and was fully phased in by 2009. The current Tier 1 standards are different between automobiles and light trucks (SUVs, pickup trucks, and minivans), but Tier 2 standards will be the same for both types of vehicles.

There are several ratings that can be given to vehicles, and a certain percentage of a manufacturer's vehicles must meet different levels in order for the company to sell its products in affected regions. Beyond Tier 1, and in order by stringency, are the following levels:

- **TLEV: Transitional Low-Emission Vehicle.** More stringent for HC than Tier 1.

- **LEV** (also known as **LEV I): Low-Emission Vehicle,** an intermediate California standard about twice as stringent as Tier 1 for HC and $NO_x$.

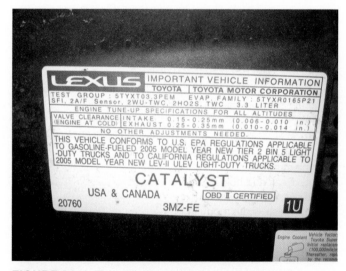

**FIGURE 24–1** The underhood decal showing that this Lexus RX-330 meets both national (Tier 2; BIN 5) and California LEV-II (ULEV) regulation standards.

- **ULEV** (also known as **ULEV I): Ultra-Low-Emission Vehicle.** A stronger California standard emphasizing very low HC emissions.

- **ULEV II: Ultra-Low-Emission Vehicle.** A cleaner-than-average vehicle certified under the Phase II LEV standard. Hydrocarbon and carbon monoxide emissions levels are nearly 50% lower than those of a LEV II-certified vehicle. ● SEE FIGURE 24–1.

- **SULEV: Super-Ultra-Low-Emission Vehicle.** A California standard even tighter than ULEV, including much lower HC and NO$_x$ emissions; roughly equivalent to Tier 2 Bin 2 vehicles.

- **ZEV: Zero-Emission Vehicle.** A California standard prohibiting any tailpipe emissions. The ZEV category is largely restricted to electric vehicles and hydrogen-fueled vehicles. In these cases, any emissions that are created are produced at another site, such as a power plant or hydrogen reforming center, unless such sites run on renewable energy.

   **NOTE: A battery-powered electric vehicle charged from the power grid will still be up to 10 times cleaner than even the cleanest gasoline vehicles over their respective lifetimes.**

- **PZEV: Partial Zero-Emission Vehicle.** Compliant with the SULEV standard; additionally has near-zero evaporative emissions and a 15 year/150,000 mile warranty on its emission control equipment.

Tier 2 standards are even more stringent. Tier 2 variations are appended with "II," such as LEV II or SULEV II. Other categories have also been created:

- **ILEV: Inherently Low-Emission Vehicle.** A vehicle certified to meet the transitional low-emission vehicle standards established by the California Air Resources Board (CARB).

- **AT-PZEV: Advanced Technology Partial Zero-Emission Vehicle.** If a vehicle meets the PZEV standards and is using high-technology features, such as an electric motor or high-pressure gaseous fuel tanks for compressed natural gas, it qualifies as an AT-PZEV. Hybrid electric vehicles such as the Toyota Prius can qualify, as can internal combustion engine vehicles that run on natural gas (CNG), such as the Honda Civic GX. These vehicles are classified as "partial" ZEV because they receive partial credit for the number of ZEV vehicles that automakers would otherwise be required to sell in California.

- **NLEV: National Low-Emission Vehicle.** All vehicles nationwide must meet this standard, which started in 2001.

**FEDERAL EPA BIN NUMBER** The higher the tier number, the newer the regulation; the lower the bin number, the cleaner the vehicle. The Toyota Prius is a very clean Bin 3, while the Hummer H2 is a dirty Bin 11. ● SEE CHARTS 24–1 THROUGH 24–3.

**SMOG EMISSION INFORMATION** New vehicles are equipped with a sticker that shows the relative level of smog-causing emissions created by the vehicle compared to others on the market. Smog-causing emissions include unburned hydrocarbons (HC) and oxides of nitrogen (NO$_x$). ● SEE FIGURE 24–2.

**CALIFORNIA STANDARDS** The pre-2004 California Air Resources Board (CARB) standards as a whole were known as LEV I. Within that, there were four possible ratings: Tier 1, TLEV, LEV, and ULEV. The newest CARB rating system (since January 1, 2004) is known as LEV II. Within that rating system there are three primary ratings: LEV, ULEV, and SULEV. States other than California are given the option to use the federal EPA standards, or they can adopt California's standards.

| CERTIFICATION LEVEL | NMOG (G/MI) | CO (G/MI) | NO$_x$ (G/MI) |
| --- | --- | --- | --- |
| Bin 1 | 0.0 | 0.0 | 0.0 |
| Bin 2 | 0.010 | 2.1 | 0.02 |
| Bin 3 | 0.055 | 2.1 | 0.03 |
| Bin 4 | 0.070 | 2.1 | 0.04 |
| Bin 5 | 0.090 | 4.2 | 0.07 |
| Bin 6 | 0.090 | 4.2 | 0.10 |
| Bin 7 | 0.090 | 4.2 | 0.15 |
| Bin 8a | 0.125 | 4.2 | 0.20 |
| Bin 8b | 0.156 | 4.2 | 0.20 |
| Bin 9a | 0.090 | 4.2 | 0.30 |
| Bin 9b | 0.130 | 4.2 | 0.30 |
| Bin 9c | 0.180 | 4.2 | 0.30 |
| Bin 10a | 0.156 | 4.2 | 0.60 |
| Bin 10b | 0.230 | 6.4 | 0.60 |
| Bin 10c | 0.230 | 6.4 | 0.60 |
| Bin 11 | 0.230 | 7.3 | 0.90 |

**CHART 24–1**

EPA Tier 2—120,000-Mile tailpipe emission limits. After January 2007, the highest allowable bin is 8. NMOG stands for non-methane organic gases, which is a measure of all gases except those often created naturally by animals.
*Source:* Data compiled from the Environmental Protection Agency (EPA).
NOTE: The bin number is determined by the type and weight of the vehicle.

| U.S. EPA VEHICLE INFORMATION PROGRAM (THE HIGHER THE SCORE, THE LOWER THE EMISSIONS) | |
| --- | --- |
| SELECTED EMISSIONS STANDARDS | SCORE |
| Bin 1 and ZEV | 10 |
| PZEV | 9.5 |
| Bin 2 | 9 |
| Bin 3 | 8 |
| Bin 4 | 7 |
| Bin 5 and LEV II cars | 6 |
| Bin 6 | 5 |
| Bin 7 | 4 |
| Bin 8 | 3 |
| Bin 9a and LEV I cars | 2 |
| Bin 9b | 2 |
| Bin 10a | 1 |
| Bin 10b and Tier 1 cars | 1 |
| Bin 11 | 0 |

**CHART 24–2**

Air Pollution score.
*Source:* Data compiled from the Environmental Protection Agency (EPA).

| MINIMUM FUEL ECONOMY (MPG) COMBINED CITY-HIGHWAY LABEL VALUE | | | | | |
|---|---|---|---|---|---|
| SCORE | GASOLINE | DIESEL | E-85 | LPG | CNG* |
| 10 | 44 | 50 | 31 | 28 | 33 |
| 9 | 36 | 41 | 26 | 23 | 27 |
| 8 | 30 | 35 | 22 | 20 | 23 |
| 7 | 26 | 30 | 19 | 17 | 20 |
| 6 | 23 | 27 | 17 | 15 | 18 |
| 5 | 21 | 24 | 15 | 14 | 16 |
| 4 | 19 | 22 | 14 | 12 | 14 |
| 3 | 17 | 20 | 12 | 11 | 13 |
| 2 | 16 | 18 | — | — | 12 |
| 1 | 15 | 17 | 11 | 10 | 11 |
| 0 | 14 | 16 | 10 | 9 | 10 |

**CHART 24-3**

Greenhouse Gas score.
*Source:* Data compiled from the Environmental Protection Agency (EPA).
*CNG assumes a gallon equivalent of 121.5 cubic feet.

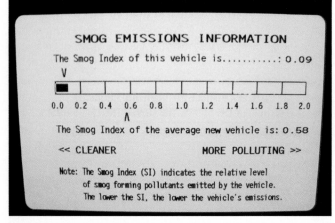

**FIGURE 24-2** This label on a Toyota Camry hybrid shows the relative smog-producing emissions, but this does not include carbon dioxide ($CO_2$), which may increase global warming.

## TIER 3 RATING

Starting in 2017, the Tier 3 sets new vehicle emissions standards and lowers the sulfur content of gasoline from the current level of 30 ppm to 10 ppm because the vehicle and its fuel is an integrated system. Tier 3 is designed to match with the California Air Resources Board (CARB) Low Emission Vehicle (LEV III) program so vehicle manufacturers can sell the same vehicles in all 50 states.

## EUROPEAN STANDARDS

Europe has its own set of standards that vehicles must meet, which includes the following tiers:

- Euro I (1992–1995)
- Euro II (1995–1999)
- Euro III (1999–2005)
- Euro IV (2005–2008)
- Euro V (2008+)

Vehicle emission standards and technological advancements have successfully reduced pollution from cars and trucks by about 90% since the 1970s. Unfortunately, there currently are more vehicles on the road, and they are being driven more miles each year, partially offsetting the environmental benefits of individual vehicle emissions reductions.

## EXHAUST ANALYSIS TESTING

The Clean Air Act Amendments require enhanced Inspection and Maintenance (I/M) programs in areas of the country that have the worst air quality and the Northeast Ozone Transport region. The states must submit to the EPA a **State Implementation Plan (SIP)** for their programs.

**FEDERAL TEST PROCEDURE (FTP)** The **Federal Test Procedure (FTP)** is the test used to certify all new vehicles before they can be sold. Once a vehicle meets these standards, it is certified by the EPA for sale in the United States. The FTP test procedure is a loaded-mode test lasting for a total duration of 505 seconds and is designed to simulate an urban driving trip. A cold start-up representing a morning start and a hot start after a soak period is part of the test. In addition to this drive cycle, a vehicle must undergo evaporative testing. Evaporative emissions are determined using the **Sealed Housing for Evaporative Determination (SHED)** test, which measures the evaporative emissions from the vehicle after a heat-up period representing a vehicle sitting in the sun. In addition, the vehicle is driven and then tested during the hot soak period.

**NOTE: A SHED is constructed entirely of stainless steel. The walls, floors, and ceiling, plus the door, are all constructed of stainless steel because it does not absorb hydrocarbons, which could offset test results.**

The FTP is a much more stringent test of vehicle emissions than is any test type that uses equipment that measures percentages of exhaust gases. The federal emission standards for each model year vehicle are the same for that model regardless of what size engine the vehicle is equipped with. This is why larger V-8 engines often are equipped with more emission control devices than smaller 4- and 6-cylinder engines.

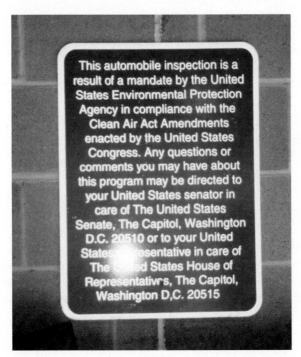

FIGURE 24–3 Photo of a sign taken at an emissions test facility.

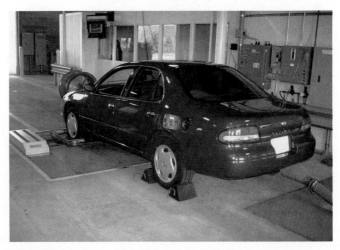

FIGURE 24–4 A vehicle being tested during an enhanced emission test.

**I/M TEST PROGRAMS** There are a variety of I/M testing programs that have been implemented by the various states. These programs may be centralized testing programs or decentralized testing programs. Each state is free to develop a testing program suitable to their needs as long as they can demonstrate to the EPA that their plan will achieve the attainment levels set by the EPA. This approach has led to a variety of different testing programs. ● **SEE FIGURE 24–3.**

**VISUAL TAMPERING CHECKS** Visual tampering checks may be part of an I/M testing program and usually include checking for the following items:

- Catalytic converter
- Fuel tank inlet restrictor
- Exhaust gas recirculation (EGR)
- Evaporative emission system
- Air-injection reaction system (AIR)
- Positive crankcase ventilation (PCV)

If any of these systems are missing, not connected, or tampered with, the vehicle will fail the emissions test and will have to be repaired/replaced by the vehicle owner before the vehicle can pass the emission test. Any cost associated with repairing or replacing these components may not be used toward the waiver amount required for the vehicle to receive a waiver.

**ONE-SPEED AND TWO-SPEED IDLE TEST** The one-speed and two-speed idle test measures the exhaust emissions from the tailpipe of the vehicle at idle and/or at 2500 RPM. This uses stand-alone exhaust gas sampling equipment that

measures the emissions in percentages. Each state chooses the standards that the vehicle has to meet in order to pass the test. The advantage to using this type of testing is that the equipment is relatively cheap and allows states to have decentralized testing programs because many facilities can afford the necessary equipment required to perform this test.

**LOADED-MODE TEST** The loaded-mode test uses a dynamometer that places a "single weight" load on the vehicle. The load applied to the vehicle varies with the speed of the vehicle. Typically, a 4-cylinder vehicle speed would be 24 mph, a 6-cylinder vehicle speed would be 30 mph, and an 8-cylinder vehicle speed would be 34 mph. Conventional stand-alone sampling equipment is used to measure HC and CO emissions. This type of test is classified as a Basic I/M test by the EPA. ● **SEE FIGURE 24–4.**

**ACCELERATION SIMULATION MODE (ASM)** The **ASM-type** of test uses a dynamometer that applies a heavy load on the vehicle at a steady-state speed. The load applied to the vehicle is based on the acceleration rate on the second simulated hill of the FTP. This acceleration rate is 3.3 mph/sec/sec (read as 3.3 mph per second per second, which is the unit of acceleration). There are different ASM tests used by different states.

The **ASM 50/15** test places a load of 50% on the vehicle at a steady 15 mph. This load represents 50% of the horsepower required to simulate the FTP acceleration rate of 3.3 mph/sec. This type of test produces relatively high levels of $NO_x$ emissions; therefore, it is useful in detecting vehicles that are emitting excessive $NO_x$.

The **ASM 25/25** test places a 25% load on the vehicle while it is driven at a steady 25 mph. This represents 25% of the load required to simulate the FTP acceleration rate of 3.3 mph/sec. Because this applies a smaller load on the vehicle at a higher speed, it will produce a higher level of HC and CO emissions than the ASM 50/15. $NO_x$ emissions will tend to be lower with this type of test.

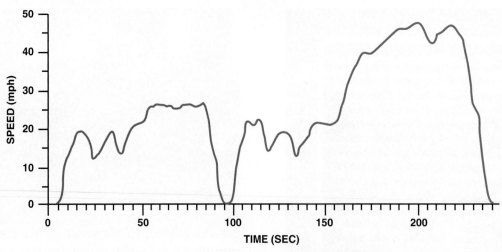

**FIGURE 24–5** Trace showing the Inspection/Maintenance 240 test. The test duplicates an urban test loop around Los Angeles, California. The first "hump" in the curve represents the vehicle being accelerated to about 20 mph, then driving up a small hill to about 30 mph and coming to a stop at 94 seconds. Then, the vehicle accelerates while climbing a hill and speeding up to about 50 mph during this second phase of the test.

## I/M 240 TEST

The **I/M 240 test** is the EPA's enhanced test. It is actually a portion of the 505-second FTP test used by the manufacturers to certify their new vehicles. The "240" stands for 240 seconds of drive time on a dynamometer. This is a loaded-mode transient test that uses constant volume sampling equipment to measure the exhaust emissions in mass just as is done during the FTP. The I/M 240 test simulates the first two hills of the FTP drive cycle. ● **FIGURE 24–5** shows the I/M 240 drive trace.

## OBD-II TESTING

In 1999, the EPA requested that states adopt OBD-II systems testing for 1996 and newer vehicles. The OBD-II system is designed to illuminate the MIL and store trouble codes any time a malfunction exists that would cause the vehicle emissions to exceed 1 1/2 times the FTP limits. If the OBD-II system is working correctly, the system should be able to detect a vehicle failure that would cause emissions to increase to an unacceptable level. The EPA has determined that the OBD-II system should detect emission failures of a vehicle even before that vehicle would fail an emissions test of the type that most states are employing. Furthermore, the EPA has determined that, as the population of OBD-II-equipped vehicles increases and the population of older non-OBD-II-equipped vehicles decreases, tailpipe testing will no longer be necessary.

The OBD-II testing program consists of a computer that can scan the vehicle OBD-II system using the DLC connector. The technician first performs a visual check of the vehicle MIL light to determine if it is working correctly. Next, the computer is connected to the vehicle's DLC connector. The computer will scan the vehicle OBD-II system and determine if there are any codes stored that are commanding the MIL light on. In addition, it will scan the status of the readiness monitors and determine if they have all run and passed. If the readiness monitors have all run and passed, it indicates that the OBD-II system has tested all the components of the emission control system. An OBD-II vehicle would fail this OBD-II test if the following occur:

- The MIL light does not come on with the key on, engine off.
- The MIL is commanded on.
- A number (varies by state) of the readiness monitors have not been run.

If none of these conditions are present, the vehicle will pass the emissions test.

## REMOTE SENSING

The EPA requires that, in high-enhanced areas, states perform on-the-road testing of vehicle emissions. The state must sample 0.5% of the vehicle population base in high-enhanced areas. This may be accomplished by using a remote sensing device. This type of sensing may be done through equipment that projects an infrared light through the exhaust stream of a passing vehicle. The reflected beam can then be analyzed to determine the pollutant levels coming from the vehicle. If a vehicle fails this type of test, the vehicle owner will receive notification in the mail that he or she must take the vehicle to a test facility to have the emissions tested.

## RANDOM ROADSIDE TESTING

Some states may implement random roadside testing that would usually involve visual checks of the emission control devices to detect tampering. Obviously, this method is not very popular, as it can lead to traffic tie-ups and delays on the part of commuters.

Exhaust analysis is an excellent tool to use for the diagnosis of engine performance concerns. In areas of the country that require exhaust testing to be able to get license plates, exhaust analysis must be able to do the following:

- Establish a baseline for failure diagnosis and service
- Identify areas of engine performance that are and are not functioning correctly
- Determine that the service and repair of the vehicle have been accomplished and are complete

**FIGURE 24–6** A partial stream sampling exhaust probe being used to measure exhaust gases in parts per million (ppm) or percent (%).

**?  FREQUENTLY ASKED QUESTION**

**What Does NMHC Mean?**

NMHC means **non-methane hydrocarbon,** and it is the standard by which exhaust emission testing for hydrocarbons is evaluated. Methane is natural gas and can come from animals, animal waste, and other natural sources. By not measuring methane gas, all background sources are eliminated, giving better results as to the true amount of unburned hydrocarbons that are present in the exhaust stream.

# EXHAUST ANALYSIS AND COMBUSTION EFFICIENCY

A popular method of engine analysis, as well as emission testing, involves the use of five-gas exhaust analysis equipment. ● **SEE FIGURE 24–6.** The five gases analyzed and their significance are discussed next.

**HYDROCARBONS**   Hydrocarbons (HC) are unburned gasoline and are measured in parts per million (ppm). A correctly operating engine should burn (oxidize) almost all the gasoline; therefore, very little unburned gasoline should be present in the exhaust. Acceptable levels of HC are 50 ppm or less. High levels of HC could be due to excessive oil consumption caused by weak piston rings or worn valve guides. The most common cause of excessive HC emissions is a fault in the ignition system. Items that should be checked include the following:

- Spark plugs
- Spark plug wires
- Distributor cap and rotor (if the vehicle is so equipped)
- Ignition timing (if possible)
- Ignition coil

**CARBON MONOXIDE**   Carbon monoxide (CO) is unstable and will easily combine with any oxygen to form stable carbon dioxide ($CO_2$). The fact that CO combines with oxygen is the reason that CO is a poisonous gas (in the lungs, it combines with oxygen to form $CO_2$ and deprives the brain of oxygen). CO levels of a properly operating engine should be less than 0.5%. High levels of CO can be caused by clogged or restricted crankcase ventilation devices such as the PCV valve, hose(s), and tubes. Other items that might cause excessive CO include the following:

- Clogged air filter
- Incorrect idle speed
- Too-high fuel-pump pressure
- Any other items that can cause a rich condition

**CARBON DIOXIDE**   Carbon dioxide ($CO_2$) is the result of oxygen in the engine combining with the carbon of the gasoline. An acceptable level of $CO_2$ is between 12% and 15%. A high reading indicates an efficiently operating engine. If the $CO_2$ level is low, the mixture may be either too rich or too lean.

**OXYGEN**   The next gas is oxygen ($O_2$). There is about 21% oxygen in the atmosphere, and most of this oxygen should be "used up" during the combustion process to oxidize all the hydrogen and carbon (hydrocarbons) in the gasoline. Levels of $O_2$ should be very low (about 0.5%). High levels of $O_2$, especially at idle, could be due to an exhaust system leak.

**NOTE: Adding 10% alcohol to gasoline provides additional oxygen to the fuel and will result in lower levels of CO and higher levels of $O_2$ in the exhaust.**

**OXIDES OF NITROGEN**   An oxide of nitrogen (NO) is a colorless, tasteless, and odorless gas when it leaves the engine, but as soon as it reaches the atmosphere and mixes with more oxygen, nitrogen oxides ($NO_2$) are formed. $NO_2$ is reddish-brown and has an acid and pungent smell. NO and $NO_2$ are grouped together and referred to as $NO_x$, where x represents any number of oxygen atoms. $NO_x$, the symbol used to represent all oxides of nitrogen, is the fifth gas commonly tested using a five-gas analyzer. The exhaust gas recirculation (EGR) system is the major controlling device limiting the formation of $NO_x$.

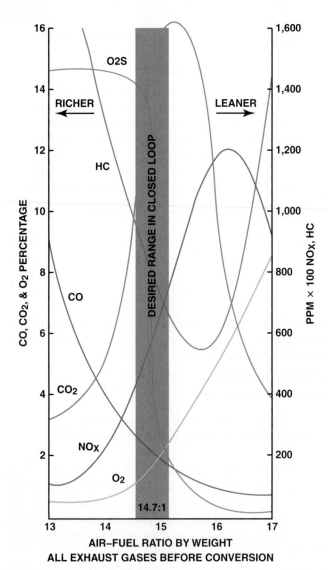

FIGURE 24–7 Exhaust emissions are very complex. When the air–fuel mixture becomes richer, some exhaust emissions are reduced, while others increase.

Acceptable exhaust emissions include the following:

| | Without Catalytic Converter | With Catalytic Converter |
|---|---|---|
| HC | 300 ppm or less | 30–50 ppm or less |
| CO | 3% or less | 0.3%–0.5% or less |
| $O_2$ | 0%–2% | 0%–2% |
| $CO_2$ | 12%–15% or higher | 12%–15% or higher |
| $NO_X$ | Less than 100 ppm at idle and less than 1000 ppm at WOT | Less than 100 ppm at idle and less than 1000 ppm at WOT |

● SEE FIGURE 24–7.

## HC TOO HIGH

High hydrocarbon exhaust emissions are usually caused by an engine misfire. What burns the fuel in an engine? The ignition system ignites a spark at the spark plug to ignite the proper mixture inside the combustion chamber. If a spark plug does not ignite the mixture, the resulting unburned fuel is pushed out of the cylinder on the exhaust stroke by the piston through the exhaust valves and into the exhaust system. Therefore, if any of the following ignition components or adjustments are not correct, excessive HC emission is likely:

1. Defective or worn spark plugs
2. Defective or loose spark plug wires
3. Defective distributor cap and/or rotor
4. Incorrect ignition timing (either too far advanced or too far retarded)
5. A lean air–fuel mixture can also cause a misfire. This condition is referred to as a lean misfire. A lean air–fuel mixture can be caused by low fuel-pump pressure, a clogged fuel filter, or a restricted fuel injector.

NOTE: To make discussion easier in future reference to these items, this list of ignition components and checks can be referred to simply as "spark stuff."

## CO TOO HIGH

Excessive carbon monoxide is an indication of too rich an air–fuel mixture. CO is the **rich indicator.** The higher the CO reading, the richer the air–fuel mixture. High concentrations of CO indicate that not enough oxygen was available for the amount of fuel. Common causes of high CO include the following:

- Too-high fuel-pump pressure
- Defective fuel-pressure regulator
- Clogged air filter or PCV valve
- Defective injectors

NOTE: One technician remembers "CO" as meaning "clogged oxygen" and always looks for restricted airflow into the engine whenever high CO levels are detected.

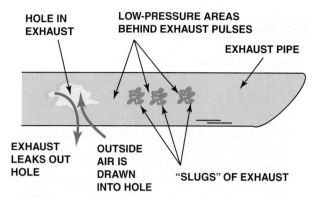

FIGURE 24-8 A hole in the exhaust system can cause outside air (containing oxygen) to be drawn into the exhaust system. This extra oxygen can be confusing to a service technician because the extra $O_2$ in the exhaust stream could be misinterpreted as a too-lean air-fuel mixture.

 **TECH TIP**

**CO Equals $O_2$**

If the exhaust is rich, CO emissions will be higher than normal. If the exhaust is lean, $O_2$ emissions will be higher than normal. Therefore, if the CO reading is the same as the $O_2$ reading, then the engine is operating correctly. For example, if both CO and $O_2$ are 0.5% and the engine develops a vacuum leak, the $O_2$ will rise. If a fuel-pressure regulator were to malfunction, the resulting richer air-fuel mixture would increase CO emissions. Therefore, if both the rich indicator (CO) and the lean indicator ($O_2$) are equal, the engine is operating correctly.

 **TECH TIP**

**How to Find a Leak in the Exhaust System**

A hole in the exhaust system can dilute the exhaust gases with additional oxygen ($O_2$). ● **SEE FIGURE 24-8.**

This additional $O_2$ in the exhaust can lead the service technician to believe that the air-fuel mixture is too lean. To help identify an exhaust leak, perform an exhaust analysis at idle and at 2500 RPM (fast idle) and compare with the following:

- If the $O_2$ is high at idle and at 2500 RPM, the mixture is lean at both idle and at 2500 RPM.
- If the $O_2$ is low at idle and high at 2500 RPM, this usually means that the vehicle is equipped with a working AIR pump.
- If the $O_2$ is high at idle, but okay at 2500 RPM, a hole in the exhaust or a small vacuum leak that is "covered up" at higher speed is indicated.

# MEASURING OXYGEN AND CARBON DIOXIDE

Two gas exhaust analyzers (HC and CO) work well, but both HC and CO are consumed (converted) inside the catalytic converter. The amount of leftover oxygen coming out of the tailpipe is an indication of leanness. The higher the $O_2$ level, the leaner the exhaust. Oxygen therefore is the **lean indicator.** Acceptable levels of $O_2$ are 0% to 2%.

NOTE: A hole in the exhaust system can draw outside air (oxygen) into the exhaust system. Therefore, to be assured of an accurate reading, carefully check the exhaust system for leaks. Using a smoke machine is an easy method to locate leaks in the exhaust system.

Carbon dioxide ($CO_2$) is a measure of efficiency. The higher the level of $CO_2$ in the exhaust stream, the more efficiently the engine is operating. Levels of 12% to 15% are considered to be acceptable. Because $CO_2$ levels peak at an air-fuel mixture of 14.7:1, a lower level of $CO_2$ indicates either a too-rich or a too-lean condition. The $CO_2$ measurement by itself does not indicate which condition is present. An example follows:

$CO_2$ = 8% (This means that efficiency is low and that the air-fuel mixture is not correct.)

Look at $O_2$ and CO levels.

A high $O_2$ indicates lean and a high CO indicates rich.

# PHOTOCHEMICAL SMOG FORMATION

Oxides of nitrogen are formed by high temperature—over 2,500°F (1,370°C)—and/or pressures inside the combustion chamber. Oxides of nitrogen contribute to the formation of photochemical **smog** when sunlight reacts chemically with $NO_x$ and unburned hydrocarbons (HC). Smog is a term derived by combining the words *smoke* and *fog*. Ground-level ozone is a constituent of smog. **Ozone** is an enriched oxygen molecule with three atoms of oxygen ($O_3$) instead of the normal two atoms of oxygen ($O_2$).

Ozone in the upper atmosphere is beneficial because it blocks out harmful ultraviolet rays that contribute to skin cancer. However, at ground level, this ozone (smog) is an irritant to the respiratory system.

### Your Nose Knows

Using the nose, a technician can often identify a major problem without having to connect the vehicle to an exhaust analyzer. An example follows:

- The strong smell of exhaust is due to excessive unburned hydrocarbon (HC) emissions. Look for an ignition system fault that could prevent the proper burning of the fuel. A vacuum leak could also cause a lean misfire and cause excessive HC exhaust emissions.
- If your eyes start to burn or water, suspect excessive oxides of nitrogen ($NO_x$) emissions. The oxides of nitrogen combine with the moisture in the eyes to form a mild solution of nitric acid. The acid formation causes the eyes to burn and water. Excessive $NO_x$ exhaust emissions can be caused by the following:
  - A vacuum leak causing higher-than-normal combustion chamber temperature
  - Overadvanced ignition timing causing higher-than-normal combustion chamber temperature
  - Lack of proper amount of exhaust gas recirculation (EGR) (This is usually noticed above idle on most vehicles.)
- Dizzy feeling or headache. This is commonly caused by excessive carbon monoxide (CO) exhaust emissions. Get into fresh air as soon as possible. A probable cause of high levels of CO is an excessively rich air–fuel mixture.

### Check for Dog Food?

A commonly experienced problem in many parts of the country involves squirrels or other animals placing dog food into the air intake ducts of vehicles. Dog food is often found packed tight in the ducts against the air filter. An air intake restriction reduces engine power and vehicle performance.

### The Case of the Retarded Exhaust Camshaft

A Toyota equipped with a double overhead camshaft (DOHC) inline 6-cylinder engine failed the state-mandated enhanced exhaust emission test for $NO_x$. The engine ran perfectly without spark knocking (ping), which is usually a major reason for excessive $NO_x$ emissions. The technician checked the following:

- The ignition timing, which was found to be set to specifications (if too far advanced, can cause excessive $NO_x$)
- The cylinders, which were decarbonized using top engine cleaner
- The EGR valve, which was inspected and the EGR passages cleaned

After all the items were completed, the vehicle was returned to the inspection station, where the vehicle again failed for excessive $NO_x$ emissions (better but still over the maximum allowable limit).

After additional hours of troubleshooting, the technician decided to go back to basics and start over again. A check of the vehicle history with the owner indicated that the only previous work performed on the engine was a replacement timing belt over a year before. The technician discovered that the exhaust cam timing was retarded two teeth, resulting in late closing of the exhaust valve. The proper exhaust valve timing resulted in a slight amount of exhaust being retained in the cylinder. This extra exhaust was added to the amount supplied by the EGR valve and helped reduce $NO_x$ emissions. After repositioning the timing belt, the vehicle passed the emissions test well within the limits.

# TESTING FOR OXIDES OF NITROGEN

Because the formation of $NO_x$ occurs mostly under load, the most efficient method to test for $NO_x$ is to use a portable exhaust analyzer that can be carried in the vehicle while the vehicle is being driven under a variety of conditions.

**SPECIFICATIONS FOR $NO_x$** From experience, a maximum reading of 1,000 parts per million (ppm) of $NO_x$ under loaded driving conditions will generally mean that the vehicle will pass an enhanced I/M roller test. A reading of over 100 ppm at idle should be considered excessive.

| Exhaust Gas Summary Chart | |
|---|---|
| Gas | Cause and Correction |
| High HC | Engine misfire or incomplete burning of fuel caused by the following:<br>1. Ignition system fault<br>2. Lean misfire<br>3. Too low an engine temperature (thermostat) |
| High CO | Rich condition caused by the following:<br>1. Leaking fuel injectors or fuel-pressure regulator<br>2. Clogged air filter or PCV system<br>3. Excessive fuel pressure |
| High HC and CO | Excessively rich condition caused by the following:<br>1. All items included under high CO<br>2. Fouled spark plugs causing a misfire to occur<br>3. Possible nonoperating catalytic converter |
| High NO$_x$ | Excessive combustion chamber temperature caused by the following:<br>1. Nonoperating EGR valve<br>2. Clogged EGR passages<br>3. Engine operating temperature too high because of cooling system restriction, worn water pump impeller, or other faults in the cooling system<br>4. Lean air–fuel mixture<br>5. High compression caused by excessive carbon buildup in the cylinders |

**REAL WORLD FIX**

**O2S Shows Rich, but Pulse Width Is Low**

A service technician was attempting to solve a driveability problem. The computer did not indicate any diagnostic trouble codes (DTCs). A check of the oxygen sensor voltage indicated a higher-than-normal reading almost all the time. The pulse width to the port injectors was lower than normal. The lower-than-normal pulse width indicates that the computer is attempting to reduce fuel flow into the engine by decreasing the amount of on-time for all the injectors.

What could cause a rich mixture if the injectors were being commanded to deliver a lean mixture? Finally, the technician shut off the engine and took a careful look at the entire fuel-injection system. Although the vacuum hose was removed from the fuel-pressure regulator, fuel was found dripping from the vacuum hose. The problem was a defective fuel-pressure regulator that allowed an uncontrolled amount of fuel to be drawn by the intake manifold vacuum into the cylinders. While the computer tried to reduce fuel by reducing the pulse width signal to the injectors, the extra fuel being drawn directly from the fuel rail caused the engine to operate with too rich an air–fuel mixture.

# SUMMARY

1. Excessive hydrocarbon (HC) exhaust emissions are created by a lack of proper combustion, such as a fault in the ignition system, too lean an air–fuel mixture, or too-cold engine operation.

2. Excessive carbon monoxide (CO) exhaust emissions are usually created by a rich air–fuel mixture.

3. Excessive oxides of nitrogen (NO$_x$) exhaust emissions are usually created by excessive heat or pressure in the combustion chamber or a lack of the proper amount of exhaust gas recirculation (EGR).

4. Carbon dioxide (CO$_2$) levels indicate efficiency. The higher the CO$_2$, the more efficient the engine operation.

5. Oxygen (O$_2$) indicates leanness. The higher the O$_2$, the leaner the air–fuel mixture.

6. A vehicle should be driven about 20 miles, especially during cold weather, to allow the engine to be fully warm before an enhanced emissions test.

# REVIEW QUESTIONS

1. List the five exhaust gases and their maximum allowable readings for a fuel-injected vehicle equipped with a catalytic converter.

2. List two causes of a rich exhaust.

3. List two causes of a lean exhaust.

4. List those items that should be checked if a vehicle fails an exhaust test for excessive NO$_x$ emissions.

1. Technician A says that high HC emission levels are often caused by a fault in the ignition system. Technician B says that high $CO_2$ emissions are usually caused by a richer-than-normal air–fuel mixture. Which technician is correct?
   a. Technician A only
   b. Technician B only
   c. Both Technicians A and B
   d. Neither Technician A nor B

2. HC and CO are high, and $CO_2$ and $O_2$ are low. This could be caused by a _____.
   a. Rich mixture
   b. Lean mixture
   c. Defective ignition component
   d. Clogged EGR passage

3. Which gas is generally considered to be the rich indicator? (The higher the level of this gas, the richer the air–fuel mixture.)
   a. HC
   b. CO
   c. $CO_2$
   d. $O_2$

4. Which gas is generally considered to be the lean indicator? (The higher the level of this gas, the leaner the air–fuel mixture.)
   a. HC
   b. CO
   c. $CO_2$
   d. $O_2$

5. Which exhaust gas indicates efficiency? (The higher the level of this gas, the more efficient the engine operates.)
   a. HC
   b. CO
   c. $CO_2$
   d. $O_2$

6. All of the gases are measured in percentages except _____.
   a. HC
   b. CO
   c. $CO_2$
   d. $O_2$

7. After the following exhaust emissions were measured, how was the engine operating?
   HC = 766 ppm    $CO_2$ = 8.2%    CO = 4.6%    $O_2$ = 0.1%
   a. Too rich
   b. Too lean

8. Technician A says that carbon inside the engine can cause excessive $NO_X$ to form. Technician B says that excessive $NO_X$ could be caused by a cooling system fault causing the engine to operate too hot. Which technician is correct?
   a. Technician A only
   b. Technician B only
   c. Both Technicians A and B
   d. Neither Technician A nor B

9. A clogged EGR passage could cause excessive _____ exhaust emissions.
   a. HC
   b. CO
   c. $NO_X$
   d. $CO_2$

10. An ignition fault could cause excessive _____ exhaust emissions.
    a. HC
    b. CO
    c. $NO_X$
    d. $CO_2$

# chapter
# 25

# EVAPORATIVE EMISSION CONTROL SYSTEMS

**LEARNING OBJECTIVES:** **After studying this chapter, the reader should be able to:** • Explain the purpose and function of the evaporative emission control (EVAP) system. • Compare enhanced and nonenhanced evaporative control (EVAP) systems. • Describe leak detection pump systems and onboard refueling vapor recovery. • Describe how to diagnose EVAP system faults. • Discuss the functions of an evaporative system monitor and interpret the EVAP diagnostic trouble codes.

**KEY TERMS:** Adsorption 318 • Evaporative control (EVAP) system 317 • Fuel tank pressure (FTP) 323 • Leak detection pump (LDP) 321 • Onboard refueling vapor recovery (ORVR) 320 • Purge valve 320 • Vent valve 321 • Volatile organic compounds (VOC) 317

## EVAPORATIVE EMISSION CONTROL SYSTEM

**PURPOSE AND FUNCTION** The purpose of the evaporative emission control system is to trap and hold gasoline vapors, also called **volatile organic compounds (VOCs).** The **evaporative control (EVAP) system** includes the charcoal canister, hoses, and valves. These vapors are routed into a charcoal canister, then into the intake airflow, where they are burned in the engine instead of being released into the atmosphere.

**COMMON COMPONENTS** The fuel tank filler caps used on vehicles with modern EVAP systems are a special design. Most EVAP fuel tank filler caps have a built-in pressure-vacuum relief valve. When pressure or vacuum exceeds a calibrated value, the valve opens. Once the pressure or vacuum has been relieved, the valve closes. If a sealed cap is used on an EVAP system that requires a pressure-vacuum relief design, a vacuum may develop in the fuel system, or the fuel tank may be damaged by fuel expansion or contraction. ● **SEE FIGURE 25–1.**

**EVAP SYSTEM OPERATION** The canister is located under the hood or underneath the vehicle and is filled with activated charcoal granules that can hold up to one-third of their own weight in fuel vapors. ● **SEE FIGURE 25–2.**

**NOTE: Some vehicles with large or dual fuel tanks may have dual canisters.**

**?** **FREQUENTLY ASKED QUESTION**

**When Filling My Fuel Tank, Why Should I Stop When the Pump Clicks Off?**

Every fuel tank has an upper volume chamber that allows for expansion of the fuel when hot. The volume of the chamber is between 10% and 20% of the volume of the tank. For example, if a fuel tank had a capacity of 20 gallons, the expansion chamber volume would be from 2 to 4 gallons. A hose is attached at the top of the chamber and vented to the charcoal canister. If extra fuel is forced into this expansion volume, liquid gasoline can be drawn into the charcoal canister. This liquid fuel can saturate the canister and create an overly rich air–fuel mixture when the canister purge valve is opened during normal vehicle operation. This extra-rich air–fuel mixture can cause the vehicle to fail an exhaust emissions test, reduce fuel economy, and possibly damage the catalytic converter. To avoid problems, simply add fuel to the next dime's worth after the nozzle clicks off. This will ensure that the tank is full yet not overfilled.

Activated charcoal is an effective vapor trap because of its great surface area. Each gram of activated charcoal has a surface area of 1,100 square meters, or more than 1/4 acre. Typical canisters hold either 300 or 625 grams of charcoal *with a surface area equivalent to 80 or 165 football fields.* By a process

**FIGURE 25–1** A capless system used on Fords. Two tabs need to be pushed outward by the nozzle to unlatch the spring-loaded flap that keeps the system sealed without the use of a removal gas cap.

**FIGURE 25–2** A charcoal canister can be located under the hood or underneath the vehicle.

called **adsorption,** the fuel vapor molecules adhere to the carbon surface. This attaching force is not strong, so the system purges the vapor molecules quite simply by sending a fresh airflow through the charcoal:

- **Vapor purging.** During engine operation, stored vapors are drawn from the canister into the engine through a hose connected to the throttle body or the air cleaner. This "purging" process mixes unburned gasoline (hydrocarbons, abbreviated HCs) vapors from the canister with the existing air–fuel charge. ● **SEE FIGURES 25–3 AND 25–4.**
- **Computer-controlled purge.** The PCM controls when the canister purges on most engines. This is done by an

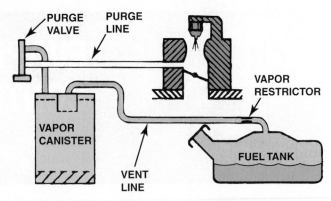

**FIGURE 25–3** The EVAP system includes all of the lines, hoses, and valves, plus the charcoal canister.

electric vacuum solenoid and one or more purge valves. Under normal conditions, most engine control systems permit purging only during closed-loop operation at cruising speeds. During other engine operation conditions, such as open-loop mode, idle, deceleration, or wide-open throttle, the PCM prevents canister purging.

**EVAPORATIVE PRESSURES** Pressures can build inside the fuel system and are usually measured in units of inches of water, abbreviated "in. $H_2O$" (28 in. $H_2O$ equals 1 pound per square inch, or 1 PSI). Pressure buildup can be caused by the following:

- Fuel evaporation rates (volatility)
- Gas tank size (fuel surface area and volume)
- Fuel level (liquid versus vapor)
- Fuel slosh (driving conditions)
- Temperature (ambient, in-tank, close to the tank)
- Returned fuel from the rail

Some scan tools display other units of measure for the EVAP system that makes understanding the system difficult. ● **SEE CHART 25–1** for pressure conversions between pounds per square inch (PSI), inches of mercury (in. Hg), and inches of water (in. $H_2O$).

## NONENHANCED EVAPORATIVE CONTROL SYSTEMS

Prior to 1996, evaporative systems were referred to as nonenhanced evaporative control (EVAP) systems. This term refers to evaporative systems that had limited diagnostic capabilities. While they are often PCM controlled, their diagnostic capability is usually limited to their ability to detect

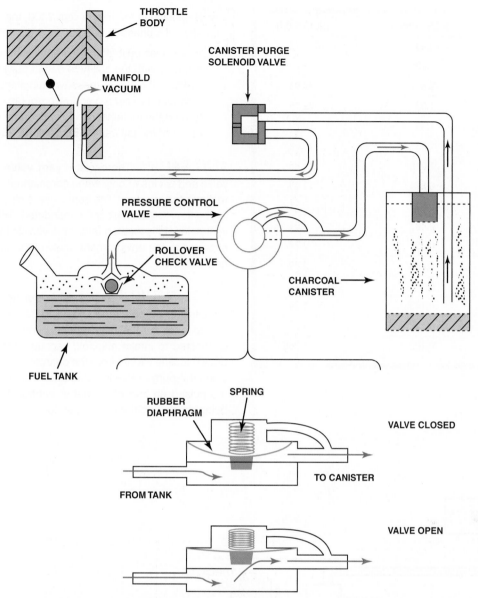

**FIGURE 25–4** A typical EVAP system. Note that when the computer turns on the canister purge solenoid valve, manifold vacuum draws any stored vapors from the canister into the engine. Manifold vacuum also is applied to the pressure control valve. When this valve opens, fumes from the fuel tank are drawn into the charcoal canister and eventually into the engine. When the solenoid valve is turned off (or the engine stops and there is no manifold vacuum), the pressure control valve is spring-loaded shut to keep vapors inside the fuel tank from escaping to the atmosphere.

if purge has occurred. Many of these older systems have a diagnostic switch that could sense if purge is occurring and set a code if no purge is detected. This system does not check for leaks. On some vehicles, the PCM also has the capability of monitoring the integrity of the purge solenoid and circuit. These systems' limitations are their ability to check the integrity of the evaporative system on the vehicle. They could not detect leaks or missing or loose gas caps that could lead to excessive evaporative emissions from the vehicle. Nonenhanced evaporative systems use either a canister purge solenoid or a vapor management valve to control purge vapor.

## ENHANCED EVAPORATIVE CONTROL SYSTEM

**BACKGROUND** Beginning in 1996 with OBD-II vehicles, manufacturers were required to install systems that are able to detect both purge flow and evaporative system leakage:

- The systems on models produced between 1996 and 2000 must be able to detect a leak as small as 0.040 inches in diameter.

| PSI | INCHES HG | INCHES H$_2$O |
|---|---|---|
| 14.7 | 29.93 | 407.19 |
| 1.0 | 2.036 | 27.7 |
| 0.9 | 1.8 | 24.93 |
| 0.8 | 1.63 | 22.16 |
| 0.7 | 1.43 | 19.39 |
| 0.6 | 1.22 | 16.62 |
| 0.5 | 1.018 | 13.85 |
| 0.4 | 0.814 | 11.08 |
| 0.3 | 0.611 | 8.31 |
| 0.2 | 0.407 | 5.54 |
| 0.1 | 0.204 | 2.77 |
| 0.09 | 0.183 | 2.49 |
| 0.08 | 0.163 | 2.22 |
| 0.07 | 0.143 | 1.94 |
| 0.06 | 0.122 | 1.66 |
| 0.05 | 0.102 | 1.385 |

**CHART 25–1**

Pressure conversions.

NOTE: 1 PSI = 28 in. H$_2$O
0.25 PSI = 7 in. H$_2$O

- Beginning in the model year 2000, the enhanced systems started a phase-in of 0.020 inch diameter leak detection.
- All vehicles built after 1995 have enhanced evaporative systems with the ability to detect purge flow and system leakage. If either of these two functions fails, the system is required to set a diagnostic trouble code (DTC) and turn on the malfunction indicator lamp (MIL) to warn the driver of the failure. ● **SEE FIGURE 25–5.**

**VENT VALVE**    The canister **vent valve** is a *normally open* valve and is closed only when commanded by the PCM during testing of the system. The vent valve is closed only during testing by the PCM as part of the mandated OBD-II standards. The vent solenoid is located under the vehicle in most cases and is exposed to the environment, making this valve subject to rust and corrosion.

**PURGE VALVE**    The canister **purge valve,** also called the *canister purge (CANP) solenoid,* is normally closed and is pulsed open by the PCM during purging. The purge valve is connected to intake manifold vacuum using a rubber hose to draw gasoline vapors from the charcoal canister into the engine when the purge valve is commanded open. Most purge valves are pulsed on and off to better control the amount of fumes being drawn into the intake manifold.

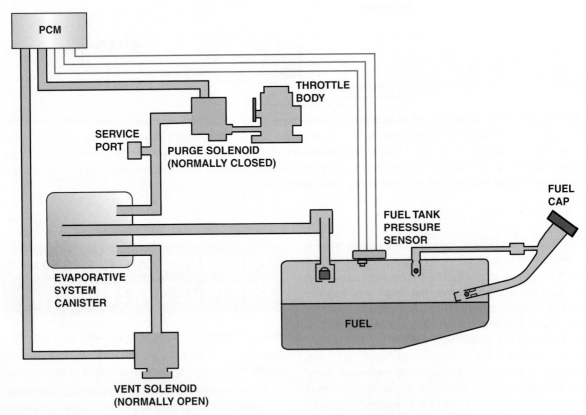

**FIGURE 25–5** An enhanced EVAP system is able to perform system and leak detection diagnosis.

**Problems after Refueling? Check the Purge Valve**

The purge valve is normally closed and opens only when the PCM is commanding the system to purge. If the purge solenoid were to become stuck in the open position, gasoline fumes would be allowed to flow directly from the gas tank to the intake manifold. When refueling, this would result in a lot of fumes being forced into the intake manifold and as a result cause a hard-to-start condition after refueling. This would also result in a rich exhaust and likely black exhaust when first starting the engine after refueling. While the purge solenoid is usually located under the hood of most vehicles and is less subject to rust and corrosion as with the vent valve, it can still fail.

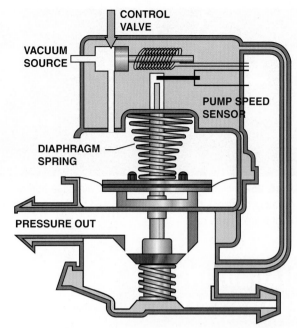

**FIGURE 25–6** A leak detection pump (LDP) used on some Chrysler and other vehicles to pressurize (slightly) the fuel system to check for leaks.

# LEAK DETECTION PUMP SYSTEM

**PURPOSE AND FUNCTION** Many vehicles use a vacuum-operated **leak detection pump (LDP)** as part of the evaporative control system diagnosis equipment. ● SEE FIGURE 25–6.

**OPERATION** The system works to test for leaks as follows:

- The purge solenoid is normally closed.
- The vent valve in the LDP is normally open. Filtered fresh air is drawn through the LDP to the canister.
- The LDP uses a spring attached to a diaphragm to apply pressure (7.5 in. $H_2O$) to the fuel tank.
- The PCM monitors the LDP switch that is triggered if the pressure drops in the fuel tank.
- The time between LDP solenoid off and LDP switch close is called the pump period. This time period is inversely proportional to the size of the leak. The shorter the pump period, the larger the leak. The longer the pump period, the smaller the leak.

  EVAP large leak (greater than 0.080 inches): less than 0.9 second

  EVAP medium leak (0.040 to 0.080 inches): 0.9 to 1.2 seconds

  EVAP small leak (0.020 to 0.040 inches): 1.2 to 6 seconds

# ONBOARD REFUELING VAPOR RECOVERY

**PURPOSE AND FUNCTION** The **onboard refueling vapor recovery (ORVR)** system was first introduced on some 1998 vehicles. Previously designed EVAP systems allowed fuel vapor to escape to the atmosphere during refueling.

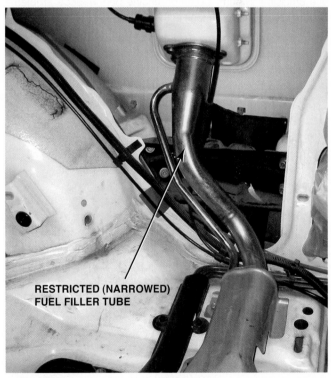

**FIGURE 25–7** A restricted fuel fill pipe shown on vehicle with the interior removed.

RESTRICTED (NARROWED) FUEL FILLER TUBE

**OPERATION** The primary feature of most ORVR systems is the restricted tank filler tube, which is about 1 inch (25 mm) in diameter. This reduced size filler tube creates an aspiration effect, which tends to draw outside air into the filler tube. During refueling, the fuel tank is vented to the charcoal canister, which captures the gas fumes, and with air flowing into the filler tube, no vapors can escape to the atmosphere. ● SEE FIGURE 25–7.

**FIGURE 25–8** Some vehicles will display a message if an evaporative control system leak is detected that could be the result of a loose gas cap.

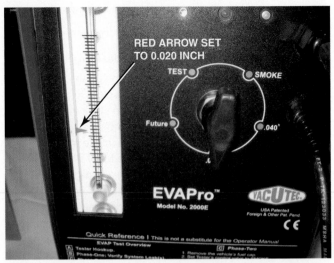

**FIGURE 25–9** To test for a leak, this tester was set to the 0.020-inch hole and turned on. The ball rose in the scale on the left, and the red arrow was moved to that location. If when testing the system for leaks the ball rises higher than the arrow, then the leak is larger than 0.02 inch. If the ball does not rise to the level of the arrow, the leak is smaller than 0.020 inch.

# DIAGNOSING THE EVAP SYSTEM

**SYMPTOMS** Before vehicle emissions testing began in many parts of the country, little service work was done on the evaporative emission system. Common engine performance problems that can be caused by a fault in this system include the following:

- **Poor fuel economy.** A leak in a vacuum-valve diaphragm can result in engine vacuum drawing in a constant flow of gasoline vapors from the fuel tank. This usually results in a drop in fuel economy of 2 to 4 miles per gallon (mpg). Use a hand-operated vacuum pump to check that the vacuum diaphragm can hold vacuum.

- **Poor performance.** A vacuum leak in the system can cause the engine to run rough. Age, heat, and time all contribute to the deterioration of rubber hoses.

**STATE EVAP TESTS** Enhanced exhaust emissions (I/M-240) testing tests the evaporative emission system. A leak in the system is tested by pressurizing the entire fuel system to a level below 1 PSI (about 14 in. $H_2O$). The system is typically pressurized with nitrogen, a nonflammable gas that makes up 78% of our atmosphere. The pressure in the system is then shut off and the pressure monitored. If the pressure drops below a set standard, then the vehicle fails the test. This test determines if there is a leak in the system.

**HINT: To help pass the evaporative section of an enhanced emissions test, arrive at the test site with less than a half tank of fuel. This means that the rest of the volume of the fuel tank is filled with air. It takes longer for the pressure to drop from a small leak when the volume of the air is greater compared to when the tank is full and the volume of air remaining in the tank is small.**

**LOCATING LEAKS IN THE SYSTEM** Leaks in the evaporative emission control system will cause the malfunction check gas cap indication lamp to light on some vehicles. ●**SEE FIGURE 25–8.**

**FIGURE 25–10** This unit is applying smoke to the fuel tank through an adapter, and the leak was easily found to be the gas cap seal.

A leak will also cause a gas smell, which would be most noticeable if the vehicle were parked in an enclosed garage. The first step is to determine if there is a leak in the system by setting the EVAP tester to rate the system for either a 0.040-inch- or a 0.020-inch-hole-size leak. ●**SEE FIGURE 25–9.**

After it has been determined that a leak exists and it is larger than specified, one of two methods can be used to check for leaks in the evaporative system:

- **Smoke machine testing.** The most efficient method of leak detection is to introduce smoke under low pressure from a machine specifically designed for this purpose. ●**SEE FIGURE 25–10.**

**FIGURE 25–11** An emission tester that uses nitrogen to pressurize the fuel system.

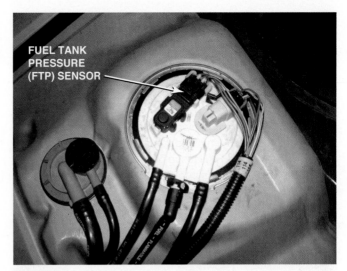

FUEL TANK PRESSURE (FTP) SENSOR

**FIGURE 25–12** The fuel tank pressure sensor (black unit with three wires) looks like a MAP sensor and is usually located on top of the fuel pump module (white unit).

- **Nitrogen gas pressurization.** This method uses nitrogen gas under a very low pressure (lower than 1 PSI) in the fuel system. The service technician then listens for the escaping air, using amplified headphones. ● **SEE FIGURE 25–11.**

# EVAPORATIVE SYSTEM MONITOR

**OBD-II REQUIREMENTS** OBD-II computer programs not only detect faults but also *periodically test various systems* and alert the driver before emissions-related components are harmed by system faults:

- Serious faults cause a blinking malfunction indicator lamp (MIL) or even an engine shutdown.
- Less serious faults may simply store a code but not illuminate the MIL.

The OBD-II requirements did not affect fuel system design. However, one new component, a fuel evaporative canister purge line pressure sensor, was added for monitoring purge line pressure during tests. The OBD-II requirements state that vehicle fuel systems are to be routinely tested *while under way* by the PCM.

All OBD-II vehicles perform a canister purge system pressure test, as commanded by the PCM. While the vehicle is being driven, the vapor line between the canister and the purge valve is monitored for pressure changes:

- When the canister purge solenoid is open, the line should be under a vacuum since vapors must be drawn from the canister into the intake system. However, when the purge solenoid is closed, there should be no vacuum in the line. The pressure sensor detects if a vacuum is present, and the information is compared to the command given to the solenoid.

- If, during the canister purge cycle, no vacuum exists in the canister purge line, a code is set indicating a possible fault, which could be caused by an inoperative or clogged solenoid or a blocked or leaking canister purge fuel line. Likewise, if vacuum exists when no command for purge is given, a stuck solenoid is evident, and a code is set. The EVAP system monitor tests for purge volume and leaks.

A typical EVAP monitor first closes off the system to atmospheric pressure and opens the purge valve during cruise operation. A **fuel tank pressure (FTP)** sensor then monitors the rate with which vacuum increases in the system. The monitor uses this information to determine the purge volume flow rate. To test for leaks, the EVAP monitor closes the purge valve, creating a completely closed system. The fuel tank pressure sensor then monitors the leak-down rate. If the rate exceeds PCM-stored values, a leak greater than or equal to the OBD-II standard of 0.04 inch (1 mm) or 0.02 inch (0.5 mm) exists. After two consecutive failed trips testing either purge volume or the presence of a leak, the PCM lights the MIL and sets a DTC.

The fuel tank pressure sensor is similar to a MAP sensor, and instead of monitoring intake manifold absolute pressure, it is used to monitor fuel tank pressure. ● **SEE FIGURE 25–12.**

**ENGINE-OFF NATURAL VACUUM** System integrity (leakage) can also be checked after the engine is shut off. The premise is that a warm evaporative system will cool down after the engine is shut off and the vehicle is stable. A slight vacuum will be created in the gas tank during this cool-down period. If a specific level of vacuum is reached and maintained, the system is said to have integrity (no leakage). Actually, the vacuum is created after a period of time because the vapor pressure tends to increase after the engine is shut off and gradually decreases over time. The PCM monitors the pressure rise and decrease over time and triggers a diagnostic trouble code (DTC) if the pressure indicates a leak in the system. ● **SEE FIGURE 25–13.**

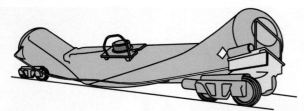

**FIGURE 25–13** A tank car was cleaned using steam, and then both the bottom drain and the top vent were closed. The next day, the tank had collapsed because of the air pressure difference when the inside cooled. The higher outside air pressure caused the tank to collapse.

 **TECH TIP**

**Always Tighten the Cap Correctly**

Many diagnostic trouble codes (DTCs) are set because the gas cap has not been properly installed. To be sure that a screw-type gas cap is properly sealed, tighten it until you hear three clicks. The clicking is a ratchet device, and the clicking does not harm the cap. Therefore, if a P0440 or similar DTC is set, check the cap. ● **SEE FIGURE 25–14.**

# TYPICAL EVAP MONITOR

The PCM will run the EVAP monitor when the following enable criteria are met:

- Barometric pressure (BARO) greater than 70 kPa (20.7 in. Hg or 10.2 PSI)
- Intake air temperature (IAT) between 39°F and 86°F (4°C to 30°C) at engine start-up
- Engine coolant temperature (ECT) between 39°F and 86°F (4°C to 30°C) at engine start-up
- ECT and IAT within 3°F of each other at engine start-up
- Fuel level within 15% to 85%
- Throttle position (TP) sensor between 9% and 35%

A typical EVAP monitor first closes off the system to atmospheric pressure and opens the purge valve during cruise operation. A fuel tank pressure (FTP) sensor then monitors the rate with which vacuum increases in the system. ● **SEE FIGURE 25–15.**

The monitor uses this information to determine the purge volume flow rate. To test for leaks, the EVAP monitor closes the purge valve, creating a completely closed system. The fuel tank pressure sensor then monitors the leak-down rate. If the rate exceeds PCM-stored values, a leak greater than or equal to the OBD-II standard of 0.04 inch (1 mm) or 0.02 inch (0.5 mm) exists.

**FIGURE 25–14** This Toyota cap warns that the check engine light will come on if not tightened until one click.

**DIFFERENTIAL PRESSURE INCHES (H2O)**

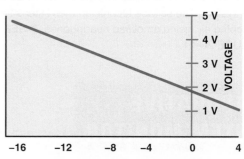

**FUEL TANK PRESSURE SENSOR VOLTAGE VS. INCHES OF WATER**

**FIGURE 25–15** To easily check the fuel tank pressure sensor, remove the cap, and the sensor should read about 1.7 volts.

**RUNNING THE EVAP MONITOR** Four tests are performed during a typical EVAP monitor. A DTC is assigned to each test:

1. **Weak vacuum test (P0440—large leak).** This test identifies gross leaks. During the monitor, the vent solenoid is closed, and the purge solenoid is duty cycled. The fuel tank pressure (FTP) should indicate a vacuum of approximately 6 to 10 in. $H_2O$.

2. **No flow during purging (P0441—no flow during purging).** This test uses the fuel tank pressure (FTP) sensor to determine that there was no change in fuel tank pressure during purging. This fault could be due to a defective purge valve or a gross leak caused by faults such as a defective or missing gas cap.

3. **Small leak test (P0442—small leak).** After the large leak test passes, the PCM checks for a small leak by keeping the vent solenoid closed and closing the purge solenoid. The system is now sealed. The PCM measures the change in FTP voltage over time.

FIGURE 25–16 The fuel level must be above 15% and below 85% before the EVAP monitor will run on most vehicles.

 **TECH TIP**

**Keep the Fuel Tank Properly Filled**

Most evaporative system monitors will not run unless the fuel level is between 15% and 85%. In other words, if a driver always runs with close to an empty tank or always tries to keep the tank full, the EVAP monitor may not run. ● **SEE FIGURE 25–16.**

## EVAP SYSTEM-RELATED DIAGNOSTIC TROUBLE CODES

| Diagnostic Trouble Code | Description | Possible Causes |
|---|---|---|
| P0440 | Evaporative system fault | • Loose gas cap <br> • Defective EVAP vent <br> • Cracked charcoal canister <br> • EVAP vent or purge vapor line problems |
| P0442 | Small leak detected | • Loose gas cap <br> • Defective EVAP vent or purge solenoid <br> • EVAP vent or purge line problems |
| P0446 | EVAP canister vent blocked | • EVAP vent or purge solenoid electrical problems <br> • Restricted EVAP canister vent line |

4. **Excess vacuum test (P0446).** This test checks for vent path restrictions. With the vent solenoid open and purge commanded, the PCM should not see excessive vacuum in the EVAP system. Typical EVAP system vacuum with the vent solenoid open is about 5 to 6 in. $H_2O$.

## SUMMARY

1. The purpose of the evaporative emission control (EVAP) system is to reduce the release of volatile organic compounds (VOCs) into the atmosphere.

2. A carbon (charcoal) canister is used to trap and hold gasoline vapors until they can be purged and run into the engine to be burned.

3. Pressures inside the EVAP system are low and are measured in inches of water (1 PSI = 28 in. $H_2O$).

4. A typical EVAP system uses a canister purge valve, which is normally closed, and a canister vent valve, which is normally open.

5. OBD-II regulation requires that the evaporative emission control system be checked for leakage and proper purge flow rates.

6. External leaks can best be located by pressurizing the fuel system with low-pressure smoke.

1. What components are used in a typical evaporative emission control system?

2. How does the computer control the purging of the vapor canister?

3. What is the difference between an enhanced and a nonenhanced evaporative control system?

4. Why is the vent valve subject to rust and corrosion?

5. What are the parameters (enable criteria) that must be met for the evaporative system monitor to run?

## CHAPTER QUIZ

1. What is the substance used in a vapor canister to absorb volatile organic compounds?
   a. Desiccant
   b. Organic absorber
   c. Pleated paper
   d. Carbon

2. Which valve(s) is (are) normally closed?
   a. Canister purge valve
   b. Canister vent valve
   c. Both canister purge and canister vent valves
   d. Neither canister purge nor canister vent valve

3. All of the following can increase the pressure in the evaporative emission control system except _____.
   a. Fuel temperature
   b. Returned fuel from the fuel-injection system
   c. Inlet fuel to the fuel pump
   d. Volatility of the fuel

4. Evaporative emission control systems operate on low pressure measured in inches of water (in. $H_2O$). One PSI is equal to how many inches of water?
   a. 1
   b. 10
   c. 18
   d. 28

5. Inadequate purge flow rate will trigger which DTC?
   a. P0440
   b. P0446
   c. P0300
   d. P0440 or P0446

6. Two technicians are discussing a state emission test. Technician A says that a vent valve that is not able to close can cause the system to fail the on-board test. Technician B says that a leaking gas cap can cause a failure of the EVAP test. Which technician is correct?
   a. Technician A only
   b. Technician B only
   c. Both Technicians A and B
   d. Neither Technician A nor B

7. Which EVAP valve is subject to rust and corrosion more than all of the others?
   a. Purge valve
   b. Vacuum control valve
   c. Vent valve
   d. Roll over check valve

8. Before an evaporative emission monitor will run, the fuel level must be where?
   a. At least 75% full
   b. Over 25%
   c. Between 15% and 85%
   d. The level of the fuel in the tank is not needed to run the monitor test

9. Technician A says that low-pressure smoke installed in the fuel system can be used to check for leaks. Technician B says that nitrogen under low pressure can be installed in the fuel system to check for leaks. Which technician is correct?
   a. Technician A only
   b. Technician B only
   c. Both Technicians A and B
   d. Neither Technician A nor B

10. A small leak is detected by the evaporative emission control system monitor that could be caused by a loose gas cap. Which DTC will likely be set?
    a. P0440
    b. P0442
    c. P0446
    d. P0440, P0441, or P0442

# chapter 26

# EXHAUST GAS RECIRCULATION SYSTEMS

**LEARNING OBJECTIVES:** After studying this chapter, the reader should be able to: • Describe the purpose and functions of exhaust gas recirculation (EGR) systems. • Explain the strategies to monitor onboard diagnostics generation II (OBD-II) exhaust gas recirculation (EGR) systems. • Understand the procedure to follow when diagnosing a defective EGR system and interpret EGR-related OBD-II diagnostic trouble codes.

**KEY TERMS:** Delta pressure feedback EGR (DPFE) sensor 331 • Detonation 328 • Digital EGR valve 330 • EGR valve position (EVP) sensor 329 • Electronic vacuum regulator valve (EVRV) 332 • Exhaust gas recirculation (EGR) 327 • Inert 327 • Linear EGR valve 330 • Nitrogen oxides ($NO_x$) 327 • Pressure feedback EGR (PFE) sensor 329

## EXHAUST GAS RECIRCULATION SYSTEMS

**INTRODUCTION** **Exhaust gas recirculation (EGR)** is an emission control system that lowers the amount of **nitrogen oxides ($NO_x$)** formed during combustion. In the presence of sunlight, $NO_x$ reacts with hydrocarbons in the atmosphere to form ozone ($O_3$) or photochemical smog, an air pollutant.

**$NO_x$ FORMATION** Nitrogen ($N_2$) and oxygen ($O_2$) molecules are separated into individual atoms of nitrogen and oxygen during the combustion process. These molecules then bond to form $NO_x$ (NO, $NO_2$). When combustion flame front temperatures exceed 2,500°F (1,370°C), $NO_x$ is formed inside the cylinder, which is then discharged into the atmosphere from the tailpipe. ● SEE FIGURE 26–1.

**CONTROLLING $NO_x$** To handle the $NO_x$ generated above 2,500°F (1,370°C), the most efficient method to meet $NO_x$ emissions without significantly affecting engine performance, fuel economy, and other exhaust emissions is to use exhaust gas recirculation (EGR). The EGR system routes small quantities, usually between 6% and 10%, of exhaust gas into the intake manifold.

Here, the exhaust gas mixes with and takes the place of some intake charge. This leaves less room for the intake charge to enter the combustion chamber. The recirculated exhaust gas is **inert** (chemically inactive) and does not enter into the combustion process. The result is a lower peak combustion temperature. When the combustion temperature is lowered, the production of oxides of nitrogen is reduced.

**FIGURE 26–1** Nitrogen oxides ($NO_x$) create a red-brown haze that often hangs over major cities.

The EGR system has some means of interconnecting the exhaust and intake manifolds. ● SEE FIGURE 26–2.

The EGR valve controls the flow of exhaust gases through the interconnecting passages:

- On V-type engines, the intake manifold crossover is used as a source of exhaust gas for the EGR system. A cast passage connects the exhaust crossover to the EGR valve. The exhaust gas is sent from the EGR valve to openings in the manifold.

- On inline-type engines, an external tube is generally used to carry exhaust gas to the EGR valve. This tube is often designed to be long so that the exhaust gas is cooled before it enters the EGR valve.

**FIGURE 26–2** When the EGR valve opens, the exhaust gases flow through the valve and into passages in the intake manifold.

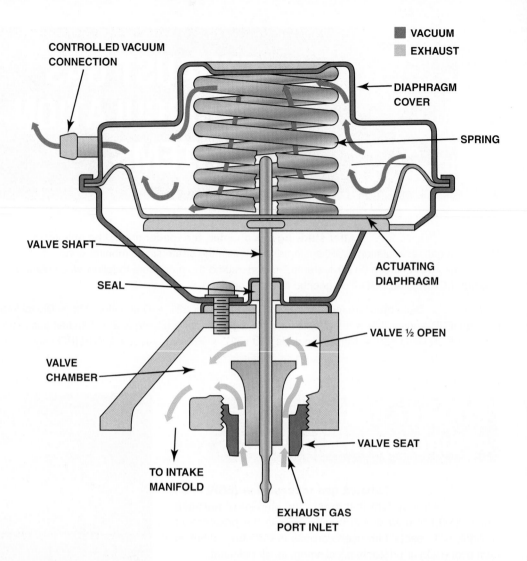

VACUUM
EXHAUST

CONTROLLED VACUUM CONNECTION

DIAPHRAGM COVER

SPRING

VALVE SHAFT

SEAL

ACTUATING DIAPHRAGM

VALVE ½ OPEN

VALVE CHAMBER

VALVE SEAT

TO INTAKE MANIFOLD

EXHAUST GAS PORT INLET

## EGR SYSTEM OPERATION
Since small amounts of exhaust are all that is needed to lower peak combustion temperatures, the orifice through which the exhaust passes is small.

EGR is usually not required during the following conditions because the combustion temperatures are low:

- During idle speed
- When the engine is cold
- At wide-open throttle (WOT) (Not allowing EGR allows the engine to provide extra power when demanded. While the $NO_x$ formation is high during these times, the overall effect of not using EGR during WOT conditions is minor.)

The level of $NO_x$ emission changes according to engine speed, temperature, and load. Many systems use a cooler to reduce the temperature of the exhaust gases before they enter the intake manifold. The cooler the exhaust gases, the more effective they are at reducing the formation of $NO_x$.

## EGR BENEFITS
In addition to lowering $NO_x$ levels, the EGR system also helps control detonation. **Detonation,** also called spark knock or ping, occurs when high pressure and heat cause the air–fuel mixture to ignite. This uncontrolled combustion can severely damage the engine.

Using the EGR system allows for greater ignition timing advance and for the advance to occur sooner without detonation problems, which increases power and efficiency.

## POSITIVE AND NEGATIVE BACK PRESSURE EGR VALVES
Some vacuum-operated EGR valves used on older engines are designed with a small valve inside that bleeds off any applied vacuum and prevents the valve from opening:

- **Positive back pressure EGR valves.** These EGR valves require a positive back pressure in the exhaust system. At low engine speeds and light engine loads, the EGR system is not needed, and the back pressure in it is also low. Without enough back pressure, the EGR valve does not open even though vacuum may be present at the EGR valve. ● **SEE FIGURE 26–3.**
- **Negative back pressure EGR valves.** On each exhaust stroke, the engine emits an exhaust "pulse." Each pulse represents a positive pressure. Behind each pulse is a small area of low pressure. Some EGR valves react to this low-pressure area by closing a small internal valve, which allows the EGR valve to be opened by vacuum.

# POSITIVE BACK PRESSURE EGR VALVE OPERATION

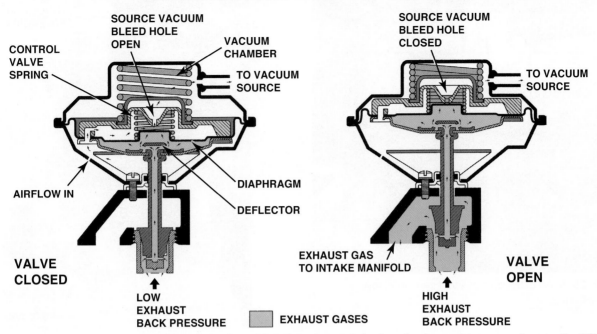

**FIGURE 26–3** Back pressure in the exhaust system is used to close the control valve, allowing engine vacuum to open the EGR valve.

The following conditions must occur before a back pressure-type vacuum-controlled EGR will operate:

1. Vacuum must be applied to the EGR valve itself. The vacuum source can be ported vacuum (above the throttle plate) or manifold vacuum (below the throttle plate) and by the computer through a solenoid valve.

2. Exhaust back pressure must be present to close an internal valve inside the EGR to allow the vacuum to move the diaphragm.

**NOTE: Installing a high-performance exhaust system could prevent a back pressure vacuum-operated EGR valve from opening. If this is occurs, excessive combustion chamber pressure may lead to severe spark knock, piston damage, or a blown head gasket.**

**COMPUTER-CONTROLLED EGR SYSTEMS** Most vehicles today use the powertrain control module (PCM) to operate the EGR system. Many PCM-controlled EGR systems have one or more solenoids controlling the EGR vacuum. The PCM controls a solenoid to shut off vacuum to the EGR valve at cold engine temperatures, idle speed, and WOT operation. If two solenoids are used, one acts as an off/on control of supply vacuum, while the second solenoid vents vacuum when EGR flow is not desired or needs to be reduced. The second solenoid is used to control a vacuum air bleed, allowing atmospheric pressure in to modulate EGR flow according to vehicle operating conditions. ● SEE FIGURE 26–4.

**EGR VALVE POSITION SENSORS** Most PCM vacuum-operated EGR systems use a sensor to indicate EGR operation.

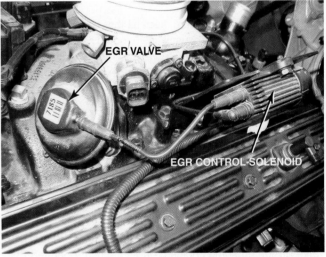

**FIGURE 26–4** Typical vacuum-operated EGR valve. The operation of the valve is controlled by the PCM by pulsing the EGR control solenoid on and off.

Onboard diagnostics generation II (OBD-II) EGR system monitors require an EGR sensor to verify that the valve opened. A linear potentiometer on the top of the EGR valve stem indicates valve position to the PCM. This is called an **EGR valve position (EVP) sensor.** Some later-model Ford EGR systems, however, use a feedback signal provided by an EGR exhaust back pressure sensor that converts the exhaust back pressure to a voltage signal. This sensor is called a **pressure feedback EGR (PFE) sensor.**

On some EGR systems, the top of the valve contains a vacuum regulator and EGR pintle-position sensor in one assembly

FIGURE 26–5 An EGR valve position sensor on top of an EGR valve.

FIGURE 26–6 Digital EGR valve as used on some older General Motors engines.

TECH TIP

**Find the Root Cause**

Excessive back pressure, such as that caused by a partially clogged exhaust system, could cause the plastic sensors on the EGR valve to melt. Always check for a restricted exhaust whenever replacing a failed EGR valve sensor.

FIGURE 26–7 A General Motors linear EGR valve.

sealed inside a nonremovable plastic cover. The pintle-position sensor provides a voltage output to the PCM, which increases as the duty cycle increases, allowing the PCM to monitor valve operation. ● **SEE FIGURE 26–5.**

**DIGITAL EGR VALVES**   General Motors used a **digital EGR valve** design on some engines. Unlike vacuum-operated EGR valves, the digital EGR valve consists of three solenoids controlled by the powertrain control module (PCM). Each solenoid controls a different size orifice in the base—small, medium, and large. The PCM controls the ground circuit of each of the solenoids individually. It can produce any of seven different flow rates, using the solenoids to open the three valves in different combinations. The digital EGR valve offers precise control and using a swivel pintle design helps prevent carbon deposit problems. ● **SEE FIGURE 26–6.**

**LINEAR EGR**   Most General Motors and many other vehicles use a **linear EGR valve** that contains a pulse-width modulated solenoid to precisely regulate exhaust gas flow and a feedback potentiometer that signals the PCM regarding the actual position of the valve. ● **SEE FIGURES 26–7 AND 26–8.**

## OBD-II EGR MONITORING STRATEGIES

**PURPOSE AND FUNCTION**   In 1996, the U.S. EPA began requiring OBD-II systems in all passenger cars and most light-duty trucks. These systems include emission system monitors that alert the driver and the technician if an emission system is malfunctioning. The OBD-II system performs this test by opening and closing the EGR valve. The PCM monitors either the oxygen or MAP sensor for a change in signal voltage. If the EGR system fails, a diagnostic trouble code (DTC) is set. If the system fails two consecutive times, the malfunction indicator lamp (MIL) is lit.

## MONITORING STRATEGIES
EGR monitoring strategies include the following:

- Some vehicle manufacturers, such as Chrysler, monitor the difference in the exhaust oxygen sensor's voltage activity as the EGR valve opens and closes. Oxygen in the exhaust decreases when the EGR valve is open and increases when the EGR valve is closed because exhaust gas is inert (contains very little oxygen) and actually displaces oxygen. The PCM sets a DTC if the sensor signal does not change.

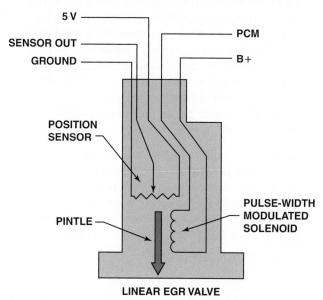

**FIGURE 26–8** The EGR valve pintle is pulse-width modulated and a three-wire potentiometer provides pintle-position information back to the PCM.

- Most Fords use an EGR monitor test sensor called a **delta pressure feedback EGR (DPFE) sensor.** This sensor measures the pressure differential between two sides of a metered orifice positioned just below the EGR valve's exhaust side. Pressure between the orifice and the EGR valve decreases when the EGR opens because it becomes exposed to the lower pressure in the intake. The DPFE sensor recognizes this pressure drop, compares it to the relatively higher pressure on the exhaust side of the orifice, and signals the value of the pressure difference to the PCM. ● **SEE FIGURE 26–9.**

- Many vehicle manufacturers use the manifold absolute pressure (MAP) sensor as the EGR monitor on some applications. After meeting the enable criteria (operating condition requirements), the EGR monitor is run. The PCM monitors the MAP sensor while it commands the EGR valve to open. The MAP sensor signal should change in response to the sudden change in manifold pressure or the fuel trim changes created by a change in the oxygen sensor voltage. If the signal value falls outside the acceptable value in the lookup table, a DTC sets. If the EGR fails on two consecutive trips, the PCM lights the MIL. ● **SEE FIGURE 26–10.**

## DIAGNOSING A DEFECTIVE EGR SYSTEM

**SYMPTOMS** If the EGR valve is not opening or the flow of the exhaust gas is restricted, then the following symptoms are likely:

- Detonation (spark knock or ping) during acceleration or during cruise (steady-speed driving)

- Excessive oxides of nitrogen ($NO_x$) exhaust emissions

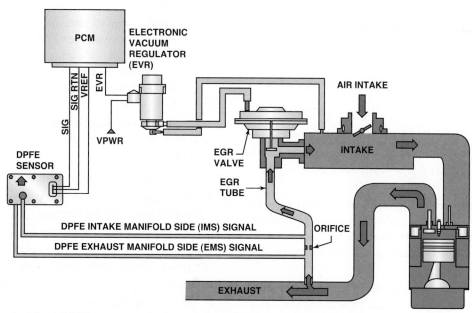

**FIGURE 26–9** A typical Ford DPFE sensor and related components.

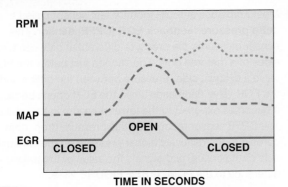

**FIGURE 26–10** An OBD-II active test. The PCM opens the EGR valve and then monitors the MAP sensor and/or engine speed (RPM) to verify that it meets acceptable values.

---

---

If the EGR valve is stuck open or partially open, then the following symptoms are likely:

- Rough idle or frequent stalling
- Poor performance/low power, especially at low engine speed

**EGR TESTING PROCEDURES** The first step in almost any diagnosis is to perform a thorough visual inspection. To check for proper operation of a vacuum-operated EGR valve, follow these steps:

**STEP 1** **Check the vacuum diaphragm of the EGR valve to see if it can hold vacuum.** Because many EGR valves require exhaust back pressure to function correctly, the engine should be running at a fast idle during this test.

**STEP 2** **Apply vacuum from a hand-operated vacuum pump and check for proper operation.** The valve itself should move when vacuum is applied, and the

---

---

engine operation should be affected. The EGR valve should be able to hold the vacuum that was applied. If the vacuum drops off, then the valve is likely to be defective.

**STEP 3** **Monitor engine vacuum drop.** Connect a vacuum gauge to an intake manifold vacuum source and monitor the engine vacuum at idle (should be 17 to 21 in. Hg at sea level). Raise the speed of the engine to 2500 RPM and note the vacuum reading (should be 17 to 21 in. Hg or higher).

Activate the EGR valve using a scan tool or vacuum pump, if vacuum controlled, and observe the vacuum gauge. The results are as follows:

- The vacuum should drop 6 to 8 in. Hg.
- If the vacuum drops less than 6 to 8 in. Hg, the valve or the EGR passages are clogged.
- If the EGR valve is able to hold vacuum but the engine is not affected when the valve is opened, then the exhaust passage(s) must be checked for restriction.

See the Tech Tip "The Snake Trick." If the EGR valve will not hold vacuum, the valve itself is likely to be defective and require replacement.

## TECH TIP

### The Snake Trick

The EGR passages on many intake manifolds become clogged with carbon, which reduces the flow of exhaust and the amount of exhaust gases in the cylinders. This reduction can cause spark knock (detonation) and increased emissions of oxides of nitrogen ($NO_x$) (especially important in areas with enhanced exhaust emissions testing).

To quickly and easily remove carbon from exhaust passages, cut an approximately 1-foot (30-cm) length from stranded wire, such as garage door guide wire or an old speedometer cable. Flare the end and place the end of the wire into the passage. Set your drill on reverse and turn it on, and the wire will pull its way through the passage, cleaning the carbon as it goes, just like a snake in a drainpipe. Some vehicles, such as Hondas, require that plugs be drilled out to gain access to the EGR passages.
● **SEE FIGURE 26–11.**

# EGR-RELATED OBD-II DIAGNOSTIC TROUBLE CODES

| Diagnostic Trouble Code | Description | Possible Causes |
|---|---|---|
| P0400 | Exhaust gas recirculation flow problems | • EGR valve<br>• EGR valve hose or electrical connection<br>• Defective PCM |
| P0401 | Exhaust gas recirculation flow insufficient | • EGR valve<br>• Clogged EGR ports or passages |
| P0402 | Exhaust gas recirculation flow excessive | • Stuck-open EGR valve<br>• Vacuum hose(s) misrouted<br>• Electrical wiring shorted |

**FIGURE 26–11** Removing the EGR passage plugs from the intake manifold on a Honda.

## SUMMARY

1. Oxides of nitrogen ($NO_x$) are formed inside the combustion chamber because of heat exceeding 2,500°F (1,370°C).

2. Recirculating 6% to 10% inert exhaust gases back into the intake system reduces peak temperature inside the combustion chamber and reduces $NO_x$ exhaust emissions.

3. EGR is usually not needed during cold engine operation, at idle speeds or during wide-open throttle conditions.

4. Vacuum-operated EGR valves are usually exhaust back pressure controlled to help match EGR flow into the intake with the load on the engine.

5. Many EGR systems use a feedback potentiometer to signal the PCM about the position of the EGR valve pintle.

6. Some EGR valves are solenoids or pulse-width modulated pintles.

7. OBD-II requires that the flow rate be tested, which can be achieved by opening the EGR valve and observing the reaction of the MAP sensor.

1. What causes the formation of oxides of nitrogen?

2. How does the use of exhaust gas reduce $NO_x$ exhaust emission?

3. How does the DPFE sensor work?

4. How does the PCM determine that the exhaust flow through the EGR system meets OBD-II regulations?

1. What causes the nitrogen and the oxygen in the air to combine and form $NO_x$?
   a. Sunlight
   b. Any spark will cause this to occur
   c. Heat above 2,500°F (1,370°C)
   d. Chemical reaction in the catalytic converter

2. Exhaust gas recirculation (EGR) is generally not needed under all the following conditions *except* _____.
   a. Idle speed
   b. Cold engine
   c. Cruise speed
   d. Wide-open throttle (WOT)

3. Technician A says that a low-restriction exhaust system could prevent a back pressure-type vacuum-controlled EGR valve from opening correctly. Technician B says restricted exhaust can cause the EGR valve position sensor to fail. Which technician is correct?
   a. Technician A only
   b. Technician B only
   c. Both Technicians A and B
   d. Neither Technician A nor B

4. EGR is used to control which exhaust emission?
   a. Unburned hydrocarbons (HC)
   b. Oxides of nitrogen ($NO_x$)
   c. Carbon monoxide (CO)
   d. Both $NO_x$ and CO

5. A typical EGR pintle-position sensor is what type of sensor?
   a. Rheostat
   b. Piezoelectric
   c. Wheatstone bridge
   d. Potentiometer

6. OBD-II regulations require that the EGR system be tested. Technician A says that the PCM can monitor the commanded position of the EGR valve to determine if it is functioning correctly. Technician B says that the PCM can open the EGR valve and monitor for a change in the MAP sensor or oxygen sensor reading to detect if the system is functioning correctly. Which technician is correct?
   a. Technician A only
   b. Technician B only
   c. Both Technicians A and B
   d. Neither Technician A nor B

7. Two technicians are discussing clogged EGR passages. Technician A says clogged EGR passages can cause excessive $NO_x$ exhaust emission. Technician B says that clogged EGR passages can cause the engine to ping (spark knock or detonation). Which technician is correct?
   a. Technician A only
   b. Technician B only
   c. Both Technicians A and B
   d. Neither Technician A nor B

8. An EGR valve that is partially stuck open would *most likely* cause what condition?
   a. Rough idle/stalling
   b. Excessive $NO_x$ exhaust emissions
   c. Ping (spark knock or detonation)
   d. Missing at highway speed

9. When testing an EGR system for proper operation using a vacuum gauge, how much should the vacuum drop when the EGR is commanded on by a scan tool?
   a. 1 to 2 in. Hg
   b. 3 to 5 in. Hg
   c. 6 to 8 in. Hg
   d. 8 to 10 in. Hg

10. A P0401 DTC (exhaust gas recirculation flow insufficient) is being discussed. Technician A says that a defective EGR valve could be the cause. Technician B says that clogged EGR passages could be the cause. Which technician is correct?
    a. Technician A only
    b. Technician B only
    c. Both Technicians A and B
    d. Neither Technician A nor B

# chapter 27

# POSITIVE CRANKCASE VENTILATION AND SECONDARY AIR-INJECTION SYSTEMS

**LEARNING OBJECTIVES:** **After studying this chapter, the reader should be able to:** • Understand the purpose and function of the positive crankcase ventilation (PCV) system and the procedure to diagnose it. • Explain the purpose and function of the secondary air-injection (SAI) system and how to diagnose faults in the system.

**KEY TERMS:** Air-injection reaction (AIR) 339 • Blowby 335 • Check valve 339 • Positive crankcase ventilation (PCV) 335 • Secondary air injection (SAI) 339 • Smog pump 339 • Thermactor pump 339

## CRANKCASE VENTILATION

**PURPOSE AND FUNCTION** The problem of crankcase ventilation has existed since the beginning of the automobile because no piston ring, new or old, can provide a perfect seal between the piston and the cylinder wall. When an engine is running, the pressure of combustion forces the piston downward. This same pressure also forces gases and unburned fuel from the combustion chamber, past the piston rings, and into the crankcase. This process of gases leaking past the rings is called **blowby,** and the gases form crankcase vapors.

These combustion by-products, particularly unburned hydrocarbons (HC) caused by blowby, must be ventilated from the crankcase. However, the crankcase cannot be vented directly to the atmosphere because the hydrocarbon vapors add to air pollution. **Positive crankcase ventilation (PCV)** systems were developed to ventilate the crankcase and recirculate the vapors to the engine's induction system so they can be burned in the cylinders. PCV systems help reduce HC and CO emissions.

All systems use the following:

1. PCV valve or calibrated orifice, or orifice and separator
2. PCV inlet air filter plus all connecting hoses

● **SEE FIGURE 27–1.**

An oil/vapor or oil/water separator is used in some systems instead of a valve or orifice, particularly with turbocharged and fuel-injected engines. The oil/vapor separator lets oil condense and drain back into the crankcase. The oil/water separator accumulates moisture and prevents it from freezing during cold engine starts.

**FIGURE 27–1** A PCV valve in a cutaway valve cover, showing the baffles that prevent liquid oil from being drawn into the intake manifold.

The air for the PCV system is drawn after the air cleaner filter, which acts as a PCV filter.

**NOTE: Some older designs drew from the dirty side of the air cleaner, where a separate crankcase ventilation filter was used.**

**PCV VALVES** The PCV valve in most systems is a one-way valve containing a spring-operated plunger that controls valve flow rate. ● **SEE FIGURE 27–2.**

Flow rate is established for each engine, and a valve for a different engine should not be substituted. The flow rate is

THIS END OF THE PCV VALVE
IS SUBJECT TO CRANKCASE
PRESSURE THAT TENDS TO
CLOSE THE VALVE.

THIS END IS SUBJECT
TO INTAKE MANIFOLD
VACUUM THAT TENDS
TO CLOSE THE VALVE.

THE SPRING FORCE OPERATES TO
OPEN THE VALVE TO MANIFOLD
VACUUM AND CRANKCASE PRESSURE.

**FIGURE 27–2** Spring force, crankcase pressure, and intake manifold vacuum work together to regulate the flow rate through the PCV valve.

AT IDLE AND LOW SPEED,
MANIFOLD VACUUM PULLS
THE VALVE TOWARD THE
RESTRICTED POSITION.

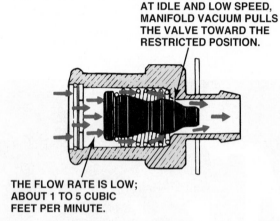

THE FLOW RATE IS LOW;
ABOUT 1 TO 5 CUBIC
FEET PER MINUTE.

**FIGURE 27–3** Air flows through the PCV valve during idle, cruising, and light-load conditions.

determined by the size of the plunger and the holes inside the valve. PCV valves usually are located in the valve cover or intake manifold.

The PCV valve regulates airflow through the crankcase under all driving conditions and speeds. When manifold vacuum is high (at idle, cruising, and light-load operation), the PCV valve restricts the airflow to maintain a balanced air–fuel ratio. ● **SEE FIGURE 27–3.**

It also prevents high intake manifold vacuum from pulling oil out of the crankcase and into the intake manifold. Under high speed or heavy loads, the valve opens and allows maximum airflow. ● **SEE FIGURE 27–4.**

If the engine backfires, the valve will close instantly to prevent a crankcase explosion. ● **SEE FIGURE 27–5.**

**ORIFICE-CONTROLLED SYSTEMS** The closed PCV system used on some 4-cylinder engines contains a calibrated orifice instead of a PCV valve. The orifice may be located in the valve cover or intake manifold or in a hose connected between the valve cover, air cleaner, and intake manifold.

AT HIGHER SPEED OR IN A HEAVY
LOAD CONDITION, MANIFOLD
VACUUM DROPS. THE SPRING MOVES
THE VALVE OPEN.

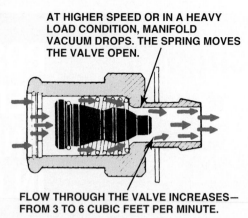

FLOW THROUGH THE VALVE INCREASES—
FROM 3 TO 6 CUBIC FEET PER MINUTE.

**FIGURE 27–4** Air flows through the PCV valve during acceleration and when the engine is under a heavy load.

IF THE ENGINE BACKFIRES DURING
CRANKING, IT CAUSES A HIGH
PRESSURE IN THE INTAKE MANIFOLD.

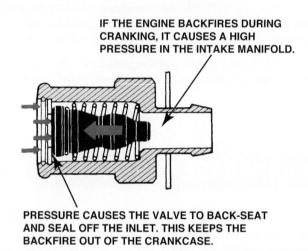

PRESSURE CAUSES THE VALVE TO BACK-SEAT
AND SEAL OFF THE INLET. THIS KEEPS THE
BACKFIRE OUT OF THE CRANKCASE.

**FIGURE 27–5** PCV valve operation in the event of a backfire.

While most orifice flow control systems work the same as a PCV valve system, they may not use fresh air scavenging of the crankcase. Crankcase vapors are drawn into the intake manifold in calibrated amounts depending on manifold pressure and the orifice size. If vapor availability is low, as during idle, air is drawn in with the vapors. During off-idle operation, excess vapors are sent to the air cleaner.

At idle, PCV flow is controlled by a 0.05 inch (1.3-mm) orifice. As the engine moves off idle, ported vacuum pulls a spring-loaded valve off of its seat, allowing PCV flow to pass through a 0.09 inch (2.3 mm) orifice.

**SEPARATOR SYSTEMS** Turbocharged and many fuel-injected engines use an oil/vapor or oil/water separator and a calibrated orifice instead of a PCV valve. In the most common applications, the air intake throttle body acts as the source for crankcase ventilation vacuum, and a calibrated orifice acts as the metering device.

## PCV SYSTEM DIAGNOSIS

**SYMPTOMS** If the PCV valve or orifice is not clogged, intake air flows freely and the PCV system functions properly. Engine design includes the air and vapor flow as a calibrated part of

**REAL WORLD FIX**

### The Whistling Engine

An older vehicle was being diagnosed for a whistling sound whenever the engine was running, especially at idle. It was finally discovered that the breather in the valve cover was plugged and caused high vacuum in the crankcase. The engine was sucking air from what was likely the rear main seal lip, making the "whistle" noise. After replacing the breather and PCV, the noise stopped.

 **TECH TIP**

### Check for Oil Leaks with the Engine Off

The owner of an older vehicle equipped with a V-6 engine complained to his technician that he smelled burning oil, but only *after* shutting off the engine. The technician found that the rocker cover gaskets were leaking. But why did the owner only notice the smell of hot oil when the engine was shut off? Because of the positive crankcase ventilation (PCV) system, engine vacuum tends to draw oil away from gasket surfaces. When the engine stops, however, engine vacuum disappears, and the oil remaining in the upper regions of the engine will tend to flow down and out through any opening. Therefore, a good technician should check an engine for oil leaks not only with the engine running but also shortly after shutdown.

 **REAL WORLD FIX**

### The Oil-Burning Chevrolet Astro Van

An automotive instructor was driving a Chevrolet Astro van to Fairbanks, Alaska, in January. It was cold, around −32°F (−36°C). As he pulled into Fairbanks and stopped at a traffic light, he smelled burning oil. He thought it was the vehicle ahead of him because it was an older model and in poor condition. However, when he stopped at the hotel he still smelled burning oil. He looked under the van and discovered a large pool of oil. After checking the oil and finding very little left, he called a local shop and was told to bring it in. The technician looked over the situation and said, "You need to put some cardboard across the grill to stop the PCV valve from freezing up." Apparently the PCV valve froze, which then caused the normal blowby gases to force several quarts out the dipstick tube. After installing the cardboard, the instructor had no further problems.

**CAUTION: Do not cover the radiator when driving unless under severe cold conditions and carefully watch the coolant temperature to avoid overheating the engine.**

the air–fuel mixture. In fact, some engines receive as much as 30% of the idle air through the PCV system. For this reason, a flow problem in the PCV system results in driveability problems.

A blocked or plugged PCV system may cause the following to occur:

- Rough or unstable idle
- Excessive oil consumption
- Oil in the air filter housing
- Oil leaks due to excessive crankcase pressure

Before attempting expensive engine repairs, check the condition of the PCV system.

**PCV SYSTEM PERFORMANCE CHECK** A properly operating positive crankcase ventilation system should be able to draw vapors from the crankcase and into the intake manifold. If the pipes, hoses, and PCV valve itself are not restricted, vacuum is applied to the crankcase. A slight vacuum is created in the crankcase (usually less than 1 in. Hg if measured at the dipstick) and is also applied to other areas of the engine. Oil drainback holes provide a path for oil to drain back into the oil pan. These holes also allow crankcase vacuum to be applied under the rocker covers

and in the valley area of most V-type engines. Several methods can be used to test a PCV system.

**RATTLE TEST** The rattle test is performed by simply removing the PCV valve and giving it a shake:

- If the PCV valve does *not* rattle, it is definitely defective and must be replaced.
- If the PCV valve *does* rattle, it does not necessarily mean that the PCV valve is good. All PCV valves contain springs that can become weaker with age and with heating and cooling cycles. Replace any PCV valve with the *exact* replacement according to the vehicle manufacturer's recommended intervals.

**THE 3 × 5 CARD TEST** Remove the oil-fill cap (where oil is added to the engine) and start the engine.

**NOTE: Use care on some overhead camshaft engines. With the engine running, oil may be sprayed from the open oil-fill opening.**

Hold a 3 × 5 card over the opening (a dollar bill or any other piece of paper can be used for this test):

- If the PCV system, including the valve and hoses, is functioning correctly, the card should be held down on the oil-fill opening by the slight vacuum inside the crankcase.
- If the card will not stay, carefully inspect the PCV valve, hose(s), and manifold vacuum port for carbon buildup (restriction). Clean or replace as necessary.

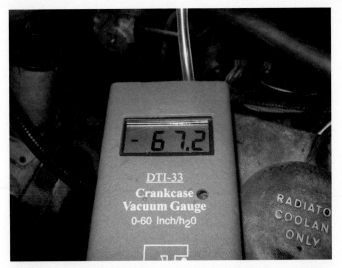

FIGURE 27–6 Using a gauge that measures vacuum in units of inches of water to test the vacuum at the dipstick tube, being sure that the PCV system is capable of drawing a vacuum on the crankcase (28 in. $H_2O$ = 1 PSI, or about 2 in. Hg of vacuum).

RIVET HOLDING
PCV VALVE
RETAINER

FIGURE 27–7 Most PCV valves used on newer vehicles are secured with fasteners, making it more difficult to disconnect and thereby less likely to increase emissions.

**NOTE: On some 4-cylinder engines, the 3 × 5 card may vibrate on the oil-fill opening when the engine is running at idle speed. This is normal because of the time intervals between intake strokes on a 4-cylinder engine.**

**SNAP-BACK TEST** The proper operation of the PCV valve can be checked by placing a finger over the inlet hole in the valve when the engine is running and removing the finger rapidly. Repeat several times. The valve should "snap back." If the valve does not snap back, replace the valve.

**CRANKCASE VACUUM TEST** Sometimes the PCV system can be checked by testing for a weak vacuum at the oil dipstick tube using an inches-of-water manometer or gauge as follows:

**STEP 1** Remove the oil-fill cap or vent PCV opening and cover the opening.

**STEP 2** Remove the oil dipstick (oil level indicator).

**STEP 3** Connect a water manometer or gauge to the dipstick tube.

**STEP 4** Start the engine and observe the gauge at idle and at 2500 RPM. ● SEE FIGURE 27–6.

The gauge should show some vacuum, especially at 2500 RPM. If not, carefully inspect the PCV system for blockages or other faults.

**PCV MONITOR** Starting with 2004 and newer vehicles, all vehicle PCMs monitor the PCV system for proper operation as part of the OBD-II system. The PCV monitor will fail if the PCM detects an opening between the crankcase and the PCV valve or between the PCV valve and the intake manifold. ● SEE FIGURE 27–7.

**FREQUENTLY ASKED QUESTION**

**What Are the Wires for at the PCV Valve?**

Ford uses an electric heater to prevent ice from forming inside the PCV valve and causing blockage.

Water is a by-product of combustion, and resulting moisture can freeze when the outside air temperature is low. General Motors and others clip a heater hose to the PCV hose to provide the heat needed to prevent an ice blockage.

# PCV-RELATED DIAGNOSTIC TROUBLE CODES

| Diagnostic Trouble Code | Description | Possible Causes |
|---|---|---|
| P0101 | MAF or airflow circuit range problem | • Defective PCV valve or hose/ connections or MAF circuit fault |
| P0505 | Idle control system problem | • Defective PCV valve or hose/ connections |

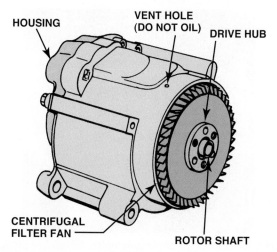

FIGURE 27–8 A typical belt-driven AIR pump. Air enters through the revolving fins behind the drive pulley. The fins act as an air filter because dirt is heavier than air, and therefore the dirt is deflected off of the fins at the same time air is being drawn into the pump.

# SECONDARY AIR-INJECTION SYSTEM

**PURPOSE AND FUNCTION** The **secondary air-injection (SAI)** system provides the air necessary for the oxidizing process either at the exhaust manifold or inside the catalytic converter.

NOTE: This system is commonly called **AIR, meaning air-injection reaction. Therefore, an AIR pump does pump air.**

**PARTS AND OPERATION** The SAI pump, also called an AIR pump, a **smog pump,** or **thermactor pump,** is mounted at the front of the engine and can be driven by a belt from the crankshaft pulley. It pulls fresh air in through an external filter and pumps the air under slight pressure to each exhaust port through connecting hoses or a manifold. The typical SAI system includes the following components:

- A belt-driven pump with inlet air filter (older models) (● SEE FIGURE 27–8.)
- An electrically driven air pump (newer models)
- One or more air distribution manifolds and nozzles
- One or more exhaust check valves
- Connecting hoses for air distribution
- Air management valves and solenoids on all newer applications

With the introduction of $NO_x$ reduction converters (also called dual-bed, three-way converters, or TWC), the output of the SAI pump is sent to the center of the converter, where the extra air can help oxidize unburned hydrocarbons (HC)

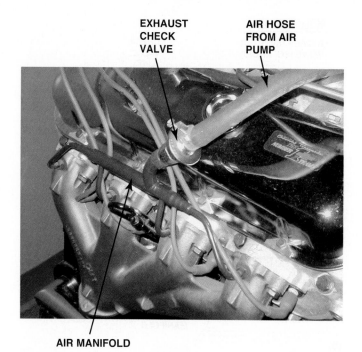

FIGURE 27–9 The external air manifold and exhaust check valve on a restored muscle car engine.

and carbon monoxide (CO) into water vapor ($H_2O$) and carbon dioxide ($CO_2$).

The computer controls the airflow from the pump by switching on and off various solenoid valves.

**AIR DISTRIBUTION MANIFOLDS AND NOZZLES** The secondary air-injection system sends air from the pump to a nozzle installed near each exhaust port in the cylinder head. This provides equal air injection for the exhaust from each cylinder and makes it available at a point in the system where exhaust gases are the hottest.

Air is delivered to the exhaust system in one of two ways:

1. An external air manifold, or manifolds, distributes the air through injection tubes with stainless-steel nozzles. The nozzles are threaded into the cylinder heads or exhaust manifolds close to each exhaust valve. This method is used primarily with smaller engines.

2. An internal air manifold distributes the air to the exhaust ports near each exhaust valve through passages cast in the cylinder head or the exhaust manifold. This method is used mainly with larger engines.

**EXHAUST CHECK VALVES** All air-injection systems use one or more one-way check valves to protect the air pump and other components from reverse exhaust flow. A **check valve** contains a spring-type metallic disc or reed that closes under exhaust back pressure. Check valves are located between the air manifold and the switching valve(s). If exhaust pressure exceeds injection pressure or if the air pump fails, the check valve spring closes the valve to prevent reverse exhaust flow. ● SEE FIGURE 27–9.

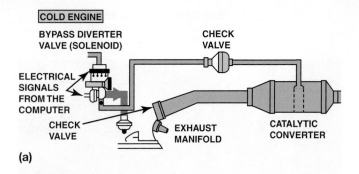

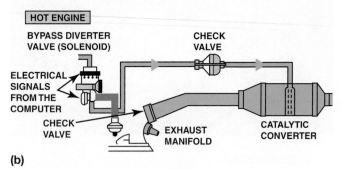

**FIGURE 27–10** (a) When the engine is cold and before the oxygen sensor is hot enough to achieve closed loop, the airflow from the air pump is directed to the exhaust manifold(s) through the one-way check valves, which keep the exhaust gases from entering the switching solenoids and the pump itself. (b) When the engine achieves closed loop, the air is directed to the catalytic converter.

**NOTE: These check valves commonly fail, resulting in excessive exhaust emissions (CO, especially). When the check valve fails, hot exhaust can travel up to and destroy the switching valve(s) and air pump itself.**

**BELT-DRIVEN AIR PUMPS**  The belt-driven air pump uses a centrifugal filter just behind the drive pulley. As the pump rotates, underhood air is drawn into the pump and slightly compressed. The system uses either vacuum- or solenoid-controlled diverter valves to air directed to the following:

- The exhaust manifold, when the engine is cold to help oxidize carbon monoxide (CO) and unburned hydrocarbons (HC) into carbon dioxide ($CO_2$) and water vapor ($H_2O$)
- The catalytic converter when the engine is warm on some models to help provide the extra oxygen needed for the efficient conversion of CO and HC into $CO_2$ and $H_2O$
- The air cleaner, during deceleration or wide-open throttle (WOT) engine operation ● **SEE FIGURE 27–10.**

**ELECTRIC MOTOR–DRIVEN AIR PUMPS**  This style of pump is generally used only during cold engine operation and is computer controlled. The secondary air-injection (SAI) system helps reduce hydrocarbons (HC) and carbon monoxide (CO).

**FIGURE 27–11**  A typical electric motor–driven SAI pump. This unit is on a Chevrolet Corvette and only works when the engine is cold.

The air flowing to the exhaust manifold also helps to warm the three-way catalytic converters quickly on engine start-up so conversion of exhaust gases may occur sooner:

- The SAI pump solenoids are controlled by the PCM. The PCM turns on the SAI pump by providing the ground to complete the circuit, which energizes the SAI pump solenoid relay. When air to the exhaust ports is desired, the PCM energizes the relay in order to turn on the solenoid and the SAI pump. ● **SEE FIGURE 27–11.**
- The PCM turns on the SAI pump during start-up any time the engine coolant temperature is above 32°F (0°C). A typical electric SAI pump operates for a maximum of four minutes or until the system enters closed-loop operation.

## SECONDARY AIR-INJECTION SYSTEM DIAGNOSIS

**SYMPTOMS**  The air pump system should be inspected if an exhaust emissions test failure occurs. In severe cases, the exhaust will enter the air cleaner assembly, resulting in a horribly running engine because the extra exhaust displaces the oxygen needed for proper combustion. With the engine running, check for normal operation. ● **SEE CHART 27–1.**

**VISUAL INSPECTION**  Carefully inspect all parts of the SAI system, including the following:

- Hoses and pipes (Any that have holes and leak air or exhaust require replacement.)
- Check valve(s), when a pump has become inoperative
- Exhaust gases, which could have gotten past the check valve and damaged the pump (Look for signs of overheated areas upstream from the check valves. In severe cases, the exhaust can enter the air cleaner assembly,

| ENGINE OPERATION | NORMAL OPERATION OF A TYPICAL SAI SYSTEM |
|---|---|
| Cold engine (open-loop operation) | Air is diverted to the exhaust manifold(s) or cylinder head |
| Warm engine (closed-loop operation) | Air is diverted to the catalytic converter |
| Deceleration | Air is diverted to the air cleaner assembly |
| Wide-open throttle | Air is diverted to the air cleaner assembly |

**CHART 27–1**

Typical SAI system operation showing location of airflow from the pump.

destroying the air filter and greatly reducing engine power.)

■ Drive belt on an engine-driven pump, for wear and proper tension (If the belt is worn or damaged, check that the AIR pump rotates.)

**FOUR-GAS EXHAUST ANALYSIS**  A SAI system can be easily tested using an exhaust gas analyzer and the following steps:

1. Start the engine and allow it to run until normal operating temperature is achieved.

2. Connect the analyzer probe to the tailpipe and observe the exhaust readings for hydrocarbons (HC) and carbon monoxide (CO).

3. Using the appropriate pinch-off pliers, shut off the airflow from the SAI system. Observe the HC and CO readings. If the SAI system is working correctly, the HC and CO levels should increase when the SAI system is shut off.

4. Record the $O_2$ reading with the SAI system still inoperative. Unclamp the pliers and watch the $O_2$ readings. If the system is functioning correctly, the $O_2$ level should increase by 1% to 4%.

# SAI-RELATED DIAGNOSTIC TROUBLE CODE

| Diagnostic Trouble Code | Description | Possible Causes |
|---|---|---|
| P0410 | SAI solenoid circuit fault | • Defective SAI solenoid<br>• Loose or corroded electrical connections<br>• Loose, missing, or defective rubber hose(s) |

# SUMMARY

1. Positive crankcase ventilation (PCV) systems use a valve or a fixed orifice to control the crankcase vapors from the crankcase back into the intake system.

2. A PCV valve regulates the flow of crankcase vapors depending on engine vacuum and seals the crankcase vent in the event of a backfire.

3. As much as 30% of the air needed by the engine at idle speed flows through the PCV system.

4. PCV tests include the rattle test, card test, snap-back test, and crankcase vacuum test.

5. The AIR system forces air at low pressure into the exhaust to reduce CO and HC exhaust emissions.

6. Exhaust check valves are used between the AIR pump and the exhaust manifold to prevent exhaust gases from flowing into and damaging the AIR pump and valves.

# REVIEW QUESTIONS

1. What exhaust emissions do the PCV valve and the SAI system control?

2. How does a PCV valve work?

3. What does the abbreviation PCV mean?

4. What does the abbreviation AIR mean?

1. The PCV system controls which exhaust emission(s)?
   a. HC
   b. CO
   c. $NO_x$
   d. Both HC and CO

2. How much of the air needed by the engine flows through the PCV system when the engine is at idle speed?
   a. 1% to 3%
   b. 5% to 10%
   c. 10% to 20%
   d. Up to 30%

3. Technician A says that if the PCV valve was defective or clogged, the engine could idle rough. Technician B says that the engine may stall. Which technician is correct?
   a. Technician A only
   b. Technician B only
   c. Both Technicians A and B
   d. Neither Technician A nor B

4. Technician A says that if a PCV valve rattles, then it is okay and does not need to be replaced. Technician B says that if a PCV valve does not rattle, it should be replaced. Which technician is correct?
   a. Technician A only
   b. Technician B only
   c. Both Technicians A and B
   d. Neither Technician A nor B

5. Technician A says that the PCV system should create a slight pressure in the crankcase at idle. Technician B says that the PCV system should create a slight vacuum in the crankcase at 2500 RPM. Which technician is correct?
   a. Technician A only
   b. Technician B only
   c. Both Technicians A and B
   d. Neither Technician A nor B

6. The SAI system is used to reduce which exhaust emission(s)?
   a. HC
   b. CO
   c. $NO_x$
   d. Both HC and CO

7. Two technicians are discussing exhaust check valves used in SAI systems. Technician A says that they are used to prevent the output from the SAI pump from entering the intake manifold. Technician B says the check valves are used to keep the exhaust from entering the AIR pump. Which technician is correct?
   a. Technician A only
   b. Technician B only
   c. Both Technicians A and B
   d. Neither Technician A nor B

8. Where is the output of the AIR pump directed when the engine is cold?
   a. Exhaust manifold
   b. Catalytic converter
   c. Air cleaner assembly
   d. To the atmosphere

9. The switching valves on the AIR pump have failed several times. Technician A says that a defective exhaust check valve could be the cause. Technician B says that a leaking exhaust system at the muffler could be the cause. Which technician is correct?
   a. Technician A only
   b. Technician B only
   c. Both Technicians A and B
   d. Neither Technician A nor B

10. When checking for the proper operation of the AIR system using an exhaust gas analyzer, how much should the oxygen ($O_2$) levels increase when the pump is allowed to function?
    a. 1% to 4%
    b. 5% to 10%
    c. 10% to 20%
    d. Up to 30%

# chapter
# 28

# CATALYTIC CONVERTERS

**LEARNING OBJECTIVES:** **After studying this chapter, the reader should be able to:** • Explain the purpose, function, construction, operation, and performance of catalytic converters. • Describe how to diagnose and replace catalytic converters.

**KEY TERMS:** Catalysts 343 • Catalytic converter 343 • Cerium 344 • Light-off temperature 344 • Light-off converter (LOC) 345 • Oxygen storage capacity (OSC) 345 • Palladium 344 • Platinum 344 • Preconverter 345 • Pup (mini) converter 345 • Rhodium 344 • Tap test 346 • Three-way converter (TWC) 344 • Washcoat 343

## CATALYTIC CONVERTERS

**PURPOSE AND FUNCTION**   A **catalytic converter** is an aftertreatment device used to reduce exhaust emissions outside of the engine. The catalytic converter uses a *catalyst:*

- A **catalyst** is a chemical that helps start a chemical reaction but does not enter into the chemical reaction.

- The catalyst materials on the surface of the material inside the converter help create a chemical reaction.

- The chemical reaction changes harmful exhaust emissions into nonharmful exhaust emissions.

- The *converter* therefore converts harmful exhaust gases into water vapor ($H_2O$) and carbon dioxide ($CO_2$).

This device is installed in the exhaust system between the exhaust manifold and the muffler and usually is positioned beneath the passenger compartment. The location of the converter is important since as much of the exhaust heat as possible must be retained for effective operation. The nearer it is to the engine, the better. ● **SEE FIGURE 28–1.**

**CATALYTIC CONVERTER CONSTRUCTION**   Most catalytic converters are constructed of a ceramic material in a honeycomb shape with square openings for the exhaust gases:

- There are approximately 400 openings per square inch (62 per sq. cm), and the wall thickness is about 0.006 inch (1.5 mm).

- The substrate is then coated with a porous aluminum material called the **washcoat,** which makes the surface rough.

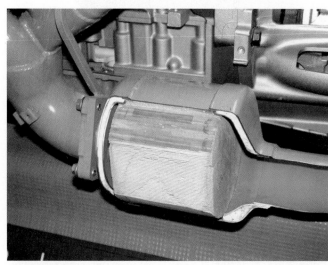

**FIGURE 28–1**  Most catalytic converters are located as close to the exhaust manifold as possible, as seen in this display of a Chevrolet Corvette.

- The catalytic materials are then applied on top of the washcoat. The substrate is contained within a round or oval shell made by welding together two stamped pieces of aluminum or stainless steel. ● **SEE FIGURE 28–2.**

The ceramic substrate in monolithic converters is not restrictive; however, the converter can be physically broken if exposed to shock or severe jolts. Monolithic converters can be serviced only as a unit.

An exhaust pipe is connected to the manifold or header to carry gases through a catalytic converter and then to the muffler or silencer. V-type engines can use dual converters or route the exhaust into one catalytic converter by using a Y-exhaust pipe.

## CATALYTIC CONVERTER OPERATION
The converter substrate contains small amounts of **rhodium, palladium,** and **platinum.** These elements act as catalysts. As mentioned, a catalyst is an element that starts a chemical reaction without becoming a part of, or being consumed in, the process. In a **three-way** (catalytic) **converter (TWC)** all three exhaust emissions ($NO_x$, HC, and CO) are converted to carbon dioxide ($CO_2$) and water ($H_2O$). As the exhaust gas passes through the catalyst, oxides of nitrogen ($NO_x$) are chemically reduced (i.e., nitrogen and oxygen are separated) in the first section of the catalytic converter. In the second section of the catalytic converter, most of the hydrocarbons and carbon monoxide remaining in the exhaust gas are oxidized to form harmless carbon dioxide ($CO_2$) and water vapor ($H_2O$). ● **SEE FIGURE 28–3.**

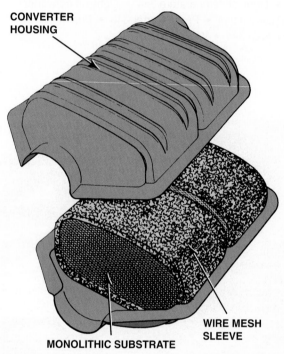

**FIGURE 28–2** A typical catalytic converter with a monolithic substrate.

CONVERTER HOUSING

WIRE MESH SLEEVE

MONOLITHIC SUBSTRATE

Since the early 1990s, many converters also contain **cerium,** an element that can store oxygen. The purpose of the cerium is to provide oxygen to the oxidation bed of the converter when the exhaust is rich and lacks enough oxygen for proper oxidation. When the exhaust is lean, the cerium absorbs the extra oxygen. For the most efficient operation, the converter should have a 14.7:1 air–fuel ratio but can use a mixture that varies slightly:

- A rich exhaust is required for reduction—stripping the oxygen ($O_2$) from the nitrogen in $NO_x$.
- A lean exhaust is required to provide the oxygen necessary to oxidize HC and CO (combining oxygen with HC and CO to form $H_2O$ and $CO_2$).

If the catalytic converter is not functioning correctly, ensure that the air–fuel mixture being supplied to the engine is correct and that the ignition system is free of defects.

## CONVERTER LIGHT-OFF TEMPERATURE
The catalytic converter does not work when cold, so it must be heated to its **light-off temperature** of close to 500°F (260°C) before it starts working at 50% effectiveness. When fully effective, the converter reaches a temperature range of 900°F to 1,600°F (482°C to 871°C). In spite of the intense heat, however, catalytic reactions do not generate a flame associated with a simple burning reaction. Because of the extreme heat (almost as hot as combustion chamber temperatures), a converter remains hot long after the engine is shut off. Most vehicles use a series of heat shields to protect the passenger compartment and other parts of the chassis from excessive heat. Vehicles have been known to start fires because of the hot converter causing tall grass or dry leaves beneath the just-parked vehicle to ignite, especially if the engine is idling. This is most likely to occur if the heat shields have been removed from the converter.

## CONVERTER USAGE
A catalytic converter must be located as close as possible to the exhaust manifold to work effectively. The farther back the converter is positioned in the exhaust system, the more the exhaust gases cool before they

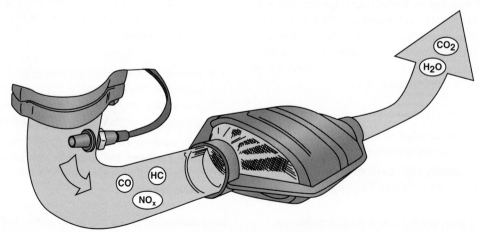

**FIGURE 28–3** The three-way catalytic converter first separates the $NO_x$ into nitrogen and oxygen and then converts the HC and CO into harmless water ($H_2O$) and carbon dioxide ($CO_2$). The nitrogen (N) passes through the converter and exits the tailpipe and enters the atmosphere which is about 78% nitrogen.

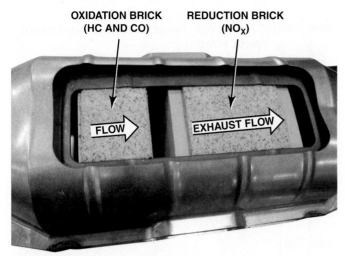

**FIGURE 28–4** The small oxidation section of the converter helps build heat for the reduction section to reduce $NO_x$ emissions in the rear brick on most newer vehicles.

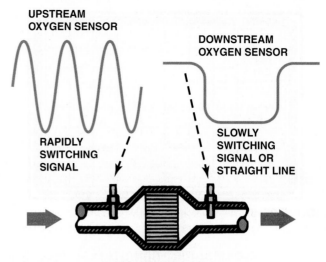

**FIGURE 28–5** The OBD-II catalytic converter monitor compares the signals of upstream and downstream oxygen sensors to determine converter efficiency.

reach the converter. Since positioning in the exhaust system affects the oxidation process, vehicle manufacturers that use only an oxidation converter generally locate it underneath the front of the passenger compartment.

Some vehicles have used a small, quick heating oxidation converter called a **preconverter** or a **pup (mini) converter** that connects directly to the exhaust manifold outlet. These have a small catalyst surface area close to the engine that heats up rapidly to start the oxidation process more quickly during cold engine warm-up. For this reason, they were often called **light-off converters (LOCs).** The larger main converter, under the passenger compartment, completes the oxidation reaction started in the LOC.

Most older vehicles used a catalytic converter that had the reduction section first to separate the oxygen from the nitrogen in NOx. The oxygen released during this action helps provide extra oxygen to help oxidize the HC and CO into harmless water ($H_2O$) and carbon dioxide ($CO_2$). However, since 2004, emission standards for oxides of nitrogen are stricter, and therefore a larger reduction section is often needed. Therefore, the reduction section is now often after the oxidation section, which is the opposite of the way it was in converters for older models.
● **SEE FIGURE 28–4.**

### OBD-II CATALYTIC CONVERTER PERFORMANCE
The PCM determines if the catalytic converter is ready for testing based on the following conditions, which may vary by vehicle make, model, and year:

- Closed-loop status achieved
- IAT sensor temperature higher than 32°F (0°C)
- ECT sensor temperature higher than 165°F (18°C)
- MAF sensor input from 15 to 32 g/sec
- Engine load less than 65% and steady
- Engine speed less than 4000 RPM
- All of the above conditions met for at least four minutes

These factors are the enable criteria that must be achieved before the OBD-II catalyst monitor will run.

## OBD-II CATALYTIC CONVERTER PERFORMANCE

With OBD-II-equipped vehicles, catalytic converter performance is monitored by a heated oxygen sensor ($HO_2S$) both before and after the converter. ● **SEE FIGURE 28–5.**

The converters used on these vehicles have **oxygen storage capacity (OSC),** due mostly to the cerium coating in the catalyst rather than the precious metals used. When the three-way converter (TWC) is operating as it should, the post-converter $HO_2S$ is far less active than the preconverter sensor. The converter stores, then releases, the oxygen during normal reduction and oxidation of the exhaust gases, smoothing out the variations in oxygen being released.

Where a cycling sensor voltage output is expected before the converter, because of the converter action, the postconverter $HO_2S$ should read a steady signal with little fluctuation.
● **SEE FIGURE 28–6.**

**NOTE: Because of more demanding exhaust emission standards for oxides of nitrogen ($NO_x$) starting in 2004, the reduction part of the converter is now usually located downstream from the oxidation section. This is opposite of the way an older style converter was constructed. With better fuel injection systems and ignition systems, the amount of CO and HC exhaust emissions is lower than in the past. Using the oxidation section first creates heat to help the reduction section better control oxides of nitrogen emissions. The rear "brick" or reduction section is also larger than the oxidation section.**

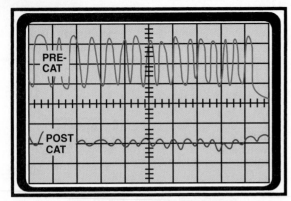

**FIGURE 28–6** The waveform of a downstream O2S sensor from a properly functioning converter shows little, if any, activity.

**CONVERTER-DAMAGING CONDITIONS** Since converters have no moving parts, they require no periodic service. Under federal law, catalyst effectiveness is warranted for 80,000 miles or eight years.

The three main causes of premature converter failure are the following:

- **Contamination.** Substances that can destroy the converter include exhaust that contains excess engine oil, antifreeze, sulfur (from poor fuel), and various other chemical substances.

- **Excessive temperatures.** Although a converter operates at high temperatures, it can be destroyed by excessive temperatures. This most often occurs either when too much unburned fuel enters the converter, or with excessively lean mixtures. Excessive temperatures may be caused by long idling periods on some vehicles since more heat develops at those times than when driving at normal highway speeds. Severe high temperatures can cause the converter to melt down, leading to the internal parts breaking apart and either clogging the converter or moving downstream to plug the muffler. In either case, the restricted exhaust flow severely reduces engine power.

- **Improper air–fuel mixtures.** Rich mixtures or raw fuel in the exhaust can be caused by engine misfiring or an excessively rich air–fuel mixture resulting from a defective coolant temp sensor or defective fuel injectors. Lean mixtures are commonly caused by intake manifold leaks. When either of these circumstances occurs, the converter can become a catalytic furnace, causing the previously described damage. For most efficient catalytic converter operation, the air–fuel mixture should be near 14.7:1. ● **SEE FIGURE 28–7.**

To avoid excessive catalyst temperatures and the possibility of fuel vapors reaching the converter, observe the following rules:

1. Do not use fuel additives or cleaners that are not converter safe.
2. Do not crank an engine for more than 40 seconds when it is flooded or misfiring.
3. Do not turn off the ignition switch when the vehicle is in motion.
4. Do not disconnect a spark plug wire for more than 30 seconds.
5. Repair engine problems such as dieseling, misfiring, or stumbling as soon as possible.

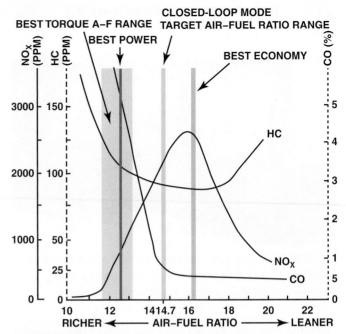

**FIGURE 28–7** The highest catalytic converter efficiency occurs when the air–fuel mixture is about 14.7:1.

**? FREQUENTLY ASKED QUESTION**

**Can a Catalytic Converter Be Defective without Being Clogged?**

Yes. Catalytic converters can fail by being chemically damaged or poisoned without being mechanically clogged. Therefore, the catalytic converter should be tested not only for physical damage (clogging) by performing a back pressure or vacuum test and a rattle test but also for temperature rise, usually with a pyrometer or propane test, to check the efficiency of the converter.

# DIAGNOSING CATALYTIC CONVERTERS

**TAP TEST** The simple **tap test** involves tapping (not pounding) on the catalytic converter using a rubber mallet. If the substrate inside the converter is broken, the converter will rattle when hit. If the converter rattles, a replacement converter is required. ● **SEE FIGURE 28–8.**

**TESTING BACK PRESSURE WITH A PRESSURE GAUGE** Exhaust system back pressure can be measured directly by installing a pressure gauge in an exhaust opening. This can be accomplished in one of the following ways:

1. To test at the oxygen sensor, remove the inside of an old, discarded oxygen sensor and thread in an adapter to convert it to a vacuum or pressure gauge.

FIGURE 28-8 A catalytic converter that rattled when tapped was removed, and the substrate, or what was left of it, fell out. This converter has to be replaced and the root cause of why it failed found and corrected.

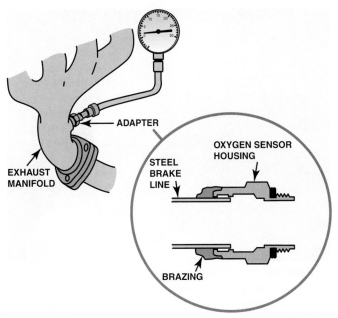

FIGURE 28-9 A back pressure tool can be made by using an oxygen sensor housing and epoxy or braze to hold the tube to the housing.

NOTE: An adapter can be easily made by inserting a metal tube or pipe into an old oxygen sensor housing. A short section of brake line works great. The pipe can be brazed to the oxygen sensor housing, or it can be glued with epoxy. An 18-mm compression gauge adapter can also be adapted to fit into the oxygen sensor opening. ● SEE FIGURE 28-9.

2. To test the exhaust back pressure at the exhaust gas recirculation (EGR) valve, remove the EGR valve and fabricate a plate equipped with a fitting for a pressure gauge.

3. To test at the secondary air-injection (SAI) check valve, remove the check valve from the exhaust tubes leading to the exhaust manifold. Use a rubber cone with a tube inside to seal against the exhaust tube. Connect the tube to a pressure gauge.

At idle, the maximum back pressure should be less than 1.5 PSI (10 kPa), and it should be less than 2.5 PSI (15 kPa) at 2500 RPM. Pressure readings higher than these indicate that the exhaust system is restricted, and further testing will be needed to determine the location of the restriction. ● SEE FIGURE 28-10.

### TESTING FOR BACK PRESSURE USING A VACUUM GAUGE
An exhaust restriction can be tested indirectly by checking the intake manifold vacuum with the engine operating at a fast idle speed (about 2500 RPM). If the exhaust is restricted, some exhaust can pass, and the effect may not be noticeable when the engine is at idle speed. However, when the engine is operating at a higher speed, the exhaust gases can build up behind the restriction and eventually will not be able to leave the combustion chamber. When some of the exhaust is left behind at the end of the exhaust stroke, the resulting pressure in the combustion chamber reduces engine vacuum. To test for an exhaust restriction using a vacuum gauge, perform the following steps:

FIGURE 28-10 This partially melted catalytic converter tested okay at idle but had excessive back pressure at idle speeds.

STEP 1 Attach a vacuum gauge to an intake manifold vacuum source.

STEP 2 Start the engine. Record the engine manifold vacuum reading. The engine vacuum should read 17 to 21 in. Hg when the engine is at idle speed.

STEP 3 Increase the engine speed to 2500 RPM and hold that speed for 60 seconds while looking at the vacuum gauge.

#### Results

■ If the vacuum reading is equal to or higher than the vacuum reading when the engine was at idle speed, the exhaust system is *not* restricted.

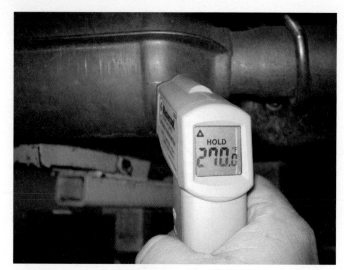

**FIGURE 28–11** The temperature of the outlet should be at least 10% hotter than the temperature of the inlet. If a converter is not working, the inlet temperature will be hotter than the outlet temperature.

- If the vacuum reading is lower than the vacuum reading when the engine was at idle speed, then the exhaust *is* restricted. Further testing will be needed to determine the location of the restriction.

### TESTING A CATALYTIC CONVERTER FOR TEMPERATURE RISE

A properly working catalytic converter should be able to reduce $NO_x$ exhaust emissions into nitrogen (N) and oxygen ($O_2$) and oxidize unburned hydrocarbon (HC) and carbon monoxide (CO) into harmless carbon dioxide ($CO_2$) and water vapor ($H_2O$). During these chemical processes, the catalytic converter should increase in temperature at least 10% if the converter is working properly. To test the converter, operate the engine at 2500 RPM for at least two minutes to fully warm the converter. Measure the inlet and the outlet temperatures using an infrared pyrometer as shown ● **SEE FIGURE 28–11.**

**NOTE: If the engine is extremely efficient, the converter may not have any excessive unburned hydrocarbons or carbon monoxide to convert! In this case, a spark plug wire could be grounded out using a vacuum hose and a test light to create some unburned hydrocarbon in the exhaust. Do not ground out a cylinder for longer than 10 seconds, or the excessive amount of unburned hydrocarbon could overheat and damage the converter.**

### CATALYTIC CONVERTER EFFICIENCY TESTS

The efficiency of a catalytic converter can be determined using an exhaust gas analyzer:

- **Oxygen level test.** With the engine warm and in closed loop, check the oxygen ($O_2$) and carbon monoxide (CO) levels. A good converter should be able to oxide the extra hydrocarbons caused by the rapid acceleration:

  - If $O_2$ is zero, go to the snap-throttle test.

 **TECH TIP**

**Aftermarket Catalytic Converters**

Some replacement aftermarket (nonfactory) catalytic converters do not contain the same amount of cerium as the original part. Cerium is the element that is used in catalytic converters to store oxygen. As a result of the lack of cerium, the correlation between the oxygen storage and the conversion efficiency may be affected enough to set a false diagnostic trouble code (P0422).

**NOTE: If an aftermarket converter is being installed, be sure that the distance between the rear of the catalyst block is the same distance from the rear oxygen sensor as the factory converter to be ensured of proper operation. Always follow the instructions that come with the replacement converter.** ● **SEE FIGURE 28–12.**

- If $O_2$ is greater than zero, check the CO level.
- If CO is greater than zero, the converter is *not* functioning correctly.
- **Snap-throttle test.** With the engine warm and in closed loop, snap the throttle to wide-open throttle (WOT) in park or neutral and observe the oxygen reading:

  - The $O_2$ reading should not exceed 1.2%; if it does, the converter is *not* working.
  - If the $O_2$ rises to 1.2%, the converter may have low efficiency.
  - If the $O_2$ remains below 1.2%, then the converter is okay.

## CATALYTIC CONVERTER REPLACEMENT GUIDELINES

Because a catalytic converter is a major exhaust gas emission control device, the U.S. Environmental Protection Agency (EPA) has strict guidelines for its replacement, including the following:

- If a converter is replaced on a vehicle with less than 80,000 miles or eight years, depending on the year of the vehicle, an original equipment catalytic converter *must* be used as a replacement.
- The replacement converter must be of the same design as the original. If the original had an AIR pump fitting, so must the replacement.
- The old converter must be kept for possible inspection by the authorities for 60 days.

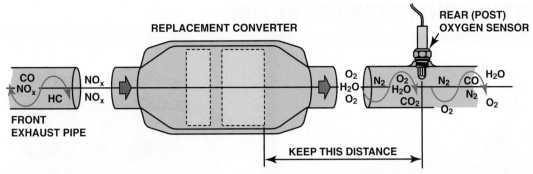

FIGURE 28–12 Whenever replacing a catalytic converter with a universal unit, first measure the distance between the rear brick and the center of the rear oxygen sensor. Be sure that the replacement unit is installed to the same dimension.

**Catalytic Converters Are Murdered**

Catalytic converters start a chemical reaction but do not enter into the chemical reaction. Therefore, catalytic converters neither wear out nor die of old age. If a catalytic converter is found to be defective (nonfunctioning or clogged), look for the *root* cause. Remember this:

"Catalytic converters do not commit suicide— they're murdered."

Items that should be checked when a defective catalytic converter is discovered include all components of the ignition and fuel systems. Excessive unburned fuel can cause the catalytic converter to overheat and fail. The oxygen sensor must be working and fluctuating from 0.5 to 5 Hz (times per second) to provide the necessary air–fuel mixture variations for maximum catalytic converter efficiency.

■ A form must be completed and signed by both the vehicle owner and a representative from the service facility. This form must state the cause of the converter failure and must remain on file for two years.

## CATALYTIC CONVERTER–RELATED DIAGNOSTIC TROUBLE CODE

| Diagnostic Trouble Code | Description | Possible Causes |
|---|---|---|
| P0420/P0422 | Catalytic converter efficiency failure | • Engine mechanical fault<br>• Exhaust leaks<br>• Fuel contaminants, such as engine oil, coolant, or sulfur |

# CATALYTIC CONVERTER OPERATION

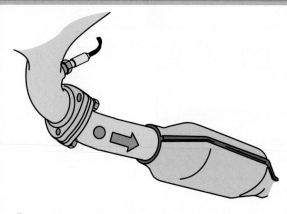

**1** Carbon monoxide leaves the engine through the exhaust valve on the exhaust stroke.

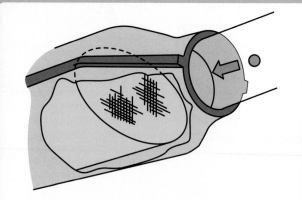

**2** The CO molecule is starting to enter the converter.

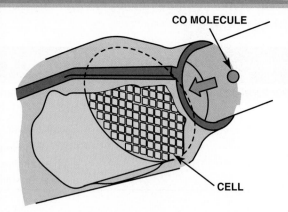

CO MOLECULE

CELL

**3** A CO molecule is ready to enter a cell. The number of cells ranges from 300 to 900 per sq. in. The substrate is cordierite (Mg, AL, Si) or foil-backed metal with a 0.002 to 0.006 inch (0.05 to 0.15 mm)-thick wall.

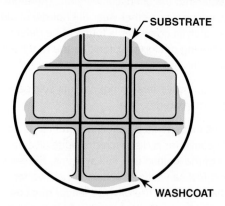

SUBSTRATE

WASHCOAT

**4** The CO molecule enters a cell. The substrate is coated with porous aluminum ($AL_2O_3$) called the washcoat. The catalytic material is sprayed onto the washcoat.

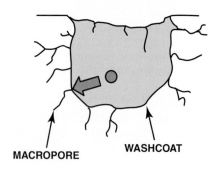

MACROPORE    WASHCOAT

**5** The CO molecule enters a micropore, which has been created in the porous washcoat.

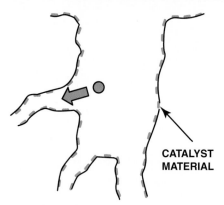

CATALYST MATERIAL

**6** The CO molecule enters a smaller micropore.

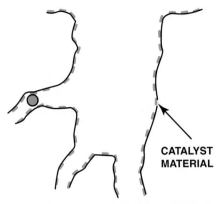

**CATALYST MATERIAL**

**7** The CO molecule is absorbed onto a catalyst side. Only a few grams of catalyst material are applied to the washcoat.

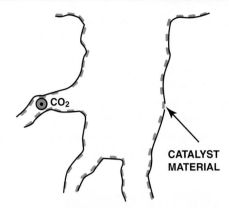

$CO_2$

**CATALYST MATERIAL**

**8** The CO molecule is converted to a $CO_2$ molecule.

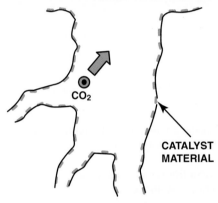

$CO_2$

**CATALYST MATERIAL**

**9** The $CO_2$ molecule is exiting the small micropore.

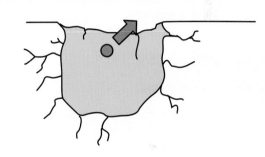

**10** The $CO_2$ molecule is exiting the larger micropore.

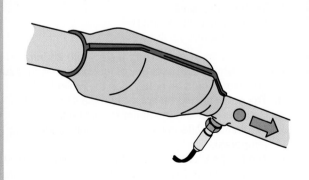

**11** The $CO_2$ molecule is exiting the converter.

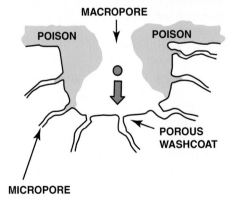

**MACROPORE**

**POISON**          **POISON**

**POROUS WASHCOAT**

**MICROPORE**

**12** A poisoned converter.

1. A catalytic converter is an after treatment device that reduces exhaust emissions outside of the engine.

2. The honeycomb shape of the catalytic converter is constructed of a ceramic material about 0.006 inch thick with small square openings.

3. A catalyst is an element that starts a chemical reaction but is not consumed in the process.

4. The catalyst materials used in a catalytic converter include rhodium, palladium, and platinum.

5. A catalytic converter has to be over 500°F (260°C) before it starts to become effective and is therefore mounted as close as possible to the exhaust parts of the engine.

6. The OBD-II system monitor compares the relative activity of a rear oxygen sensor to the precatalytic oxygen sensor to determine catalytic converter efficiency.

7. Catalytic converters can be tested for restriction and for efficiency.

## REVIEW QUESTIONS

1. What are the three most commonly used catalysts in a catalytic converter?

2. How does a catalytic converter reduce $NO_x$ to nitrogen and oxygen?

3. Why must a catalytic converter be mounted close to the exhaust ports of the engine?

4. How does the computer monitor catalytic converter performance?

5. What tests can be performed by a service technician to test the catalytic converter?

## CHAPTER QUIZ

1. What is applied to the ceramic substrate to make the surface porous?
   a. Honeycomb filler
   b. Washcoat
   c. Aluminum
   d. Cerium

2. Two technicians are discussing catalytic converters. Technician A says that the exhaust mixture must fluctuate between rich and lean for the best efficiency. Technician B says that the air–fuel mixture must be leaner than 14.7:1 for best performance from a three-way catalytic converter. Which technician is correct?
   a. Technician A only
   b. Technician B only
   c. Both Technicians A and B
   d. Neither Technician A nor B

3. A catalytic converter has to be at least how hot before it starts to work?
   a. 500°F (260°C)
   b. 1,000°F (540°C)
   c. 1,500°F (815°C)
   d. 2,000°F (1,100°C)

4. What two primary sensors does the PCM use to check the catalytic converter?
   a. Catalytic converter temperature sensor and rear oxygen sensor
   b. Precat and postcat oxygen sensor
   c. Precat oxygen sensor and MAF
   d. MAP and TP

5. A catalytic converter can be harmed by _____.
   a. Excessive engine oil
   b. Antifreeze
   c. Sulfur from poor-quality fuel
   d. Any of the above

6. Two technicians are discussing testing a catalytic converter. Technician A says that a vacuum gauge can be used and observed to see if the vacuum drops with the engine at 2500 RPM for 60 seconds. Technician B says that a pressure gauge can be used to check for back pressure. Which technician is correct?
   a. Technician A only
   b. Technician B only
   c. Both Technicians A and B
   d. Neither Technician A nor B

7. A catalytic converter is being tested with an infrared pyrometer. Which is an acceptable (good converter) result?
   a. The inlet should be hotter than the outlet by 10%.
   b. The outlet should be hotter than the inlet by 10%.
   c. Both the inlet and the outlet should be the same temperature after the converter reaches operating temperature.
   d. The temperature of a catalytic converter is the best test to perform to locate a restricted (clogged) unit.

8. Which exhaust gas reading indicates a good catalytic converter?
   a. O$_2$ is zero
   b. CO is zero
   c. Both a and b
   d. Neither a nor b

9. A P0422 (catalytic converter efficiency failure) is set. What is a possible cause?
   a. Engine mechanical fault
   b. Exhaust leak
   c. Fuel contamination
   d. Any of the above

10. Technician A says that the catalytic converter is warranted for eight years or 80,000 miles, whichever comes first. Technician B says that after replacing the catalytic converter, the old converter must be kept for possible inspection for 60 days. Which technician is correct?
   a. Technician A only
   b. Technician B only
   c. Both Technicians A and B
   d. Neither Technician A nor B

**LEARNING OBJECTIVES:** **After studying this chapter, the reader should be able to:** • Understand how the ignition system and ignition coils work. • Discuss crankshaft position sensors and the operation of pickup coils. • Explain the operation of waste-spark and coil-on-plug ignition systems. • Discuss ignition system diagnosis. • Understand the construction and operation of different types of spark plugs and discuss how to inspect spark plug wires. • List the steps necessary to check and/or adjust ignition timing on engines equipped with a distributor.

**KEY TERMS:** Coil-on-plug (COP) ignition 355 • Companion cylinders 361 • Detonation 365 • Distributor ignition 355 • Electronic ignition 355 • Electromagnetic induction (EMI) 355 • Firing order 360 • Hall effect 357 • ICM 356 • Ignition coil 355 • Ignition timing 375 • Ion-sensing ignition 365 • Iridium spark plugs 372 • Knock sensors 365 • Magnetic pulse generator 357 • Pickup coil 357 • Ping 365 • Platinum spark plugs 372 • Primary ignition circuit 356 • Primary winding 355 • Schmitt trigger 358 • Secondary ignition circuit 356 • Secondary winding 355 • Spark knock 365 • Spark plugs 372 • Spark tester 367 • Switching 356 • Track 370 • Transistor 357 • Trigger 357 • Turns ratio 355 • Waste-spark system 355

## IGNITION SYSTEM

**PURPOSE AND FUNCTION** The ignition system includes components and wiring necessary to create and distribute a high voltage (up to 40,000 volts or more) and send to the spark plug. A high-voltage arc occurs across the gap of a spark plug at the right time inside the combustion chamber. The spark raises the temperature of the air–fuel mixture and starts the combustion process inside the cylinder.

**BACKGROUND** All ignition systems apply battery voltage (close to 12 volts) to the positive side of the ignition coil(s) and pulse the negative side to ground:

- **Early ignition systems.** Before the mid-1970s, ignition systems used a mechanically opened set of contact points to make and break the electrical connection to ground. A cam lobe, located in and driven by the distributor, opened the points. There was one lobe for each cylinder. The points used a rubbing block that was lubricated by applying a thin layer of grease on the cam lobe at each service interval. Each time the points opened, a high voltage was created in the ignition coil. The high-voltage then traveled to each spark plug through the distributor cap and rotor. The distributor was used twice in the creation of the spark, as follows:

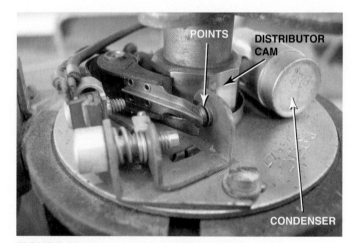

**FIGURE 29–1** A point-type distributor from a hot rod.

1. It was connected to the camshaft, which rotated the distributor cam, causing the points to open and close.

2. It used a rotor to send the high voltage from the coil entering the center of the distributor cap to inserts connected to spark plug wires to each cylinder.

● **SEE FIGURE 29–1.**

- **Electronic ignition.** Since the mid-1970s, ignition systems have used sensors, such as a pickup coil and reluctor (trigger wheel), to trigger or signal an electronic module that switches the primary ground circuit of the ignition coil. **Distributor ignition** is the term specified by the Society of Automotive Engineers (SAE) for an ignition system that uses a distributor. **Electronic ignition** is the term specified by the SAE for an ignition system that does not use a distributor. Electronic ignition system types include the following:

1. **Waste-spark system.** This type of system uses one ignition coil to fire the spark plugs for two cylinders at the same time.

2. **Coil-on-plug (COP) system.** This type of system uses a single ignition coil for each cylinder with the coil placed above or near the spark plug.

## IGNITION COIL CONSTRUCTION

The heart of any ignition system is the **ignition coil.** When the coil negative lead is grounded, the primary (low-voltage) circuit of the coil is complete, and a magnetic field is created around the coil windings. When the circuit is opened, the magnetic field collapses and induces a high voltage in the secondary winding of the ignition coil.

The coil creates a high-voltage spark by electromagnetic induction. Many ignition coils contain two separate but electrically connected windings of copper wire. Other coils are true transformers in which the primary and secondary windings are not electrically connected. ● **SEE FIGURE 29–2.**

The center of an ignition coil contains a core of laminated soft iron (thin strips of soft iron). This core increases the magnetic strength of the coil:

■ **Secondary coil winding.** Surrounding the laminated core are approximately 20,000 turns of fine wire (approximately 42 gauge). The winding is called the **secondary winding.**

■ **Primary coil winding.** Surrounding the secondary windings are approximately 150 turns of heavy wire (approximately 21 gauge). The winding is called the **primary winding.** The secondary winding has about 100 times the number of turns of the primary winding, referred to as the **turns ratio** (approximately 100:1).

In older coils, these windings are surrounded with a thin metal shield and insulating paper and placed into a metal container filled with transformer oil to help cool the coil windings. Other coil designs use an air-cooled, epoxy-sealed E coil. The *E coil* is so named because the laminated, soft iron core is E shaped, with the coil wire turns wrapped around the center "finger" of the E and the primary winding wrapped inside the secondary winding. ● **SEE FIGURES 29–3 AND 29–4.**

## IGNITION COIL OPERATION

All ignition systems use electromagnetic induction to produce a high-voltage spark from the ignition coil. **Electromagnetic induction (EMI)** means that a current can be created in a conductor (coil winding) by a moving magnetic field. The magnetic field in an ignition coil is produced by current flowing through the primary winding of

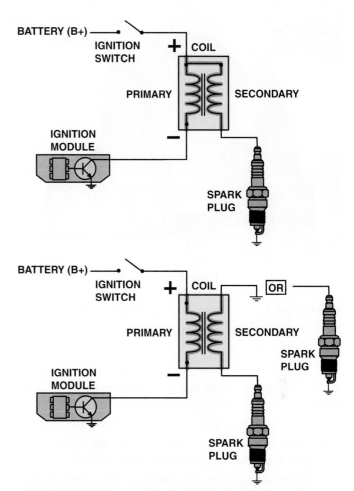

**FIGURE 29–2** Some ignition coils are electrically connected, called married (top figure), whereas others use separate primary and secondary windings, called divorced (lower figure). The polarity (positive or negative) of a coil is determined by the direction in which the coil is wound.

**FIGURE 29–3** The steel lamination used in an E coil helps increase the magnetic field strength, which helps the coil produce higher energy output for a more complete combustion in the cylinders.

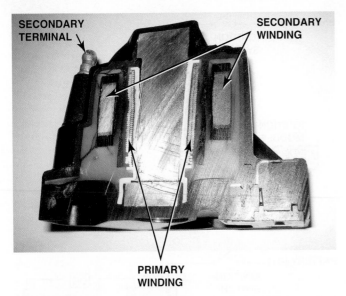

FIGURE 29–4 The primary windings are inside the secondary windings on this General Motors coil.

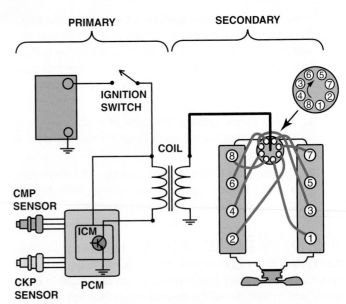

FIGURE 29–5 The primary ignition system is used to trigger and therefore create the secondary (high-voltage) spark from the ignition coil.

the coil. An ignition coil is able to increase battery voltage to 40,000 volts or more in the following way:

- Battery voltage is applied to the primary winding.

- A ground is provided to the primary winding by the **ignition control module (ICM),** igniter, or PCM.

- Current (approximately 2 to 6 amperes) flows in the primary coil creating a magnetic field in the primary winding.

- When the ground is opened by the ICM, the built-up magnetic field collapses.

- The movement of the collapsing magnetic field induces a voltage of 250 to 400 volts in the primary winding and 20,000 to 40,000 volts or more in the secondary winding with a current of 0.020 to 0.080 ampere.

- The high voltage created in the secondary winding is high enough to jump the air gap at the spark plug.

- The electrical arc at the spark plug ignites the air–fuel mixture in the combustion chamber of the engine.

- For each spark that occurs, the coil must be charged with a magnetic field and then discharged.

 **WARNING**

The spark from an ignition coil is strong enough to cause physical injury. Always follow the exact service procedure and avoid placing hands near the secondary ignition components when the engine is running.

The ignition components that regulate the current in the coil primary winding by turning it on and off are collectively known as the **primary ignition circuit.** When the primary circuit is carrying current, the secondary circuit is off. When the primary circuit is turned off, the secondary circuit has high voltage. The components necessary to create and distribute the high voltage produced in the secondary windings of the coil are called the **secondary ignition circuit.** ● **SEE FIGURE 29–5.**

These circuits include the following components:

- Primary ignition circuit
  1. Battery
  2. Ignition switch
  3. Primary windings of coil
  4. Pickup coil (crankshaft position sensor)
  5. Ignition control module (igniter)
- Secondary ignition circuit
  1. Secondary windings of coil
  2. Distributor cap and rotor (if the vehicle is so equipped)
  3. Spark plug wires
  4. Spark plugs

# IGNITION SWITCHING AND TRIGGERING

**SWITCHING** For any ignition system to function, the primary current must be turned on to charge the coil and off to allow the coil to discharge, creating a high-voltage spark. This turning on and off of the primary circuit is called **switching.** The unit that does

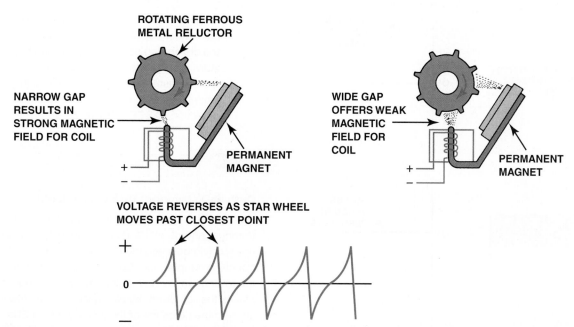

**FIGURE 29–6** Operation of a typical pulse generator (pickup coil). At the bottom is a line drawing of a typical scope pattern of the output voltage of a pickup coil. The module receives this voltage from the pickup coil and opens the ground circuit to the ignition coil when the voltage starts down from its peak (just as the reluctor teeth start moving away from the pickup coil).

the switching is an electronic switch, such as a power transistor. This power transistor can be found in the following locations:

- Ignition control module (ICM) or igniter
- PCM (computer)

**NOTE: On some coil-on-plug (COP) systems, the ICM is part of the ignition coil itself and is serviced as an assembly.**

### TRIGGERING
The device that signals the switching of the coil on and off or just on in most instances is called the **trigger.** A trigger is typically a pickup coil in some distributor-type ignitions and a crankshaft position sensor (CKP) on electronic systems (waste spark and coil on plug). There are three types of devices used for triggering:

1. Magnetic sensor
2. Hall-effect switch
3. Optical sensor

### PRIMARY CIRCUIT OPERATION
To get a spark out of an ignition coil, the primary coil circuit must be turned on and off. The primary circuit current switching is controlled by a **transistor** (electronic switch) inside the ignition module (or igniter) or PCM and is controlled by one of several devices, including the following:

- **Magnetic sensor.** A simple and common ignition electronic switching device is the magnetic pulse generator system. This is a type of magnetic sensor, often called a **magnetic pulse generator** or **pickup coil,** and is installed in the distributor housing. The pulse generator consists of a trigger wheel (reluctor) and a pickup coil. The pickup coil consists of an iron core wrapped with fine wire, in a coil at one end and attached to a permanent magnet at the other end. The center of the coil is called the pole piece. The pickup coil signal

triggers the transistor inside the module and is also used by the PCM for piston position information and engine speed (RPM). The reluctor is shaped so that the magnetic strength changes enough to create a usable varying signal for use by the module to trigger the coil. ● **SEE FIGURE 29–6.**

*Magnetic crankshaft position sensors* use the changing strength of the magnetic field surrounding a coil of wire to signal the module and computer. This signal is used by the electronics in the module and computer to determine piston position and engine speed (RPM). This sensor operates similarly to the distributor magnetic pickup coil. The crankshaft position sensor uses the strength of the magnetic field surrounding a coil of wire to signal the ICM. The rotating crankshaft has notches cut into it that trigger the magnetic position sensor, which change the strength of the magnetic field as the notches pass by the position sensor. ● **SEE FIGURE 29–7.**

- **Hall-effect switch.** This switch also uses a stationary sensor and rotating trigger wheel (shutter). Unlike the magnetic pulse generator, the Hall-effect switch requires a small input voltage to generate an output or signal voltage. **Hall effect** has the ability to generate a voltage signal in semiconductor material (gallium arsenate crystal) by passing current through it in one direction and applying a magnetic field to it at a right angle to its surface. If the input current is held steady and the magnetic field fluctuates, an output voltage is produced that changes in proportion to field strength. Most Hall-effect switches in distributors have the following:

1. Hall element or device
2. Permanent magnet
3. Rotating ring of metal blades (shutters) similar to a trigger wheel (Another method uses a stationary sensor with a rotating magnet.) ● **SEE FIGURE 29–8.**

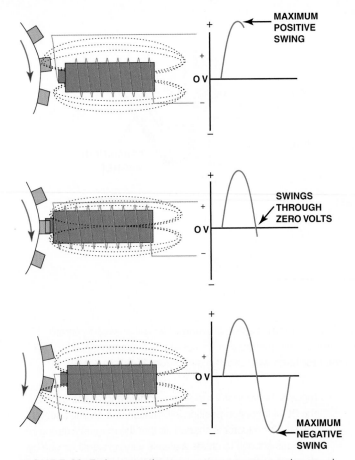

MAXIMUM
POSITIVE
SWING

0 V

SWINGS
THROUGH
ZERO VOLTS

0 V

0 V

MAXIMUM
NEGATIVE
SWING

**FIGURE 29–7** A magnetic sensor uses a permanent magnet surrounded by a coil of wire. The notches of the crankshaft (or camshaft) creates a variable magnetic field strength around the coil and create an analog signal when the engine rotates. When a metallic section is close to the sensor, the magnetic field is stronger because metal is a better conductor of magnetic lines of force than air.

Some blades are designed to hang down, typically found in Bosch and Chrysler systems, while others may be on a separate ring on the distributor shaft, typically found in General Motors and Ford Hall-effect distributors. There are two types of Hall-effect sensors used:

- When the shutter blade enters the gap between the magnet and the Hall element, it creates a magnetic shunt that changes the field strength through the Hall element.

- This analog signal is sent to a **Schmitt trigger** inside the sensor itself, which converts the analog signal into a digital signal. A digital (on or off) voltage signal is created at a varying frequency to the ignition module or onboard computer. ● SEE FIGURE 29–9.

- **Optical sensors.** These use light from an LED and a phototransistor to signal the computer. An interrupter disc between the LED and the phototransistor has slits that allow the light from the LED to

trigger the phototransistor on the other side of the disc. Most optical sensors (usually located inside the distributor) use two rows of slits to provide individual cylinder recognition (low resolution) and precise distributor angle recognition (high resolution) signals that are used for cylinder misfire detection. ● SEE FIGURE 29–10.

### TECH TIP

#### Optical Distributors Do Not Like Light

Optical distributors use the light emitted from LEDs to trigger phototransistors. Most optical distributors use a shield between the distributor rotor and the optical interrupter ring. Sparks jump the gap from the rotor tip to the distributor cap inserts. This shield blocks the light from the electrical arc from interfering with the detection of the light from the LEDs.

If this shield is not replaced during service, the light signals are reduced, and the engine may not operate correctly. ● SEE FIGURE 29–11.

This can be difficult to detect because nothing looks wrong during a visual inspection. Remember that all optical distributors must be shielded between the rotor and the interrupter ring.

### TECH TIP

#### The Tachometer Trick

When diagnosing a no-start or intermediate missing condition, check the operation of the tachometer. If the tachometer does not indicate engine speed (no-start condition) or drops toward zero (engine missing), then the problem is due to a defect in the *primary* ignition circuit. The tachometer gets its signal from the pulsing of the primary winding of the ignition coil. The following components in the primary circuit could cause the tachometer to not work when the engine is cranking:

- Pickup coil
- Crankshaft position sensor
- Ignition module (igniter)
- Coil primary wiring

If the vehicle is not equipped with a tachometer, use a scan tool to look at engine RPM. The results are as follows:

- No or an unstable engine RPM reading means the problem is in the primary ignition circuit.
- A steady engine RPM reading means the problem is in the secondary ignition circuit or is a fuel-related problem.

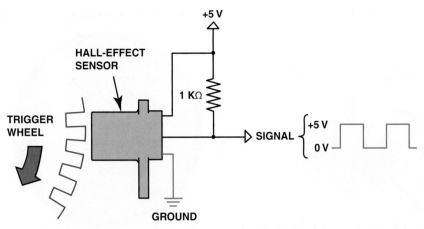

**FIGURE 29–8** A Hall-effect sensor produces a digital on-off voltage signal whether it is used with a blade or a notched wheel.

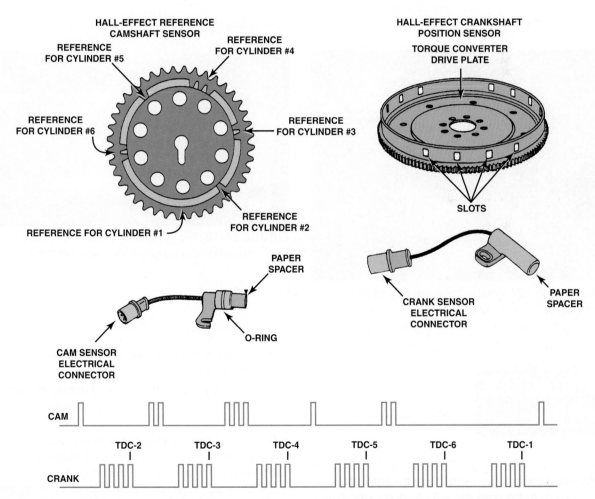

**FIGURE 29–9** Some Hall-effect sensors look like magnetic sensors. This Hall-effect camshaft reference sensor and crankshaft position sensor have an electronic circuit built in that creates a 0- to 5-volt signal as shown at the bottom. These Hall-effect sensors have three wires: a power supply (8 volts) from the computer (controller), a signal (0 to 5 volts), and a signal ground.

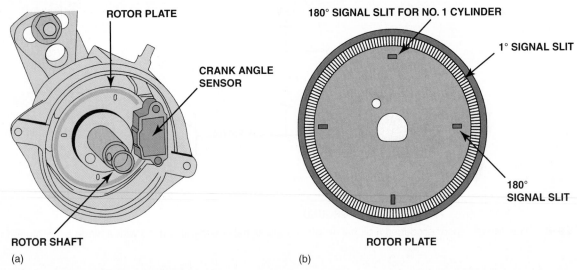

FIGURE 29–10 (a) Typical optical distributor. (b) Cylinder I slit signals the computer the piston position for cylinder I. The 1-degree slits provide accurate engine speed information to the PCM. Optical sensors generate digital (on and off) signals.

FIGURE 29–11 A light shield is being installed on an optical distributor before the rotor is attached.

FIGURE 29–12 The firing order is cast or stamped on the intake manifold on most engines that have a distributor ignition.

## DISTRIBUTOR IGNITION

**PURPOSE AND FUNCTION**  The purpose of a distributor is to distribute the high-voltage spark from the output terminal of the ignition coil to the spark plugs for each cylinder. A gear or shaft drives the distributor that is connected to the camshaft and is driven at camshaft speed. Most distributor ignition systems also use a sensor to trigger the ignition control module.

**OPERATION OF DISTRIBUTOR IGNITION**  The distributor is used twice in most ignition systems that use a distributor:

- First, to trigger the ignition control module by the use of the rotating distributor shaft
- Second, by rotating the rotor to distribute the high-voltage spark to the individual spark plugs

**FIRING ORDER**  **Firing order** means the order that the spark is distributed to the correct spark plug at the right time. The firing order of an engine is determined by crankshaft and camshaft design. The firing order is determined by the location of the spark plug wires in the distributor cap of an engine equipped with a distributor. The firing order is often cast into the intake manifold for easy reference. ● **SEE FIGURE 29–12.**

Service information also shows the firing order and the direction of the distributor rotor rotation as well as the location of the spark plug wires on the distributor cap.

**CAUTION:** Ford V-8s use two different firing orders depending on whether the engine is high output or standard. Using the incorrect firing order can cause the engine to backfire and could cause engine damage or personal injury. General Motors V-6 engines use different firing orders and different locations for cylinder 1 between the 60-degree V-6 and the 90-degree V-6. Using the incorrect firing order or cylinder number location chart could result in poor engine operation or a no start. Firing order is also important for waste-spark-type ignition systems. The spark plug wire can often be installed on the wrong coil pack, which can create a no-start condition or poor engine operation.

# WASTE-SPARK IGNITION SYSTEMS

**PARTS INVOLVED** Waste-spark ignition is another name for distributorless ignition system (DIS) or electronic ignition. Waste-spark ignition was introduced in the mid-1980s and uses the ignition control module (ICM) and/or the powertrain control module (PCM) to fire the ignition coils. A 4-cylinder engine uses two ignition coils, and a 6-cylinder engine uses three ignition coils. Each coil is a true transformer because the primary winding and secondary winding are not electrically connected. Each end of the secondary winding is connected to a cylinder exactly opposite the other in the firing order, which is called a **companion** (paired) **cylinder.**  **SEE FIGURE 29–13.**

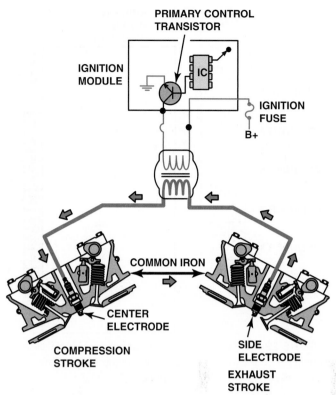

**FIGURE 29–13** A waste-spark system fires one cylinder while its piston is on the compression stroke and into paired or companion cylinders while it is on the exhaust stroke. In a typical engine, it requires only about 2 to 3 kV to fire the cylinder on the exhaust stroke. The remaining coil energy is available to fire the spark plug under compression (typically about 8 to 12 kV).

**? FREQUENTLY ASKED QUESTION**

**How Can You Determine the Companion Cylinder?**

Companion cylinders are two cylinders in the same engine that both reach top dead center (TDC) at the same time:

- One cylinder is on the compression stroke.
- The other cylinder is on the exhaust stroke.

To determine which two cylinders are companion cylinders in the engine, follow these steps:

**STEP 1** Determine the firing order (such as 165432 for a typical V-6 engine).

**STEP 2** Write the firing order and then place the second half under the first half:

$$\frac{165}{432}$$

**STEP 3** The cylinder numbers above and below each other are companion or paired cylinders. In this case 1 and 4, 6 and 3, and 5 and 2 are companion cylinders.

**🔧 TECH TIP**

**Odds Fire Straight**

Waste-spark ignition systems fire two spark plugs at the same time. Most vehicle manufacturers use a waste-spark system that fires the odd number cylinders (1, 3, and 5) by straight polarity (current flow from the top of the spark plug through the gap and to the ground electrode). The even number cylinders (2, 4, and 6) are fired reverse polarity, meaning that the spark jumps from the side electrode to the center electrode. Some vehicle manufacturers equip their vehicles with platinum plugs that have the expensive platinum alloy on only one electrode, as follows:

- On odd number cylinders (1, 3, 5), the platinum is on the center electrode.
- On even number cylinders (2, 4, 6), the platinum is on the ground electrode.

Replacement spark plugs use platinum on both electrodes (double platinum) and, therefore, can be placed in any cylinder location.

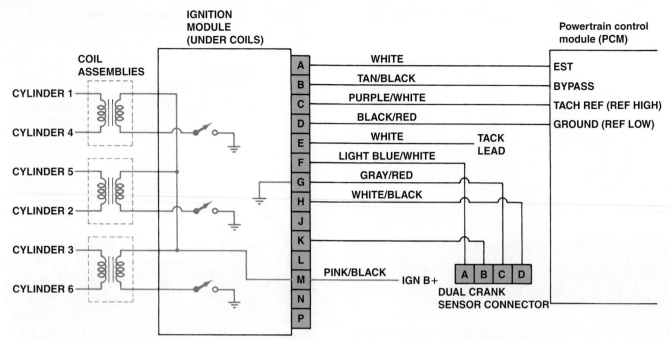

**FIGURE 29-14** Typical wiring diagram of a GM V-6 waste-spark ignition system. The PCM uses input data from all of the engine sensors and determines the optimum ignition timing, then triggers the primary ignition circuit to fire the spark plug or sends the trigger signal to the ICM, if equipped.

**WASTE-SPARK SYSTEM OPERATION** *Both* spark plugs fire at the same time (within nanoseconds of each other):

- When one cylinder (e.g., cylinder number 6) is on the compression stroke, the other cylinder (number 3) is on the exhaust stroke.

- The spark that occurs on the exhaust stroke is called the *waste spark* because it does no useful work and is only used as a ground path for the secondary winding of the ignition coil. The voltage required to jump the spark plug gap on cylinder 3 (the exhaust stroke) is only 2 to 3 kV.

- The cylinder on the compression stroke uses the remaining coil energy.

- One spark plug of each pair always fires straight polarity (center to side electrode), and the other cylinder always fires reverse polarity (side to center electrode). Spark plug life is not greatly affected by the reverse polarity. If there is only one defective spark plug wire or spark plug, two cylinders may be affected.

The coil polarity is determined by the direction the coil is wound (left-hand rule for conventional current flow) and cannot be changed.

Each spark plug for a particular cylinder always will be fired either with straight or reversed polarity, depending on its location in the engine and how the coils are wired. However, the compression and waste-spark condition flip-flops. When one cylinder is on compression, such as cylinder 1, then the paired cylinder (number 4) is on the exhaust stroke. During the next rotation of the crankshaft, cylinder 4 is on the compression stroke, and cylinder 1 is on the exhaust stroke:

**Cylinder 1**  Always fires straight polarity (from the center electrode to the ground electrode), one time, requiring 10 to 12 kV, and one time, requiring 3 to 4 kV.

**Cylinder 4**  Always fires reverse polarity (from the ground electrode to the center electrode), one time, requiring 10 to 12 kV, and one time, requiring 3 to 4 kV.

Waste-spark ignitions require a sensor (usually a crankshaft sensor) to trigger the coils at the correct time. ● SEE FIGURE 29-14.

The crankshaft sensor cannot be moved to adjust ignition timing because ignition timing is not adjustable. The slight adjustment of the crankshaft sensor is designed to position the sensor exactly in the middle of the rotating metal disc for maximum clearance.

**COMPRESSION-SENSING WASTE-SPARK IGNITION** Some waste-spark ignition systems, such as those used on Saturns and others, use the voltage required to fire the cylinders to determine cylinder position. It requires a higher voltage to fire a spark plug under compression than it does when the spark plug is being fired on the exhaust stroke. The electronics in the coil and the PCM can detect which of the two companion (paired) cylinders that are fired at the same time requires the higher firing voltage and a slightly delayed break-over voltage, which indicates the cylinder that is on the compression stroke. For example, a typical 4-cylinder engine equipped with a waste-spark ignition system will fire both cylinders 1 and 4. If cylinder 4 requires a higher voltage to fire, as determined by the electronics connected to the

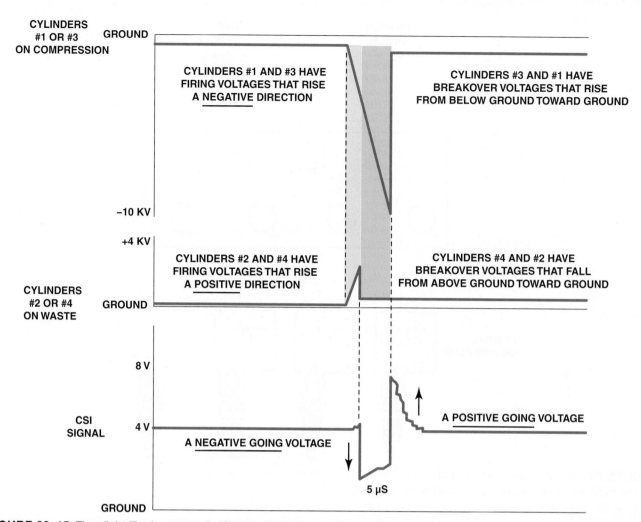

CYLINDERS #1 OR #3 ON COMPRESSION

GROUND

CYLINDERS #1 AND #3 HAVE FIRING VOLTAGES THAT RISE A NEGATIVE DIRECTION

CYLINDERS #3 AND #1 HAVE BREAKOVER VOLTAGES THAT RISE FROM BELOW GROUND TOWARD GROUND

−10 KV

+4 KV

CYLINDERS #2 AND #4 HAVE FIRING VOLTAGES THAT RISE A POSITIVE DIRECTION

CYLINDERS #4 AND #2 HAVE BREAKOVER VOLTAGES THAT FALL FROM ABOVE GROUND TOWARD GROUND

CYLINDERS #2 OR #4 ON WASTE

GROUND

8 V

CSI SIGNAL

4 V

A POSITIVE GOING VOLTAGE

A NEGATIVE GOING VOLTAGE

5 μS

GROUND

**FIGURE 29–15** The slight (5 microsecond) difference in the firing of the companion cylinders is enough time to allow the PCM to determine which cylinder is firing on the compression stroke. The compression sensing ignition (CSI) signal is then processed by the PCM which then determines which cylinder is on the compression stroke.

coil, then the PCM assumes that cylinder 4 is on the compression stroke. Engines equipped with compression-sensing ignition systems do not require the use of a camshaft position sensor to determine specific cylinder numbers. ● SEE FIGURE 29–15.

## COIL-ON-PLUG IGNITION

**TERMINOLOGY**   Coil-on-plug (COP) ignition uses one ignition coil for each spark plug. This system is also called *coil-by-plug, coil-near-plug,* or *coil-over-plug ignition.* ● SEE FIGURES 29–16 AND 29–17.

**ADVANTAGES**   The COP system eliminates the spark plug wires that are often the source of electromagnetic interference (EMI) that can cause problems to some computer signals. The vehicle computer controls the timing of the spark. Ignition timing also can be changed (retarded or advanced) on a cylinder-by-cylinder basis for maximum performance and to respond to knock sensor signals.

**TYPES OF COP SYSTEMS**   There are two basic types of COP ignition systems:

- **Two primary wires.** This design uses the vehicle computer to control the firing of the ignition coil. The two wires include the ignition voltage feed and the pulse ground wire, which is controlled by the computer. The ignition control module is located in the PCM, which handles all ignition timing and coil on-time control.

- **Three primary wires.** This design includes an ignition module at each coil. The three wires include the following:
    - Ignition voltage
    - Ground
    - Pulse from the computer to the built-in ignition module

  Vehicles use a variety of COP-type ignition systems, including the following:

    - Many General Motors V-8 engines use a coil-near-plug system with individual coils and modules for each individual cylinder that are placed on the valve covers. Short secondary ignition spark plug wires are used to connect the output terminal of the ignition coil to the

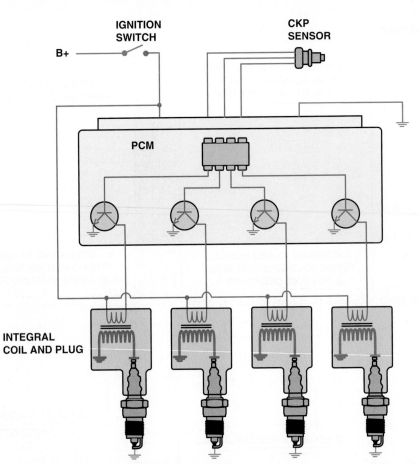

**FIGURE 29–16** A typical two wire coil-on-plug (COP) ignition system showing the triggering and the switching being performed by the PCM from input from the crankshaft position sensor.

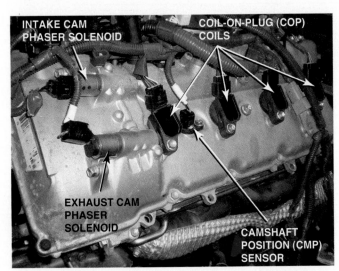

**FIGURE 29–17** An overhead camshaft engine equipped with variable valve timing on both the intake and exhaust camshafts and coil-on-plug (COP) ignition.

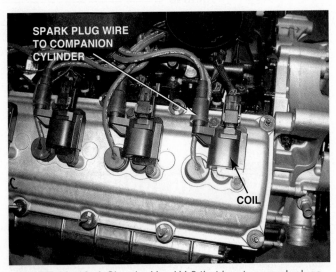

**FIGURE 29–18** A Chrysler Hemi V-8 that has two spark plugs per cylinder. The coil on top of one spark fires that plug and, through a spark plug wire, fires a plug in the companion cylinder.

spark plug, and therefore this system is called a *coil-near-plug* system.

■ In a combination of COP and waste-spark systems, the systems fire a spark plug attached to the coil and use a spark plug wire attached to the other secondary terminal

of the coil to fire another spark plug of the companion cylinder. This type of system is used in some Chrysler Hemi V-8 and Toyota V-6 engines. ● **SEE FIGURE 29–18.**

Most new engines use coil-on-plug-type ignition systems. Each coil is controlled by the PCM, which can vary the ignition

timing separately for each cylinder based on signals the PCM receives from the knock sensor(s). For example, if the knock sensor detects that a spark knock has occurred after firing cylinder 3, then the PCM will continue to monitor cylinder 3 and retard timing on just this one cylinder if necessary to prevent engine-damaging detonation.

### ION-SENSING IGNITION

In an **ion-sensing ignition** system, the spark plug itself becomes a sensor. An ion-sensing ignition uses a COP design where the ignition control module (ICM) applies a DC voltage across the spark plug gap *after* the ignition event to sense the ionized gases (called plasma) inside the cylinder. Ion-sensing ignition is used in the Saab four and six cylinder engines and on many Harley-Davidson motorcycles.
● SEE FIGURE 29–19.

The secondary coil discharge voltage (10 to 15 kV) is electrically isolated from the ion-sensing circuit. The combustion flame is ionized and will conduct some electricity, which can be accurately measured at the spark plug gap. The purpose of this circuit includes the following:

- Misfire detection (required by OBD-II regulations)
- Knock detection (eliminates the need for a knock sensor)
- Ignition timing control (to achieve the best spark timing for maximum power with lowest exhaust emissions)
- Exhaust gas recirculation (EGR) control
- Air–fuel ratio control on an individual cylinder basis

Ion-sensing ignition systems still function the same as conventional COP designs, but the engine does not need to be equipped with a camshaft position sensor for misfire detection or a knock sensor because both of these faults are achieved using the electronics inside the ignition control circuits.

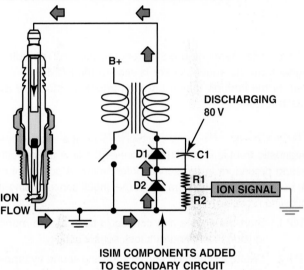

**FIGURE 29–19** A DC voltage is applied across the spark plug gap after the plug fires, and the circuit can determine if the correct air–fuel ratio was present in the cylinder and if knock occurred. The applied voltage for ion-sensing does not jump the spark plug gap but determines the conductivity of the ionized gases left over from the combustion process.

## KNOCK SENSORS

### PURPOSE AND FUNCTION

**Knock sensors** are used to detect abnormal combustion, often called **ping, spark knock,** or **detonation.** Whenever abnormal combustion occurs, a rapid pressure increase occurs in the cylinder, creating a vibration in the engine block. It is this vibration that is detected by the knock sensor. The signal from the knock sensor is used by the PCM to retard the ignition timing until the knock is eliminated, thereby reducing the damaging effects of the abnormal combustion on pistons and other engine parts.

Inside the knock sensor is a piezoelectric element that is a type of crystal that produces a voltage when pressure or a vibration is applied to the unit. The knock sensor is tuned to the engine knock frequency, which is a range from

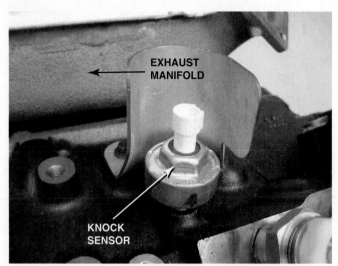

**FIGURE 29–20** A typical knock sensor on the side of the block. Some are located in the "V" of a V-type engine and are not noticeable until the intake manifold has been removed.

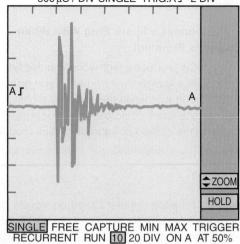

A 50 V AC 1:1 PROBE   B 200 mV  OFF 1:1 PROBE
500 µS / DIV SINGLE  TRIG:A ⌐–2 DIV

SINGLE FREE CAPTURE MIN MAX TRIGGER
RECURRENT RUN 10 20 DIV ON A AT 50%

**FIGURE 29–21** A typical waveform from a knock sensor during a spark knock event. This signal is sent to the computer, which in turn retards the ignition timing. This timing retard is accomplished by an output command from the computer to either a spark advance control unit or directly to the ignition module.

5 to 10 kHz, depending on the engine design. The voltage signal from the knock sensor is sent to the PCM. The PCM retards the ignition timing until the knocking stops. ● **SEE FIGURE 29–20.**

**DIAGNOSING THE KNOCK SENSOR**  If a knock sensor diagnostic trouble code (DTC) is present, follow the specified testing procedure in the service information. A scan tool can be used to check the operation of the knock sensor, using the following procedure:

**STEP 1**  Start the engine and connect a scan tool to monitor ignition timing and/or knock sensor activity.

**STEP 2**  Create a simulated engine knocking sound by tapping on the engine block or cylinder head with a soft-faced mallet or small ball peen hammer.

**STEP 3**  Observe the scan tool display. The vibration from the tapping should have been interpreted by the knock sensor as a knock, resulting in a knock sensor signal and a reduction in the spark advance.

A knock sensor also can be tested using a digital storage oscilloscope. ● **SEE FIGURE 29–21.**

**NOTE: Some engine computers are programmed to ignore knock sensor signals when the engine is at idle speed to avoid having the noise from a loose accessory drive belt or other accessory interpreted as engine knock. Always follow the vehicle manufacturer's recommended testing procedure.**

**REPLACING A KNOCK SENSOR**  If replacing a knock sensor, be sure to purchase the exact replacement needed, because they often look the same, but the frequency range can

  **REAL WORLD FIX**

**The Low-Power Toyota**

A technician talked about the driver of a Toyota who complained about poor performance and low fuel economy. The technician checked everything and even replaced all secondary ignition components. Then the technician connected a scan tool and noticed that the knock sensor was commanding the timing to be retarded. Careful visual inspection revealed a "chunk" missing from the serpentine belt, which caused a "noise" similar to a spark knock. Apparently the knock sensor was "hearing" the accessory drive belt noise and kept retarding the ignition timing. After replacing the accessory drive belt, a test drive confirmed that normal engine power was restored.

Other items that can fool the knock sensor to retard the ignition timing include the following:
- Loose valve lifter adjustment
- Engine knocks
- Loose accessory brackets such as air-conditioning compressor, power steering pumps, or alternator

vary according to engine design and location on the engine. Always tighten the knock sensor using a torque wrench and tighten to the specified torque to avoid causing damage to the piezoelectric element inside the sensor.

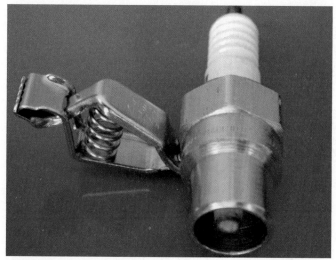

FIGURE 29–22 A spark tester looks like a regular spark plug with an alligator clip attached to the shell. This tester has a specified gap that requires at least 25,000 volts (25 kV) to fire.

FIGURE 29–23 A close-up showing the recessed center electrode on a spark tester. It is recessed 3/8 inch into the shell, and the spark must then jump another 3/8 inch to the shell for a total gap of 3/4 inch.

## IGNITION SYSTEM DIAGNOSIS

**CHECKING FOR SPARK** In the event of a no-start condition, the first step should be to check for secondary voltage out of the ignition coil or to the spark plugs. If the engine is equipped with a separate ignition coil, remove the coil wire from the center of the distributor cap, install a **spark tester** and crank the engine. See the Tech Tip "Always Use a Spark Tester." A good coil and ignition system should produce a blue spark at the spark tester. ● **SEE FIGURES 29–22 AND 29–23.**

If the ignition system being tested does not have a separate ignition coil, disconnect any spark plug wire from a spark plug and, while cranking the engine, test for spark available at the spark plug wire, again using a spark tester.

NOTE: An intermittent spark should be considered a no-spark condition.

Typical causes of a no-spark (intermittent spark) condition include the following:

1. Weak ignition coil
2. Low or no voltage to the primary (positive) side of the coil
3. High resistances, open coil wire, or spark plug wire
4. Negative side of the coil not being pulsed by the ignition module
5. Defective pickup coil or crankshaft position sensor
6. Defective ignition control module (ICM)
7. Defective main relay (can be labeled Main, EFI, ASD on Chrysler products; EEC on Ford vehicle relays)

The triggering sensor has to work to create a spark from the ignition coil(s). If there is a no-spark condition, check for triggering by using a scan tool and check for engine RPM while cranking the engine:

■ If the engine speed (RPM) shows zero or almost zero while cranking, the most likely cause is a defective triggering sensor or sensor circuit fault.

■ If the engine speed (RPM) is shown on the scan tool while cranking the engine, then the triggering sensor is working (in most cases).

Check service information for the exact procedure to follow for testing triggering sensors.

**IGNITION COIL TESTING USING AN OHMMETER** If an ignition coil is suspected of being defective, a simple ohmmeter check can be performed to test the resistance of the primary and secondary windings inside the coil. For accurate resistance measurements, the wiring to the coil should be removed before testing. To test the primary coil winding resistance, take the following steps: ● **SEE FIGURE 29–24.**

STEP 1 Set the meter to read low ohms.

STEP 2 Measure the resistance between the positive terminal and the negative terminal of the ignition coil. Most coils will give a reading between less than 1 and 3 ohms. Check the manufacturer's specifications for the exact resistance values.

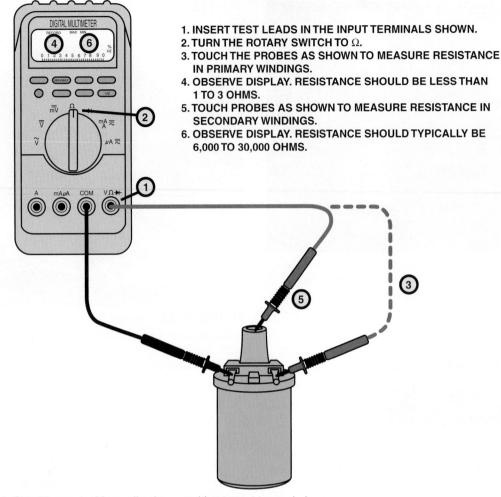

1. INSERT TEST LEADS IN THE INPUT TERMINALS SHOWN.
2. TURN THE ROTARY SWITCH TO Ω.
3. TOUCH THE PROBES AS SHOWN TO MEASURE RESISTANCE IN PRIMARY WINDINGS.
4. OBSERVE DISPLAY. RESISTANCE SHOULD BE LESS THAN 1 TO 3 OHMS.
5. TOUCH PROBES AS SHOWN TO MEASURE RESISTANCE IN SECONDARY WINDINGS.
6. OBSERVE DISPLAY. RESISTANCE SHOULD TYPICALLY BE 6,000 TO 30,000 OHMS.

**FIGURE 29–24** Checking an ignition coil using a multimeter set to read ohms.

 **TECH TIP**

### Always Use a Spark Tester

A spark tester looks like a spark plug except it has a recessed center electrode and no side electrode. The tester commonly has an alligator clip attached to the shell so that it can be clamped on a good ground connection on the engine. A good ignition system should be able to cause a spark to jump this wide gap at atmospheric pressure. Without a spark tester, a technician might assume that the ignition system is okay because it can spark across a normal, grounded spark plug. The voltage required to fire a standard spark plug when it is out of the engine and not under pressure is about 3,000 volts or less. An electronic ignition spark tester requires a minimum of 25,000 volts to jump the 3/4-inch gap. Therefore, never assume that the ignition system is okay because it fires a spark plug—always use a spark tester. *Remember that an intermittent spark across a spark tester should be interpreted as a no-spark condition.*

To test the secondary coil winding resistance, follow these steps:

**STEP 1** Set the meter to read kilohms (kΩ).

**STEP 2** Measure the resistance either between the primary terminal and the secondary coil tower or between the secondary towers. The normal resistance of most coils ranges between 6,000 and 30,000 ohms. Check the manufacturer's specifications for the exact resistance values.

**MAGNETIC SENSOR TESTING** Magnetic sensor such as the pickup coil, located under the distributor cap on many electronic ignition engines, can cause a no-spark condition if defective. The sensor must generate an AC voltage pulse to the ignition module so that the module can pulse the ignition coil.

The sensor contains a coil of wire, and the resistance of this coil should be within the range specified by the manufacturer.

Some common tests for pickup coils and magnetic crankshaft position sensors include the following:

- **Resistance.** Usually between 150 and 1,500 ohms, but check service information for the exact specifications.
  ● **SEE FIGURE 29–25.**
- **Coil shorted to ground.** Check that the coil windings are insulated from ground by checking for continuity

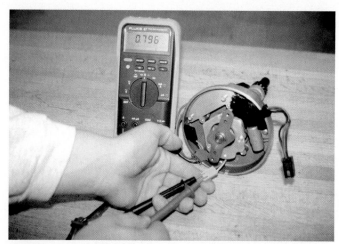

**FIGURE 29–25** Measuring the resistance of an HEI pickup coil using a digital multimeter set to the ohms position. The reading on the face of the meter is 0.796 kΩ, or 796 ohms in the middle of the 500- to 1,500-ohm specifications.

using an ohmmeter. With one ohmmeter lead attached to ground, touch the other lead of the ohmmeter to the pickup coil terminal. The ohmmeter should read OL (over limit) with the ohmmeter set on the high scale. If the sensor resistance is not within the specified range or if it has continuity to ground, replace the pickup coil assembly.

- **AC voltage output.** The sensor also can be tested for proper voltage output. During cranking, most sensors should produce a minimum of 0.25 volt AC.

**TESTING HALL-EFFECT SENSORS** As with any other sensor, the output of the Hall-effect sensor should be tested first. Using a digital voltmeter, check for the following:

- Power and ground to the sensor
- Changing voltage (pulsed on and off or digital DC voltage) when the engine is being cranked
- Waveform, using an oscilloscope ● **SEE FIGURE 29–26.**

**TESTING OPTICAL SENSORS** Optical sensors will not operate if they are dirty or covered in oil. Perform a thorough visual inspection and look for an oil leak that could cause dirty oil to get on the LED or phototransistor. Also be sure that the light shield is securely fastened and that the seal is lightproof. An optical sensor also can be checked using an oscilloscope. ● **SEE FIGURE 29–27.**

Because of the speed of the engine and the number of slits in the optical sensor disk, a scope is one of the only tools that can capture useful information. For example, a Nissan has 360 slits, and if it is running at 2000 RPM, a signal is generated 720,000 times per minute, or 12,000 times per second.

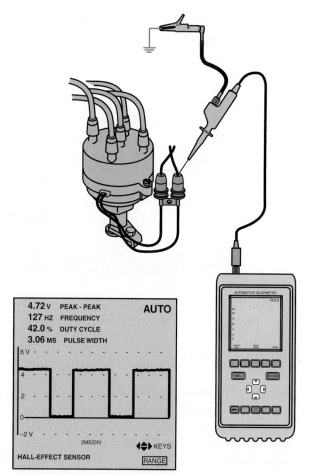

**FIGURE 29–26** The connection required to test a Hall-effect sensor. A typical waveform from a Hall-effect sensor.

## SPARK PLUG WIRE INSPECTION

Spark plug wires should be visually inspected for cuts or defective insulation. Faulty spark plug wire insulation can cause hard starting or no starting in rainy or damp weather conditions. When removing a spark plug wire, be sure to rotate the boot of the wire at the plug before pulling it off the spark plug. This will help prevent damaging the wire, as many wires are stuck to the spark plug and are often difficult to remove.

**VISUAL INSPECTION** A thorough visual inspection should include a look at the following items:

- Check all spark plug wires for proper routing. All plug wires should be in the factory wiring separators and be clear of any metallic object that could damage the insulation and cause a short-to-ground fault.
- Check that all spark plug wires are securely attached to the spark plugs and to the distributor cap or ignition coil(s).

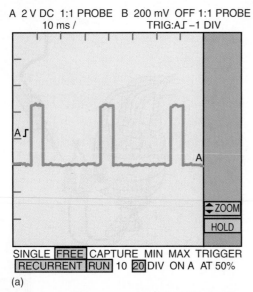

A 2 V DC  1:1 PROBE   B  200 mV  OFF 1:1 PROBE
10 ms /                           TRIG:A⌐ –1 DIV

A ⌐

A

ZOOM
HOLD

SINGLE FREE CAPTURE MIN MAX TRIGGER
RECURRENT RUN 10 20 DIV ON A AT 50%

(a)

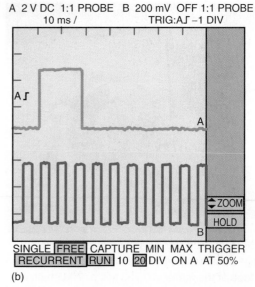

A 2 V DC  1:1 PROBE   B  200 mV  OFF 1:1 PROBE
10 ms /                           TRIG:A⌐ –1 DIV

A ⌐

A

B

ZOOM
HOLD

SINGLE FREE CAPTURE MIN MAX TRIGGER
RECURRENT RUN 10 20 DIV ON A AT 50%

(b)

**FIGURE 29–27** (a) The low-resolution signal has the same number of pulses as the engine has cylinders. (b) A dual-trace pattern showing both the low-resolution signal and the high-resolution signals that usually represent 1 degree of rotation.

 **TECH TIP**

### Bad Wire? Replace the Coil!

When performing engine testing (such as a compression test), always ground the coil wire. Never allow the coil to discharge without a path to ground for the spark. High-energy ignition systems can produce 40,000 volts or more of electrical pressure. If the spark cannot spark to ground, the coil energy can (and usually does) arc inside the coil itself, creating a low-resistance path to the primary windings or the steel laminations of the coil. ● SEE FIGURE 29–28.

This low-resistance path is called a **track** and could cause an engine miss under load even though all of the remaining component parts of the ignition system are functioning correctly. Often these tracks do not show up on any coil test, including most scopes. Because the track is a lower resistance path to ground than normal, it requires that the ignition system be put under a load for it to be detected, and even then, the misfire may be intermittent. If a misfire was the result of an open circuit in the secondary circuit, the coil is ruined and must be replaced.

When disabling an ignition system, perform one of the following procedures to prevent possible ignition coil damage:

1. Remove the power source wire from the ignition system to prevent any ignition operation.
2. On distributor-equipped engines, remove the secondary coil wire from the center of the distributor cap and connect a jumper wire between the disconnected coil wire and a good engine ground. This ensures that the secondary coil energy will be safely grounded and prevents high-voltage coil damage.

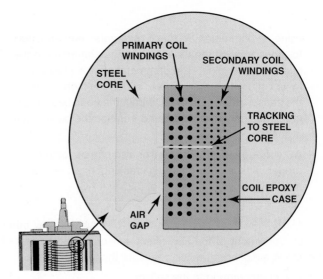

**FIGURE 29–28** A track inside an ignition coil is not a short but a low-resistance path or hole that has been burned through from the secondary wiring to the steel core.

- Check that all spark plug wires are clean and free from excessive dirt or oil. Check that all protective covers normally covering the coil and/or distributor cap are in place and not damaged.
- Carefully check the cap and distributor rotor for faults or coil secondary terminal on waste spark coils. ● SEE FIGURE 29–29.

Visually check the wires and boots for damage. ● SEE FIGURE 29–30.

Check all spark plug wires with an ohmmeter for proper resistance. Good spark plug wires should measure less than 10,000 ohms per foot of length. ● SEE FIGURE 29–31.

**FIGURE 29–29** Corroded terminals on a waste-spark coil can cause misfire diagnostic trouble codes to be set.

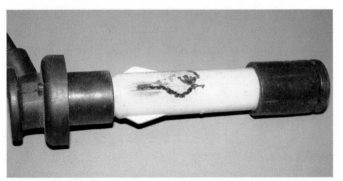

**FIGURE 29–30** This spark plug boot on an overhead camshaft engine has been arcing to the valve cover causing a misfire to occur.

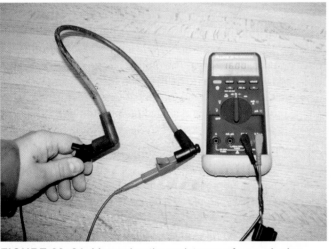

**FIGURE 29–31** Measuring the resistance of a spark plug wire with a multimeter set to the ohms position. The reading of 16.03 kΩ (16,030 ohms) is okay because the wire is about 2 feet long. Maximum allowable resistance for a spark plug wire this long would be 20 kΩ (20,000 ohms).

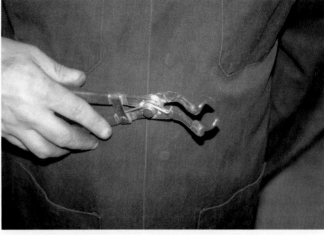

**FIGURE 29–32** This spark plug wire boot pliers is a handy addition to any tool box.

**TECH TIP**

**Spark Plug Wire Pliers Are a Good Investment**

Spark plug wires are often difficult to remove. Using good-quality spark plug wire pliers, as shown in
●**FIGURE 29–32,** saves time and reduces the chance of harming the wire during removal.

**TECH TIP**

**Route the Wires Right!**

High voltage is present through spark plug wires when the engine is running. Surrounding the spark plug wires is a magnetic field that can affect other circuits or components of the vehicle. For example, if a spark plug wire is routed too closely to the signal wire from a mass airflow (MAF) sensor, the induced signal from the ignition wire could create a false MAF signal to the computer. The computer, not able to detect that the signal was false, would act on the MAF signal and command the appropriate amount of fuel based on the false MAF signal.

To prevent any problems associated with high-voltage spark plug wires, be sure to route them using all of the factory holding brackets and wiring combs.
●**SEE FIGURE 29–33.**

If the factory method is unknown, most factory service information shows the correct routing.

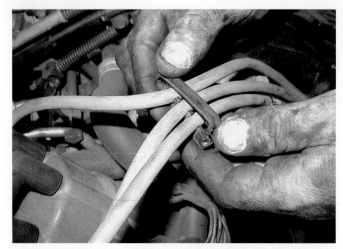

**FIGURE 29–33** Always take the time to install spark plug wires back into the original holding brackets (wiring combs).

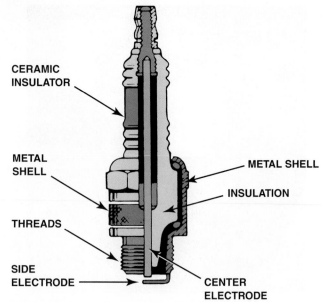

**FIGURE 29–34** Parts of a spark plug.

# SPARK PLUGS

**SPARK PLUG CONSTRUCTION** **Spark plugs** are manufactured from ceramic insulators inside a steel shell. The threads of the shell are rolled and a seat is formed to create a gastight seal with the cylinder head. ● **SEE FIGURE 29–34.**

The physical differences in spark plugs include the following:

- **Reach.** This is the length of the threaded part of the plug.
- **Heat range.** This refers to how rapidly the heat created at the tip is transferred to the cylinder head. A spark plug with a long ceramic insulator path will run hotter at the tip than one that has a shorter path because the heat must travel farther. ● **SEE FIGURE 29–35.**
- **Type of seat.** Some spark plugs use a gasket, and others rely on a tapered seat to seal.

**RESISTOR SPARK PLUGS** Most spark plugs include a resistor in the center electrode, which helps to reduce electromagnetic noise or radiation from the ignition system. The closer the resistor is to the actual spark or arc, the more effective it becomes. The value of the resistor is usually between 2,500 and 7,500 ohms.

**PLATINUM SPARK PLUGS** **Platinum spark plugs** have a small amount of the precious metal platinum included on the end of the center electrode as well as on the ground or side electrode. Platinum is a gray-white metal that does not react with oxygen and, therefore, will not erode away as can occur with conventional nickel alloy spark plug electrodes. Platinum is also used as a catalyst in catalytic converters, where it is able to start a chemical reaction without itself being consumed.

**IRIDIUM SPARK PLUGS** Iridium is a white precious metal and is the most corrosion-resistant metal known. Most **iridium spark plugs** use a small amount of iridium welded onto the tip of a small center electrode, 0.0015 to 0.002 inch (0.4 to 0.6 mm)

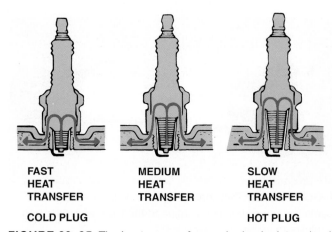

**FIGURE 29–35** The heat range of a spark plug is determined by distance the heat flows from the tip to the cylinder head.

in diameter. The small diameter reduces the voltage required to jump the gap between the center and the side electrode, thereby reducing possible misfires. The ground or side electrode is usually tipped with platinum to help reduce electrode gap wear.

Spark plugs should be inspected when an engine performance problem occurs and should be replaced at specified intervals to ensure proper ignition system performance:

- Nonplatinum spark plugs have a service life of over 20,000 miles (32,000 km).
- Platinum-tipped original equipment spark plugs have a typical service life of 60,000 to 100,000 miles (100,000 to 160,000 km) or longer.

Used spark plugs should *not* be cleaned and reused unless absolutely necessary. The labor required to remove and replace (R & R) spark plugs is the same whether the spark plugs are replaced or cleaned. Although cleaning spark plugs often restores proper engine operation, the service life of cleaned spark plugs is definitely shorter than that of new spark plugs.

NOTE: Platinum-tipped spark plugs should not be re-gapped on one that has been used in an engine be-fore. The engine heat makes the platinum brittle, and the center electrode can be easily broken if regapping the plug is attempted. Using a gapping tool can break the platinum after it has been used in an engine. Check service information regarding the recommended type of spark plugs and the specified service procedures.

## SPARK PLUG SERVICE
When replacing spark plugs, per-form the following steps:

**STEP 1** **Check service information.** Check for the exact spark plug to use and the specified instructions and/or technical service bulletins that affect the number of plug to be used or a revised replacement procedure.

**STEP 2** **Allow the engine to cool before removing spark plugs.** This is true especially on engines with alumi-num cylinder heads.

**STEP 3** **Use compressed air or a brush to remove dirt from around the spark plug before removal.** This step helps prevent dirt from getting into the cylinder of an engine while removing a spark.

**STEP 4** **Check the spark plug gap and correct as needed.** Be careful not to damage the tip on the center electrode if adjusting a platinum or iridium type of spark plug.

**STEP 5** **Install the spark plugs by hand.** After tightening by hand, use a torque wrench and tighten the spark plugs to factory specifications. ● **SEE FIGURES 29–36 AND 29–37.**

Spark plugs are the windows to the inside of the combustion chamber. A thorough visual inspection of the spark plugs often can lead to the root cause of an engine performance problem. Two indications on spark plugs and their possible root causes in engine performance include the following:

1. **Carbon fouling.** If the spark plug(s) has *dry black carbon* (soot), the usual causes include the following:
   - Excessive idling
   - Overly rich air–fuel mixture due to a fuel system fault
   - Weak ignition system output

2. **Oil fouling.** If the spark plug has wet, oily deposits with lit-tle electrode wear, oil may be getting into the combustion chamber from the following:
   - Worn or broken piston rings
   - Worn valve guides
   - Defective or missing valve stem seals

When removing spark plugs, place them in order so that they can be inspected to check for engine problems that might af-fect one or more cylinders. All spark plugs should be in the same condition, and the color of the center insulator should be light tan or gray. If all the spark plugs are black or dark, the engine should be checked for conditions that could cause an overly rich air–fuel mixture or possible oil burning. If only one or a few spark plugs are black, check those cylinders for proper firing (possible defec-tive spark plug wire) or an engine condition affecting only those particular cylinders. ● **SEE FIGURES 29–38 THROUGH 29–41.**

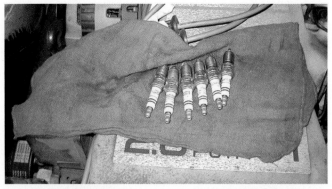

**FIGURE 29–36** When removing spark plugs, it is wise to arrange them in the order by the cylinder they were removed from so that they can be compared and any problem can be identified with a particular cylinder.

**FIGURE 29–37** A spark plug thread chaser is a low-cost tool that hopefully will not be used often but is necessary to use to clean the threads before new spark plugs are installed.

If all spark plugs are white, check for possible overad-vanced ignition timing or a vacuum leak causing a lean air–fuel mixture. If only one or a few spark plugs are white, check for a vacuum leak or injector fault affecting the air–fuel mixture only to those particular cylinders.

NOTE: The engine computer "senses" rich or lean air-fuel ratios by means of input from the oxygen sensor(s). If one cylinder is lean, the PCM may make all other cyl-inders richer to compensate.

Inspect all spark plugs for wear by first checking the condi-tion of the center electrode. As a spark plug wears, the center electrode becomes rounded. If the center electrode is rounded, higher ignition system voltage is required to fire the spark plug.

When installing spark plugs, always use the correct tight-ening torque to ensure proper heat transfer from the spark plug shell to the cylinder head. ● **SEE CHART 29–1.**

NOTE: General Motors does not recommend the use of antiseize compound on the threads of spark plugs be-ing installed in an aluminum cylinder head because the spark plug will be overtightened. This excessive tighten-ing torque places the threaded portion of the spark plug too far into the combustion chamber, where carbon can accumulate and result in the spark plugs being difficult to remove. If antiseize compound is used on spark plug threads, reduce the tightening torque by 40%. Always follow the vehicle manufacturer's recommendations.

**FIGURE 29–38** A normally worn spark plug that uses a tapered platinum-tipped center electrode.

**FIGURE 29–40** A spark plug from an engine that had a blown head gasket. The white deposits could be from the aluminum of the piston or from the additives in the coolant.

**FIGURE 29–39** A worn spark plug showing fuel and/or oil deposits.

**FIGURE 29–41** A platinum tipped spark plug that is fuel soaked indicating a fault with the fuel system or the ignition system causing the spark plug to not fire.

| SPARK PLUG TYPE | TORQUE WITH TORQUE WRENCH (LB-FT) | | TORQUE WITHOUT TORQUE WRENCH (TURNS AFTER SEATED) | |
| --- | --- | --- | --- | --- |
| | CAST-IRON HEAD | ALUMINUM HEAD | CAST-IRON HEAD | ALUMINUM HEAD |
| Gasket | 26–30 | 18–22 | 1/4 | 1/4 |
| 14 mm | 32–38 | 28–34 | 1/4 | 1/4 |
| 18 mm | | | | |
| Tapered seat | 7–15 | 7–15 | 1/16 (snug) | 1/16 (snug) |
| 14 mm | 15–20 | 15–20 | 1/16 (snug) | 1/16 (snug) |
| 18 mm | | | | |

Typical spark plug installation torque.

### TECH TIP

**Two-Finger Trick**

To help prevent overtightening a spark plug when a torque wrench is not available, simply use two fingers on the ratchet handle. Even the strongest service technician cannot overtighten a spark plug by using two fingers.

## IGNITION TIMING

**PURPOSE**   **Ignition timing** refers to when the spark plug fires in relation to piston position. The time when the spark occurs depends on engine speed and, therefore, must be advanced (spark plugs fire sooner) as the engine rotates faster. The ignition in the cylinder takes a certain amount of time, usually 30 ms (30/1,000 of a second), and remains constant regardless of engine speed. Therefore, to maintain the most efficient combustion, the ignition sequence has to occur sooner as the engine speed increases. For maximum efficiency from the expanding gases inside the combustion chamber, the burning of the air–fuel mixture should end by about 10 degrees after top dead center (ATDC). If the burning of the mixture is still occurring after that point, the expanding gases do not exert much force on the piston because the gases are "chasing" the piston as it moves downward.

Therefore, to achieve the goal of having the air–fuel mixture be completely burned by the time the piston reaches 10 degrees ATDC, the spark must be advanced (occur sooner) as the engine speed increases. This timing advance is determined and controlled by the PCM on most vehicles. ● **SEE FIGURES 29–42 AND 29–43.**

If the engine is equipped with a distributor, it may be possible to adjust the base or the initial timing. The initial timing is usually set to fire the spark plug between zero degrees (TDC)

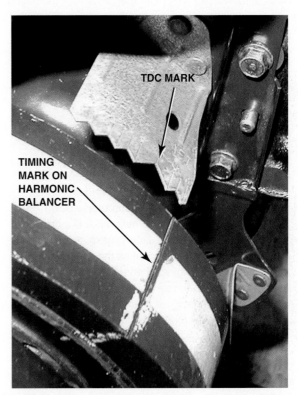

**FIGURE 29–42** Ignition timing marks are found on the harmonic balancers on engines equipped with distributors that can be adjusted for timing.

or slightly before TDC (BTDC). Ignition timing changes as mechanical wear occurs to the following:

- Timing chain
- Distributor gear
- Camshaft drive gear

**CHECKING IGNITION TIMING**   To be assured of the proper ignition timing, follow exactly the timing procedure indicated on the underhood vehicle emission control information (VECI) decal. ● **SEE FIGURE 29–44.**

**NOTE: The ignition timing for waste-spark and coil-on-plug ignition systems cannot be adjusted.**

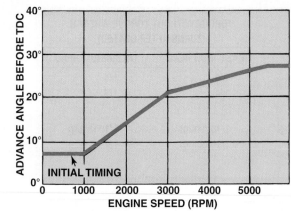

FIGURE 29–43 The initial (base) timing is where the spark plug fires at idle speed. The PCM then advances the timing based primarily on engine speed.

## TECH TIP

### Two Marks Are the Key to Success

When a distributor is removed from an engine, always mark where the rotor is pointing to ensure that the distributor is reinstalled in the correct position. Because of the helical cut on the distributor drive gear, the rotor rotates as the distributor is being removed from the engine. To help reinstall a distributor without any problems, simply make another mark where the rotor is pointing just as the distributor is lifted out of the engine. Then to reinstall, simply line up the rotor to the second mark and lower the distributor into the engine. The rotor should then line up with the original mark as a double check.

## TECH TIP

### Use a Water Spray Bottle to Check the Secondary Ignition Circuit

To check for breaks in the insulation in the secondary circuit perform the following steps:

**STEP 1** Start the engine and allow to reach normal operating temperature.

**STEP 2** Using a water spray bottle set to mist, spray the secondary ignition components and listen for any change in operation.

**STEP 3** If the engine operation is affected at all, there is a break in the secondary ignition circuit insulation. Continue spraying to pinpoint the exact location.

(a)

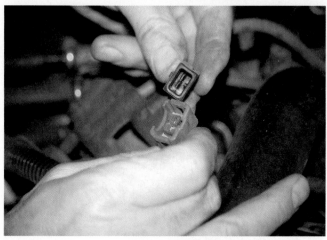

(b)

FIGURE 29–44 (a) Typical SPOUT connector as used on many Ford engines equipped with distributor ignition. (b) The connector must be opened (disconnected) to check and/or adjust the ignition timing. On DIS/EDIS systems, the connector is called SPOUT/SAW (spark output/spark angle word).

## IGNITION SYSTEM SYMPTOM GUIDE

| Problem | Possible Causes and/or Solutions |
|---|---|
| No spark out of the coil | ▪ Open in the ignition switch circuit or theft deterrent system fault<br>▪ Defective ignition control module<br>▪ Defective triggering device (magnetic sensor, Hall effect or optical sensor) |
| Weak spark out of the coil | ▪ High-resistance coil wire or spark plug wire<br>▪ Poor ground between the distributor or ignition control module and the engine block |
| Engine misfire | ▪ Defective (open) spark plug wire<br>▪ Worn or fouled spark plugs<br>▪ Defective ignition control module (ICM) |

1. All inductive ignition systems supply battery voltage to the positive side of the ignition coil and pulse the negative side of the coil on and off to ground to create a high-voltage spark.

2. If an ignition system uses a distributor, it is a distributor ignition system.

3. If an ignition system does not use a distributor, it is an electronic ignition system.

4. A waste-spark ignition system fires two spark plugs at the same time.

5. A coil-on-plug ignition system uses an ignition coil for each spark plug.

6. A thorough visual inspection should be performed on all ignition components when diagnosing an engine performance problem.

7. Platinum spark plugs should not be regapped after use in an engine.

## REVIEW QUESTIONS

1. How can 12 volts from a battery be changed to 40,000 volts for ignition?

2. How does a magnetic sensor work?

3. How does a Hall-effect sensor work?

4. How does a waste-spark ignition system work?

5. Why should a spark tester be used to check for spark rather than a standard spark plug?

6. How do you test a pickup coil for resistance and AC voltage output?

7. What harm can occur if the engine is cranked or run with an open (defective) spark plug wire?

## CHAPTER QUIZ

1. The primary (low-voltage) ignition system must be working correctly before any spark occurs from a coil. Which component is not in the primary ignition circuit?
   a. Spark plug wiring
   b. Ignition module (igniter)
   c. Pickup coil (pulse generator)
   d. Ignition switch

2. The ignition module has direct control over the firing of the coil(s) of an ignition system. Which component(s) triggers (controls) the module?
   a. Pickup coil     c. Crankshaft sensor
   b. Computer     d. All of the above

3. Distributor ignition systems can be triggered by a _____.
   a. Hall-effect sensor     c. Spark sensor
   b. Magnetic sensor     d. Either a or b

4. Ignition coil primary resistance is usually _____ ohms.
   a. 6,000 to 30,000     c. Less than 1 to 3
   b. 150 to 1,500     d. Zero

5. Coil polarity is determined by the _____.
   a. Direction of rotation of the coil windings
   b. Turn ratio
   c. Direction of laminations
   d. Saturation direction

6. A compression-sensing ignition system uses a _____ type of ignition.
   a. Distributor     c. Waste-spark
   b. Coil-on-plug     d. All of the above

7. The pulse generator _____.
   a. Fires the spark plug directly
   b. Signals the ignition control module (ICM)
   c. Signals the computer that fires the spark plug directly
   d. Is used as a tachometer reference signal by the computer and has no other function

8. Two technicians are discussing coil-on-plug ignition systems. Technician A says that they can be called coil-near-plug or coil-by-plug ignition systems. Technician B says that some can use ion sensing. Which technician is correct?
   a. Technician A only
   b. Technician B only
   c. Both Technicians A and B
   d. Neither Technician A nor B

9. A waste-spark-type ignition system fires _____.
   a. Two spark plugs at the same time
   b. One spark plug with reverse polarity
   c. One spark plug with straight polarity
   d. All of the above

10. Technician A says that a defective crankshaft position sensor can cause a no-spark condition. Technician B says that a faulty ignition control module can cause a no-spark condition. Which technician is correct?
    a. Technician A only
    b. Technician B only
    c. Both Technicians A and B
    d. Neither Technician A nor B

# chapter 30

# SCAN TOOLS AND ENGINE PERFORMANCE DIAGNOSIS

**LEARNING OBJECTIVES:** **After studying this chapter, the reader should be able to:** • Identify the steps of a diagnostic process. • Identify the types of scan tools. • Describe the troubleshooting procedures to follow when a diagnostic trouble code is set. • Explain the procedure for diagnosing and testing an onboard diagnostics generation II (OBD II) system. • Explain the methods that are used to reprogram (reflash) a vehicle computer. • Discuss the diagnostic routines and the procedures for resetting the PCM.

**KEY TERMS:** Data link connector (DLC) 383 • Drive cycle 395 • Flash codes 386 • Paper test 381 • Pending code 382 • Smoke machine 381 • Technical service bulletin (TSB) 382 • Trip 391

## THE EIGHT-STEP DIAGNOSTIC PROCEDURE

It is important that all automotive service technicians know how to diagnose and troubleshoot engine computer systems. The diagnostic process is a strategy that eliminates known-good components or systems in order to find the root cause of automotive engine performance problems. All vehicle manufacturers recommend a diagnostic procedure, and the plan suggested in this chapter combines most of the features of these plans plus additional steps developed over years of real-world problem solving.

Many different things can cause an engine performance problem or concern. The service technician has to narrow the possibilities to find the cause of the problem and correct it. A funnel is a way of visualizing a diagnostic procedure. ● **SEE FIGURE 30–1.** At the wide top are the symptoms of the problem; the funnel narrows as possible causes are eliminated until the root cause is found and corrected at the bottom of the funnel.

All problem diagnosis deals with symptoms that could be the result of many different causes. The wide range of possible solutions must be narrowed to the most likely, and these must eventually be further narrowed to the actual cause. The following section describes eight steps the service technician can take to narrow the possibilities to one cause.

**STEP 1 VERIFY THE PROBLEM (CONCERN)** Before a minute is spent on diagnosis, be certain that a problem exists. If the problem cannot be verified, it cannot be solved or tested to verify that the repair was complete. ● **SEE FIGURE 30–2.**

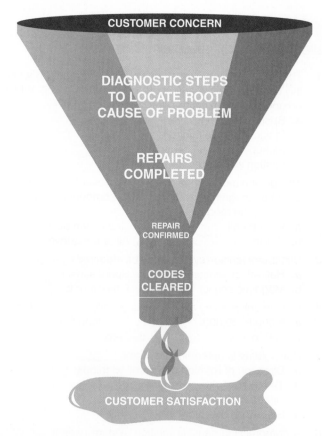

**FIGURE 30–1** A funnel is one way to visualize the diagnostic process. The purpose is to narrow the possible causes of a concern until the root cause is determined and corrected.

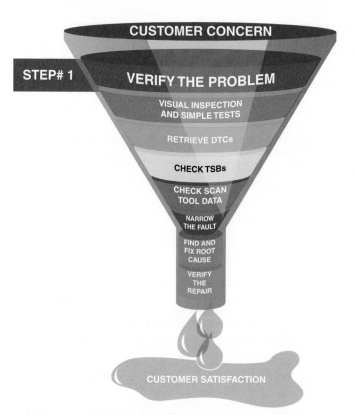

FIGURE 30–2 Step #1 is to verify the customer concern or problem. If the problem cannot be verified, then the repair cannot be verified.

**"Original Equipment" Is Not a Four-Letter Word**

To many service technicians, an original-equipment part is considered to be only marginal, and to get the really "good stuff" an aftermarket (renewal market) part has to be purchased. However, many problems can be traced to the use of an aftermarket part that has failed early in its service life. Technicians who work at dealerships usually begin their diagnosis with an aftermarket part identified during a visual inspection. It has been their experience that simply replacing the aftermarket part with the factory original-equipment (OE) part often solves the problem.

Original equipment parts are *required* to pass quality and durability standards and tests at a level not required of aftermarket parts. The technician should be aware that the presence of a new part does not necessarily mean that the part is good.

The driver of the vehicle knows much about the vehicle and how it is driven. *Before* diagnosis, always ask the following questions:

- Is the malfunction indicator light (check engine) on?
- What was the temperature outside?
- Was the engine warm or cold?
- Was the problem during starting, acceleration, cruise, or some other condition?
- How far had the vehicle been driven?
- Were any dash warning lights on? If so, which one(s)?
- Has there been any service or repair work performed on the vehicle lately?

**NOTE: This last question is very important. Many engine performance faults are often the result of something being knocked loose or a hose falling off during repair work. Knowing that the vehicle was just serviced before the problem began may be an indicator as to where to look for the solution to a problem.**

After the nature and scope of the problem are determined, the complaint should be verified before further diagnostic tests are performed. A sample form that customers could fill out with details of the problem is shown in ● **FIGURE 30–3.**

**NOTE: Because drivers differ, it is sometimes the best policy to take the customer on the test drive to verify the concern.**

## STEP 2 PERFORM A THOROUGH VISUAL INSPECTION AND BASIC TESTS
The visual inspection is the most important aspect of diagnosis! Most experts agree that between 10% and 30% of all engine performance problems can be found simply by performing a *thorough* visual inspection. The inspection should include the following:

- **Check for obvious problems (basics, basics, basics).**
  Fuel leaks
  Vacuum hoses that are disconnected or split
  Corroded connectors
  Unusual noises, smoke, or smell
  Check the air cleaner and air duct (squirrels and other small animals can build nests or store dog food in them) ● **SEE FIGURE 30–4.**
- **Check everything that does and does not work.** This step involves turning things on and observing that everything is working properly.
- **Look for evidence of previous repairs.** Any time work is performed on a vehicle, there is always a risk that something will be disturbed, knocked off, or left disconnected.
- **Check oil level and condition.** Another area for visual inspection is oil level and condition.
  **Oil level.** Oil should be to the proper level.

# ENGINE PERFORMANCE DIAGNOSIS WORKSHEET

(To Be Filled Out By the Vehicle Owner)

Name: _____ Mileage: _____ Date: _____

Make: _____ Model: _____ Year: _____ Engine: _____

| (Please Circle All That Apply in All Categories) | |
|---|---|
| **Describe Problem:** | |
| **When Did the Problem First Occur?** | • Just Started • Last Week • Last Month • Other _____ |
| **List Previous Repairs in the Last 6 Months:** | |
| **Starting Problems** | • Will Not Crank • Cranks, but Will Not Start • Starts, but Takes a Long Time |
| **Engine Quits or Stalls** | • Right after Starting • When Put into Gear • During Steady Speed Driving • Right after Vehicle Comes to a Stop • While Idling • During Acceleration • When Parking |
| **Poor Idling Conditions** | • Is Too Slow at All Times • Is Too Fast • Intermittently Too Fast or Too Slow • Is Rough or Uneven • Fluctuates Up and Down |
| **Poor Running Conditions** | • Runs Rough • Lacks Power • Bucks and Jerks • Poor Fuel Economy • Hesitates or Stumbles on Acceleration • Backfires • Misfires or Cuts Out • Engine Knocks, Pings, Rattles • Surges • Dieseling or Run-On |
| **Auto. Transmission Problems** | • Improper Shifting (Early/Late) • Changes Gear Incorrectly • Vehicle Does Not Move when in Gear • Jerks or Bucks |
| **Usually Occurs** | • Morning • Afternoon • Anytime |
| **Engine Temperature** | • Cold • Warm • Hot |
| **Driving Conditions During Occurrence** | • Short—Less Than 2 Miles • 2–10 Miles • Long—More Than 10 Miles • Stop and Go • While Turning • While Braking • At Gear Engagement • With A/C Operating • With Headlights On • During Acceleration • During Deceleration • Mostly Downhill • Mostly Uphill • Mostly Level • Mostly Curvy • Rough Road |
| **Driving Habits** | • Mostly City Driving • Highway • Park Vehicle Inside • Park Vehicle Outside **Drive Per Day:** • Less Than 10 Miles • 10–50 • More Than 50 |
| **Gasoline Used** | **Fuel Octane:** • 87 • 89 • 91 • More Than 91 **Brand:** _____ |
| **Temperature when Problem Occurs** | • 32–55° F • Below Freezing (32° F) • Above 55° F |
| **Check Engine Light/ Dash Warning Light** | • Light on Sometimes • Light on Always • Light Never On |
| **Smells** | • "Hot" • Gasoline • Oil Burning • Electrical |
| **Noises** | • Rattle • Knock • Squeak • Other |

**FIGURE 30–3** A form that the customer should fill out if there is a driveability concern to help the service technician more quickly find the root cause.

**FIGURE 30-4** This is what was found when removing an air filter from a vehicle that had a lack-of-power concern. Obviously the nuts were deposited by squirrels or some other animal, blocking a lot of the airflow into the engine.

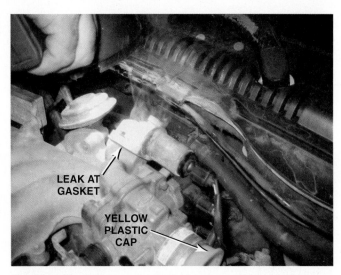

LEAK AT GASKET

YELLOW PLASTIC CAP

**FIGURE 30-5** Using a bright light makes seeing where the smoke is coming from easier. In this case, smoke was added to the intake manifold with the inlet blocked with a yellow plastic cap, and smoke was seen escaping past a gasket at the idle air control.

## TECH TIP

### Smoke Machine Testing

Vacuum (air) leaks can cause a variety of driveability problems and are often difficult to locate. One good method is to use a machine that generates a stream of smoke. Connecting the outlet of the **smoke machine** to the hose that was removed from the vacuum brake booster allows smoke to enter the intake manifold. Any vacuum leaks will be spotted by observing smoke coming out of the leak. ● **SEE FIGURE 30-5.**

**Oil condition.** Using a match or lighter, try to light the oil on the dipstick; if the oil flames up, gasoline is present in the engine oil. Drip some engine oil from the dipstick onto the hot exhaust manifold. If the oil bubbles or boils, coolant (water) is present in the oil. Check for grittiness by rubbing the oil between your fingers.

**NOTE: Gasoline in the oil will cause the engine to run rich by drawing fuel through the positive crankcase ventilation (PCV) system.**

■ **Check coolant level and condition.** Many mechanical engine problems are caused by overheating. The proper operation of the cooling system is critical to the life of any engine.

**NOTE: Check the coolant level in the radiator only if the radiator is cool. If the radiator is hot and the radiator cap is removed, the drop in pressure above the coolant will cause the coolant to boil immediately, which can cause severe burns because the coolant expands explosively upward and outward from the radiator opening.**

■ **Use the paper test.** A sound engine should produce even and steady exhaust flow at the tailpipe when running. For the **paper test**, hold a piece of paper (even a dollar bill works) or a 3-by-5-inch card within 1 inch (2.5 cm) of the tailpipe with the engine running at idle. The paper should blow evenly away from the end of the tailpipe without "puffing" or being drawn inward toward the end of the tailpipe. If the paper is at times drawn *toward* the tailpipe, the valves in one or more cylinders could be burned. Other reasons why the paper might be drawn toward the tailpipe include the following:

1. The engine could be misfiring because of a lean condition that could occur normally when the engine is cold.

2. Pulsing of the paper toward the tailpipe could also be caused by a hole in the exhaust system. If exhaust escapes through a hole in the exhaust system, air could be drawn—in the intervals between the exhaust puffs—from the tailpipe to the hole in the exhaust, causing the paper to be drawn toward the tailpipe.

■ **Ensure adequate fuel level.** Make certain that the fuel tank is at least one-fourth to one-half full; if the fuel level is low, it is possible that any water or alcohol at the bottom of the fuel tank is more concentrated and can be drawn into the fuel system.

■ **Check the battery voltage.** The voltage of the battery should be at least 12.4 volts, and the charging voltage (engine running) should be 13.5 to 15.0 volts at 2000 RPM on most vehicles. Low battery voltage can cause a variety of problems, including reduced fuel economy and incorrect (usually too high) idle speed. Higher-than-normal battery voltage can also cause the powertrain control module (PCM) problems and could cause damage to electronic modules.

■ **Check the spark using a spark tester.** Remove one spark plug wire and attach the removed plug wire to the spark tester. Attach the grounding clip of the spark tester to a good clean engine ground, start or crank the engine

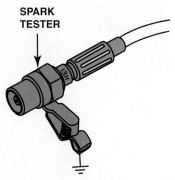

FIGURE 30–6 A spark tester connected to a spark plug wire or coil output. A typical spark tester will fire only if at least 25,000 volts is available from the coil, making a spark tester a very useful tool. Do not use one that just lights when a spark is present because they do not require more than about 2,000 volts to light.

and observe the spark tester. ● SEE FIGURE 30–6. The spark at the spark tester should be steady and consistent. If an intermittent spark occurs, then this condition should be treated as a no-spark condition. If this test does not show satisfactory spark, carefully inspect and test all components of the primary and secondary ignition systems.

NOTE: Do not use a standard spark plug to check for proper ignition system voltage. An electronic ignition spark tester is designed to force the spark to jump about 0.75 inch (19 mm). This amount of gap requires between 25,000 and 30,000 volts (25 to 30 kV) at atmospheric pressure, which is enough voltage to ensure that a spark can occur under compression inside an engine.

- **Check the fuel-pump pressure.** Checking the fuel-pump pressure is relatively easy on many port-fuel-injected engines. Often the cause of intermittent engine performance is due to a weak electric fuel pump or clogged fuel filter. Checking fuel pump pressure early in the diagnostic process eliminates low fuel pressure as a possibility.

## STEP 3 RETRIEVE THE DIAGNOSTIC TROUBLE CODES
If a diagnostic trouble code (DTC) is present in the computer memory, it may be signaled by illuminating a malfunction indicator lamp (MIL), commonly labeled "check engine" or "service engine soon." ● SEE FIGURE 30–7. Any code(s) that is displayed on a scan tool when the MIL is *not* on is called a **pending code**. Because the MIL is not on, it indicates that the fault has not repeated to cause the PCM to turn on the MIL. Although this pending code is helpful to the technician to know that a fault has, in the past, been detected, further testing will be needed to find the root cause of the problem.

## STEP 4 CHECK FOR TECHNICAL SERVICE BULLETINS (TSBs)
Check for corrections or repair procedures in **technical service bulletins (TSBs)** that match the symptoms. ● SEE FIGURE 30–8. According to studies performed by automobile manufacturers, as many as 30% of vehicles can be repaired following the information, suggestions, or replacement parts

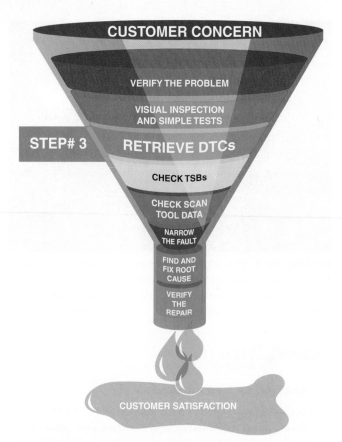

FIGURE 30–7 Step 3 in the diagnostic process is to retrieve any stored diagnostic trouble codes (DTCs).

FIGURE 30–8 After checking for stored diagnostic trouble codes (DTCs), the wise technician checks service information for any technical service bulletins that may relate to the vehicle being serviced.

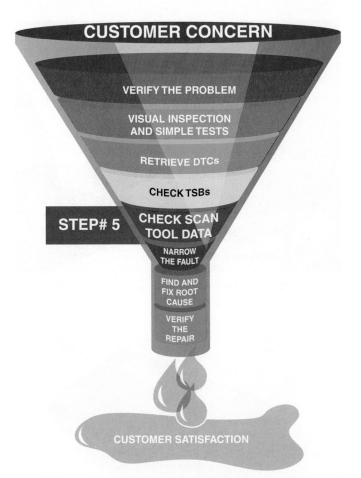

**FIGURE 30–9** Looking carefully at the scan tool data is very helpful in locating the source of a problem.

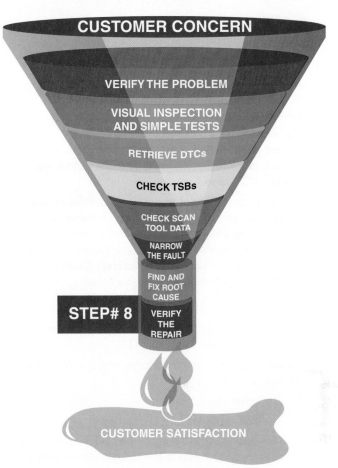

**FIGURE 30–10** Step 8 is very important. Be sure that the customer's concern has been corrected.

found in a service bulletin. DTCs must be known before searching for service bulletins because bulletins often include information on solving problems that involve a stored DTC.

### STEP 5 LOOK CAREFULLY AT SCAN TOOL DATA

Vehicle manufacturers have been giving the technician more and more data on a scan tool connected to the **data link connector (DLC).** ● **SEE FIGURE 30–9.** Beginning technicians are often observed scrolling through scan data without a real clue about what they are looking for. When asked, they usually reply that they are looking for something unusual, as if the screen will flash a big message "LOOK HERE—THIS IS NOT CORRECT." That statement does not appear on scan tool displays. The best way to look at scan data is in a definite sequence and with specific, selected bits of data that can tell the most about the operation of the engine, such as the following:

- Engine coolant temperature (ECT) is the same as intake air temperature (IAT) after the vehicle sits for several hours.
- Idle air control (IAC) valve is being commanded to an acceptable range.
- Oxygen sensor (O2S) is operating properly:
  1. Readings below 200 mV at times
  2. Readings above 800 mV at times
  3. Rapid transitions between rich and lean

### STEP 6 NARROW THE PROBLEM TO A SYSTEM OR CYLINDER
Narrowing the focus to a system or individual cylinder is the hardest part of the entire diagnostic process:

- Perform a cylinder power balance test.
- If a weak cylinder is detected, perform a compression and a cylinder leakage test to determine the probable cause.

### STEP 7 REPAIR THE PROBLEM AND DETERMINE THE ROOT CAUSE
The repair or part replacement must be performed following vehicle manufacturer's recommendations and be certain that the root cause of the problem has been found. Also follow the manufacturer's recommended repair procedures and methods.

### STEP 8 VERIFY THE REPAIR AND CLEAR ANY STORED DTCS
● **SEE FIGURE 30–10.**

- Test-drive to verify that the original problem (concern) is fixed.
- Verify that no additional problems have occurred during the repair process.
- Check for and then clear all DTCs. (This step ensures that the computer will not make any changes based on a stored DTC but should not be performed if the vehicle is

**One Test Is Worth 1,000 "Expert" Opinions**

Whenever any vehicle has an engine performance or driveability concern, certain people always say the following:

"Sounds like it's a bad injector."

"I'll bet you it's a bad computer."

"I had a problem just like yours yesterday and it was a bad EGR valve."

Regardless of the skills and talents of those people, it is still more accurate to perform tests on the vehicle than to rely on feelings or opinions of others who have not even seen the vehicle. Even your own opinion should not sway your thinking. Follow a plan and perform tests, and the test results will lead to the root cause.

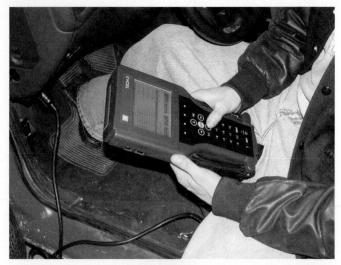

**FIGURE 30–11** A TECH 2 scan tool is the factory scan tool used on General Motors vehicles.

going to be tested for emissions because all of the monitors will need to be run and pass.)

- Return the vehicle to the customer and double-check the following:

  1. The vehicle is clean.

  2. The radio is turned off.

  3. The clock is set to the right time, and the radio stations have been restored if the battery was disconnected during the repair procedure.

## SCAN TOOLS

Scan tools are the workhorse for any diagnostic work on all vehicles. Scan tools can be divided into two basic groups:

1. **Factory scan tools.** These are the scan tools required by all dealers that sell and service the brand of vehicle. Examples of factory scan tools include the following:

   - **General Motors**—Tech 2. ● **SEE FIGURE 30–11.**
   - **Ford**—New Generation Star (NGS) and IDS (Integrated Diagnostic Software).
   - **Chrysler**—DRB III, Star Scan or wiTECH for CAN-equipped vehicles.
   - **Honda**—HDS or Master Tech
   - **Toyota**—Master Tech

   All factory scan tools are designed to provide bidirectional capability, which allows the service technician the opportunity to operate components using the scan tool, thereby confirming that the component is able to work when commanded. Also, all factory scan tools are capable of displaying all factory parameters.

2. **Aftermarket scan tools.** These scan tools are designed to function on more than one brand of vehicle. Examples of aftermarket scan tools include the following:

   - **Snap-on** (various models including the MT2500 and Modis)

**FIGURE 30–12** A Bluetooth adapter that plugs into the DLC and transmits global OBD II information to a smart phone that has a scan tool app installed.

- **OTC** (various models including Pegasus, Genisys, and Task Master)
- **AutoEnginuity** and other programs that use a laptop or handheld computer for the display

While many aftermarket scan tools can display most if not all of the parameters of the factory scan tool, there can be a difference when trying to troubleshoot some faults. ● **SEE FIGURE 30–12.**

## RETRIEVAL OF DIAGNOSTIC INFORMATION

To retrieve diagnostic information from the PCM, a scan tool is needed. If a factory or factory-level scan tool is used, then all of the data can be retrieved. If a global (generic) only type scan tool is used, only the emissions-related data can be retrieved.

To retrieve diagnostic information from the PCM, use the following steps:

**STEP 1**  Locate and gain access to the data link connector (DLC).

**STEP 2**  Connect the scan tool to the DLC and establish communication.

> NOTE: **If no communication is established, follow the vehicle manufacturer's specified instructions.**

**STEP 3**  Follow the on-screen instructions of the scan tool to correctly identify the vehicle.

**STEP 4**  Observe the scan data as well as any DTCs.

**STEP 5**  Check to see that all of the monitors have run and passed. A diagnostic trouble code (DTC) will not be set unless the monitor for that system has been run.

**STEP 6**  Follow vehicle manufacturer's instructions if any DTCs are stored. If no DTCs are stored, compare all sensor values with a factory acceptable range chart to see if any sensor values are out of range.

| Parameter Identification (PID) | | |
|---|---|---|
| **Scan Tool Parameter** | **Units Displayed** | **Typical Data Value** |
| Engine Idling/Radiator Hose Hot/Closed Throttle/ Park or Neutral/Closed Loop/Accessories Off/ Brake Pedal Released | | |
| 3X Crank Sensor | RPM | Varies |
| 24X Crank Sensor | RPM | Varies |
| Actual EGR Position | Percent | 0 |
| BARO | kPa/Volts | 65–110 kPa/ 3.5–4.5 Volts |
| CMP Sensor Signal Present | Yes/No | Yes |
| Commanded Fuel Pump | On/Off | On |
| Cycles of Misfire Data | Counts | 0–99 |
| Desired EGR Position | Percent | 0 |
| ECT | °C/°F | Varies |
| EGR Duty Cycle | Percent | 0 |
| Engine Run Time | Hr: Min: Sec | Varies |
| EVAP Canister Purge | Percent | Low and Varying |
| EVAP Fault History | No Fault/ Excess Vacuum/ Purge Valve Leak/ Small Leak/ Weak Vacuum | No Fault |
| Fuel Tank Pressure | Inches of $H_2O$/ Volts | Varies |
| $HO_2S$ Sensor 1 | Ready/Not Ready | Ready |
| $HO_2S$ Sensor 1 | Millivolts | 0–1,000 and Varying |

| **Scan Tool Parameter** | **Units Displayed** | **Typical Data Value** |
|---|---|---|
| $HO_2S$ Sensor 2 | Millivolts | 0–1,000 and Varying |
| $HO_2S$ X Counts | Counts | Varies |
| IAC Position | Counts | 15–25 preferred |
| IAT | °C/°F | Varies |
| Knock Retard | Degrees | 0 |
| Long Term FT | Percent | 0–10 |
| MAF | Grams per second | 3–7 |
| MAF Frequency | Hz | 1,200–3,000 (depends on altitude and engine load) |
| MAP | kPa/Volts | 20–48 kPa/ 0.75–2 Volts (depends on altitude) |
| Misfire Current Cyl. 1–10 | Counts | 0 |
| Misfire History Cyl. 1–10 | Counts | 0 |
| Short-term FT | Percent | 0–10 |
| Start Up ECT | °C/°F | Varies |
| Start Up IAT | °C/°F | Varies |
| Total Misfire Current Count | Counts | 0 |
| Total Misfire Failures | Counts | 0 |
| Total Misfire Passes | Counts | 0 |
| TP Angle | Percent | 0 |
| TP Sensor | Volts | 0.20–0.74 |
| Vehicle Speed | MPH/Km/h | 0 |

**Note:** Viewing the PID screen on the scanner is useful in determining if a problem is occurring at the present time

# TROUBLESHOOTING USING DTC DIAGNOSTIC TROUBLE CODES

Pinning down causes of the actual problem can be accomplished by trying to set the opposite code. For example, if a code indicates an open throttle position (TP) sensor (high resistance), clear the code and create a shorted (low-resistance) condition. This can be accomplished by using a jumper wire and connecting the signal terminal to the 5-volt reference terminal. This should set a DTC:

■ **If the opposite code sets,** this indicates that the wiring and connector for the sensor is okay and the sensor itself is defective (open).

■ **If the same code sets,** this indicates that the wiring or electrical connection is open (has high resistance) and is the cause of the setting of the DTC.

## METHODS FOR CLEARING DIAGNOSTIC TROUBLE CODES
Clearing DTCs from a vehicle computer sometimes needs to be performed. There are three methods that can be used to clear stored DTCs:

**CAUTION: Clearing diagnostic trouble codes (DTCs) also will clear all of the noncontinuous monitors.**

■ **Clearing codes—Method 1.** The preferred method of clearing codes is by using a scan tool. This is the method recommended by most vehicle manufacturers if the procedure can be performed on the vehicle. The computer of some vehicles cannot be cleared with a scan tool.

■ **Clearing codes—Method 2.** If a scan tool is not available or a scan tool cannot be used on the vehicle being serviced, the power to the computer can be disconnected:

1. Disconnect the fusible link (if so equipped) that feeds the computer.

2. Disconnect the fuse or fuses that feed the computer.

**NOTE: The fuse may not be labeled as a computer fuse. For example, many Toyotas can be cleared by disconnecting the fuel-injection fuse. Some vehicles require that two fuses be disconnected to clear any stored codes.**

■ **Clearing codes—Method 3.** If the other two methods cannot be used, the negative battery cable can be disconnected to clear stored DTCs.

**NOTE: Because of the adaptive learning capacity of the computer, a vehicle may fail an exhaust emissions test if the vehicle is not driven enough to allow the computer to run all of the monitors.**

**CAUTION: By disconnecting the battery, the radio presets and clock information will be lost. They should be reset before returning the vehicle to the customer. If the radio has a security code, the code must be entered before the radio will function. Before disconnecting the battery, always check with the vehicle owner to be sure that the code is available.**

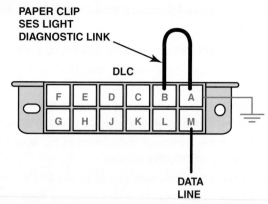

**FIGURE 30–13** To retrieve flash codes from an OBD-I General Motors vehicle, without a scan tool, connect terminals A and B with the ignition on–engine off. The M terminal is used to retrieve data from the sensors to a scan tool.

🔧 **TECH TIP**

**Do Not Lie to a Scan Tool!**
Because computer calibration may vary from year to year, using the incorrect year for the vehicle while using a scan tool can cause the data retrieved to be incorrect or inaccurate.

# RETRIEVING CODES PRIOR TO 1996

**FLASH CODES** Most vehicles from the early 1980s through 1995 used some method to retrieve diagnostic trouble codes. For example, General Motors diagnostic trouble codes could be retrieved by using a metal tool and contacting terminals A and B of the 12-pin DLC. ● **SEE FIGURE 30–13.**

**FIGURE 30–14** Diagnostic trouble codes (DTCs) from Chrysler and Dodge vehicles can be retrieved by turning the ignition switch to on and then off three times.

This method is called flash code retrieval because the MIL will flash to indicate diagnostic trouble codes. The steps are as follows:

1. Turn the ignition switch to on (engine off). The "check engine" light or "service engine soon" light should be on. If the amber malfunction indicator light (MIL) is not on, a problem exists within the light circuit.

2. Connect terminals A and B at the DLC.

3. Observe the MIL. A code 12 (one flash, then a pause, then two flashes) reveals that there is no engine speed indication to the computer. Because the engine is not running, this simply indicates that the computer diagnostic system is working correctly.

**RETRIEVAL METHODS** Check service information for the exact procedures to follow to retrieve diagnostic trouble codes. Depending on the exact make, model, and year of manufacture, the procedure can include the use of one or more of the following:

- scan tool
- special tester
- fused jumper wire
- test light

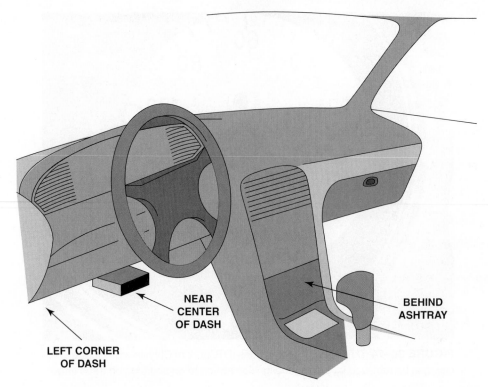

**FIGURE 30–15** The data link connector (DLC) can be located in various locations but always within 12 inches (30 cm) of the center of the vehicle.

NEAR
CENTER
OF DASH

BEHIND
ASHTRAY

LEFT CORNER
OF DASH

## DLC LOCATIONS

The data link connector (DLC) is a standardized 16-cavity connector where a scan tool can be connected to retrieve diagnostic information from the vehicle's computers.

The normal location is under the dash on the driver's side but it can be located within 12 inches (30cm) of the center of the vehicle. It can be covered but if it is, then the cover has to be able to be removed without the use of a tool such as when located underneath the ash tray. ● **SEE FIGURE 30–15.**

**FIGURE 30–16** A typical OBD-II data link connector (DLC). The location varies with make and model and may even be covered, but a tool is not needed to gain access. Check service information for the exact location if needed.

## OBD-II DIAGNOSIS

Starting with the 1996 model year, all vehicles sold in the United States must use the same type of 16-pin data link connector (DLC) and must monitor emission-related components. ● **SEE FIGURE 30–16.**

**RETRIEVING OBD-II CODES**   A scan tool is required to retrieve DTCs from most OBD-II vehicles. Every OBD-II scan tool will be able to read all generic Society of Automotive Engineers (SAE) DTCs from any vehicle.

## Fuel and Air Metering System

| | |
|---|---|
| P0100 | Mass or Volume Airflow Circuit Problem |
| P0101 | Mass or Volume Airflow Circuit Range or Performance Problem |
| P0102 | Mass or Volume Airflow Circuit Low Input |
| P0103 | Mass or Volume Airflow Circuit High Input |
| P0105 | Manifold Absolute Pressure or Barometric Pressure Circuit Problem |
| P0106 | Manifold Absolute Pressure or Barometric Pressure Circuit Range or Performance Problem |
| P0107 | Manifold Absolute Pressure or Barometric Pressure Circuit Low Input |
| P0108 | Manifold Absolute Pressure or Barometric Pressure Circuit High Input |
| P0110 | Intake Air Temperature Circuit Problem |
| P0111 | Intake Air Temperature Circuit Range or Performance Problem |
| P0112 | Intake Air Temperature Circuit Low Input |
| P0113 | Intake Air Temperature Circuit High Input |
| P0115 | Engine Coolant Temperature Circuit Problem |
| P0116 | Engine Coolant Temperature Circuit Range or Performance Problem |
| P0117 | Engine Coolant Temperature Circuit Low Input |
| P0118 | Engine Coolant Temperature Circuit High Input |
| P0120 | Throttle Position Circuit Problem |
| P0121 | Throttle Position Circuit Range or Performance Problem |
| P0122 | Throttle Position Circuit Low Input |
| P0123 | Throttle Position Circuit High Input |
| P0125 | Excessive Time to Enter Closed-Loop Fuel Control |
| P0128 | Coolant Temperature Below Thermostat Regulating Temperature |
| P0130 | O2 Sensor Circuit Problem (Bank 1* Sensor 1) |
| P0131 | O2 Sensor Circuit Low Voltage (Bank 1* Sensor 1) |
| P0132 | O2 Sensor Circuit High Voltage (Bank 1* Sensor 1) |
| P0133 | O2 Sensor Circuit Slow Response (Bank 1* Sensor 1) |
| P0134 | O2 Sensor Circuit No Activity Detected (Bank 1* Sensor 1) |
| P0135 | O2 Sensor Heater Circuit Problem (Bank 1* Sensor 1) |
| P0136 | O2 Sensor Circuit Problem (Bank 1* Sensor 2) |
| P0137 | O2 Sensor Circuit Low Voltage (Bank 1* Sensor 2) |
| P0138 | O2 Sensor Circuit High Voltage (Bank 1* Sensor 2) |
| P0139 | O2 Sensor Circuit Slow Response (Bank 1* Sensor 2) |
| P0140 | O2 Sensor Circuit No Activity Detected (Bank 1* Sensor 2) |
| P0141 | O2 Sensor Heater Circuit Problem (Bank 1* Sensor 2) |
| P0142 | O2 Sensor Circuit Problem (Bank 1* Sensor 3) |
| P0143 | O2 Sensor Circuit Low Voltage (Bank 1* Sensor 3) |
| P0144 | O2 Sensor Circuit High Voltage (Bank 1* Sensor 3) |
| P0145 | O2 Sensor Circuit Slow Response (Bank 1* Sensor 3) |
| P0146 | O2 Sensor Circuit No Activity Detected (Bank 1* Sensor 3) |
| P0147 | O2 Sensor Heater Circuit Problem (Bank 1* Sensor 3) |
| P0150 | O2 Sensor Circuit Problem (Bank 2 Sensor 1) |
| P0151 | O2 Sensor Circuit Low Voltage (Bank 2 Sensor 1) |
| P0152 | O2 Sensor Circuit High Voltage (Bank 2 Sensor 1) |
| P0153 | O2 Sensor Circuit Slow Response (Bank 2 Sensor 1) |
| P0154 | O2 Sensor Circuit No Activity Detected (Bank 2 Sensor 1) |
| P0155 | O2 Sensor Heater Circuit Problem (Bank 2 Sensor 1) |
| P0156 | O2 Sensor Circuit Problem (Bank 2 Sensor 2) |
| P0157 | O2 Sensor Circuit Low Voltage (Bank 2 Sensor 2) |
| P0158 | O2 Sensor Circuit High Voltage (Bank 2 Sensor 2) |
| P0159 | O2 Sensor Circuit Slow Response (Bank 2 Sensor 2) |
| P0160 | O2 Sensor Circuit No Activity Detected (Bank 2 Sensor 2) |
| P0161 | O2 Sensor Heater Circuit Problem (Bank 2 Sensor 2) |
| P0162 | O2 Sensor Circuit Problem (Bank 2 Sensor 3) |
| P0163 | O2 Sensor Circuit Low Voltage (Bank 2 Sensor 3) |
| P0164 | O2 Sensor Circuit High Voltage (Bank 2 Sensor 3) |
| P0165 | O2 Sensor Circuit Slow Response (Bank 2 Sensor 3) |
| P0166 | O2 Sensor Circuit No Activity Detected (Bank 2 Sensor 3) |
| P0167 | O2 Sensor Heater Circuit Problem (Bank 2 Sensor 3) |
| P0170 | Fuel Trim Problem (Bank 1*) |
| P0171 | System Too Lean (Bank 1*) |
| P0172 | System Too Rich (Bank 1*) |
| P0173 | Fuel Trim Problem (Bank 2) |
| P0174 | System Too Lean (Bank 2) |
| P0175 | System Too Rich (Bank 2) |
| P0176 | Fuel Composition Sensor Circuit Problem |
| P0177 | Fuel Composition Sensor Circuit Range or Performance |
| P0178 | Fuel Composition Sensor Circuit Low Input |
| P0179 | Fuel Composition Sensor Circuit High Input |
| P0180 | Fuel Temperature Sensor Problem |
| P0181 | Fuel Temperature Sensor Circuit Range or Performance |
| P0182 | Fuel Temperature Sensor Circuit Low Input |
| P0183 | Fuel Temperature Sensor Circuit High Input |

## Fuel and Air Metering (Injector Circuit)

| | |
|---|---|
| P0201 | Injector Circuit Problem—Cylinder 1 |
| P0202 | Injector Circuit Problem—Cylinder 2 |
| P0203 | Injector Circuit Problem—Cylinder 3 |
| P0204 | Injector Circuit Problem—Cylinder 4 |
| P0205 | Injector Circuit Problem—Cylinder 5 |
| P0206 | Injector Circuit Problem—Cylinder 6 |
| P0207 | Injector Circuit Problem—Cylinder 7 |
| P0208 | Injector Circuit Problem—Cylinder 8 |
| P0209 | Injector Circuit Problem—Cylinder 9 |
| P0210 | Injector Circuit Problem—Cylinder 10 |
| P0211 | Injector Circuit Problem—Cylinder 11 |
| P0212 | Injector Circuit Problem—Cylinder 12 |
| P0213 | Cold Start Injector 1 Problem |
| P0214 | Cold Start Injector 2 Problem |

## Ignition System or Misfire

| | |
|---|---|
| P0300 | Random Misfire Detected |
| P0301 | Cylinder 1 Misfire Detected |
| P0302 | Cylinder 2 Misfire Detected |
| P0303 | Cylinder 3 Misfire Detected |
| P0304 | Cylinder 4 Misfire Detected |
| P0305 | Cylinder 5 Misfire Detected |

*(continued)*

| | |
|---|---|
| P0306 | Cylinder 6 Misfire Detected |
| P0307 | Cylinder 7 Misfire Detected |
| P0308 | Cylinder 8 Misfire Detected |
| P0309 | Cylinder 9 Misfire Detected |
| P0310 | Cylinder 10 Misfire Detected |
| P0311 | Cylinder 11 Misfire Detected |
| P0312 | Cylinder 12 Misfire Detected |
| P0320 | Ignition or Distributor Engine Speed Input Circuit Problem |
| P0321 | Ignition or Distributor Engine Speed Input Circuit Range or Performance |
| P0322 | Ignition or Distributor Engine Speed Input Circuit No Signal |
| P0325 | Knock Sensor 1 Circuit Problem |
| P0326 | Knock Sensor 1 Circuit Range or Performance |
| P0327 | Knock Sensor 1 Circuit Low Input |
| P0328 | Knock Sensor 1 Circuit High Input |
| P0330 | Knock Sensor 2 Circuit Problem |
| P0331 | Knock Sensor 2 Circuit Range or Performance |
| P0332 | Knock Sensor 2 Circuit Low Input |
| P0333 | Knock Sensor 2 Circuit High Input |
| P0335 | Crankshaft Position Sensor Circuit Problem |
| P0336 | Crankshaft Position Sensor Circuit Range or Performance |
| P0337 | Crankshaft Position Sensor Circuit Low Input |
| P0338 | Crankshaft Position Sensor Circuit High Input |

## Auxiliary Emission Controls

| | |
|---|---|
| P0400 | Exhaust Gas Recirculation Flow Problem |
| P0401 | Exhaust Gas Recirculation Flow Insufficient Detected |
| P0402 | Exhaust Gas Recirculation Flow Excessive Detected |
| P0405 | Air Conditioner Refrigerant Charge Loss |
| P0410 | Secondary Air Injection System Problem |
| P0411 | Secondary Air Injection System Insufficient Flow Detected |
| P0412 | Secondary Air Injection System Switching Valve or Circuit Problem |
| P0413 | Secondary Air Injection System Switching Valve or Circuit Open |
| P0414 | Secondary Air Injection System Switching Valve or Circuit Shorted |
| P0420 | Catalyst System Efficiency below Threshold (Bank 1*) |
| P0421 | Warm Up Catalyst Efficiency below Threshold (Bank 1*) |
| P0422 | Main Catalyst Efficiency below Threshold (Bank 1*) |
| P0423 | Heated Catalyst Efficiency below Threshold (Bank 1*) |
| P0424 | Heated Catalyst Temperature below Threshold (Bank 1*) |
| P0430 | Catalyst System Efficiency below Threshold (Bank 2) |
| P0431 | Warm Up Catalyst Efficiency below Threshold (Bank 2) |
| P0432 | Main Catalyst Efficiency below Threshold (Bank 2) |
| P0433 | Heated Catalyst Efficiency below Threshold (Bank 2) |
| P0434 | Heated Catalyst Temperature below Threshold (Bank 2) |
| P0440 | Evaporative Emission Control System Problem |
| P0441 | Evaporative Emission Control System Insufficient Purge Flow |

| | |
|---|---|
| P0442 | Evaporative Emission Control System Leak Detected |
| P0443 | Evaporative Emission Control System Purge Control Valve Circuit Problem |
| P0444 | Evaporative Emission Control System Purge Control Valve Circuit Open |
| P0445 | Evaporative Emission Control System Purge Control Valve Circuit Shorted |
| P0446 | Evaporative Emission Control System Vent Control Problem |
| P0447 | Evaporative Emission Control System Vent Control Open |
| P0448 | Evaporative Emission Control System Vent Control Shorted |
| P0450 | Evaporative Emission Control System Pressure Sensor Problem |
| P0451 | Evaporative Emission Control System Pressure Sensor Range or Performance |
| P0452 | Evaporative Emission Control System Pressure Sensor Low Input |
| P0453 | Evaporative Emission Control System Pressure Sensor High Input |

## Vehicle Speed Control and Idle Control

| | |
|---|---|
| P0500 | Vehicle Speed Sensor Problem |
| P0501 | Vehicle Speed Sensor Range or Performance |
| P0502 | Vehicle Speed Sensor Low Input |
| P0505 | Idle Control System Problem |
| P0506 | Idle Control System RPM Lower Than Expected |
| P0507 | Idle Control System RPM Higher Than Expected |
| P0510 | Closed Throttle Position Switch Problem |

## Computer Output Circuit

| | |
|---|---|
| P0600 | Serial Communication Link Problem |
| P0605 | Internal Control Module (Module Identification Defined by J1979) |

## Transmission

| | |
|---|---|
| P0703 | Brake Switch Input Problem |
| P0705 | Transmission Range Sensor Circuit Problem (PRNDL Input) |
| P0706 | Transmission Range Sensor Circuit Range or Performance |
| P0707 | Transmission Range Sensor Circuit Low Input |
| P0708 | Transmission Range Sensor Circuit High Input |
| P0710 | Transmission Fluid Temperature Sensor Problem |
| P0711 | Transmission Fluid Temperature Sensor Range or Performance |
| P0712 | Transmission Fluid Temperature Sensor Low Input |
| P0713 | Transmission Fluid Temperature Sensor High Input |
| P0715 | Input or Turbine Speed Sensor Circuit Problem |
| P0716 | Input or Turbine Speed Sensor Circuit Range or Performance |
| P0717 | Input or Turbine Speed Sensor Circuit No Signal |
| P0720 | Output Speed Sensor Circuit Problem |
| P0721 | Output Speed Sensor Circuit Range or Performance |
| P0722 | Output Speed Sensor Circuit No Signal |
| P0725 | Engine Speed Input Circuit Problem |
| P0726 | Engine Speed Input Circuit Range or Performance |

| P0727 | Engine Speed Input Circuit No Signal |
| P0730 | Incorrect Gear Ratio |
| P0731 | Gear 1 Incorrect Ratio |
| P0732 | Gear 2 Incorrect Ratio |
| P0733 | Gear 3 Incorrect Ratio |
| P0734 | Gear 4 Incorrect Ratio |
| P0735 | Gear 5 Incorrect Ratio |
| P0736 | Reverse Incorrect Ratio |
| P0740 | Torque Converter Clutch System Problem |
| P0741 | Torque Converter Clutch System Performance or Stuck Off |
| P0742 | Torque Converter Clutch System Stuck On |
| P0743 | Torque Converter Clutch System Electrical |
| P0745 | Pressure Control Solenoid Problem |
| P0746 | Pressure Control Solenoid Performance or Stuck Off |
| P0747 | Pressure Control Solenoid Stuck On |
| P0748 | Pressure Control Solenoid Electrical |
| P0750 | Shift Solenoid A Problem |
| P0751 | Shift Solenoid A Performance or Stuck Off |
| P0752 | Shift Solenoid A Stuck On |
| P0753 | Shift Solenoid A Electrical |
| P0755 | Shift Solenoid B Problem |
| P0756 | Shift Solenoid B Performance or Stuck Off |
| P0757 | Shift Solenoid B Stuck On |
| P0758 | Shift Solenoid B Electrical |
| P0760 | Shift Solenoid C Problem |
| P0761 | Shift Solenoid C Performance or Stuck Off |
| P0762 | Shift Solenoid C Stuck On |
| P0763 | Shift Solenoid C Electrical |
| P0765 | Shift Solenoid D Problem |
| P0766 | Shift Solenoid D Performance or Stuck Off |
| P0767 | Shift Solenoid D Stuck On |
| P0768 | Shift Solenoid D Electrical |
| P0770 | Shift Solenoid E Problem |
| P0771 | Shift Solenoid E Performance or Stuck Off |
| P0772 | Shift Solenoid E Stuck On |
| P0773 | Shift Solenoid E Electrical |

* The side of the engine where number one cylinder is located.

## OBD-II ACTIVE TESTS

The vehicle computer must run tests on the various emission-related components and turn on the malfunction indicator lamp (MIL) if faults are detected. OBD-II is an *active* computer analysis system because it actually tests the operation of the oxygen sensors, exhaust gas recirculation system, and so forth whenever conditions permit. It is the purpose and function of the PCM to monitor these components and perform these active tests.

For example, the PCM may open the EGR valve momentarily to check its operation while the vehicle is decelerating. A change in the manifold absolute pressure (MAP) sensor signal will indicate to the computer that the exhaust gas is, in fact, being introduced into the engine. Because these tests are active and certain conditions must be present before these tests can be run, the computer uses its internal diagnostic program to keep track of all the various conditions and to schedule active tests so that they will not interfere with each other.

**OBD-II DRIVE CYCLE**   The vehicle must be driven under a variety of operating conditions for all active tests to be performed. A **trip** is defined as an engine-operating drive cycle that contains the necessary conditions for a particular test to be performed. For example, for the EGR test to be performed, the engine has to be at normal operating temperature and decelerating for a minimum amount of time. Some tests are performed when the engine is cold, whereas others require that the vehicle be cruising at a steady highway speed.

**TYPES OF OBD-II CODES**   Not all OBD-II DTCs are of the same importance for exhaust emissions. Each type of DTC has different requirements for it to set, and the computer will turn on the MIL only for emissions-related DTCs.

**TYPE A CODES.**   A type A DTC is emission related and will cause the MIL to be turned on at the *first trip* if the computer has detected a problem. Engine misfire or a very rich or lean air–fuel ratio, for example, would cause a type A DTC. These codes alert the driver to an emissions problem that may cause damage to the catalytic converter.

**TYPE B CODES.**   A type B code will be stored as a pending code in the PCM, and the MIL will be turned on only after the second consecutive trip, alerting the driver to the fact that a diagnostic test was performed and failed.

**NOTE: Type A and Type B codes are emission related and will cause the lighting of the malfunction indicator lamp, usually labeled "check engine" or "service engine soon."**

**TYPE C AND TYPE D CODES.**   Type C and type D codes are for use with non-emission-related diagnostic tests. They will cause the lighting of a "service" lamp (if the vehicle is so equipped).

**OBD-II FREEZE-FRAME**   To assist the service technician, OBD-II requires the computer to take a "snapshot" or freeze-frame of all data at the instant an emission-related DTC is set. A scan tool is required to retrieve this data. CARB and EPA regulations require that the controller store specific freeze-frame (engine-related) data when the first emission related fault is detected. The data stored in freeze-frame can be replaced only by data from a trouble code with a higher priority, such as a trouble related to a fuel system or misfire monitor fault.

**NOTE: Although OBD-II requires that just one freeze-frame of data be stored, the instant an emission-related DTC is set, vehicle manufacturers usually provide expanded data about the DTC beyond that required. However, retrieving enhanced data usually requires the use of an enhanced or factory-level scan tool.**

The freeze-frame has to contain data values that occurred at the time the code was set (these values are provided in standard units of measurement). Freeze-frame data is recorded during the first trip on a two-trip fault. As a result, OBD-II systems record the data present at the time an emission-related code is recorded and the MIL activated. This data can be accessed and displayed on a scan tool. Freeze-frame data is one frame or one instant in time. Freeze-frame data is not updated (refreshed) if the same monitor test fails a second time.

### REQUIRED FREEZE-FRAME DATA ITEMS

- Code that triggered the freeze-frame
- A/F ratio, airflow rate, and calculated engine load
- Base fuel injector pulse width
- ECT, IAT, MAF, MAP, TP, and VS sensor data
- Engine speed and amount of ignition spark advance
- Open- or closed-loop status
- Short-term and long-term fuel trim values
- For misfire codes—identify the cylinder that misfired

**NOTE: All freeze-frame data will be lost if the battery is disconnected, power to the PCM is removed, or the scan tool is used to erase or clear trouble codes.**

### DIAGNOSING INTERMITTENT MALFUNCTIONS

Of all the different types of conditions that you will see, the hardest to accurately diagnose and repair are intermittent malfunctions. These conditions may be temperature related (occur only when the vehicle is hot or cold) or humidity related (occur only when it is raining). Regardless of the conditions that will cause the malfunction to occur, you must diagnose and correct the condition.

When dealing with an intermittent concern, you should determine the conditions when the malfunction occurs and then try to duplicate those conditions. If a cause is not readily apparent to you, ask the customer when the symptom occurs. Ask if there are any conditions that seem to be related to or cause the concern.

Another consideration when working on an OBD-II-equipped vehicle is whether a concern is intermittent or whether it occurs only when a specific diagnostic test is performed by the PCM. Since OBD-II systems conduct diagnostic tests only under very precise conditions, some tests may be run only once during an ignition cycle. Additionally, if the requirements needed to perform the test are not met, the test will not run during an ignition cycle. This type of on-board diagnostics could be mistaken as "intermittent" when, in fact, the tests are only infrequent (depending on how the vehicle is driven). Examples of this type of diagnostic test are $HO_2S$ heaters, evaporative canister purge, catalyst efficiency, and EGR flow. When diagnosing intermittent concerns on an OBD-II-equipped vehicle, a logical diagnostic strategy is essential. The use of stored freeze-frame information can also be very useful when diagnosing an intermittent malfunction if a code has been stored.

## SERVICE/FLASH PROGRAMMING

**PURPOSE** Designing a program that allows an engine to meet strict air quality and fuel economy standards while providing excellent performance is no small feat. However, this is only part of the challenge facing engineers assigned with the task of developing OBD-II software. The reason for this is the countless variables involved with running the diagnostic monitors. Although programmers do their best to factor in any and all operating conditions when writing this complex code, periodic revisions are often required.

Reprogramming consists of downloading new calibration files from a scan tool, personal computer (PC), or modem into the PCM's electronically erasable programmable read-only memory (EEPROM). This can be done on or off the vehicle using the appropriate equipment. Since reprogramming is not an OBD-II requirement however, many vehicles will need a new PCM in the event software changes become necessary. Physically removing and replacing the PROM chip is no longer possible.

The following are three industry-standard methods used to reprogram the EEPROM:

- Remote programming
- Direct programming
- Off-board programming

**REMOTE PROGRAMMING** Remote programming uses the scan tool to transfer data from the manufacturer's shop PC to the vehicle's PCM. This is accomplished by performing the following steps:

- Connect the scan tool to the vehicle's DLC. ● **SEE FIGURE 30–17.**
- Enter the vehicle information into the scan tool through the programming application software incorporated in the scan tool. ● **SEE FIGURE 30–18.**
- Download VIN and current EEPROM calibration using a scan tool.
- Disconnect the scan tool from the DLC and connect the tool to the shop PC.
- Download the new calibration from the PC to the scan tool. ● **SEE FIGURE 30–19.**
- Reconnect the scan tool to the vehicle's DLC and download the new calibration into the PCM.

**CAUTION: Before programming, the vehicle's battery must be between 11 and 14 volts. Do not attempt to program while charging the battery unless using a special battery charger that does not produce excessive ripple voltage, such as the Midtronics PSC-300 (30 amp) or PSC-550 (55 amp) or similar as specified by the vehicle manufacturer.**

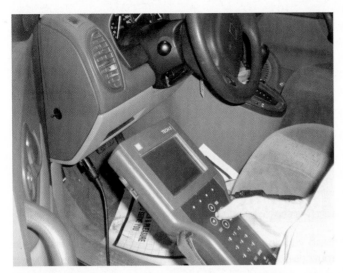

**FIGURE 30–17** The first step in the reprogramming procedure is to determine the current software installed using a scan tool. Not all scan tools can be used. In most cases using the factory scan tool is needed for reprogramming unless the scan tool is equipped to handle reprogramming.

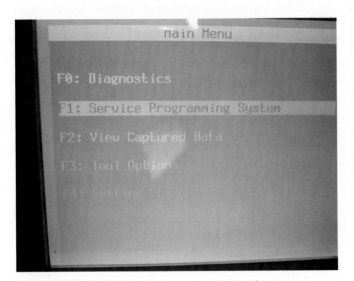

**FIGURE 30–18** Follow the on-screen instructions.

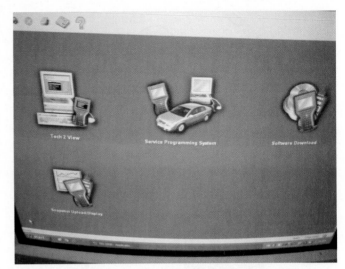

**FIGURE 30–19** An Internet connection is usually needed to perform updates, although some vehicle manufacturers use CDs that are updated regularly at a cost to the shop.

**FIGURE 30–20** Connecting cables and a computer to perform off-board programming.

**DIRECT PROGRAMMING** Direct programming does utilize a connection between the shop PC and the vehicle DLC.

**OFF-BOARD PROGRAMMING** Off-board programming is used if the PCM must be programmed away from the vehicle. This is preformed using the off-board programming adapter. ● **SEE FIGURE 30–20.**

**J2534 REPROGRAMMING** Legislation has mandated that vehicle manufacturers meet the SAE J2534 standards for all emissions-related systems on all new vehicles starting with model year 2004. This standard enables independent service repair operators to program or reprogram emissions-related ECMs from a wide variety of vehicle manufacturers with a single tool. ● **SEE FIGURE 30–21.** A J2534-compliant pass-through system is a standardized programming and diagnostic system. It uses a PC plus a standard interface to a software device driver and a hardware vehicle communication interface. The interface connects to a PC and to a programmable ECM on a vehicle through the J1962 data link connector (DLC). This system allows programming of all vehicle manufacturer ECMs using a single set of programming hardware. Programming software made available by the vehicle manufacturer must be functional with a J2534-compliant pass-through system.

The software for a typical pass-through application consists of two major components, including the following:

- The part delivered by the company that furnishes the hardware for J2534 enables the pass-through vehicle communication interface to communicate with the PC and provides

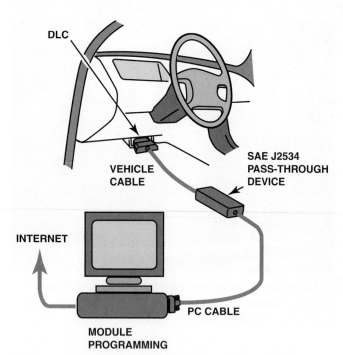

FIGURE 30–21 The J2534 pass-through reprogramming system does not need a scan tool to reflash the PCM on most 2004 and newer vehicles.

FIGURE 30–22 A typical J2534 universal reprogrammer that uses the J2534 standards.

for all Vehicle Communication Protocols as required by SAE J2534. It also provides for the software interface to work with the software applications as provided for by the vehicle manufacturers. ● **SEE FIGURE 30–22.**

■ The second part of the pass-through enabling software is provided for by the vehicle manufacturers. This is normally a subset of the software used with their original equipment manufacturer (OEM) tools, and their website will indicate how to obtain this software and under what conditions it can be used. Refer to the National Automotive Service

Task Force (NASTF) website for the addresses for all vehicle manufacturers' service information and cost, *www.NASTF.org.*

Since the majority of vehicle manufacturers make this software available in downloadable form, having an Internet browser (Explorer/Netscape) and connection is a must.

## MANUFACTURER'S DIAGNOSTIC ROUTINES

Each vehicle manufacturer has established its own diagnostic routines and they should be followed. Most include the following steps:

**STEP 1**  Retrieve DTCs.

**STEP 2**  Check for all technical service bulletins that could be related to the stored DTC.

**STEP 3**  If there are multiple DTCs, the diagnostic routine may include checking different components or systems instead of when only one DTC was stored.

**STEP 4**  Perform system checks.

**STEP 5**  Perform the necessary service or repair

**STEP 6**  Perform a road test matching the parameters recorded in the freeze-frame to check that the repair has corrected the malfunction.

**STEP 7**  Repeat the road test to cause the MIL to be extinguished.

**NOTE: Do not clear codes (DTCs) unless instructed by the service information.**

Following the vehicle manufacturer's specific diagnostic routines will ensure that the root cause is found and the repair verified. This is important for customer satisfaction.

## COMPLETING SYSTEM REPAIRS

After the repair has been successfully completed, the vehicle should be driven under similar conditions that caused the original concern. Verify that the problem has been corrected. To perform this test drive, it is helpful to have a copy of the freeze-frame parameters that were present when the DTC was set. By driving under similar conditions, the PCM may perform a test of the system and automatically extinguish the MIL. This is the method preferred by most vehicle manufacturers. The DTC can be cleared using a scan tool, but then that means that monitors will have to be run, and the vehicle may fail an emission inspection if driven directly to the testing station.

## PROCEDURES FOR RESETTING THE PCM

The PCM can be reset or cleared of previously set DTCs and freeze-frame data in the following ways:

1. **Driving the vehicle.** Drive the vehicle under similar conditions that were present when the fault occurred. If the conditions are similar and the PCM performed the noncontinuous monitor test and it passed three times, then the PCM will extinguish the MIL. This is the method preferred by most vehicle manufacturers, however, this method could be time consuming. If three passes cannot be achieved, the owner of the vehicle will have to be told that even though the check engine light (MIL) is on, the problem has been corrected, and the MIL should go out in a few days of normal driving.

2. **Clear DTCs using a scan tool.** A scan tool can be used to clear the DTC, which will also delete all of the freeze-frame data. The advantage of using a scan tool is that the check engine (MIL) will be out and the customer will be happy that the problem (MIL on) has been corrected. Do not use a scan tool to clear a DTC if the vehicle is going to be checked soon at a test station for state-mandated emission tests.

3. **Battery disconnect.** Disconnecting the negative battery cable will clear the DTCs and freeze-frame on many vehicles but not all. Besides clearing the DTCs, disconnecting the battery for about 20 minutes will also erase radio station presets and other memory items in many cases. Most vehicle manufacturers do not recommend that the battery be disconnected to clear DTCs, and it may not work on some vehicles.

## ROAD TEST (DRIVE CYCLE)

Use the freeze-frame data and test-drive the vehicle so that the vehicle is driven to match the conditions displayed on the freeze-frame. If the battery has been disconnected, then the vehicle may have to be driven under conditions that allow the PCM to conduct monitor tests. This drive pattern is called a **drive cycle.** The drive cycle is different for each vehicle manufacturer, but a universal drive cycle may work in many cases. In many cases performing a universal drive cycle will reset most monitors in most vehicles.

### UNIVERSAL DRIVE CYCLE

**PRECONDITIONING: Phase 1**

> MIL must be off.
> No DTCs present.
> Fuel fill between 15% and 85%.
> Cold start—Preferred = 8-hour soak at 68°F to 86°F.
> Alternative = ECT below 86°F.

1. With the ignition off, connect scan tool.
2. Start engine and drive between 20 and 30 mph for 22 minutes, allowing speed to vary.
3. Stop and idle for 40 seconds, gradually accelerate to 55 mph.
4. Maintain 55 mph for 4 minutes using a steady throttle input.
5. Stop and idle for 30 seconds, then accelerate to 30 mph.
6. Maintain 30 mph for 12 minutes.
7. Repeat steps 4 and 5 four times.

> Using scan tool, check readiness. If insufficient readiness set, continue to universal drive trace phase II.

> **Important: (Do not shut off engine between phases).**

**Phase II:**

1. Vehicle at a stop and idle for 45 seconds, then accelerate to 30 mph.
2. Maintain 30 mph for 22 minutes.
3. Repeat steps 1 and 2 three times.
4. Bring vehicle to a stop and idle for 45 seconds, then accelerate to 35 mph.
5. Maintain speed between 30 and 35 mph for 4 minutes.
6. Bring vehicle to a stop and idle for 45 seconds, then accelerate to 30 mph.
7. Maintain 30 mph for 22 minutes.
8. Repeat steps 6 and 7 five times.
9. Using scan tool, check readiness.

1. Funnel diagnostics—Visual approach to a diagnostic procedure:
   **Step 1** Verify the problem (concern)
   **Step 2** Perform a thorough visual inspection and basic tests
   **Step 3** Retrieve the diagnostic trouble codes (DTCs)
   **Step 4** Check for technical service bulletins (TSBs)
   **Step 5** Look carefully at scan tool data
   **Step 6** Narrow the problem to a system or cylinder
   **Step 7** Repair the problem and determine the root cause
   **Step 8** Verify the repair and check for any stored DTCs

2. A thorough visual inspection is important during the diagnosis and troubleshooting of any engine performance problem or electrical malfunction.

3. If the MIL is on, retrieve the DTC and follow the manufacturer's recommended procedure to find the root cause of the problem.

4. OBD-II vehicles use a 16-pin DLC and common DTCs.

# REVIEW QUESTIONS

1. Explain the procedure to follow when diagnosing a vehicle with stored DTCs using a scan tool.

2. Discuss what the PCM does during a drive cycle to test emission-related components.

3. Explain the difference between a type A and type B OBD-II DTC.

4. List three things that should be checked as part of a thorough visual inspection.

5. List the eight-step funnel diagnostic procedure.

6. Explain why a bulletin search should be performed after stored DTCs are retrieved.

7. List the three methods that can be used to reprogram a PCM.

# CHAPTER QUIZ

1. Technician A says that the first step in the diagnostic process is to verify the problem (concern). Technician B says the second step is to perform a thorough visual inspection. Which technician is correct?
   a. Technician A only
   b. Technician B only
   c. Both Technicians A and B
   d. Neither Technician A nor B

2. Which item is *not* important to know before starting the diagnosis of an engine performance problem?
   a. List of previous repairs
   b. The brand of engine oil used
   c. The type of gasoline used
   d. The temperature of the engine when the problem occurs

3. A paper test can be used to check for a possible problem with _____.
   a. The ignition system (bad spark plug wire)
   b. A faulty injector on a multiport engine
   c. A burned valve
   d. All of the above

4. Which step should be performed *last* when diagnosing an engine performance problem?
   a. Checking for any stored diagnostic trouble codes (DTCs)
   b. Checking for any technical service bulletins (TSBs)
   c. Performing a thorough visual inspection
   d. Verify the repair

5. Technician A says that if the opposite DTC can be set, the problem is the component itself. Technician B says if the opposite DTC cannot be set, the problem is with the wiring or grounds. Which technician is correct?
   a. Technician A only
   b. Technician B only
   c. Both Technicians A and B
   d. Neither Technician A nor B

6. The preferred method to clear DTCs is to _____.
   a. Disconnect the negative battery cable for 10 seconds
   b. Use a scan tool
   c. Remove the computer (PCM) power feed fuse
   d. Cycle the ignition key on and off 40 times

7. Which is the factory scan tool for Chrysler brand vehicles equipped with CAN?
   a. Star Scan
   b. Tech 2
   c. NGS
   d. Master Tech

8. Technician A says that reprogramming a PCM using the J2534 system requires a factory scan tool. Technician B says that reprogramming a PCM using the J2534 system requires Internet access. Which technician is correct?
   a. Technician A only
   b. Technician B only
   c. Both Technicians A and B
   d. Neither Technician A nor B

9. Technician A says that knowing if there are any stored DTCs may be helpful when checking for related TSBs. Technician B says that only a factory scan tool should be used to retrieve DTCs. Which technician is correct?
   a. Technician A only
   b. Technician B only
   c. Both Technicians A and B
   d. Neither Technician A nor B

10. Which method can be used to reprogram a PCM?
    a. Remote
    b. Direct
    c. Off-board
    d. All of the above

# chapter 31

# HYBRID SAFETY AND SERVICE PROCEDURES

**LEARNING OBJECTIVES:** **After studying this chapter, the reader should be able to:** • Identify the safety equipment to be used with high-voltage circuits. • Explain how to de-power high-voltage systems. • Explain the procedure to move and tow a hybrid electric vehicle (HEV). • Discuss the steps to perform for routine services on hybrid electric vehicles.

**KEY TERMS:** ANSI 397 • ASTM 397 • CAT III 398 • DMM 398 • Floating ground 400 • HV 397 • HV cables 397 • IEC 398 • Lineman's gloves 397 • NiMH 403 • OSHA 397 • Service plug 401

## HIGH-VOLTAGE SAFETY

**NEED FOR CAUTION**    There have been electrical systems on vehicles for over 100 years. Technicians have been repairing vehicle electrical systems without fear of serious injury or electrocution. However, when working with hybrid electric vehicles, this is no longer true. It is now possible to be seriously injured or electrocuted (killed) if proper safety procedures are not followed.

Hybrid electric vehicles and all electric vehicles use **high-voltage (HV)** circuits that if touched with an unprotected hand could cause serious burns or even death.

**IDENTIFYING HIGH-VOLTAGE CIRCUITS**    **High-voltage cables** are identified by color of the plastic conduit and include the following:

- **Blue or yellow.** 42 volts (not a shock hazard but an arc will be maintained if a circuit is opened)
- **Orange.** 144 to 600 volts or higher

> ☠ **WARNING**
>
> Touching circuits or wires containing high voltage can cause severe burns or death.

## HIGH-VOLTAGE SAFETY EQUIPMENT

**RUBBER GLOVES**    Before working on the high-voltage system of a hybrid electric vehicle, be sure that high-voltage **lineman's gloves** are available. Be sure that the gloves are rated at least 1,000 volts and class "0" by ANSI/ASTM. The **American National Standards Institute (ANSI)** is a private, nonprofit organization that administers and coordinates the U.S. voluntary standardization and conformity assessment system. ASTM International, originally known as the **American Society for Testing and Materials (ASTM),** was formed over a century ago to address the need for component testing in industry. The **Occupational Safety and Health Administration (OSHA)** requirements specify that the HV gloves get inspected every six months by a qualified glove inspection laboratory. Use an outer leather glove to protect the HV rubber gloves. Inspect the gloves carefully before each use. High voltage and current (amperes) in combination is fatal. ● **SEE FIGURES 31–1 AND 31–2.**

**NOTE: The high-voltage insulated safety gloves must be recertified every six months to remain within Occupational Safety and Health Administration (OSHA) guidelines.**

Before using the rubber gloves, they should be tested for leaks using the following procedure:

1. Roll the glove up from the open end until the lower portion of the glove begins to balloon from the resulting air pressure. Be sure to "lean" into the sealed glove to raise the internal air pressure. If the glove leaks any air, discard the gloves. ● **SEE FIGURE 31–3.**

2. The gloves should not be used if they show any signs of wear and tear.

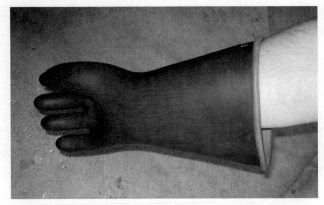

**FIGURE 31–1** Rubber lineman's gloves protect the wearer from a shock hazard.

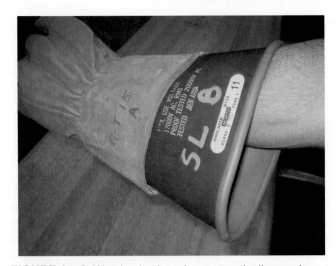

**FIGURE 31–2** Wearing leather gloves over the lineman's gloves helps protect the rubber gloves from damage.

> ☠ **WARNING**
>
> Cables and wiring are orange in color. High-voltage insulated safety gloves and a face shield must be worn when carrying out any diagnostics involving the high-voltage systems or components.

**CAT III-RATED DIGITAL MULTIMETER** Hybrid electric vehicles are equipped with electrical systems whose voltages can exceed 600 volts DC. A CAT III-certified **digital multimeter (DMM)** is required for making measurements on these high-voltage systems.

The **International Electrotechnical Commission (IEC)** has several categories of voltage standards for meter and meter leads. These categories are ratings for over-voltage protection and are rated CAT I, CAT II, CAT III, and CAT IV. The higher the category (CAT) rating, the greater the protection to the technician when measuring high-energy voltage. Under each category there are various voltage ratings:

**FIGURE 31–3** Checking rubber lineman's gloves for pinhole leaks.

**CAT I** Typically, a CAT I meter is used for low-voltage measurements, such as voltage measurements at wall outlets in the home. Meters with a CAT I rating are usually rated at 300 to 800 volts. CAT I is for relatively low-energy levels, and while the voltage level is high enough for use when working on a hybrid electric vehicle, the protective energy level is lower than what is needed.

**CAT II** A higher-rated meter that would be typically used for checking voltages at the circuit-breaker panel in the home. Meters with a CAT II rating are usually rated at 300 to 600 volts. CAT II rated meters have similar voltage ratings as the other CAT ratings, but the energy level of protection is higher with a CAT II compared to a CAT I.

**CAT III** **CAT III** is the minimum-rated meter that should be used for hybrid vehicles. Meters with a CAT III rating are usually rated at 600 to 1,000 volts and the highest energy level that is needed to protect the service technician.

**CAT IV** CAT IV meters are for clamp-on meters only. A clamp-on meter is used to measure current (amperes) in a circuit by placing the clamp around the wire carrying the current. If a clamp-on meter also has meter leads for voltage measurements, that part of the meter will be rated as CAT III.

**INSULATION TESTER** An electrical insulation tester, such as the Fluke 1587, is used to test for electrical continuity between the high-voltage wires or components and the body of the vehicle. If a hybrid electric vehicle has been involved in any type of collision or any other incident where damage could occur to the insulation, the high-voltage system should be checked. An insulation tester is more expensive than a digital meter. This means that an individual service technician often does not purchase one, but any technician or service shop that works on hybrid electric vehicles should have one available.

FIGURE 31–4 Be sure to only use a meter that is CAT III rated when taking electrical voltage measurements on a hybrid electric or electric vehicle.

FIGURE 31–5 The meter leads should also be CAT III rated when checking voltages on a hybrid electric vehicle.

**FIBERGLASS POLE** Ford requires that a 10 foot insulated fiberglass pool be available outside the safety zone to be used to pull a technician away from the vehicle in the unlikely event of an accident where the technician is shocked or electrocuted.

**FREQUENTLY ASKED QUESTION**

**Is It the Voltage Rating That Determines the CAT Rating?**

Yes and no. The voltages stated for the various CAT ratings are important, but the potential harm to a technician due to the energy level is what is most important. For example, some CAT II-rated meters may have a stated voltage higher than a meter that has a CAT III rating. Always use a meter that has a CAT III rating when working on a hybrid electric vehicle. ● **SEE FIGURES 31–4 AND 31–5.**

**EYE PROTECTION** Eye protection should be worn when testing for high voltage, which is considered by many experts to be over 60 volts. Eye protection should include the following features:

1. Plastic frames (avoid metal frames as these are conductive and could cause a shock hazard)
2. Side shields
3. Meet the standard ANSI Z87.1

Most hybrid electric systems use voltages higher than this threshold. If the system has not been powered down or has not had the high-voltage system disabled, a shock hazard is always possible. Even when the high-voltage system has been disconnected, there is still high voltage in the HV battery box.

NOTE: Some vehicle manufactures specify that full face shields be worn instead of safety glasses when working with high voltage circuits or components.

**SAFETY CONES** Ford requires that cones be placed at the four corners of any hybrid electric vehicle when service work on the high-voltage system is being performed. They are used to establish a safety zone around the vehicles so that other technicians will know that a possible shock hazard may be present.

## ELECTRIC SHOCK POTENTIAL

**LOCATIONS WHERE SHOCKS CAN OCCUR** Accidental and unprotected contact with any electrically charged ("hot" or "live") high-voltage component can cause serious injury or death. However, receiving an electric shock from a hybrid vehicle is highly unlikely because of the following:

1. Contact with the battery module or other components inside the battery box can occur only if the box is damaged and the contents are exposed or the box is opened without following proper precautions.
2. Contact with the electric motor can occur only after one or more components are removed.
3. The high-voltage cables can be easily identified by their distinctive orange color, and contact with them can be avoided.
4. The system main relays (SMRs) disconnect power from the cables the moment the ignition is turned off.

**LOCATIONS OF AUXILIARY BATTERIES** ● **SEE CHART 31–1** for a summary of the locations of auxiliary batteries.

## DEPOWERING THE HIGH-VOLTAGE SYSTEM

**THE NEED TO DEPOWER THE HV SYSTEM** During routine vehicle service work there is no need to go through any procedures needed to depower or to shut off the high-voltage circuits.

## HYBRID VEHICLE AUXILIARY BATTERY CHART

| VEHICLE | AUXILIARY BATTERY TYPE | AUXILIARY BATTERY LOCATION |
|---|---|---|
| Honda Insight Hybrid | Flooded lead acid | Underhood; center near bulkhead |
| Honda Civic Hybrid | Flooded lead acid | Underhood; driver's side |
| Honda Accord Hybrid | Flooded lead acid | Underhood; driver's side |
| Ford Escape Hybrid | Flooded lead acid | Underhood; driver's side |
| Toyota Prius Hybrid (2001–2003) | Absorbed glass mat (AGM) | Trunk; driver's side |
| Toyota Prius Hybrid (2004–2007) | Absorbed glass mat (AGM) | Trunk; passenger side |
| Toyota Highlander Hybrid | Flooded lead acid | Underhood; passenger side |
| Toyota Camry Hybrid | Absorbed glass mat (AGM) | Trunk; passenger side |
| Lexus RX 400h Hybrid | Flooded lead acid | Underhood; passenger side |
| Lexus GS 450h Hybrid | Absorbed glass mat (AGM) | Trunk; driver's side |
| Chevrolet/GMC Hybrid Pickup Truck | Flooded lead acid | Underhood; driver's side |

**CHART 31–1**

As a rule of thumb, the auxiliary battery is usually a flood type if it is located under the hood and an AGM type if it is in the trunk area.

**FIGURE 31–6** The Ford Escape Hybrid instrument panel showing the vehicle in park and the tachometer on "EV" instead of 0 RPM. This means that the gasoline engine could start at any time depending on the state-of-charge of the high-voltage batteries and other factors.

### TECH TIP

**Silence Is NOT Golden**

Never assume the vehicle is shut off just because the engine is off. When working with a Toyota or Lexus hybrid electric vehicle, always look for the **READY** indicator status on the dash display. The vehicle is shut off when the **READY** indicator is off.

The vehicle may be powered by the following:

1. The electric motor only
2. The gasoline engine only
3. A combination of both the electric motor and the gasoline engine

The vehicle computer determines the mode in which the vehicle operates to improve fuel economy and reduce emissions. The driver cannot manually select the mode. ● **SEE FIGURE 31–6.**

### TECH TIP

**High Voltage Is Insulated from the Vehicle Body**

Both positive and negative high-voltage power cables are isolated from the metal chassis, so there is no possibility of shock by touching the metal chassis. This design is called a **floating ground.**

A ground fault monitor continuously monitors for high-voltage leakage to the metal chassis while the vehicle is running. If a malfunction is detected, the vehicle computer will illuminate the master warning light in the instrument cluster and the hybrid warning light in the LCD display. The HV battery pack relays will automatically open to stop electricity flow in a collision sufficient to activate the SRS airbags.

### WARNING

Power remains in the high-voltage electrical system for up to 10 minutes after the HV battery pack is shut off. Never touch, cut, or open any orange high-voltage power cable or high-voltage component without confirming that the high-voltage has been completely discharged.

However, if work is going to be performed on any of the following components, then service information procedures must be followed to prevent possible electrical shock and personal injury:

- The high-voltage (HV) battery pack
- Any of the electronic controllers that use orange cables such as the inverter and converters
- The air-conditioning compressor if electrically driven and has orange cables attached

To safely depower the vehicle always follow the instructions found in service information for the exact vehicle being serviced. The steps usually include the following:

**STEP 1**   Turn the ignition off and remove the key (if equipped) from the ignition.

> **CAUTION: If a push-button start is used, remove the key fob at least 15 feet (5 m) from the vehicle to prevent the vehicle from being powered up.**

**STEP 2**   Remove the 12-volt power source to the HV controller. This step could involve the following:
- Removing a fuse or a relay
- Disconnecting the negative battery cable from the auxiliary 12-volt battery

**STEP 3**   Remove the high-voltage (HV) fuse or **service plug** or switch.

### WARNING

Even if all of the above steps are followed, there is still a risk for electrical shock at the high-voltage batteries. Always follow the vehicle manufacturer's instructions exactly and wear high-voltage gloves and other specified personal protective equipment (PPE).

## COLLISION AND REPAIR INDUSTRY ISSUES

**JUMP-STARTING**   The 12-volt auxiliary battery may be jump-started if the vehicle does not start. The 12-volt auxiliary battery is located under the hood or in the cargo (trunk) area of some HEVs. Using a jump box or jumper cable from another vehicle, make the connections to the positive and negative battery terminals. ● **SEE FIGURE 31-7.**

On the 2004+ Toyota Prius vehicles, there is a stud located under the hood that can be used to jump-start the auxiliary battery, which is located in the truck. The trunk has an electric latch and cannot be opened if the auxiliary battery is dead. ● **SEE FIGURE 31-8.**

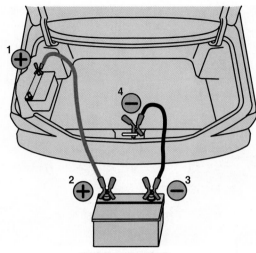

**RESCUE VEHICLE**

**FIGURE 31-7** Jump-starting a 2001–2003 Toyota Prius using a 12-volt supply to boost the 12-volt auxiliary battery in the trunk.

**FIGURE 31-8** The underhood 12-volt jump-start terminal on this 2004+ Toyota Prius has a red plastic cover with a "+" sign. The positive booster cable clamp will attach directly to the vertical metal bracket.

**NOTE: The high-voltage HV battery pack cannot be jump-started on most HEVs. One exception is the Ford Escape/ Mercury Mariner hybrids that use a special "jump-start" button located behind the left kick panel. When this button is pushed, the auxiliary battery is used to boost the HV battery through a DC-DC converter.**

## MOVING AND TOWING A HYBRID

**TOWING**   If a disabled vehicle needs to be moved a short distance (to the side of the road, for example) and the vehicle can still roll on the ground, the easiest way is to shift the transmission into neutral and manually push the vehicle. To transport a vehicle

FIGURE 31–9 Using a warning cover over the steering wheel helps others realize that work is being performed on the high-voltage system and that no one is to attempt to start or move the vehicle.

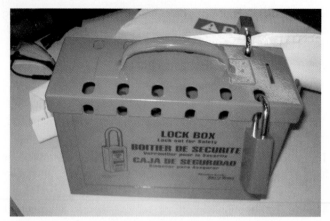

FIGURE 31–10 A lock box is a safe location to keep the ignition keys of a hybrid electric vehicle while it is being serviced.

away from an emergency location, a flatbed truck should be used if the vehicle might be repaired. If a flatbed is not available, the vehicle should be towed by wheel-lift equipment with the front wheels off the ground (FWD hybrid electric vehicles only). Do not use sling-type towing equipment. In the case of 4WD HEVs such as the Toyota Highlander, only a flatbed vehicle should be used.

### MOVING THE HYBRID VEHICLE IN THE SHOP
After an HEV has been serviced, it may be necessary to push the vehicle to another part of the shop or outside as parts are ordered. Make sure to tape any orange cable ends that were disconnected during the repair procedure. Permanent magnets are used in all the drive motors and generators, and it is possible that a high-voltage arc could occur as the wheels turn and produce voltage. Another way to prevent this is to use wheel dollies. A sign that says "HIGH VOLTAGE—DO NOT TOUCH" could also be added to the roof of the vehicle. Remove the keys from the vehicle and keep in a safe location. ● SEE FIGURES 31–9 AND 31–10.

 FREQUENTLY ASKED QUESTION

**When Do I Need to DePower the High-Voltage System?**

During routine service work, there is no need for a technician to depower the high-voltage system. The only time when this process is needed is if service repairs or testing is being performed on any circuit that has an orange cable attached. These include the following:

- AC compressor if electrically powered
- High-voltage battery pack or electronic controllers

The electric power steering system usually operates on 12 volts or 42 volts, and neither is a shock hazard. However, an arc will be maintained if a 42-volt circuit is opened. Always refer to service information if servicing the electric power steering system or any other system that may contain high voltage.

## REMOVING THE HIGH-VOLTAGE BATTERIES

### PRECAUTIONS
The HV battery box should always be removed as an assembly, placed on a rubber-covered work bench, and handled carefully. Every other part, especially the capacitors, should be checked for voltage reading while wearing HV rubber gloves. Always check for voltage as the components become accessible before proceeding. When removing high-voltage components, it is wise to use insulated tools. ● SEE FIGURE 31–11.

### STORING THE HIGH-VOLTAGE BATTERIES
If a hybrid is to be stored for any length of time, the state of charge of the HV batteries must be maintained. If possible, start the vehicle every month and run it for at least 30 minutes to help recharge the HV batteries. This is necessary because NiMH batteries suffer from self-discharge over time. High-voltage battery chargers are expensive and may be hard to find. If the HV battery SOC was over 60% when it was put into storage, the batteries may be stored for about a month without a problem. If, however, the SOC is less than 60%, a problem with a discharged HV battery may result.

### HOISTING A HYBRID VEHICLE
When hoisting or using a floor jack, pay attention to the lift points. Orange cables run under the vehicle just inside the frame rails on most hybrids. ● SEE FIGURE 31–12.

Some Honda hybrid vehicles use an aluminum pipe painted orange that includes three HV cables for the starter/generator and also three more cables for the HV air-conditioning compressor. If any damage occurs to any high-voltage cables, the MIL will light up, and a no-start will result if the PCM senses a fault. The cables are not repairable and are expensive. The cables can be identified by an orange outer casing, but in some cases, the orange casing is not exposed until a black plastic underbelly shield is removed first.

### HV BATTERY DISPOSAL
The hybrid electric vehicle manufacturers are set up to ship NiMH battery packs to a recycling

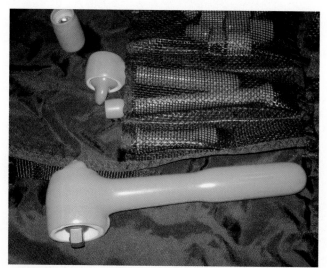

FIGURE 31–11 Insulated tools, such as this socket set, would provide an additional margin of safety to the service technician when working around high-voltage components and systems.

FIGURE 31–12 The high-voltage wiring on this Honda hybrid is colored orange for easy identification.

center. There is an 800 number located under the hood or on the HV battery pack that can be used to gain information on how to recycle these batteries.

Always follow the proper safety procedures, and then minor service to hybrid vehicles can be done with a reasonable level of safety.

## ROUTINE SERVICE PROCEDURES

**DIAGNOSIS PROCEDURES** Hybrid electric vehicles should be diagnosed the same as any other type of vehicle. This means following a diagnostic routine, which usually includes the following steps:

**STEP 1** Verify the customer concern.

**STEP 2** Check for diagnostic trouble codes (DTCs). An enhanced or factory level scan tool may be needed to get access to codes and subcodes.

**STEP 3** Perform a thorough visual inspection. If a DTC is stored, carefully inspect those areas that might be the cause of the trouble code.

**STEP 4** Check for technical service bulletins (TSBs) that may relate to the customer concern.

**STEP 5** Follow service information specified steps and procedures. This could include checking scan tool data for sensors or values that are not within normal range.

**STEP 6** Determine and repair the root cause of the problem.

**STEP 7** Verify the repair and clear any stored diagnostic trouble codes unless in an emission testing area. If in an emission test area, drive the vehicle until the powertrain control module (PCM) passes the fault and turns off the malfunction indicator lamp (MIL), thereby allowing the vehicle to pass the inspection.

**STEP 8** Complete the work order and record the "three Cs" (complaint, cause, and correction).

## OIL CHANGE
Performing an oil change is similar to changing oil in any vehicle equipped with an internal combustion engine. However, there are several items to know when changing oil in a hybrid electric vehicle, including the following:

- **Use vehicle manufacturer's recommended hoisting locations.** Use caution when hoisting a hybrid electric vehicle and avoid placing the pads on or close to the orange high-voltage cables that are usually located under the vehicle.

- **Always use the specified oil viscosity.** Most hybrid electric vehicles require either of the following:

  SAE 0W-20

  SAE 5W-20

  Using the specified oil viscosity is important because the engine stops and starts many times, and using the incorrect viscosity not only can cause a decrease in fuel economy but also could cause engine damage.

- **Always follow the specified procedures.** Be sure that the internal combustion engine (ICE) is off and that the "READY" lamp is off. If there is a smart key or the vehicle has a push-button start, be sure that the key fob is at least 15 feet (5 m) away from the vehicle to help prevent the engine from starting accidentally.

## COOLING SYSTEM SERVICE
Performing cooling system service is similar to performing this service in any vehicle equipped with an internal combustion engine. However, there are several items to know when servicing the cooling system on a hybrid electric vehicle, including the following:

- **Always check service information for the exact procedure to follow.** The procedure will include the following:

  1. **The specified coolant.** Most vehicle manufacturers will recommend using premixed coolant because using water (half of the coolant) that has minerals could cause corrosion issues.

  2. **The specified coolant replacement interval.** While this may be similar to the coolant replacement interval for a conventional vehicle, always check to be sure that this service is being performed at the specified time or mileage interval.

  3. **The specified precautions.** Some Toyota Prius HEVs use a coolant storage bottle that keeps the coolant hot for up to three days. Opening a coolant hose could cause the release of this hot coolant and can cause serious burns to the technician.

  4. Always read, understand, and follow all of the service information instructions when servicing the cooling system on a hybrid electric vehicle.

## AIR FILTER SERVICE
Performing air filter service is similar to performing this service in any vehicle equipped with an internal combustion engine. However, there are several items to know when servicing the air filter on a hybrid electric vehicle, including the following:

**REAL WORLD FIX**

### A Bad Day Changing Oil
A shop owner was asked by a regular customer, who had just bought a Prius, if the oil could be changed there. The owner opened the hood, made sure the filter was in stock (it is a standard Toyota filter used on other models), and said yes. A technician with no prior knowledge of hybrids drove the warmed-up vehicle into the service bay. The internal combustion engine never started, as it was in electric (stealth) mode at the time. Not hearing the engine running, the technician hoisted the vehicle into the air, removed the drain bolt, and drained the oil into the oil drain unit. When the filter was removed, oil started to fly around the shop. The engine was in "standby" mode during the first part of the oil change. When the voltage level dropped, the onboard computer started the engine so that the HV battery could recharge. The technician should have removed the key to keep this from happening. Be sure that the "ready" light is off before changing the oil or doing any other service work that may cause personal harm or harm to the vehicle if the engine starts.

1. Always follow the service information recommended air filter replacement interval.

2. For best results use the factory-type and quality air filter.

   **CAUTION: Using a "high-performance" cotton-type air filter that uses oil on the cotton to trap dirt can cause problems with the mass airflow sensor. If a MAF sensor diagnostic trouble code (DTC) is set, check to see if an aftermarket air filter that has been overoiled could be the cause.**

3. Double-check that all of the air ducts are securely fastened after checking or replacing the air filter.

## AIR-CONDITIONING SERVICE
Performing air-conditioning system service is similar to performing this service in any vehicle equipped with an internal combustion engine. However, there are several items to know when servicing the air-conditioning system on a hybrid electric vehicle, including the following:

1. Many hybrid electric vehicles use an air-conditioning compressor that uses high voltage from the high-voltage (HV) battery pack to operate the compressor either all of the time, such as many Toyota/Lexus models, or during idle stop periods, such as on Honda hybrids.

2. If the system is electrically driven, then special refrigerant oil is used that is nonconductive. This means that a separate recovery machine should be used to avoid the possibility of mixing regular refrigerant oils in with the oil used in hybrids.

3. Always read, understand, and follow all of the service information instructions when servicing the air-conditioning system on a hybrid electric vehicle.

## STEERING SYSTEM SERVICE
Performing steering system service is similar to performing this service in any vehicle equipped with an internal combustion engine. However, there are several items to know when servicing the steering system on a hybrid electric vehicle, including the following:

1. Check service information for any precautions that are specified to be followed when servicing the steering system on a hybrid electric vehicle.

2. Most hybrid electric vehicles use an electric power steering system. These can be powered by one of two voltages:
   - **12 volts.** These systems can be identified by the red or black wiring conduit and often use an inverter that increases the voltage to operate the actuator motor (usually to 42 volts). While this higher voltage is contained in the controller and should not create a shock hazard, always follow the specified safety precautions and wear protective high-voltage gloves as needed.
   - **42 volts.** These systems use a yellow or blue plastic conduit over the wires to help identify the possible hazards from this voltage level. This voltage level is not a shock hazard but can maintain an arc if a circuit carrying 42 volts is opened.

3. Always read, understand, and follow all of the service information instructions when servicing the steering system on a hybrid electric vehicle.

## BRAKING SYSTEM SERVICE
Performing braking system service is similar to performing this service in any vehicle equipped with an internal combustion engine. However, there are several items to know when servicing the braking system on a hybrid electric vehicle, including the following:

1. Check service information for any precautions that are specified to be followed when servicing the braking system on a hybrid electric vehicle.

2. All hybrid electric vehicles use a regenerative braking system, which captures the kinetic energy of the moving vehicle and converts it to electrical energy and is sent to the high-voltage battery pack. The amount of current produced during hard braking can exceed 100 amperes. This current is stored in the high-voltage battery pack and is then used as needed to help power the vehicle.

3. The base brakes used on hybrid electric vehicles are the same as any other conventional vehicle except for the master cylinder and related control systems. There is no high-voltage circuits associated with the braking system, as the regeneration occurs inside the electric drive (traction) motors and is controlled by the motor controller.

4. The base brakes on many hybrid vehicles are often found to be stuck or not functioning correctly because the brakes are not doing much work and can rust.

**NOTE: Always check the base brakes whenever there is a poor fuel economy complaint heard from an owner of a hybrid vehicle. Often when a disc brake caliper sticks, the brakes drag but the driver is not aware of any performance problems but the fuel economy drops.**

5. Always read, understand, and follow all of the service information instructions when servicing the braking system on a hybrid electric vehicle.

## TIRES
Performing tire-related service is similar to performing this service in any vehicle equipped with an internal combustion engine. However, there are several items to know when servicing tires on a hybrid electric vehicle, including the following:

1. Tire pressure is very important to not only the fuel economy but also on the life of the tire. Lower inflation pressure increases rolling resistance and reduces load-carrying capacity and tire life. Always inflate the tires to the pressure indicated on the door jamb sticker or found in service information or the owner's manual.

2. All tires create less rolling resistance as they wear. This means that even if the same identical tire is used as a replacement, the owner may experience a drop in fuel economy.

3. Tires can have a big effect on fuel economy. It is best to warn the owner that replacement of the tires can and often will cause a drop in fuel economy, even if low rolling resistance tires are selected.

4. Try to avoid using tires that are larger than used from the factory. The larger the tire, the heavier it is and it takes more energy to rotate, resulting in a decrease in fuel economy.

5. Follow normal tire inspections and tire rotation intervals as specified by the vehicle manufacturer.

## AUXILIARY BATTERY TESTING AND SERVICE
Performing auxiliary battery service is similar to performing this service in any vehicle equipped with an internal combustion engine. However, there are several items to know when servicing the auxiliary battery on a hybrid electric vehicle, including the following:

1. Auxiliary 12-volt batteries used in hybrid electric vehicles are located in one of two general locations.
   - **Under the hood.** If the 12-volt auxiliary battery is under the hood, it is generally a flooded-type lead-acid battery and should be serviced the same way as any conventional battery.
   - **In the passenger or trunk area.** If the battery is located in the passenger or truck area of the vehicle, it is usually of the absorbed glass mat (AGM) design. This type of battery requires that a special battery charger that limits the charging voltage be used.

2. The auxiliary 12-volt battery is usually smaller than a battery used in a conventional vehicle because it is not used to actually start the engine unless under extreme conditions on Honda hybrids only.

3. The 12-volt auxiliary battery can be tested and serviced the same as any battery used in a conventional vehicle.

4. Always read, understand, and follow all of the service information instructions when servicing the auxiliary battery on a hybrid electric vehicle.

# HV GLOVE USE

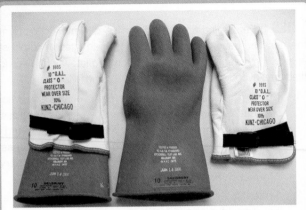

**1** The cuff of the rubber glove should extend at least 1/2 in. beyond the cuff of the leather protector.

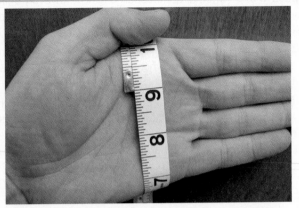

**2** To determine correct glove size, use a soft tape to measure around the palm of the hand. A measurement of 9 inches would correspond with a glove size of 9.

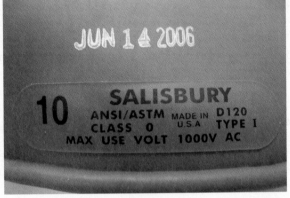

**3** The glove rating and the date of the last test should be stamped on the glove cuff.

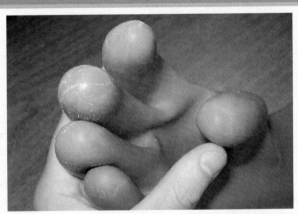

**4** Start with a visual inspection of the glove fingertips, making sure that no cuts or other damage is present.

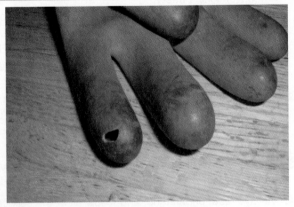

**5** The damage on this glove was easily detected with a simple visual inspection. Note that the rubber glove material can be damaged by petroleum products, detergents, certain hand soaps, and talcum powder.

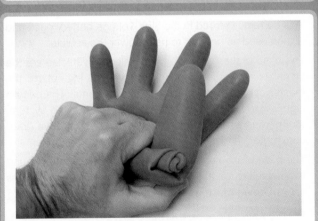

**6** Manually inflate the glove to inspect for pinhole leaks. Starting at the cuff, roll up the glove and trap air at the finger end. Listen and watch carefully for deflation of the glove. If a leak is detected, the glove must be discarded.

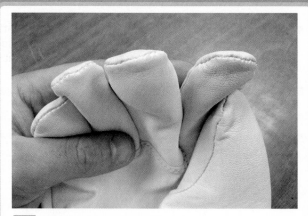

**7** Petroleum on the leather protector's surfaces will damage the rubber glove underneath.

**8** Glove powder (glove dust) should be used to absorb moisture and reduce friction.

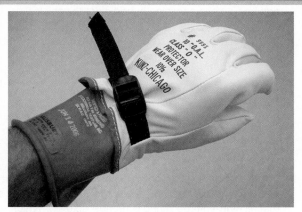

**9** Put on the gloves and tighten the straps on the back of the leather protectors.

**10** Technicians MUST wear HV gloves and leather protectors whenever working around the high-voltage areas of a hybrid electric vehicle.

**11** HV gloves and leather protectors should be placed in a canvas storage bag when not in use. Note the ventilation hole at the bottom of this bag.

**12** Make sure that the rubber gloves are not folded when placed in the canvas bag. Folding increases mechanical stress on the rubber and can lead to premature failure of the glove material.

1. Personal protective equipment (PPE) for work on hybrid electric vehicles includes the wearing of high-voltage rubber gloves rated at 1,000 volts or more worn with outer leather gloves to help protect the rubber gloves.

2. A digital meter that meets CAT III standards should be used when working around the high-voltage section of a hybrid electric vehicle.

3. Safety glasses and a face shield should be worn whenever working around the high-voltage circuits of a hybrid electric vehicle.

4. The high-voltage system can be shut off at the battery pack by simply being certain that the ignition is off. Disconnecting the 12-volt battery is additional security that the high-voltage circuits are depowered.

5. When servicing a hybrid electric vehicle, always observe safety procedures.

## REVIEW QUESTIONS

1. What are the recommended items that should be used when working with the high-voltage circuits of a hybrid electric vehicle?

2. What actions are needed to disable the high-voltage circuit?

3. What are the precautions that service technicians should adhere to when servicing hybrid electric vehicles?

## CHAPTER QUIZ

1. Rubber gloves should be worn whenever working on or near the high-voltage circuits or components of a hybrid electric vehicle. Technician A says that the rubber gloves should be rated at 1,000 volts or higher. Technician B says that leather gloves should be worn over the high-voltage rubber gloves. Which technician is correct?
   a. Technician A only
   b. Technician B only
   c. Both Technicians A and B
   d. Neither Technician A nor B

2. A CAT III-certified DMM should be used whenever measuring high-voltage circuits or components. The CAT III rating relates to _____.
   a. High voltage
   b. High energy
   c. High electrical resistance
   d. Both a and b

3. All of the following will shut off the high voltage to components and circuits, except _____.
   a. Opening the driver's door
   b. Turning the ignition off
   c. Disconnecting the 12-volt auxiliary battery
   d. Removing the main fuse, relay, or HV plug

4. If the engine is not running, Technician A says that the high-voltage circuits are depowered. Technician B says that all high-voltage wiring is orange-colored. Which technician is correct?
   a. Technician A only
   b. Technician B only
   c. Both Technicians A and B
   d. Neither Technician A nor B

5. Which statement is false about high-voltage wiring?
   a. Connects the battery pack to the electric controller
   b. Connects the controller to the motor/generator
   c. Is electrically grounded to the frame (body) of the vehicle
   d. Is controlled by a relay that opens if the ignition is off

6. What routine service procedure could result in lower fuel economy, which the owner may discover?
   a. Using the wrong viscosity engine oil
   b. Replacing tires
   c. Replacing the air filter
   d. Either a or b

7. Two technicians are discussing jump-starting a hybrid electric vehicle. Technician A says that the high-voltage batteries can be jumped on some HEV models. Technician B says that the 12-volt auxiliary battery can be jumped using a conventional jump box or jumper Which technician is correct?
   a. Technician A only
   b. Technician B only
   c. Both Technicians A and B
   d. Neither Technician A nor B

8. What can occur if a hybrid electric vehicle is pushed in the shop?
   a. The high-voltage battery pack can be damaged
   b. The tires will be locked unless the ignition is on
   c. Damage to the electronic controller can occur
   d. High voltage will be generated by the motor/generator

9. Nickel-metal hydride (NiMH) batteries can be damaged if exposed to temperatures higher than about _____.
   a. 150°F (66°C)
   b. 175°F (79°C)
   c. 200°F (93°C)
   d. 225°F (107°C)

10. How should NiMH batteries be disposed?
    a. In regular trash
    b. Call an 800 number shown under the hood of the vehicle for information
    c. Submerged in water and then disposed of in regular trash
    d. Burned at an EPA-certified plant

# chapter 32

# FUEL CELLS AND ADVANCED TECHNOLOGIES

**LEARNING OBJECTIVES:** **After studying this chapter, the reader should be able to:** • Understand the technology of fuel cells. • Explain fuel-cell vehicle systems. • Discuss hydraulic hybrid storage systems. • Explain homogeneous charge compression ignition (HCCI). • Discuss plug-in hybrid electric vehicles (PHEVs).

**KEY TERMS:** Double-layer technology 415 • Electrolysis 410 • Electrolyte 412 • Energy carrier 410 • Energy density 413 • Electric vehicles 413 • Farads 415 • Fuel cell 410 • Fuel-cell hybrid vehicle (FCHV) 411 • Fuel-cell stack 412 • Fuel-cell vehicle (FCV) 411 • Homogeneous charge compression ignition (HCCI) 419 • Hydraulic power assist (HPA) 419 • Inverter 416 • Low-grade heat 414 • Membrane electrode assembly (MEA) 412 • Plug-in hybrid electric vehicle (PHEV) 420 • Polymer electrolyte fuel cell (PEFC) 412 • Proton exchange membrane (PEM) 412 • Range 421 • Specific energy 410 • Ultracapacitor 414 • Wheel motors 416 • Wind farms 422

## FUEL-CELL TECHNOLOGY

**WHAT IS A FUEL CELL?** A **fuel cell** is an electrochemical device in which the chemical energy of hydrogen and oxygen is converted into electrical energy. The principle of the fuel cell was first discovered in 1839 by Sir William Grove, a Welsh physician. In the 1950s, NASA put this principle to work in building devices for powering space exploration vehicles. In the present day, fuel cells are being developed to power homes and vehicles while producing low or zero emissions. ● **SEE FIGURE 32–1.**

The chemical reaction in a fuel cell is the opposite of **electrolysis.** Electrolysis is the process in which electrical current is passed through water in order to break it into its components, hydrogen and oxygen. While energy is required to bring about electrolysis, this same energy can be retrieved by allowing hydrogen and oxygen to reunite in a fuel cell. It is important to note that while hydrogen can be used as a fuel, it is *not* an energy source. Instead, hydrogen is only an **energy carrier,** as energy must be expended to generate the hydrogen and store it so it can be used as a fuel.

In simple terms, a fuel cell is a hydrogen-powered battery. Hydrogen is an excellent fuel because it has a very high **specific energy** when compared to an equivalent amount of fossil fuel. One kilogram (kg) of hydrogen has three times the energy content as one kilogram of gasoline. Hydrogen is the most abundant element on earth, but it does not exist by itself in nature. This is because its natural tendency is to react with oxygen in the atmosphere to form water ($H_2O$). Hydrogen is also found in many other compounds, most notably hydrocarbons,

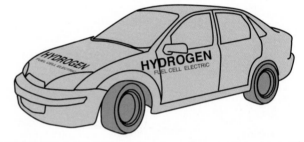

**FIGURE 32–1** Ford Motor Company has produced a number of demonstration fuel-cell vehicles based on the Ford Focus.

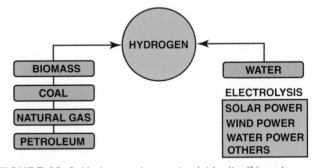

**FIGURE 32–2** Hydrogen does not exist by itself in nature. Energy must be expended to separate it from other, more complex materials.

such as natural gas or crude oil. In order to store hydrogen for use as a fuel, processes must be undertaken to separate it from these materials. ● **SEE FIGURE 32–2.**

**BENEFITS OF A FUEL CELL** A fuel cell can be used to move a vehicle by generating electricity to power electric drive motors as well as powering the remainder of the vehicle's

**FIGURE 32–3** The Mercedes-Benz B-Class fuel-cell car was introduced in 2005.

**FIGURE 32–4** The Toyota FCHV is based on the Highlander platform and uses much of Toyota's Hybrid Synergy Drive (HSD) technology in its design.

electrical system. Since they are powered by hydrogen and oxygen, fuel cells by themselves do not generate carbon emissions such as $CO_2$. Instead, their only emissions are water vapor and heat, and this makes the fuel cell an ideal candidate for a zero-emission vehicle (ZEV).

A fuel cell is also much more energy efficient than a typical internal combustion engine. While a vehicle powered by an internal combustion engine (ICE) is anywhere from 15% to 20% efficient, a fuel-cell vehicle can achieve efficiencies upwards of 40%. Another major benefit of fuel cells is that they have very few moving parts and have the potential to be very reliable. A number of OEMs have spent many years and millions of dollars in order to develop a low-cost, durable, and compact fuel cell that will operate satisfactorily under all driving conditions. ● **SEE FIGURE 32–3.**

A **fuel-cell vehicle (FCV)** uses the fuel cell as its only source of power, whereas a **fuel-cell hybrid vehicle (FCHV)** would also have an electrical storage device that can be used to power the vehicle. Most new designs of fuel-cell vehicles are now based on a hybrid configuration because of the significant increase in efficiency and driveability that can be achieved with this approach. ● **SEE FIGURE 32–4.**

**FUEL-CELL CHALLENGES** While major automobile manufacturers continue to build demonstration vehicles and work on improving fuel-cell system design, no vehicle powered by a fuel cell has been placed into mass production. There are a number of reasons for this, including the following:

- High cost
- Lack of refueling infrastructure
- Safety perception
- Insufficient vehicle range
- Lack of durability
- Freeze starting problems
- Insufficient power density

All of these problems are being actively addressed by researchers, and significant improvements are being made. Once cost and performance levels meet that of current vehicles, fuel cells will be adopted as a mainstream technology. ● **SEE CHART 32–1.**

|  | PAFC (PHOSPHORIC ACID FUEL CELL) | PEM (POLYMER ELECTROLYTE MEMBRANE) | MCFC (MOLTEN CARBONATE FUEL CELL) | SOFC (SOLID OXIDE FUEL CELL) |
|---|---|---|---|---|
| *Electrolyte* | Orthophosphoric acid | Sulfonic acid in polymer | Li and K carbonates | Yttrium-stabilized zirconia |
| *Fuel* | Natural gas, hydrogen | Natural gas, hydrogen, methanol | Natural gas, synthetic gas | Natural gas, synthetic gas |
| *Operating Temp (F) (C)* | 360–410°F | 176–212°F | 1,100–1,300°F | 1,200–3,300°F |
|  | 180–210°C | 80–100°C | 600–700°C | 650–1,800°C |
| *Electric Efficiency* | 40% | 30%–40% | 43%–44% | 50%–60% |
| *Manufacturers* | ONSI Corp. | Avista, Ballard, Energy Partners, H-Power, International, Plug Power | Fuel Cell Energy, IHI, Hitachi, Siemens | Honeywell, Siemens-Westinghouse, Ceramic |
| *Applications* | Stationary power | Vehicles, portable power, small stationary power | Industrial and institutional power | Stationary power, military vehicles |

**CHART 32–1**

Fuel cell types and their operating temperature range.

## TYPES OF FUEL CELLS

There are a number of different types of fuel cells, and these are differentiated by the type of **electrolyte** that is used in their design. Some electrolytes operate best at room temperature, whereas others are made to operate at up to 1,800°F. See the accompanying chart showing the various fuel-cell types and applications.

The fuel-cell design that is best suited for automotive applications is the **proton exchange membrane (PEM).** A PEM fuel cell must have hydrogen for it to operate, and this may be stored on the vehicle or generated as needed from another type of fuel.

# PEM FUEL CELLS

## DESCRIPTION AND OPERATION

The Proton Exchange Membrane fuel cell is also known as a **polymer electrolyte fuel cell (PEFC).** The PEM fuel cell is known for its lightweight and compact design as well as its ability to operate at ambient temperatures. This means that a PEM fuel cell can start quickly and produce full power without an extensive warm-up period. The PEM is a simple design based on a membrane that is coated on both sides with a catalyst such as platinum or palladium. There are two electrodes, one located on each side of the membrane. These are responsible for distributing hydrogen and oxygen over the membrane surface, removing waste heat, and providing a path for electrical current flow. The part of the PEM fuel cell that contains the membrane, catalyst coatings, and electrodes is known as the **membrane electrode assembly (MEA).**

The negative electrode (anode) has hydrogen gas directed to it, while oxygen is sent to the positive electrode (cathode). Hydrogen is sent to the negative electrode as H2 molecules, which break apart into $H^+$ ions (protons) in the presence of the catalyst. The electrons ($e^-$) from the hydrogen atoms are sent through the external circuit, generating electricity that can be utilized to perform work. These same electrons are then sent to the positive electrode, where they rejoin the $H^+$ ions that have passed through the membrane and have reacted with oxygen in the presence of the catalyst. This creates $H_2O$ and waste heat, which are the only emissions from a PEM fuel cell. **●SEE FIGURE 32–5.**

**NOTE: It is important to remember that a fuel cell generates direct current (DC) electricity as electrons flow in only one direction (from the anode to the cathode).**

## FUEL-CELL STACKS

A single fuel cell by itself is not particularly useful, as it will generate less than 1 volt of electrical potential. It is more common for hundreds of fuel cells to be built together in a **fuel-cell stack.** In this arrangement, the fuel cells are connected in series so that total voltage of the stack is the sum of the individual cell voltages. The fuel cells are placed end to end in the stack, much like slices in a loaf of bread. Automotive fuel-cell stacks contain more than 400 cells in their construction. **●SEE FIGURE 32–6.**

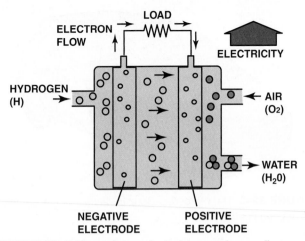

**FIGURE 32–5** The polymer electrolyte membrane allows only $H^+$ ions (protons) to pass through it. This means that electrons must follow the external circuit and pass through the load to perform work.

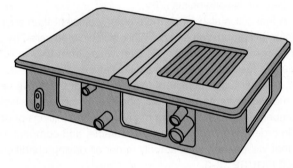

**FIGURE 32–6** A fuel-cell stack is made up of hundreds of individual cells connected in series.

  TECH TIP

### CO Poisons the PEM Fuel-Cell Catalyst

Purity of the fuel gas is critical with PEM fuel cells. If more than 10 parts per million (ppm) of carbon monoxide is present in the hydrogen stream being fed to the PEM anode, the catalyst will be gradually poisoned, and the fuel cell will eventually be disabled. This means that the purity must be "five nines" (99.999% pure). This is a major concern in vehicles where hydrogen is generated by reforming hydrocarbons such as gasoline because it is difficult to remove all CO from the hydrogen during the reforming process. In these applications, some means of hydrogen purification must be used to prevent CO poisoning of the catalyst.

The total voltage of the fuel-cell stack is determined by the number of individual cells incorporated into the assembly. The current-producing ability of the stack, however, is dependent on the surface area of the electrodes. Since output of the fuel-cell

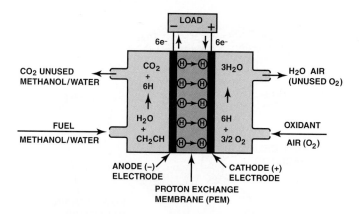

**FIGURE 32–7** A direct methanol fuel cell uses a methanol/water solution for fuel instead of hydrogen gas.

**FIGURE 32–8** A direct methanol fuel cell can be refueled similar to a gasoline-powered vehicle.

stack is related to both voltage and current (voltage × current = power), increasing the number of cells or increasing the surface area of the cells will increase power output. Some fuel-cell vehicles will use more than one stack, depending on power output requirements and space limitations.

### DIRECT METHANOL FUEL CELLS
High-pressure cylinders are one method of storing hydrogen onboard a vehicle for use in a fuel cell. This is a simple and lightweight storage method but often does not provide sufficient vehicle driving range. Another approach has been to fuel a modified PEM fuel cell with liquid methanol instead of hydrogen gas. ● **SEE FIGURE 32–7.**

Methanol is most often produced from natural gas and has a chemical symbol of $CH_3OH$. It has a higher **energy density** than gaseous hydrogen because it exists in a liquid state at normal temperatures and is easier to handle since no compressors or other high-pressure equipment is needed. This means that a fuel-cell vehicle can be refueled with a liquid instead of high-pressure gas, which makes the refueling process simpler and produces a greater vehicle driving range. ● **SEE FIGURE 32–8.**

Unfortunately, direct methanol fuel cells suffer from a number of problems, not the least of which is the corrosive nature of methanol itself. This means that methanol cannot be stored in existing tanks and thus requires a separate infrastructure for handling and storage. Another problem is "fuel crossover," in

**?** FREQUENTLY ASKED QUESTION

#### What Is the Role of the Humidifier in a PEM Fuel Cell?

The polymer electrolyte membrane assembly in a PEM fuel cell acts as conductor of positive ions and as a gas separator. However, it can perform these functions effectively only if it is kept moist. A fuel-cell vehicle uses an air compressor to supply air to the positive electrodes of each cell, and this air is sometimes sent through a humidifier first to increase its moisture content. The humid air then comes in contact with the membrane assembly and keeps the electrolyte damp and functioning correctly.

which methanol makes its way across the membrane assembly and diminishes performance of the cell. Direct methanol fuel cells also require much greater amounts of catalyst in their construction, which leads to higher costs. These challenges are leading researchers to look for alternative electrolyte materials and catalysts to lower cost and improve cell performance.

**NOTE: Direct methanol fuel cells are not likely to see service in automotive applications. However, they are well suited for low-power applications, such as cell phones or laptop computers.**

## FUEL-CELL VEHICLE SYSTEMS

**HUMIDIFIERS** Water management inside a PEM fuel cell is critical. Too much water can prevent oxygen from making contact with the positive electrode; too little water can allow the electrolyte to dry out and lower its conductivity. The amount of water and where it resides in the fuel cell is also critical in determining at how low a temperature the fuel cell will start because water freezing in the fuel cell can prevent it from starting. The role of the humidifier is to achieve a balance where it is providing sufficient moisture to the fuel cell by recycling water that is evaporating at the cathode. The humidifier is located in the air line leading to the cathode of the fuel-cell stack. ● **SEE FIGURE 32–9.**

Some newer PEM designs manage the water in the cells in such a way that there is no need to prehumidify the incoming reactant gases. This eliminates the need for the humidifier assembly and makes the system simpler overall.

**FUEL-CELL COOLING SYSTEMS** Heat is generated by the fuel cell during normal operation. Excess heat can lead to a breakdown of the polymer electrolyte membrane, so a liquid cooling system must be utilized to remove waste heat from the

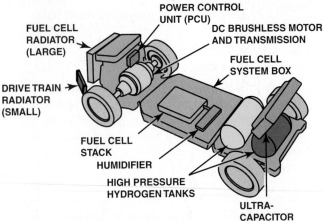

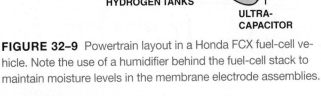

**FIGURE 32–9** Powertrain layout in a Honda FCX fuel-cell vehicle. Note the use of a humidifier behind the fuel-cell stack to maintain moisture levels in the membrane electrode assemblies.

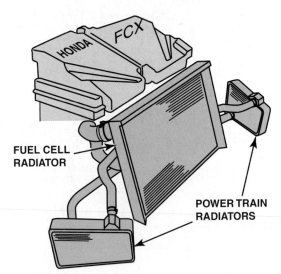

**FIGURE 32–10** The Honda FCX uses one large radiator for cooling the fuel cell and two smaller ones on either side for cooling drivetrain components.

**FIGURE 32–11** Space is limited at the front of the Toyota FCHV engine compartment, so an auxiliary heat exchanger is located under the vehicle to help cool the fuel-cell stack.

### ? FREQUENTLY ASKED QUESTION

**When Is Methanol Considered to Be a "Carbon-Neutral" Fuel?**

Most of the methanol in the world is produced by reforming natural gas. Natural gas is a hydrocarbon but does not increase the carbon content of our atmosphere as long as it remains in reservoirs below the earth's surface. However, natural gas that is used as a fuel causes extra carbon to be released into the atmosphere, which is said to contribute to global warming. Natural gas is not a carbon-neutral fuel, and neither is methanol, which is made from natural gas.

Fortunately, it is possible to generate methanol from biomass and wood waste. Methanol made from renewable resources is carbon neutral because no extra carbon is being released into the earth's atmosphere than what was originally absorbed by the plants used to make the methanol.

fuel-cell stack. One of the major challenges for engineers in this regard is the fact that the heat generated by the fuel cell is classified as **low-grade heat.** This means that there is only a small difference between the temperature of the coolant and that of the ambient air. Heat transfers very slowly under these conditions, so heat exchangers with a much larger surface area must be utilized. ● **SEE FIGURE 32–10.**

In some cases, heat exchangers may be placed in other areas of the vehicle when available space at the front of the engine compartment is insufficient. In the case of the Toyota FCHV, an auxiliary heat exchanger is located underneath the vehicle to increase the cooling system heat-rejection capacity. ● **SEE FIGURE 32–11.**

An electric water pump and a fan drive motor are used to enable operation of the fuel cell's cooling system. These and other support devices use electrical power that is generated by the fuel cell and therefore tend to decrease the overall efficiency of the vehicle.

**AIR SUPPLY PUMPS** Air must be supplied to the fuel-cell stack at the proper pressure and flow rate to enable proper performance under all driving conditions. This function is performed by an onboard air supply pump that compresses atmospheric air and supplies it to the fuel cell's positive electrode (cathode). This pump is often driven by a high-voltage electric drive motor.

**FUEL-CELL HYBRID VEHICLES** Hybridization tends to increase efficiency in vehicles with conventional drivetrains, as energy that was once lost during braking and otherwise normal operation is instead stored for later use in a high-voltage battery or **ultracapacitor.** This same advantage can be gained

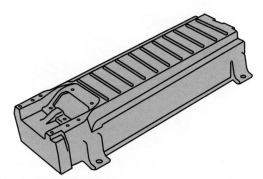

**FIGURE 32–12** The secondary battery in a fuel-cell hybrid vehicle is made up of many individual cells connected in series, much like a fuel-cell stack.

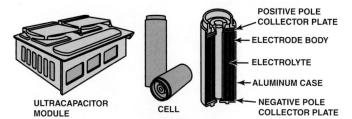

ULTRACAPACITOR
MODULE
CELL

POSITIVE POLE
COLLECTOR PLATE
ELECTRODE BODY
ELECTROLYTE
ALUMINUM CASE
NEGATIVE POLE
COLLECTOR PLATE

**FIGURE 32–13** The Honda ultracapacitor module and construction of the individual cells.

by applying the hybrid design concept to fuel-cell vehicles. Whereas the fuel cell is the only power source in a fuel-cell vehicle, the fuel-cell hybrid vehicle (FCHV) relies on both the fuel cell and an electrical storage device for motive power. Driveability is also enhanced with this design, as the electrical storage device is able to supply energy immediately to the drive motors and overcome any "throttle lag" on the part of the fuel cell.

**SECONDARY BATTERIES** All hybrid vehicle designs require a means of storing electrical energy that is generated during regenerative braking and other applications. In most FCHV designs, a high-voltage nickel-metal hydride (NiMH) battery pack is used as a secondary battery. This is most often located near the back of the vehicle, either under or behind the rear passenger seat. ● **SEE FIGURE 32–12.** The secondary battery is built similar to a fuel-cell stack because it is made up of many low-voltage cells connected in series to build a high-voltage battery.

**ULTRACAPACITORS** An alternative to storing electrical energy in batteries is to use ultracapacitors. A capacitor is best known as an electrical device that will block DC current but allow AC to pass. However, a capacitor can also be used to store electrical energy, and it is able to do this without a chemical reaction. Instead, a capacitor stores electrical energy using the principle of electrostatic attraction between positive and negative charges.

Ultracapacitors are built very different from conventional capacitors. Ultracapacitor cells are based on **double-layer technology,** in which two activated-carbon electrodes are immersed in an organic electrolyte. The electrodes have a very large surface area and are separated by a membrane that allows ions to migrate but prevents the electrodes from touching. ● **SEE FIGURE 32–13.** Charging and discharging occurs

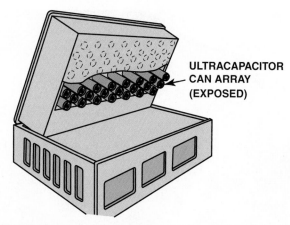

ULTRACAPACITOR
CAN ARRAY
(EXPOSED)

**FIGURE 32–14** An ultracapacitor can be used in place of a high-voltage battery in a hybrid electric vehicle. This example is from the Honda FCX fuel-cell hybrid vehicle.

as ions move within the electrolyte, but no chemical reaction takes place. Ultracapacitors can charge and discharge quickly and efficiently, making them especially suited for electric assist applications in fuel-cell hybrid vehicles.

Ultracapacitors that are used in fuel-cell hybrid vehicles are made up of multiple cylindrical cells connected in parallel. ● **SEE FIGURE 32–14.** This results in the total capacitance being the sum of the values of each individual cell. For example, 10 1.0-**farad** capacitors connected in parallel will have a total capacitance of 10.0 farads. Greater capacitance means greater electrical storage ability, and this contributes to greater assist for the electric motors in a fuel-cell hybrid vehicle.

Ultracapacitors have excellent cycle life, meaning that they can be fully charged and discharged many times without degrading their performance. They are also able to operate over a wide temperature range and are not affected by low temperatures to the same degree as many battery technologies. The one major downside of ultracapacitors is a lack of specific energy, which means that they are best suited for sudden bursts of energy as opposed to prolonged discharge cycles. Research is being conducted to improve this and other aspects of ultracapacitor performance.

**FUEL-CELL TRACTION MOTORS** Much of the technology behind the electric drive motors being used in fuel-cell vehicles was developed during the early days of the California ZEV mandate. This was a period when battery-powered electric vehicles were being built by the major vehicle manufacturers in an effort to meet a legislated quota in the state of California. The ZEV mandate rules were eventually relaxed to allow other types of vehicles to be substituted for credit, but the technology that had been developed for pure electric vehicles was now put to work in these other vehicle designs.

The electric traction motors used in fuel-cell hybrid vehicles are very similar to those being used in current hybrid electric vehicles. The typical drive motor is based on an AC synchronous design, which is sometimes referred to as a DC brushless motor. This design is very reliable, as it does not use a commutator or brushes but instead has a three-phase stator

FIGURE 32–15 Drive motors in fuel-cell hybrid vehicles often use stator assemblies similar to ones found in Toyota hybrid electric vehicles. The rotor turns inside the stator and has permanent magnets on its outer circumference.

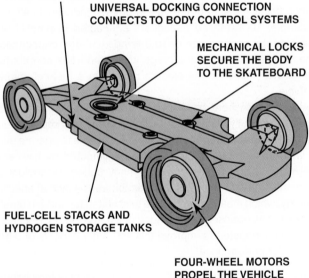

HEAT VENTS TO DISSIPATE HEAT GENERATED BY THE FUEL CELL AND ELECTRONICS

UNIVERSAL DOCKING CONNECTION CONNECTS TO BODY CONTROL SYSTEMS

MECHANICAL LOCKS SECURE THE BODY TO THE SKATEBOARD

FUEL-CELL STACKS AND HYDROGEN STORAGE TANKS

FOUR-WHEEL MOTORS PROPEL THE VEHICLE

FIGURE 32–16 The General Motors "Skateboard" concept uses a fuel-cell propulsion system with wheel motors at all four corners.

and a permanent magnet rotor. ● SEE FIGURE 32–15. An electronic controller (inverter) is used to generate the three-phase high-voltage AC current required by the motor. While the motor itself is very simple, the electronics required to power and control it are complex.

Some fuel-cell hybrid vehicles use a single electric drive motor and a transaxle to direct power to the vehicle's wheels. It is also possible to use **wheel motors** to drive individual wheels. While this approach adds a significant amount of unsprung weight to the chassis, it allows for greater control of the torque being applied to each individual wheel. ● SEE FIGURE 32–16.

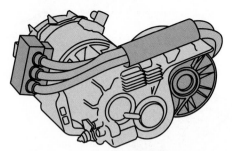

FIGURE 32–17 The electric drive motor and transaxle assembly from a Toyota FCHV. Note the three orange cables, indicating that this motor is powered by high-voltage three-phase alternating current.

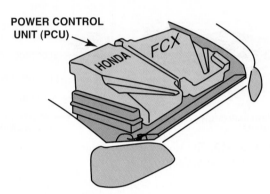

POWER CONTROL UNIT (PCU)

FIGURE 32–18 The power control unit (PCU) on a Honda FCX fuel-cell hybrid vehicle is located under the hood.

**TRANSAXLES** Aside from the hydrogen fueling system, fuel-cell hybrid vehicles are effectively pure electric vehicles in that their drivetrain is electrically driven. Electric motors work very well for automotive applications because they produce high torque at low RPMs and are able to maintain a consistent power output throughout their entire RPM range. This is in contrast to vehicles powered by internal combustion engines, which produce very little torque at low RPMs and have a narrow range where significant horsepower is produced.

ICE-powered vehicles require complex transmissions with multiple speed ranges in order to accelerate the vehicle quickly and maximize the efficiency of the ICE. Fuel-cell hybrid vehicles use electric drive motors that require only a simple reduction in their final drive and a differential to send power to the drive wheels. No gear shifting is required, and mechanisms such as torque converters and clutches are done away with completely. A reverse gear is not required either, as the electric drive motor is simply powered in the opposite direction. The transaxles used in fuel-cell hybrid vehicles are extremely simple with few moving parts, making them extremely durable, quiet, and reliable. ● SEE FIGURE 32–17.

**POWER CONTROL UNITS** The drivetrain of a fuel-cell hybrid vehicle is controlled by a power control unit (PCU), which controls fuel-cell output and directs the flow of electricity between the various components. One of the functions of the PCU is to act as an **inverter,** which changes direct current from the fuel-cell stack into three-phase alternating current for use in the vehicle drive motor(s). ● SEE FIGURE 32–18.

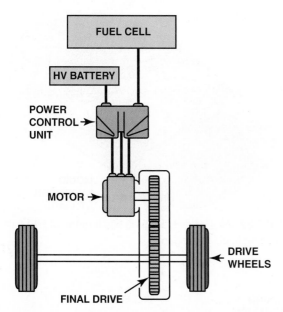

FIGURE 32–19 Toyota's FCHV uses a power control unit that directs electrical energy flow between the fuel cell, battery, and drive motor.

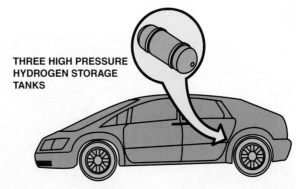

FIGURE 32–20 This GM fuel-cell vehicle uses compressed hydrogen in three high-pressure storage tanks.

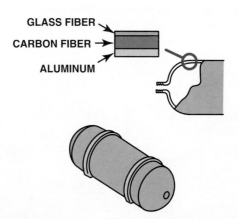

FIGURE 32–21 The Toyota FCHV uses high-pressure storage tanks that are rated at 350 bar. This is the equivalent of 5,000 pounds per square inch.

Power to and from the secondary battery is directed through the power control unit, which is also responsible for maintaining the battery pack's state of charge and for controlling and directing the output of the fuel-cell stack. ● SEE FIGURE 32–19.

During regenerative braking, the electric drive motor acts as a generator and converts kinetic (moving) energy of the vehicle into electricity for recharging the high-voltage battery pack. The PCU must take the three-phase power from the motor (generator) and convert (or *rectify*) this into DC voltage to be sent to the battery. DC power from the fuel cell will also be processed through the PCU for recharging the battery pack.

A DC-to-DC converter is used in hybrid-electric vehicles for converting the high voltage from the secondary battery pack into the 12 volts required for the remainder of the vehicle's electrical system. Depending on the vehicle, there may also be 42 volts required to operate accessories such as the electric-assist power steering. In fuel-cell hybrid vehicles, the DC-to-DC converter function may be built into the power control unit, giving it full responsibility for the vehicle's power distribution.

## HYDROGEN STORAGE
One of the pivotal design issues with fuel-cell hybrid vehicles is how to store sufficient hydrogen onboard to allow for reasonable vehicle range. Modern drivers have grown accustomed to having a minimum of 300 miles between refueling stops, a goal that is extremely difficult to achieve when fueling the vehicle with hydrogen. Hydrogen has a very high energy content on a pound-for-pound basis, but its energy density is less than that of conventional liquid fuels. This is because gaseous hydrogen, even at high pressure, has a very low physical density (mass per unit volume). ● SEE FIGURE 32–20.

A number of methods of hydrogen storage are being considered for use in fuel-cell hybrid vehicles. These include high-pressure compressed gas, liquefied hydrogen, and solid storage in metal hydrides. Efficient hydrogen storage is one of the technical issues that must be solved in order for fuel cells to be adopted for vehicle applications. Much research is being conducted to solve the issue of onboard hydrogen storage.

**HIGH-PRESSURE COMPRESSED GAS** Most current fuel-cell hybrid vehicles use compressed hydrogen that is stored in tanks as a high-pressure gas. This approach is the least complex of all the storage possibilities but also has the least energy density. Multiple small storage tanks are often used rather than one large one in order to fit them into unused areas of the vehicle. One drawback with this approach is that only cylinders can be used to store gases at the required pressures. This creates a good deal of unused space around the outside of the cylinders and leads to further reductions in hydrogen storage capacity. It is common for a pressure of 5,000 PSI (350 bar) to be used, but technology is available to store hydrogen at up to 10,000 PSI (700 bar). ● SEE FIGURE 32–21.

**FIGURE 32–22** The high-pressure fitting used to refuel a fuel-cell hybrid vehicle.

**FIGURE 32–23** Note that high-pressure hydrogen storage tanks must be replaced in 2020.

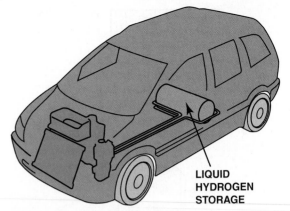

**FIGURE 32–24** GM's Hydrogen3 has a range of 249 miles when using liquid hydrogen.

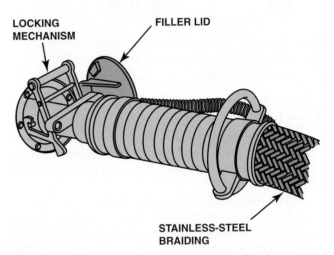

**FIGURE 32–25** Refueling a vehicle with liquid hydrogen.

The tanks used for compressed hydrogen storage are typically made with an aluminum liner wrapped in several layers of carbon fiber and an external coating of fiberglass. In order to refuel the compressed hydrogen storage tanks, a special high-pressure fitting is installed in place of the filler neck used for conventional vehicles. ● SEE FIGURE 32–22. There is also a special electrical connector that is used to enable communication between the vehicle and the filling station during the refueling process. ● SEE FIGURE 32–23.

The filling station utilizes a special coupler to connect to the vehicle's high-pressure refueling fitting. The coupler is placed on the vehicle fitting, and a lever on the coupler is rotated to seal and lock it into place.

**LIQUID HYDROGEN** Hydrogen can be liquefied in an effort to increase its energy density, but this requires that it be stored in cryogenic tanks at −423°F (−253°C). This increases vehicle range but impacts overall efficiency, as a great deal of energy is required to liquefy the hydrogen, and a certain amount of the liquid hydrogen will "boil off" while in storage.

One liter of liquid hydrogen only has one-fourth the energy content of 1 liter of gasoline. ● SEE FIGURES 32–24 AND 32–25.

**SOLID STORAGE OF HYDROGEN** One method discovered to store hydrogen in solid form is as a metal hydride, similar to how a nickel-metal hydride (NiMH) battery works.

A demonstration vehicle features a lightweight fiber-wrapped storage tank under the body that stores 3 kg (about 6.6 pounds) of hydrogen as a metal hydride at low pressure. The vehicle can travel almost 200 miles with this amount of fuel. One kilogram of hydrogen is equal to 1 gallon of gasoline. Three gallons of water will generate 1 kilogram of hydrogen.

A metal hydride is formed when gaseous hydrogen molecules disassociate into individual hydrogen atoms and bond with the metal atoms in the storage tank. This process uses powdered metallic alloys capable of rapidly absorbing hydrogen to make this occur.

**FIGURE 32–26** Carbon deposits, such as these, are created by incomplete combustion of a hydrocarbon fuel.

 **TECH TIP**

**Hydrogen Fuel = No Carbon**

Most fuels contain hydrocarbons or molecules that contain both hydrogen and carbon. During combustion, the first element that is burned is the hydrogen. If combustion is complete, then all of the carbon is converted to carbon dioxide gas and exits the engine in the exhaust. However, if combustion is not complete, carbon monoxide is formed, leaving some unburned carbon to accumulate in the combustion chamber. ● **SEE FIGURE 32–26**.

# HYDRAULIC HYBRID STORAGE SYSTEM

Ford Motor Co. is experimenting with a system it calls **hydraulic power assist (HPA)**. This system converts kinetic energy to hydraulic pressure and then uses that pressure to help accelerate the vehicle. It is currently being tested on a four-wheel-drive (4WD) Lincoln Navigator with a 4.0-L V-8 engine in place of the standard 5.4-L engine.

A variable-displacement hydraulic pump/motor is mounted on the transfer case and connected to the output shaft that powers the front driveshaft. The HPA system works with or without 4WD engaged. A valve block mounted on the pump contains solenoid valves to control the flow of hydraulic fluid. A 14-gallon, high-pressure accumulator is mounted behind the rear axle, with a low-pressure accumulator right behind it to store hydraulic fluid. The master cylinder has a "deadband," meaning the first few fractions of an inch of travel do not pressurize the brake system. When the driver depresses the brake pedal, a pedal movement sensor signals the control unit, which then operates solenoid valves to send hydraulic fluid from the low-pressure reservoir to the pump. The pumping action slows the vehicle, similar to engine compression braking, and the fluid is pumped into the high-pressure reservoir. Releasing the brake and pressing on the accelerator signals the control unit

to send that high-pressure fluid back to the pump, which then acts as a hydraulic motor and adds torque to the driveline. The system can be used to launch the vehicle from a stop and/or add torque for accelerating from any speed.

While the concept is simple, the system itself is very complicated. Additional components include the following:

- Pulse suppressors
- Filters
- An electric circulator pump for cooling the main pump/motor

Potential problems with this system include leakage problems with seals and valves, getting air out of the hydraulic fluid system, and noise. In prototype stages the system demands different driving techniques. Still, this system was built to prove the concept, and engineers believe that these problems can be solved and that a control system can be developed that will make HPA transparent to the driver. A 23% improvement in fuel economy and improvements in emissions reduction were achieved using a dynamometer set for a 7,000-pound vehicle. While the HPA system could be developed for any type of vehicle with any type of drivetrain, it does add weight and complexity, which would add to the cost.

# HCCI

**Homogeneous charge compression ignition (HCCI)** is a combustion process. HCCI is the combustion of a very lean gasoline air–fuel mixture without the use of a spark ignition. It is a low-temperature, chemically controlled (flameless) combustion process. ● **SEE FIGURE 32–27**.

HCCI combustion is difficult to control and extremely sensitive to changes in temperature, pressure, and fuel type. While the challenges of HCCI are difficult, the advantages include having a gasoline engine being able to deliver 80% of diesel efficiency (a 20% increase in fuel economy) for 50% of the cost. A diesel engine using HCCI can deliver gasoline-like emissions. Spark and injection timing are no longer a factor as they are in a conventional port-fuel injection system.

While much research and development needs to be performed using this combustion process, it has been shown to give excellent performance from idle to midload and from ambient to warm operating temperatures as well as cold-start and run capability. Because an engine only operates in HCCI mode at light throttle settings, such as during cruise conditions at highway speeds, engineers need to improve the transition in and out of the HCCI mode. Work is also being done on piston and combustion chamber shape to reduce combustion noise and vibration that is created during operation in the HCCI operating mode.

Ongoing research is focusing on improving fuel economy under real-world operating conditions as well as controlling costs.

DIESEL ENGINE
(COMPRESSION IGNITION)

FUEL
INJECTOR →

GASOLINE ENGINE
(SPARK IGNITED)

HCCI ENGINE
(HOMOGENEOUS CHARGE
COMPRESSION IGNITION)

HOT REGIONS
CREATES NO$_X$ AND
SOOT (PM) EMISSIONS

HOT REGIONS
CREATES NO$_X$
EMISSIONS

LOW-TEMPERATURE
COMBUSTION RESULTS
IN REDUCED EMISSIONS

**FIGURE 32–27** Both diesel and conventional gasoline engines create exhaust emissions due to high peak temperatures created in the combustion chamber. The lower combustion temperatures during HCCI operation result in high efficiency with reduced emissions.

# PLUG-IN HYBRID ELECTRIC VEHICLES

**PRINCIPLES** A **plug-in hybrid electric vehicle (PHEV)** is a vehicle that is designed to be plugged into an electrical outlet at night to charge the batteries. By charging the batteries in the vehicle, it can operate using electric power alone (stealth mode) for a longer time, thereby reducing the use of the internal combustion engine (ICE). The less the ICE is operating, the less fuel is consumed and the lower the emissions.

**PHEV BATTERY CAPACITY** The size or capacity of the battery pack used determines how far that the vehicle can travel without using the ICE, commonly called the electric vehicle or EV range.

- A lower kilowatt-hour (kWh)-rated battery weighs less and is less expensive but the range that the vehicle can travel on battery power alone is limited.

- A higher kWh rating battery means that the battery is capable of propelling the vehicle for a greater distance before the ICE is used. This reduces the fuel used but the larger battery weighs and costs more.

- Therefore, a plug-in vehicle is a compromise between cost and weight of the battery.

**PHEV EXAMPLES** A standard Toyota Prius has a 1.3 kWh battery pack, whereas the plug-in version has a larger 4.4 kWh battery, allowing electric-only travel of about 10 miles before the ICE is used. When the battery pack SOC has been depleted, the vehicle operates as a standard HEV with the ICE and the electric motor, both used to propel the vehicle.

A Chevrolet Volt has a larger 16 kWh battery pack and, as a result, can travel over 30 miles on electric power alone,

without using the ICE until the battery has been discharged to about 25% to 35%. At this stage, the ICE is operated to keep the battery pack at a level high enough to keep propelling the vehicle. ● SEE FIGURE 32–28.

**CHARGING A PHEV** After the battery pack has been discharged propelling the vehicle, the ICE is used to keep the battery charged enough to propel the vehicle, but it does not fully recharge the battery pack. To fully charge the high-voltage battery pack in a plug-in hybrid electric vehicle requires that it be plugged into either a 120-volt or a 240-volt outlet. ● SEE FIGURE 32–29.

# ELECTRIC VEHICLES

**PRINCIPLES** Electric vehicles (EV) use a high-voltage battery pack to supply electrical energy to an electric motor(s) to propel the vehicle under all driving conditions. The capacity of the battery pack in kilowatt-hours (kWh) determines the range of the vehicle and has to be plugged in to recharge the battery before it can be driven further.

**COLD WEATHER CONCERNS** Past models of electric vehicles, such as the General Motors electric vehicle (EV1), were restricted to locations such as Arizona and southern California that had a warm climate. Cold weather is a major disadvantage to the use of electric vehicles for the following reasons:

- Cold temperatures reduce battery efficiency.
- Additional electrical power from the batteries is needed to heat the batteries themselves to be able to achieve reasonable performance.
- Passenger compartment heating is a concern for an electric vehicle because it requires the use of resistance units or other technology that reduces the range of the vehicle.

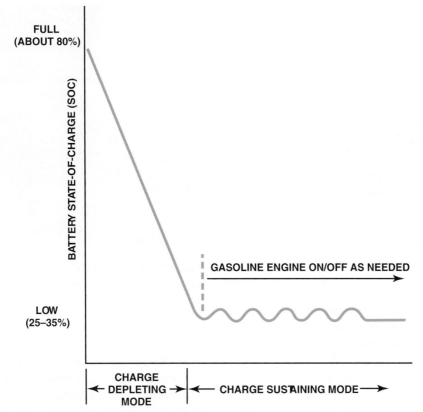

FULL
(ABOUT 80%)

BATTERY STATE-OF-CHARGE (SOC)

GASOLINE ENGINE ON/OFF AS NEEDED

LOW
(25–35%)

CHARGE
DEPLETING
MODE

CHARGE SUSTAINING MODE

**FIGURE 32–28** After the Chevrolet Volt has been charged, it uses the electrical power stored in the high-voltage battery to propel the vehicle and provide heating and cooling for 25 to 50 miles (40 to 80 km). Then the gasoline engine starts and maintains the SOC between 25% and 35%. The gasoline engine cannot fully charge the high-voltage batteries but rather the vehicle has to be plugged in to provide a higher SOC level.

(a)

(b)

**FIGURE 32–29** a) The Chevrolet Volt is charged using a standard SAE 1772 connector using either 110 or 220 volts. (b) After connecting the charging plug, a light on the top of the dash turns green and the dash display shows the estimated time when the high-voltage battery will be fully charged and the estimated current range using battery power alone.

**HOT WEATHER CONCERNS** Batteries do not function well at high temperatures, and therefore, some type of battery cooling system must be added to the vehicle to allow for maximum battery performance. This results in a reduction of vehicle range due to the use of battery power needed just to keep the batteries work-ing properly. Besides, the batteries also have to supply the power needed to keep the interior and the other accessories of the vehicle cool. These combined electrical loads represent a huge battery drain and reduce the range of the vehicle.

**Batteries Like the Same Temperature Range as Humans**

Batteries work best when they are kept within a temperature range that is also the most comfortable for humans. Most people are comfortable when the temperature is between 68°F and 78°F (20°C and 26°C).

- Below 68°F (20°C), most people want heat.
- Above 78°F (26°C), most people want cooling.

Batteries perform best when they too are exposed to the same temperature range. Therefore, a proper heating and cooling system must be used to keep the batteries within this fairly narrow temperature range for best performance.

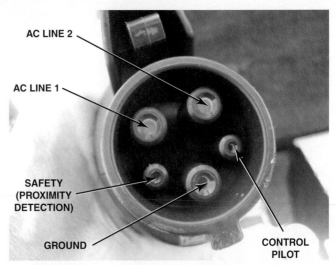

**FIGURE 32–30** The SAE J 1772 plug is used on most electric and plug-in hybrid electric vehicles and is designed to work with Level 1 (110 to 120 volt) and Level 2 (220 to 240 volt) charging.

**RANGE** How far an electric vehicle can travel on a full battery charge is called its **range**. The range of an electric vehicle depends on many factors, including:

- Battery energy storage capacity
- Vehicle weight
- Outside temperature
- Terrain (driving in hilly or mountainous areas requires more energy from the battery)
- Use of air conditioning and other electrical devices

**BATTERY CAPACITY AND RANGE EXAMPLES** ●SEE **CHART 32–2** for some examples of the battery capacity and the range of selected electric vehicles:

| VEHICLE | BATTERY CAPACITY IN KWH | RANGE (MILES/KM) |
|---------|------------------------|------------------|
| Nissan Leaf | 24 | 84/135 |
| Ford Focus EV | 23 | 76/122 |
| Chevrolet Spark Electric | 21 | 82/132 |
| Fiat 500 EV | 24 | 87/140 |
| Honda Fit EV | 20 | 82/132 |
| Mitsubishi i-MiEV | 16 | 62/100 |
| Tesla | 60 | 208/335 |
|  | 85 | 265/426 |

**CHART 32–2**

Selected electric vehicles and their battery capacity and published range.

**SAE STANDARD CHARGER PLUG** Most electric vehicles and plug-in hybrid electric vehicles, such as the Chevrolet Volt and Toyota Prius, use a standard charger plug. The standard charger plug meets the specification as designated by SAE standard J1772 (updated in 2009). ● **SEE FIGURE 32–30.**

Because electric vehicles have a relatively short range, charging stations must be made available in areas where these vehicles are driven. For example, when the state of California mandated the sale of zero-emission vehicles (ZEV), charging stations were set up in many areas, usually in parking lots of businesses and schools. The parking spaces near the charging stations are designated for electric vehicles only and can be used for free to recharge electric vehicles.

**CHARGING LEVELS** There are three levels of chargers that can be used to charge a plug-in hybrid electric vehicle (PHEV) or electric vehicle (EV). They are:

- **Level 1** uses 110- to 120-volt standard electric outlets and is capable of charging a Chevrolet Volt extended range electric vehicle in 10 hours or more. The advantage is that there is little if any installation cost as most houses are equipped with 110-volt outlets and can charge up to 16 amperes.

- **Level 2** chargers use 220 to 240 volts to charge the same vehicle in about 4 hours. Level 2 chargers can be added to most houses, making recharging faster (up to 80 amperes) when at home, and are the most commonly used charging stations available at stores and colleges. Adding a Level 2 charging outlet to the garage or parking location can cost $2,000 or more depending on the location and the wiring of the house or apartment.

- **Level 3** charging stations use 440 volts and can charge most electric vehicles to 80% charge in less than 30 minutes. This high-charge rate may be harmful to battery life. Always follow the charging instructions and recommendations as stated in the owner's manual of the vehicle being charged. Level 3 chargers charge the vehicle using direct current (DC) at a rate up to 125 amperes. A Level

FIGURE 32–31 A Nissan Leaf electric vehicle charging ports located at the front of the vehicle under a hinged door for easy access.

FIGURE 32–32 A typical wind generator that is used to generate electricity.

 FREQUENTLY ASKED QUESTION

**What Is a "CHAdeMO" Connector?**

CHAdeMO is a Japanese trade name of a quick charging method for Level 3 charging using DC electricity at a high rate. "CHAdeMO" is an abbreviation of "CHArge de Move," which can be translated to mean "charge for moving." ● SEE FIGURE 32–31.

3 charger station can cost $50,000 or more, making this type of charger most suitable where facilities will be selling the service of rapidly charging the vehicle. The Tesla "superchargers" are Level 3 chargers and are free to use by Tesla owners.

FIGURE 32–33 The Hoover Dam in Nevada/Arizona is used to create electricity for use in the southwestern United States.

# WIND POWER

Wind power is used to help supplement electric power generation in many parts of the country. Because AC electricity cannot be stored, this energy source is best used to reduce the use of natural gas and coal to help reduce $CO_2$ emissions. Wind power is most economical if the windmills are located where the wind blows consistently above 8 miles per hour (13 km/h). Locations include the eastern slope of mountain ranges, such as the Rocky Mountains, or on high ground in many states. Often, wind power is used as supplemental power in the evenings when it is most needed and is allowed to stop rotating in daylight hours when the power is not needed. Windmills are usually grouped together to form **wind farms,** where the product is electrical energy. Energy from wind farms can be used to charge plug-in hybrid vehicles as well as for domestic lighting and power needs. ● SEE FIGURE 32–32.

# HYDROELECTRIC POWER

Hydroelectric power is limited to locations where there are dammed rivers and hydroelectric plants. However, electricity can and is transmitted long distances—so that electricity generated at the Hoover Dam can be used in California and other remote locations. Electrical power from hydroelectric sources can be used to charge plug-in hybrid electric vehicles, thereby reducing emissions that would normally be created by burning coal or natural gas to create electricity. However, hydroelectric plants are limited as to the amount of power they can produce, and constructing new plants is extremely expensive. ● SEE FIGURE 32–33.

1. The chemical reaction inside a fuel cell is the opposite of electrolysis in that electricity is created when hydrogen and oxygen are allowed to combine in the fuel cell.

2. A fuel cell produces electricity and releases heat and water as the only by-products.

3. The major disadvantages of fuel cells include the following:
   - High cost
   - Lack of hydrogen refueling stations
   - Short range
   - Freezing-temperature starting problems

4. Types of fuel cells include PEM (the most commonly used), PAFC, MCFC, and SOFC.

5. Ultracapacitors are an alternative to batteries for the storage of electrical energy.

6. A gasoline-powered engine can be more efficient if it uses a homogeneous charge compression ignition (HCCI) combustion process.

7. Plug-in hybrid electric vehicles could expand the range of hybrid vehicles by operating on battery power alone.

8. Wind power and hydroelectric power are being used to recharge plug-in hybrids and provide electrical power for all uses, without harmful emissions.

## REVIEW QUESTIONS

1. How does a fuel cell work?

2. What are the advantages and disadvantages of fuel cells?

3. What are the uses of the various types of fuel cells?

4. How does an ultracapacitor work?

5. What are the advantages and disadvantages of using hydrogen?

6. What alternative power sources could be used for vehicles?

## CHAPTER QUIZ

1. A fuel cell produces electricity from _____ and _____.
   - a. Gasoline/oxygen
   - b. Nitrogen/hydrogen
   - c. Hydrogen/oxygen
   - d. Water/oxygen

2. What are the by-products (emissions) from a fuel cell?
   - a. Water
   - b. $CO_2$
   - c. CO
   - d. Nonmethane hydrocarbon

3. Which type of fuel cell is the most likely to be used to power vehicles?
   - a. PAFC
   - b. PEM
   - c. MCFC
   - d. SOFC

4. Which liquid fuel could be used to directly power a fuel cell?
   - a. Methanol
   - b. Ethanol
   - c. Biodiesel
   - d. Unleaded gasoline

5. Which is *not* a function of an ultracapacitor?
   - a. Can pass AC current
   - b. Can be charged with DC current
   - c. Discharges DC current
   - d. Can pass DC current

6. Hydrogen is commonly stored at what pressure?
   - a. 100,000 PSI
   - b. 50,000 PSI
   - c. 5,000 PSI
   - d. 1,000 PSI

7. Hydrogen storage tanks are usually constructed from _____.
   - a. Steel
   - b. Aluminum
   - c. Carbon fiber
   - d. Both b and c

8. HCCI is a process that eliminates what parts or components in a gasoline engine?
   - a. Fuel tank
   - b. Battery
   - c. Fuel injectors
   - d. Ignition system

9. A plug-in hybrid is different from a conventional hybrid electric vehicle because it has _____.
   - a. A built-in battery charger
   - b. Li Ox batteries
   - c. More batteries
   - d. Bigger motor/generator

10. Which energy source(s) is (are) currently being used to help reduce the use of fossil fuels?
    - a. Hydrogen
    - b. Wind power
    - c. Hydroelectric power
    - d. Both b and c

# SAMPLE ASE A8 CERTIFICATION-TYPE TEST

| CONTENT AREA | QUESTIONS IN TEST | PERCENTAGE OF TEST | COVERED IN CHAPTER |
|---|---|---|---|
| A. General Engine Diagnosis | 12 | 24% | 10 |
| B. Ignition System Diagnosis and Repair | 8 | 16% | 29 |
| C. Fuel, Air Induction, and Exhaust Systems Diagnosis and Repair | 9 | 18% | 8, 9, 19, 20,23 |
| D. Emissions Control Systems Diagnosis and Repair | 8 | 16% | 24, 25, 26, 27, 28 |
|    1. Positive Crankcase Ventilation (1) | | | |
|    2. Exhaust Gas Recirculation (2) | | | |
|    3. Secondary Air Injection (AIR) and Catalytic Converter (2) | | | |
|    4. Evaporative Emissions Controls (3) | | | |
| E. Computerized Engine Controls Diagnosis and Repair (Including OBD II) | 13 | 26% | 11, 12,13, 14, 15, 16, 17, 18, |
| TOTAL | 50 | 100% | |

## A. GENERAL ENGINE DIAGNOSIS

1. A blown head gasket is suspected on a five-year-old vehicle. The service technician should perform which of the following tests to confirm the problem?
   a. Running compression test, vacuum test
   b. Leak-down test, compression test
   c. Oil pressure test, leak-down test
   d. Check for DTCs, Check for TSBs

2. Two technicians are discussing oil leaks. Technician A says that an oil leak can be found using a fluorescent dye in the oil with a black light to check for leaks. Technician B says that a white spray powder can be used to locate oil leaks. Which technician is correct?
   a. Technician A only
   b. Technician B only
   c. Both Technicians A and B
   d. Neither Technician A nor B

3. A smoothly operating engine depends on _____.
   a. High compression on most cylinders
   b. Equal compression among cylinders
   c. Cylinder compression levels above 100 PSI (700 kPa) and within 70 PSI (500 kPa) of each other
   d. Compression levels below 100 PSI (700 kPa) on most cylinders

4. A good reading for a cylinder leakage test would be _____.
   a. Within 20% among cylinders
   b. All cylinders below 20% leakage
   c. All cylinders above 20% leakage
   d. All cylinders above 70% leakage and within 7% of each other

5. Technician A says that during a power balance test, the cylinder that causes the biggest RPM drop is the weak cylinder. Technician B says that if one spark plug wire is grounded out and the engine speed does not drop, a weak or dead cylinder is indicated. Which technician is correct?
   a. Technician A only
   b. Technician B only
   c. Both Technicians A and B
   d. Neither Technician A nor B

6. White exhaust can be caused by _____
   a. Coolant entering the combustion chamber
   b. Engine oil getting past the piston rings
   c. A vacuum leak at the intake manifold gasket
   d. Any of the above

7. An engine is misfiring. A power-balance test indicates that when the spark to cylinder #4 is grounded, there is no change in the engine speed. Technician A says that a burned valve is a possible cause. Technician B says that a defective cylinder #4 injector or spark plug wire could be the cause. Which technician is correct?
   a. Technician A only
   b. Technician B only
   c. Both Technicians A and B
   d. Neither Technician A nor B

8. Two technicians are discussing an engine vibration during acceleration. Technician A says that a defective (collapsed) mount can cause an engine or driveline vibration. Technician B says that the some mounts are fluid filled and should be checked for leakage. Which technician is correct?
   a. Technician A only
   b. Technician B only
   c. Both Technicians A and B
   d. Neither Technician A nor B

9. An engine uses an excessive amount of oil and the exhaust is blue but only at first engine start in the morning. What is the most likely cause?
   a. Leaking fuel injector
   b. Worn valve stem seals
   c. Leaking valve (cylinder head) cover
   d. Overfilled oil level

10. Two technicians are discussing the diagnosis of a lack-of-power problem. Technician A says that a clogged catalytic converter could be the cause. Technician B says that a restricted or clogged muffler could be the cause. Which technician is correct?
    a. Technician A only
    b. Technician B only
    c. Both Technicians A and B
    d. Neither Technician A nor B

11. A compression test gave the following results:

    cylinder #1 = 155, cylinder #2 = 140, cylinder #3 = 110, cylinder #4 = 105

    Technician A says that a defective (burned) valve is the most likely cause. Technician B says that a leaking head gasket could be the cause. Which technician is correct?
    a. Technician A only
    b. Technician B only
    c. Both Technicians A and B
    d. Neither Technician A nor B

12. Two technicians are discussing a compression test. Technician A says that the engine should be cranked over with the pressure gauge installed for "4 puffs." Technician B says that the maximum difference between the highest-reading cylinder and the lowest-reading cylinder should be 20%. Which technician is correct?
    a. Technician A only
    b. Technician B only
    c. Both Technicians A and B
    d. Neither Technician A nor B

## B. IGNITION SYSTEM DIAGNOSIS AND REPAIR

13. How should a service technician test for spark?
    a. Hold the plug wire ¼ in. from the block
    b. Use a spark tester
    c. Pull the spark plug wire away from the plug ½ in.
    d. Measure the output with a meter set to kV

14. An engine will not start and a check of the ignition system output indicates no spark to any of the spark plugs. Technician A says that a defective crankshaft position sensor (CKP) could be the cause. Technician B says a defective ignition switch could be the cause. Which technician is correct?
    a. Technician A only
    b. Technician B only
    c. Both Technicians A and B
    d. Neither Technician A nor B

15. A spark plug wire is 2.5 feet long. If it is okay, its resistance should be less than _____
    _____.
    a. 25 k ohms
    b. 200,000 ohms
    c. 250,000 ohms
    d. 2.50 k ohms

16. A P0300 (random misfire detected) diagnostic trouble code (DTC) was being diagnosed and a defective (open) spark plug wire was found on a waste-spark-type ignition system. Technician A says that the companion cylinder spark plug wire should also be carefully inspected and replaced if necessary. Technician B says that the ignition coil should be replaced because the bad wire could have caused the coil to become damaged (tracked) internally. Which technician is correct?
    a. Technician A only
    b. Technician B only
    c. Both Technicians A and B
    d. Neither Technician A nor B

17. What can be adjusted to set the ignition timing on a waste-spark or coil-on-plug-type ignition system?
    a. Crankshaft position (CKP) sensor
    b. Camshaft position (CMP) sensor
    c. Either a or b depending on make and model
    d. None of the above

18. An engine produces less than normal power and is slow to accelerate when the throttle is opened. Technician A says that the exhaust system could be restricted. Technician B says the valve timing may be retarded. Which technician is correct?
    a. Technician A only
    b. Technician B only
    c. Both Technicians A and B
    d. Neither Technician A nor B

19. Which is *least likely* to cause a weak spark at the spark plug?
    a. A partially shorted primary winding in the ignition coil
    b. A 12.2-volt battery voltage
    c. A high-resistance spark plug wire(s)
    d. A voltage drop across the ignition switch

20. Which is *most likely* to cause an engine miss on one cylinder?
    a. An open spark plug wire
    b. A high-resistance spark plug wire
    c. Excessive rotor gap
    d. A clogged fuel filter

## C. FUEL, AIR INDUCTION, AND EXHAUST SYSTEMS DIAGNOSIS AND REPAIR

21. An engine equipped with a turbocharger is burning oil (blue exhaust smoke all the time). Technician A says that a defective wastegate could be the cause. Technician B says that a clogged PCV system could be the cause. Which technician is correct?
    a. Technician A only
    b. Technician B only
    c. Both Technicians A and B
    d. Neither Technician A nor B

22. A vehicle equipped with a mass airflow sensor will stumble or stall when in "drive," but operate normally when driven in reverse. What is the most likely cause?
    a. A split or crack in the air intake hose
    b. A clogged fuel filter
    c. A restricted air filter
    d. A leaking fuel injector

23. A poor fuel economy concern is being discussed. Technician A says that a pinched fuel return line could be the cause. Technician B says that a partially clogged fuel filter could be the cause. Which technician is correct?
    a. Technician A only
    b. Technician B only
    c. Both Technicians A and B
    d. Neither Technician A nor B

24. An engine equipped with return–type electronic port fuel injection is hard to start and emits black exhaust smoke when being started when hot. What is the *most likely* cause?
    a. A defective fuel pressure regulator
    b. A shorted fuel injector
    c. A clogged fuel filter
    d. A clogged air filter

25. A vehicle fails an enhanced emission test for excessive carbon monoxide (CO) emission. Which is the most likely cause?
    a. Clogged fuel injector(s)
    b. A stuck open fuel pressure regulator
    c. A stuck idle air control (IAC)
    d. A clogged fuel return line

26. Technician A says that the exhaust system can be checked for restriction by using a vacuum gauge attached to manifold vacuum and operating the engine at idle speed. Technician B says the exhaust is restricted if the vacuum increases at 2000 RPM. Which technician is correct?
    a. Technician A only
    b. Technician B only
    c. Both Technicians A and B
    d. Neither Technician A nor B

27. A fuel pump should be tested for all of the following except:
    a. Pressure
    b. Volume
    c. Current draw
    d. Resistance

28. An engine idles roughly and stalls occasionally when hot. This can be caused by _____.
    a. A partially clogged air filter
    b. A partially clogged fuel filter
    c. Using winter-blended gasoline in warm weather
    d. A loose gas cap

29. Technician A says that black exhaust smoke is an indication of excessive oil consumption. Technician B says that blue smoke is an indication of excessive amount of fuel being burned in the engine. Which technician is correct?
    a. Technician A only
    b. Technician B only
    c. Both Technicians A and B
    d. Neither Technician A nor B

## D. EMISSIONS CONTROL SYSTEMS DIAGNOSIS AND REPAIR

30. Two technicians are discussing positive crankcase ventilation (PCV) valves. Technician A says that if the valve rattles, it is good. Technician B says the PCV valve may still require replacement even if it rattles. Which technician is correct?
    a. Technician A only
    b. Technician B only
    c. Both Technicians A and B
    d. Neither Technician A nor B

31. Technician A says that a defective one-way exhaust check valve could cause the air pump to fail. Technician B says that the airflow to the exhaust manifold when the engine is warm can cause a drivability problem. Which technician is correct?
    a. Technician A only
    b. Technician B only
    c. Both Technicians A and B
    d. Neither Technician A nor B

32. A vehicle is running rich. Technician A says that overfilling the fuel tank can cause the carbon canister to become saturated with gasoline, which can cause a rich running condition. Technician B says that an exhaust leak upstream from the O2S could be the cause. Which technician is correct?
    a. Technician A only
    b. Technician B only
    c. Both Technicians A and B
    d. Neither Technician A nor B

33. Technician A says that a partially clogged EGR passage can cause the vehicle to fail due to excessive NOx emissions. Technician B says the vehicle could fail for excessive CO if the EGR passage were clogged. Which technician is correct?
    a. Technician A only
    b. Technician B only
    c. Both Technicians A and B
    d. Neither Technician A nor B

34. Technician A says the catalytic converter must be replaced if it rattles when tapped. Technician B says a catalytic converter can be defective and not be working yet not be clogged. Which technician is correct?
    a. Technician A only
    b. Technician B only
    c. Both Technicians A and B
    d. Neither Technician A nor B

35. Used catalytic converters must be kept for possible inspection by the EPA for how long?
   a. 30 days
   b. 60 days
   c. 90 days
   d. 6 months

36. A vehicle fails an emission test for excessive NOx. Which exhaust control device has the greatest effect on the amount of NOx produced by the engine?
   a. PCV
   b. Air pump
   c. Carbon (charcoal) canister
   d. EGR

37. The oxygen sensor of a vehicle has a constant voltage output of about 750 mV. Which exhaust emission control device could be damaged if the vehicle is not repaired to operate correctly?
   a. PCV
   b. Carbon (charcoal) canister
   c. Catalytic converter
   d. EGR

38. Two technicians are discussing the evaporative control system. Technician A says that the carbon (charcoal) canister should be replaced regularly as part of routine maintenance. Technician B says the carbon (charcoal) inside of the EVAP canister can dissolve in gasoline and leave a yellow deposit in the engine when burned. Which technician is correct?
   a. Technician A only
   b. Technician B only
   c. Both Technicians A and B
   d. Neither Technician A nor B

39. An EGR valve is stuck partially open. What is the most likely result?
   a. Pinging (spark knock)
   b. Rough idle—runs normally at highway speeds
   c. Fast idle
   d. Lack of power at highway speeds

## E. COMPUTERIZED ENGINE CONTROLS DIAGNOSIS AND REPAIR (INCLUDING OBD II)

40. An oxygen sensor (O2S) is being tested and the O2S voltage is fluctuating between 800 millivolts and 200 millivolts. Technician A says the engine is operating too lean. Technician B says the engine is operating too rich. Which technician is correct?
   a. Technician A only
   b. Technician B only
   c. Both Technicians A and B
   d. Neither Technician A nor B

41. An oxygen sensor in a fuel-injected engine is slow to react to changes in air–fuel mixture. Technician A says that the O2S may need to be replaced. Technician B says that driving the vehicle at highway speeds may restore proper operation of the O2S. Which technician is correct?
   a. Technician A only
   b. Technician B only

c. Both Technicians A and B
d. Neither Technician A nor B

42. Technician A says that OBD II SAE (generic) codes are the same for all OBD II vehicles. Technician B says that the DLC is located under the hood on all OBD II vehicles. Which technician is correct?
   a. Technician A only
   b. Technician B only
   c. Both Technicians A and B
   d. Neither Technician A nor B

43. The IAC counts are zero. Technician A says that the engine may have a vacuum leak or a stuck throttle cable. Technician B says the throttle plate(s) may be dirty or partially clogged. Which technician is correct?
   a. Technician A only
   b. Technician B only
   c. Both Technicians A and B
   d. Neither Technician A nor B

44. An engine is operating at idle speed with all accessories off and the gear selector in Park. Technician A says that a scan tool should display injector pulse width between 1.5 and 3.5 milliseconds. Technician B says that the oxygen sensor activity as displayed on a scan tool should indicate over 800 millivolts and less than 200 millivolts. Which technician is correct?
   a. Technician A only
   b. Technician B only
   c. Both Technicians A and B
   d. Neither Technician A nor B

45. Two technicians are discussing fuel trim. Technician A says that oxygen sensor activity determines short-term fuel trim numbers. Technician B says that a positive (+) long-term fuel trim means that the computer is adding fuel to compensate for a lean exhaust. Which technician is correct?
   a. Technician A only
   b. Technician B only
   c. Both Technicians A and B
   d. Neither Technician A nor B

46. An engine will not go into closed loop. Which sensor is the most likely to be at fault?
   a. Oxygen sensor (O2S)
   b. Intake air temperature (IAT)
   c. MAP sensor
   d. BARO sensor

47. A technician is looking at scan data with the engine at idle speed and in Park and notices that the MAP sensor voltage reading is about 1.0 volt (18 in. Hg). Technician A says that the reading is normal. Technician B says that the reading indicates a possible MAP sensor fault. Which technician is correct?
   a. Technician A only
   b. Technician B only
   c. Both Technicians A and B
   d. Neither Technician A nor B

48. The voltage output of a zirconia oxygen sensor is low (close to zero volts). Technician A says the engine is oper-

ating too lean. Technician B says the engine is operating too rich. Which technician is correct?
a. Technician A only
b. Technician B only
c. Both Technicians A and B
d. Neither Technician A nor B

49. A typical TP sensor used in electronic throttle control (ETC) systems includes _____.
a. One standard three wire TP sensor
b. Two TP sensors in one with one producing an increase in voltage as the other one produces a decreasing voltage as the throttle plate moves toward wide open.

c. Three TP sensors in one with two producing an increase in voltage as the other one produces a decreasing voltage as the throttle plate moves toward wide open.
d. Either b or c

50. Which of the following describes acceptable oxygen sensor activity as measured with a multimeter set to read DC volts?
a. 0.350 to 0.550 volts
b. 0.150 to 0.950 volts
c. 0.450 to 0.850 volts
d. 0.450 volts and steady

## ANSWERS

| | | | | | | | |
|---|---|---|---|---|---|---|---|
| 1. | b | 14. | c | 27. | d | 40. | d |
| 2. | c | 15. | a | 28. | c | 41. | c |
| 3. | b | 16. | c | 29. | d | 42. | a |
| 4. | b | 17. | d | 30. | c | 43. | a |
| 5. | b | 18. | c | 31. | c | 44. | c |
| 6. | a | 19. | b | 32. | c | 45. | c |
| 7. | c | 20. | a | 33. | a | 46. | a |
| 8. | c | 21. | b | 34. | c | 47. | a |
| 9. | b | 22. | a | 35. | b | 48. | a |
| 10. | c | 23. | a | 36. | d | 49. | b |
| 11. | b | 24. | a | 37. | c | 50. | b |
| 12. | c | 25. | d | 38. | d | | |
| 13. | b | 26. | a | 39. | b | | |

# 2013 NATEF CORRELATION CHART

**MLR-** Maintenance & Light Repair
**AST-** Auto Service Technology (Includes MLR)
**MAST-** Master Auto Service Technology (Includes MLR and AST)

## ENGINE PERFORMANCE (A8)

| TASK | PRIORITY | MLR | AST | MAST | TEXT PAGE # | TASK PAGE # |
|------|----------|-----|-----|------|-------------|-------------|
| **A. GENERAL: ENGINE DIAGNOSIS** | | | | | | |
| 1. Identify and interpret engine performance concerns; determine necessary action. | P-1 | | ✔ | ✔ | 135–139 | 146 |
| 2. Research applicable vehicle and service information, vehicle service history, service precautions, and technical service bulletins. | P-1 | ✔ | ✔ | ✔ | 1–4 | 4–7 |
| 3. Diagnose abnormal engine noises or vibration concerns; determine necessary action. | P-3 | | ✔ | ✔ | 138–139 | 36 |
| 4. Diagnose the cause of excessive oil consumption, coolant consumption, unusual exhaust color, odor, and sound; determine necessary action. | P-2 | | ✔ | ✔ | 135 | 37 |
| 5. Perform engine absolute (vacuum/boost) manifold pressure tests; determine necessary action. | P-1 | ✔ | ✔ | ✔ | 144–146 | 38 |
| 6. Perform cylinder power balance test; determine necessary action. | P-1 | ✔ | ✔ | ✔ | 144 | 39 |
| 7. Perform cylinder cranking and running compression tests; determine necessary action. | P-1 | ✔ | ✔ | ✔ | 140–143 | 40 |
| 8. Perform cylinder leakage test; determine necessary action. | P-1 | ✔ | ✔ | ✔ | 143 | 41 |
| 9. Diagnose engine mechanical, electrical, electronic, fuel, and ignition concerns; determine necessary action | P-2 | | ✔ | ✔ | 136–148; 286–376 | 33 |
| 10. Verify engine operating temperature; determine necessary action. | P-1 | ✔ | ✔ | ✔ | 180 | 43 |
| 11. Verify correct camshaft timing. | P-1 | | ✔ | ✔ | 131 | 145 |
| **B. COMPUTERIZED CONTROLS DIAGNOSIS AND REPAIR** | | | | | | |
| 1. Retrieve and record diagnostic trouble codes, OBD monitor status, and freeze frame data; clear codes when applicable. | P-1 | | ✔ | ✔ | 385–394 | 46;47 |
| 2. Access and use service information to perform step-by-step (troubleshooting) diagnosis. | P-1 | | ✔ | ✔ | 378 | 146 |
| 3. Perform active tests of actuators using a scan tool; determine necessary action. | P-2 | | ✔ | ✔ | 332 | 153 |
| 4. Describe the importance of running all OBDII monitors for repair verification | P-1 | ✔ | ✔ | ✔ | 331 | 154 |
| 5. Diagnose the causes of emissions or drivability concerns with stored or active diagnostic trouble codes; obtain, graph, and interpret scan tool data. | P-1 | | ✔ | ✔ | 385–394 | 147 |

| TASK | PRIORITY | MLR | AST | MAST | TEXT PAGE # | TASK PAGE # |
|------|----------|-----|-----|------|-------------|-------------|
| 6. Diagnose emissions or drivability concerns without stored diagnostic trouble codes; determine necessary action. | P-1 | | | ✔ | 392 | 49; 155 |
| 7. Inspect and test computerized engine control system sensors, powertrain/engine control module (PCM/ECM), actuators, and circuits using a graphing multimeter (GMM)/digital storage oscilloscope (DSO); perform necessary action. | P-2 | | | ✔ | 175; 184; 287; 346 | 59; 65; 72; 76; 79; 112 |
| 8. Diagnose driveability and emissions problems resulting from malfunctions of interrelated systems (cruise control, security alarms, suspension controls, traction controls, A/C, automatic transmissions, non-OEM installed accessories, or similar systems); determine necessary action. | P-3 | | | ✔ | 379 | 155 |

### C. IGNITION SYSTEM DIAGNOSIS AND REPAIR

| TASK | PRIORITY | MLR | AST | MAST | TEXT PAGE # | TASK PAGE # |
|------|----------|-----|-----|------|-------------|-------------|
| 1. Diagnose (troubleshoot) ignition system related problems such as no-starting, hard starting, engine misfire, poor drivability, spark knock, power loss, poor mileage, and emissions concerns; determine necessary action. | P-2 | | ✔ | ✔ | 367–374 | 139–145 |
| 2. Inspect and test crankshaft and camshaft position sensor(s); perform necessary action. | P-1 | | ✔ | ✔ | 357–360 | 145 |
| 3. Inspect, test, and/or replace ignition control module, powertrain/engine control module; reprogram as necessary. | P-3 | | ✔ | ✔ | 355 | 140 |
| 4. Remove and replace spark plugs; inspect secondary ignition components for wear and damage. | P-1 | ✔ | ✔ | ✔ | 373–374 | 142; 143 |

### D. FUEL, AIR INDUCTION, AND EXHAUST SYSTEMS DIAGNOSIS AND REPAIR

| TASK | PRIORITY | MLR | AST | MAST | TEXT PAGE # | TASK PAGE # |
|------|----------|-----|-----|------|-------------|-------------|
| 1. Diagnose (troubleshoot) hot or cold no-starting, hard starting, poor drivability, incorrect idle speed, poor idle, flooding, hesitation, surging, engine misfire, power loss, stalling, poor mileage, dieseling, and emissions problems; determine necessary action. | P-2 | | | ✔ | 286–299 | 49; 50 |
| 2. Check fuel for contaminants; determine necessary action. | P-2 | | ✔ | ✔ | 90–93 | 18; 19; 20 |
| 3. Inspect and test fuel pumps and pump control systems for pressure, regulation, and volume; perform necessary action. | P-1 | | ✔ | ✔ | 248–253 | 83–87 |
| 4. Replace fuel filter(s). | P-1 | ✔ | ✔ | ✔ | 248 | 88 |
| 5. Inspect, service, or replace air filters, filter housings, and intake duct work. | P-1 | ✔ | ✔ | ✔ | 114 | 68; 93 |
| 6. Inspect throttle body, air induction system, intake manifold and gaskets for vacuum leaks and/or unmetered air. | P-2 | | ✔ | ✔ | 298 | 92 |
| 7. Inspect and test fuel injectors. | P-2 | | ✔ | ✔ | 291–298 | 109–113 |
| 8. Verify idle control operation. | P-1 | | ✔ | ✔ | 295 | 91 |
| 9. Inspect integrity of the exhaust manifold, exhaust pipes, muffler(s), catalytic converter(s), resonator(s), tail pipe(s), and heat shields; perform necessary action. | P-1 | ✔ | ✔ | ✔ | 119–120; 346–347 | 29 |
| 10. Inspect condition of exhaust system hangers, brackets, clamps, and heat shields; repair or replace as needed. | P-1 | ✔ | ✔ | ✔ | 120 | 29 |

| TASK | PRIORITY | MLR | AST | MAST | TEXT PAGE # | TASK PAGE # |
|---|---|---|---|---|---|---|
| 11. Perform exhaust system back-pressure test; determine necessary action. | P-2 | | | ✔ | 346–347 | 135 |
| 12. Check and refill diesel exhaust fluid (DEF). | P-3 | ✔ | ✔ | ✔ | 74 | 15 |
| 13. Test the operation of turbocharger/supercharger systems; determine necessary action. | P-3 | | | ✔ | 125–129 | - |

## E. EMISSIONS CONTROL SYSTEMS DIAGNOSIS AND REPAIR

| TASK | PRIORITY | MLR | AST | MAST | TEXT PAGE # | TASK PAGE # |
|---|---|---|---|---|---|---|
| 1. Diagnose oil leaks, emissions, and drivability concerns caused by the positive crankcase ventilation (PCV) system; determine necessary action. | P-3 | | ✔ | ✔ | 336–338 | 34 |
| 2. Inspect, test, and service positive crankcase ventilation (PCV) filter/breather cap, valve, tubes, orifices, and hoses; perform necessary action. | P-2 | ✔ | ✔ | ✔ | 336–338 | 130; 131 |
| 3. Diagnose emissions and drivability concerns caused by the exhaust gas recirculation (EGR) system; determine necessary action. | P-3 | | ✔ | ✔ | 331–333 | 126 |
| 4. Diagnose emissions and drivability concerns caused by the secondary air injection and catalytic converter systems; determine necessary action. | P-2 | | ✔ | ✔ | 339–341 | 134 |
| 5. Diagnose emissions and drivability concerns caused by the evaporative emissions control system; determine necessary action. | P-2 | | | ✔ | 332–325 | 120–125 |
| 6. Inspect and test electrical/electronic sensors, controls, and wiring of exhaust gas recirculation (EGR) systems; perform necessary action. | P-2 | | | ✔ | 329–331 | 128; 129 |
| 7. Inspect, test, service, and replace components of the EGR system including tubing, exhaust passages, vacuum/pressure controls, filters, and hoses; perform necessary action | P-2 | | ✔ | ✔ | 332–333 | 127 |
| 8. Inspect and test electrical/electronically-operated components and circuits of air injection systems; perform necessary action. | P-3 | | ✔ | ✔ | 340–341 | 134 |
| 9. Inspect and test catalytic converter efficiency. | P-2 | | ✔ | ✔ | 348 | 138 |
| 10. Inspect and test components and hoses of the evaporative emissions control system; perform necessary action. | P-1 | | ✔ | ✔ | 320–325 | 120–124 |
| 11. Interpret diagnostic trouble codes (DTCs) and scan tool data related to the emissions control systems; determine necessary action. | P-1 | | ✔ | ✔ | 324; 332; 341 | 125 |

# GLOSSARY

**AC coupling** A selection that can be made to observe a waveform.

**AC/DC clamp-on DMM** A type of meter that has a clamp that is placed around the wire to measure current.

**Acceleration simulation mode** Uses a dynamometer that applies a heavy load on the vehicle at a steady-state speed.

**Accelerator pedal position (APP) sensor** A sensor that is used to monitor the position and rate of change of the accelerator pedal.

**Accumulator** A temporary location for fluid under pressure.

**Actuator** An electrical or mechanical device that converts electrical energy into a mechanical action, such as adjusting engine idle speed, altering suspension height, or regulating fuel metering.

**Adjustable wrench** A wrench that has a movable jaw to allow it to fit many sizes of fasteners.

**Adsorption** Attaches the fuel vapor molecules to the carbon surface.

**AFV** Alternative-fuel vehicle.

**AGST** Aboveground storage tank, used to store used oil.

**Air–fuel ratio** The ratio of air to fuel in an intake charge as measured by weight.

**Air reference chamber** An internal chamber of a wide-band oxygen sensor that is exposed to ambient air.

**Air-fuel ratio sensor** A term used to describe an oxygen sensor that is capable of measuring the air-fuel ratio in the exhaust.

**Air-injection reaction (AIR)** A term used to describe a secondary air-injection system that provides the air necessary for the oxidizing process either at the exhaust manifold or inside the catalytic converter.

**Ambient air electrode** The electrode inside a wide-band oxygen sensor that is exposed to outside (ambient) air.

**Ambient side electrode** The electrode inside a wide-band oxygen sensor that is exposed to outside (ambient) air.

**Analog-to-digital (AD) converter** An electronic circuit that converts analog signals into digital signals that can then be used by a computer.

**Anhydrous ethanol** A type of ethanol that has almost zero absorbed water.

**Annealing** A heat-treating process that takes out the brittle hardening of the casting to reduce the chance of cracking from the temperature changes.

**ANSI** American National Standards Institute.

**Antiknock Index (AKI)** The pump octane.

**API gravity** An arbitrary scale expressing the gravity or density of liquid petroleum products devised jointly by the American Petroleum Institute and the National Bureau of Standards.

**Asbestosis** A health condition where asbestos causes scar tissue to form in the lungs causing shortness of breath.

**ASD** Automatic Shutdown Relay.

**ASM 50/15 test** Places a load of 50% on the vehicle at a steady 15 mph. This load represents 50% of the horsepower required to simulate the FTP acceleration rate of 3.3 mph/sec.

**ASM 25/25 test** Places a 25% load on the vehicle while it is driven at a steady 25 mph. This represents 25% of the load required to simulate the FTP acceleration rate of 3.3 mph/sec.

**ASTM** American Society for Testing Materials.

**B20** A blend of 20% biodiesel with 80% petroleum diesel.

**Back pressure** The exhaust system's resistance to flow. Measured in pounds per square inch (PSI).

**Baffle** A plate or shield used to direct the flow of a liquid or gas.

**Bar** When air is pumped into the cylinder, the combustion chamber receives an increase of air pressure known as boost and is measured in pounds per square inch (PSI), atmospheres (ATM), or bar.

**BARO sensor** A sensor used to measure barometric pressure.

**Base timing** The timing of the spark before the computer advances the timing.

**Baud rate** The speed at which bits of computer information are transmitted on a serial data stream. Measured in bits per second (bps).

**BCI** Battery Council International.

**Bench grinder** An electric-powered grinding stone usually combined with a wire wheel and mounted to a bench.

**Bias voltage** A weak signal voltage applies to an oxygen sensor by the PCM. This weak signal voltage is used by the PCM to detect when the oxygen sensor has created a changing voltage and for diagnosis of the oxygen senor circuit.

**Binary system** A computer system that uses a series of zeros and ones to represent information.

**Biodiesel** A renewable fuel manufactured from vegetable oils, animal fats, or recycled restaurant grease.

**Biomass** Nonedible farm products, such as cornstalks, cereal straws, and plant wastes from industrial processes, such as sawdust and paper pulp, used in making ethanol.

**Block** The foundation of any engine. All other parts are either directly or indirectly attached to the block of an engine.

**BMAP sensor** A sensor that has individual circuits to measure barometric and manifold pressure. This input not only allows the computer to adjust for changes in atmospheric pressure due to weather, but also is the primary sensor used to determine altitude.

**BNC connector** A miniature standard coaxial cable connector.

**BOB** Break-out box.

**Bolts** A threaded fastener use to attach two parts. The threaded end can be installed into a casting such as an engine block or a nut used to join two parts.

**Boost** An increase in air pressure above atmospheric. Measured in pounds per square inch (PSI).

**Bore** The inside diameter of the cylinder in an engine.

**BOV** Also called a dump valve or vent valve, the BOV features an adjustable spring design that keeps the valve closed until a sudden release of the throttle.

**Boxer** A type of engine design that is flat and has opposing cylinders. Called a boxer because the pistons on one side resemble a boxer during engine operation. Also called a pancake engine.

**Breaker bar** A handle used to rotate a socket; also called a flex handle.

**Breakout box (BOB)** A piece of test equipment that allows access to the terminals while connecting to the vehicle, using a scan tool.

**British thermal unit**  A unit of heat measurement.

**Bump cap**  A hat that is made of plastic and is hard enough to protect the head from bumps.

**Burn kV**  Spark line voltage.

**BUS**  A term used to describe a communication network.

**Bypass ignition**  Commonly used on General Motors engines equipped with distributor ignition (DI), as well as those equipped with waste-spark ignition.

**Bypass valve**  Allows intake air to flow directly into the intake manifold bypassing the supercharger.

**CAA**  Clean Air Act. Federal legislation passed in 1970 that established national air quality standards.

**Calibration codes**  Codes used on many powertrain control modules.

**California Air Resources Board**  A state of California agency that regulates the air quality standards for the state.

**Cam-in-block design**  An engine where the crankshaft is located in the block rather than in the cylinder head.

**Campaign**  A recall where vehicle owners are contacted to return a vehicle to a dealer for corrective action.

**Camshaft**  A shaft in an engine that is rotated by the crankshaft by a belt or chain and used to open valves.

**CAN**  A type of serial data transmission.

**Cap screw**  A bolt that is threaded into a casting.

**Casting number**  An identification code cast into an engine block or other large cast part of a vehicle.

**CAT III**  An electrical measurement equipment rating created by the International Electrotechnical Commission (IEC). CAT III indicates the lowest level of instrument protection that should be in place when performing electrical measurements on hybrid electric vehicles.

**Catalysts**  Platinum and palladium used in the catalytic converter to combine oxygen ($O_2$) with hydrocarbons (HC) and carbon monoxide (CO) to form nonharmful tailpipe emissions of water ($H_2O$) and carbon dioxide ($CO_2$).

**Catalytic converter**  An emission control device located in the exhaust system that changes HC and CO into harmless $H_2O$ and $CO_2$. If a three-way catalyst, $NO_x$ is also separated into harmless, separate N and O.

**Catalytic cracking**  Breaking hydrocarbon chains using heat in the presence of a catalyst.

**CCM**  Comprehensive Component Monitor.

**Cellulose ethanol**  Ethanol produced from biomass feedstock such as agricultural and industrial plant wastes.

**Cellulosic biomass**  Composed of cellulose and lignin, with smaller amounts of proteins, lipids (fats, waxes, and oils), and ash.

**Cerium**  An element that can store oxygen.

**Cetane number**  A measure of the ease with which the fuel can be ignited.

**CFR**  Code of Federal Regulations.

**Cheater bar**  A pipe or other object used to lengthen the handle of a ratchet or breaker bar. Not recommended to be used as the extra force can cause the socket or ratchet to break.

**Check valve**  A one-way valve used in the fuel line to keep fuel from leaking if the vehicle rolls over or in involved in a collision.

**Chisels**  A type of hand tool used to mark or cut strong material such as steel.

**Chrysler Collision Detection (CCD)**  A type of multiplex network used by Chrysler for scan tool and module communications. The "collision" in the Chrysler Collision detection BUS communications refers to the program that avoids conflicts of information exchange within the BUS and does not refer to airbags or other accident-related circuits of the vehicle.

**CID**  Component Identification.

**CKP**  Crankshaft position sensor.

**Class 2**  A type of BUS communication used in General Motors vehicles.

**Clean Air Act Amendments (CAAA)**  All vehicles sold in the United States must meet Tier 1 standards that went into effect in 1994 and are the least stringent.

**Clock generator**  A crystal that determines the speed of computer circuits.

**Close-end wrench**  A type of hand tool that is closed at both ends.

**Closed-loop operation**  A phase of computer-controlled engine operation in which oxygen sensor feedback is used to calculate air–fuel mixture.

**Cloud point**  The low-temperature point at which the waxes present in most diesel fuel tend to form wax crystals that clog the fuel filter.

**CMP**  Camshaft position sensor.

**CNG**  Compressed natural gas.

**Coal-to-liquid**  A refining process in which coal is converted to liquid fuel.

**Coast-down stall**  A condition that results in the engine stalling when coasting to a stop .

**Coil-on-plug ignition**  An ignition system without a distributor, where each spark plug is integrated with an ignition coil.

**Combination wrench**  A wrench that is open ended at one end and has a box end at the other end.

**Combustion chamber**  The space left within the cylinder when the piston is at the top of its combustion chamber.

**Combustion**  The rapid burning of the air–fuel mixture in the engine cylinders, creating heat and pressure.

**Companion cylinders**  Two cylinders that share an ignition coil on a waste-spark-type ignition system.

**Component identification (CID)**  A component module that communicates with the PCM and is identified using mode $06 data using a scan tool.

**Comprehensive component monitor (CCM)**  This monitor watches the sensors and actuators in the OBD-II system. Sensor values are constantly compared with known-good values stored in the PCM's memory.

**Compressed natural gas (CNG)**  An alternative fuel that uses natural gas compressed at high pressures and used as a vehicle fuel.

**Compression ratio**  The ratio of the volume in the engine cylinder with the piston at bottom dead center (BDC) to the volume at top dead center (TDC).

**Compression-sensing ignition**  Does not require the use of a camshaft position sensor to determine cylinder number.

**Compression test**  An engine test that helps determine the condition of an engine based on how well each cylinder is able to compress the air on the compression stroke.

**Compressor bypass valve**  This type of relief valve routes the pressurized air to the inlet side of the turbocharger for reuse and is quiet during operation.

**Connecting rod**   Connects the pistons to the crankshaft.

**Continuity light**   A test light that has a battery and lights if there is continuity (electrical connection) between the two points that are connected to the tester.

**Controller**   A term that is usually used to refer to a computer or an electronic control unit (ECU).

**CPS**   Canister purge solenoid.

**CPU**   Central processor unit.

**Cracking**   A refinery process in which hydrocarbons with high boiling points are broken into hydrocarbons with low boiling points.

**Cranking vacuum test**   Measuring the amount of manifold vacuum during cranking.

**Crankshaft**   The part of an engine that transfers the up and down motion of the pistons to rotary motion.

**Crest**   The outside diameter of a bolt measured across the threads.

**Cross counts**   The number of times an oxygen sensor changes voltage from high to low (from low to high voltage is not counted) in 1 second (or 1.25 second, depending on scan tool and computer speed).

**CRT**   Cathode ray tube.

**Cup design**   A term used to describe the type of oxygen sensor element that is shaped like a thimble.

**Cycle**   A series of events such as the operation of the four strokes of an engine that repeats.

**Cycle life**   The number of times a battery can be charged and discharged without suffering significant degradation in its performance.

**Cylinder**   The part of an engine that is round and houses the piston.

**Cylinder head temperature (CHT) sensor-**   A temperature sensor mounted on the cylinder head and used by the PCM to determine fuel delivery.

**Cylinder leakage test**   A test that involves injecting air under pressure into the cylinders one at a time. The amount and location of any escaping air helps the technician determine the condition of the engine.

**Data link connector (DLC)**   The connector usually located under the dash used to communicate with a scan tool.

**DC coupling**   A selection that can be made to observe a waveform.

**DDS**   Demand delivery system.

**Default position**   The position of the throttle plate in an electronic throttle control without any signals from the controller.

**Delivery system**   A system that includes all of the parts and componentry need to deliver clean fuel tor e engine under pressure.

**Delta Pressure Feedback EGR sensor**   This sensor measures the pressure differential between two sides of a metered orifice positioned just below the EGR valve's exhaust side.

**Demand delivery system (DDS)**   A type of electronic fuel injection system.

**Detonation**   A violent explosion in the combustion chamber created by uncontrolled burning of the air–fuel mixture; often causes a loud, audible knock. Also known as spark knock or ping.

**DI**   Distributor ignition.

**Diagnostic executive**   Software program designed to manage the operation of all OBD-II monitors by controlling the sequence of steps necessary to execute the diagnostic tests and monitors.

**Diesel exhaust fluid (DEF)**   Urea also called Adblue used in some diesel engines' exhaust system to reduce emissions.

**Diesel exhaust particulate filter (DPF)**   filter that traps PM (soot) located in the exhaust system of most 2007 and newer diesel engines.

**Diesel oxidation catalyst**   Consists of a flow-through honeycomb-style substrate structure that is washcoated with a layer of catalyst materials, similar to those used in a gasoline engine catalytic converter.

**Diesohol**   Standard #2 diesel fuel combined with up to 15% ethanol.

**Diffusion chamber**   A section inside a wide-band oxygen sensor that is exposed to the exhaust gases.

**Digital computer**   A computer that uses on and off signals only. Uses an A to D converter to change analog signals to digital before processing.

**Digital EGR valve**   A digital EGR valve consists of three solenoids controlled by the powertrain control module (PCM). Each solenoid controls a different size orifice in the base—small, medium, and large. The PCM controls the ground circuit of each of the solenoids individually. It can produce any of seven different flow rates, using the solenoids to open the three valves in different combinations.

**Direct injection**   A fuel-injection system design in which gasoline is injected directly into the combustion chamber.

**DIS**   Distributorless ignition system. Also called direct-fire ignition system.

**Displacement**   The total volume displaced or swept by the cylinders in an internal combustion engine.

**Distillation**   The process of purification through evaporation and then condensation of the desired liquid.

**Distillation curve**   A graph that plots the temperatures at which the various fractions of a fuel evaporate.

**Distributor cap**   Provides additional space between the spark plug connections to help prevent crossfire.

**Distributor ignition**   A term specified by the Society of Automotive Engineers (SAE) for an ignition system that uses a distributor.

**Division**   A block.

**Divorced coil**   Used by most waste-spark ignition coils to keep both the primary and secondary winding separated.

**DMM**   Digital multimeter. A digital multimeter is capable of measuring electrical current, resistance, and voltage.

**Double overhead camshaft (DOHC)**   An engine design that has two overhead camshafts. One camshaft operates the intake valves and the other for the exhaust valves.

**Double-layer technology**   Technology used to build ultracapacitors. Involves the use of two carbon electrodes separated by a membrane.

**DPS**   Differential pressure sensor.

**Driveability index (DI)**   A calculation of the various boiling temperatures of gasoline that once complied can indicate the fuels ability to perform well at low temperatures.

**Drive-by-wire**   A term used to describe an engine equipped with an electronic throttle control (ETC) system.

**Drive cycle**   A series of driving conditions that will reset most monitors in most vehicles.

**Drive size**   The size in fractions of an inch of the square drive for sockets.

**Dry System**   A type of nitrous system that does not include additional gasoline.

**DSO**   Digital storage oscilloscope, takes samples of the signals that can be stopped or stored.

**Dual cell** A planar-type wide-band oxygen sensor is made like a conventional planar O2S and is labeled Nernst cell.

**Dual overhead camshaft** An engine design with two camshafts above each line of cylinders—one for the exhaust valves and one for the intake valves.

**Dump valve** Features an adjustable spring design that keeps the valve closed until a sudden release of the throttle. The resulting pressure increase opens the valve and vents the pressurized air directly into the atmosphere.

**Duty cycle** Refers to the percentage of on-time of the signal during one complete cycle.

**DVOM** Digital volt-ohm-millimeter.

**Dwell section** The amount of time that the current is charging the coil from the transistor-on point to the transistor-off point.

**Dwell** The number of degrees of distributor cam rotation that the points are closed.

**Dynamic compression test** A compression test done with the engine running rather than during engine cranking as is done in a regular compression test.

**E & C** Entertainment and comfort.

**E10** A fuel blend of 10% ethanol and 90% gasoline.

**E2PROM** Electrically erasable programmable read-only memory.

**E85** A fuel blend of 85% ethanol and 15% gasoline.

**ECA** Electronic Control Module. The name used by Ford to describe the computer used to control spark and fuel on older-model vehicles.

**ECM** Electronic control module on a vehicle.

**ECT** Engine coolant temperature.

**ECU** Electronic control unit on a vehicle.

**E-diesel** Standard #2 diesel fuel combined with up to 15% ethanol. Also known as diesohol.

**EECS** Evaporative Emission Control System.

**EEPROM** Electronically erasable programmable read-only memory.

**EGR valve position** A linear potentiometer on the top of the EGR valve stem indicates valve position for the computer.

**Electrolysis** The process in which electric current is passed through water in order to break it into hydrogen and oxygen gas.

**Electrolyte** The insulation between two different parts of a battery or fuel cell.

**Electromagnetic interference** An undesirable electronic signal. It is caused by a magnetic field building up and collapsing, creating unwanted electrical interference on a nearby circuit.

**Electronic air control** The idle air control valve.

**Electronic ignition** General term used to describe any of the various types of ignition systems that use electronic instead of mechanical components, such as contact points.

**Electronic returnless fuel system** A fuel delivery system that does not return fuel to the tank.

**Electronic spark timing** The computer controls spark timing advance.

**Electronic throttle control (ETC)** A system that moves the throttle plate using an electric motor instead of a mechanical linkage from the accelerator pedal.

**Electronic vacuum regulator valve (EVRV)** A type of EGR valve controlled by a solenoid and controlled by the PCM.

**EMI** A term used to describe a situation where a current can be created in a conductor (coil winding) by a moving magnetic field.

**Enable criteria** Operating condition requirements.

**Energy carrier** Any medium that is utilized to store or transport energy. Hydrogen is an energy carrier because energy must be used to generate hydrogen gas that is used as a fuel.

**Energy density** A measure of the amount of energy that can be stored in a battery relative to the volume of the battery container. Energy density is measured in terms of watt-hours per liter (Wh/L).

**Engine coolant temperature** A sensor constructed of a semiconductor material that decreases in resistance as the temperature of the sensor increases.

**Engine fuel temperature (EFT)** sensor- A temperature sensor located on the fuel rail that measures the temperature of the fuel entering the engine.

**Engine mapping** A computer program that uses engine test data to determine the best fuel–air ratio and spark advance to use at each speed of the engine for best performance.

**EPA** Environmental Protection Agency.

**Ethyl alcohol** See *ethanol*.

**Ethyl tertiary butyl ether** An octane enhancer for gasoline. It is also a fuel oxygenate that is manufactured by reacting isobutylene with ethanol. The resulting ether is high octane and low volatility. ETBE can be added to gasoline up to a level of approximately 13%.

**Ethanol** Grain alcohol that is blended with gasoline to produce motor fuel. Also known as ethyl alcohol.

**Evaporative control (EVAP) system** This system includes the charcoal canister, hoses, and valves. These vapors are routed into a charcoal canister, then into the intake airflow, where they are burned in the engine instead of being released into the atmosphere.

**Exhaust gas recirculation (EGR)** An emission control device to reduce $NO_x$ (oxides of nitrogen).

**Exhaust side electrode** The electrode inside a wide-band oxygen sensor that is exposed to the exhaust stream.

**Exhaust valve** The valve in an engine that opens to allow the exhaust to escape into the exhaust manifold.

**Exponentially weighted moving average (EWMA)** This monitor is a mathematical method used to determine performance. This method smooths out any variables in the readings over time and results in a running average. This method is used by some vehicle manufacturers for two monitors: the catalyst monitor and EGR monitor.

**Extension** A socket wrench tool used between a ratchet or breaker bar and a socket.

**External combustion engine** A type of engine that burns fuel from outside the engine itself such as a steam engine.

**External trigger** Occurs when the trace starts when a signal is received from another (external) source.

**Eye wash station** A water fountain designed to rinse the eyes with a large volume of water.

**False air** A term used to describe air that enters the engine without being measured by the mass air flow sensor.

**False lean indication** Occurs when an oxygen sensor reads low as a result of a factor besides a lean mixture.

**False rich indication** A high oxygen sensor voltage reading that is not the result of a rich exhaust. Some common causes for this false rich indication include a contaminated oxygen sensor and having the signal wire close to a high-voltage source such as a spark plug wire.

**Fail-safe position**   A term used to describe the default position for the throttle plate in an electronic throttle control (ETC) system.

**Farads**   A unit of capacitance.

**FCHV**   Fuel-cell hybrid vehicle.

**FCV**   Fuel-cell vehicle.

**Federal Test Procedure (FTP)**   This is the test used to certify all new vehicles before they can be sold. Once a vehicle meets these standards, it is certified by the EPA for sale in the United States.

**Federal Test Procedure (FTP)**   This is the test used to certify all new vehicles before they can be sold. Once a vehicle meets these standards, it is certified by the EPA for sale in the United States.

**FFV**   Flexible fuel vehicle.

**Files**   A hand tool that is used to smooth rough or sharp edges from metal.

**Filter basket**   A fine mesh screen filter located in the top of a fuel injector.

**Finger design**   A term used to describe the type of oxygen sensor element that is shaped like a thimble.

**Fire blanket**   A fireproof wool blanket used to cover a person who is on fire and smother the fire.

**Fire extinguisher classes**   The types of fires that a fire extinguisher is designed to handle are referred to as fire classes.

**Firing line**   The leftmost vertical (upward) line.

**Firing order**   The order that the spark is distributed to the correct spark plug at the right time.

**Fischer-Tropsch**   A method to create synthetic liquid fuel from coal. Flammable and can form explosive mixtures with air. It is slightly soluble.

**Flare**   An increase and then decrease in engine speed.

**Flare-nut wrench**   A type of wrench used to remove fuel, brake, or air-conditioning lines.

**Flash point**   The temperature at which the vapors on the surface of the fuel will ignite if exposed to an open flame.

**Flex fuel**   Flex-fuel vehicles are capable of running on straight gasoline or gasoline/ethanol blends.

**Floating ground**   Both positive and negative high-voltage power cables are isolated from the

**Flow gauge**   Tests for proper airflow in the EVAP system.

**Flyback voltage**   The inductive kick created when the primary field collapses is used by the PCM to monitor secondary ignition performance.

**Forced induction systems**   A term used to describe a turbocharger or supercharger.

**Formaldehyde**   Formed when RFG is burned, and the vehicle exhaust has a unique smell when reformulated gasoline is used.

**Four-stroke cycle**   An engine design that requires four stokes to complete one cycle with each stroke requiring 180 degrees of crankshaft rotation.

**Freeze-frame**   A snapshot of all of the engine data at the time the DTC was set.

**Frequency**   The number of times a waveform repeats in one second, measured in hertz (Hz), frequency band.

**FTD**   Fischer-Tropsch diesel.

**FTP**   Federal Test Procedure.

**Fuel cell**   An electrochemical device that converts the energy stored in hydrogen gas into electricity, water, and heat.

**Fuel compensation sensor**   A sensor used in flex-fuel vehicles that provides information to the PCM on the ethanol content and temperature of the fuel as it is flowing through the fuel delivery system.

**Fuel-cell stack**   A collection of individual fuel cells, which are stacked end-to-end into one compact package.

**Fuel rail**   A term used to describe the tube that delivers the fuel from the fuel line to the individual fuel injectors.

**Fuel tank pressure (FTP)**   A sensor that monitors the pressure inside the fuel tank and is used to monitor if the EVAP system is functioning correctly.

**Fuel-cell hybrid vehicle (FCHV)**   A vehicle that uses a fuel cell to create electrical energy and a high-voltage battery to store electrical energy used by an electric motor to propel the vehicle.

**Fuel-cell vehicle (FCV)**   A vehicle that uses a fuel cell to create electrical energy used by an electric motor to propel the vehicle.

**Functionality**   Refers to PCM inputs checking the operation of the outputs.

**Fungible**   A term used to describe a product that has the same grade or meets the same specifications, such as oil, that can be interchanged with another product without any affect.

**Gang fired**   Pulsing injectors in groups.

**Gasoline direct injection**   A fuel-injection system design in which gasoline is injected directly into the combustion chamber.

**Gasoline**   Refined petroleum product that is used primarily as a motor fuel. Gasoline is made up of many different hydrocarbons and also contains additives for enhancing its performance in an ICE.

**Gas-to-liquid**   A refining process in which natural gas is converted into liquid fuel.

**GAWR**   Gross axle weight rating. A rating of the load capacity of a vehicle and included on placards on the vehicle and in the owner's manual.

**Gerotor**   A type of positive displacement pump that is often used in engine oil pumps. It uses the meshing of internal and external gear teeth to pressurize the fuel.

**Glow plug**   A heating element that uses 12 volts from the battery and aids in the starting of a cold engine.

**GMLAN**   GM local area network. A type of serial data transmission by General Motors.

**GMM**   Graphing multimeter. A cross between a digital meter and a digital storage oscilloscope.

**Grade**   The strength rating of a bolt.

**Grain alcohol**   See *ethanol*.

**Graticule**   The grid lines on the scope screen.

**GVWR**   Gross vehicle weight rating. The total weight of the vehicle including the maximum cargo.

**Hacksaws**   A type of hand tool that is used to cut metal or other hard materials.

**Hall-effect switch**   A semiconductor moving relative to a magnetic field, creating a variable voltage output. Used to determine position. A type of electromagnetic sensor used in electronic ignition and other systems. Named for Edwin H. Hall, who discovered the Hall effect in 1879.

**Hammers**   A hand tool that is used to apply force by swinging.

**Hangers**   Made of rubberized fabric with metal ends that hold the muffler and tailpipe in position so that they do not touch any metal part. This helps to isolate the exhaust noise from the rest of the vehicle.

**Hazardous waste material**   A classification of materials that can.

**Heat of compression**   Air is compressed until its temperature reaches about 1000°F.

**Helmholtz resonator**   Used on the intake duct between the air filter and the throttle body to reduce air intake noise during engine acceleration.

**HEPA vacuum**   High-efficiency particulate air filter vacuum used to clean brake dust.

**Hertz**   The measurement of frequency.

**HEUI**   Hydraulic Electronic Unit Injection.

**HEV**   an abbreviation for hybrid electric vehicle.

**High Energy Ignition**   General Motors' name for their electronic ignition.

**High-impedance meter**   Measures the total internal resistance of the meter circuit due to internal coils, capacitors, and resistors.

**High-pressure common rail**   Diesel fuel under high pressure, over 20,000 PSI (138,000 kPa), is applied to the injectors, which are opened by a solenoid controlled by the computer. Because the injectors are computer controlled, the combustion process can be precisely controlled to provide maximum engine efficiency with the lowest possible noise and exhaust emissions.

**Homogeneous charge compression ignition**   A low-temperature combustion process that involves air–fuel mixtures being burned without the use of spark ignition.

**Homogeneous mode**   In this mode of operation, the air-fuel mixture is the same throughout the cylinder.

**HV cables**   Vehicle cables that carry high voltage.

**HV**   High voltage. Applies to any voltage above 50 volts.

**Hydraulic power assist**   A hybrid vehicle configuration that utilizes hydraulic pumps and accumulators for energy regeneration.

**Hydrocracking**   A refinery process that converts hydrocarbons with a high boiling point into ones with low boiling points.

**Hydrokinetic pump**   This design of pump rapidly moves the fuel to create pressure.

**I/M 240 test**   It is a portion of the 505-second FTP test used by the manufacturers to certify their new vehicles. The "240" stands for 240 seconds of drive time on a dynamometer.

**IAC**   Idle air control.

**ICE**   Internal combustion engine.

**IEC**   International Electrotechnical Commission.

**Idle air control (IAC)**   An electronic stepper motor or pulse-width modulated solenoid used to maintain the correct idle speed.

**Idle air control counts**   The calculated relative idle air control valve position as displayed on a scan tool.

**Idle speed control (ISC) motor**   A motor, usually a stepper motor used to move a pintle that allows more or less air past the throttle plate thereby controlling idle speed.

**Idle vacuum test**   A test performed using a vacuum gauge attached to the intake manifold of an engine running at idle speed.

**Igniter**   Ignition Control Module.

**Ignition coil**   An electrical device that consists of two separate coils of wire: a primary and a secondary winding. The purpose of an ignition coil is to produce a high-voltage (20,000 to 40,000 volts), low-amperage (about 80 mA) current necessary for spark ignition.

**Ignition control module**   Controls (turns on and off) the primary ignition current of an electronic ignition system.

**Ignition control**   Igniter.

**Ignition timing**   The exact point of ignition in relation to piston position.

**Inches of Mercury**   A measurement of vacuum; pressure below atmospheric pressure.

**Indirect injection**   Fuel is injected into a small prechamber, which is connected to the cylinder by a narrow opening. The initial combustion takes place in this prechamber. This has the effect of slowing the rate of combustion, which tends to reduce noise.

**Inductive ammeter**   A type of ammeter that is used as a Hall-effect sensor in a clamp that surrounds a conductor carrying a current.

**Inductive reactance**   An opposing current created in a conductor whenever there is a charging current flow in a conductor.

**Inert**   Chemically inactive.

**Inertia switch**   Turns off the electric fuel pump in an accident.

**Initial timing**   Where the spark plug fires at idle speed. The computer then advances the timing based off engine speed and other factors.

**Injection pump**   Delivers fuel to the injectors at a high pressure and at timed intervals. Each injector sprays fuel into the combustion chamber at the precise moment required for efficient combustion.

**Input conditioning**   What the computer does to the input signals to make them useful; usually includes an analog to digital converter and other electronic circuits that eliminate electrical noise.

**Input**   Information on data from sensors to an electronic controller is called input. Sensors and switches provide the input signals.

**Intercooler**   Similar to a radiator, wherein outside air can pass through, cooling the pressurized heated air.

**Intermediate oscillations**   Also called the "ringing" of the coil as it is pulsed.

**Inverter**   An electronic device used to convert DC (direct current) into AC (alternating current).

**Ion-sensing ignition**   A type of coil-on-plug ignition that uses a signal voltage across the spark plug gap after the plug has fired to determine the air-fuel mixture and if spark knock occurred.

**Iridium spark plugs**   Use a small amount of iridium welded onto the tip of a small center electrode 0.0015 to 0.002 inch (0.4 to 0.6 mm) in diameter. The small diameter reduces the voltage required to jump the gap between the center and the side electrode, thereby reducing possible misfires. The ground or side electrode is usually tipped with platinum to help reduce electrode sap wear.

**ISC**   Idle speed control motor.

**KAM**   Keep-alive memory.

**Keyword**   A type of network communications used in many General Motors vehicles.

**Kilo**   Means 1,000; abbreviated k or K.

**Knock sensor**   A sensor that can detect engine spark knock.

**Leak defection pump**   Chrysler uses an electric pump that pressurizes the fuel system to check for leaks by having the PCM monitor the fuel tank pressure sensor.

**Lean indicator**   Oxygen.

**LED test light**   Uses an LED instead of a standard automotive bulb for a visual indication of voltage.

**Lift pump**   The diesel fuel is drawn from the fuel tank by the lift pump and delivers the fuel to the injection pump.

**Light-off temperature**   The temperature where a catalytic converter starts to work.

**Light-off time (LOT)**   The fast heating of a wide-band oxygen sensor helps improve fuel economy and reduces cold-start exhaust emissions.

**Linear air–fuel ratio sensor**   See lean air–fuel ratio sensor.

**Linear EGR**  Contains a stepper motor to precisely regulate exhaust gas flow and a feedback potentiometer that signals to the computer the actual position of the valve.

**Lineman's gloves**  Type of gloves worn by technicians when working around high-voltage circuits. Usually includes a rubber inner glove rated at 1,000 volts and a protective leather outer glove when used for hybrid electric vehicle service.

**Linesman's gloves**  Type of gloves worn by technicians when working around high-voltage circuits. Usually includes a rubber inner glove rated at 1,000 volts and a protective leather outer glove when used for hybrid electric vehicle service.

**Liquefied petroleum gas**  Sold as compressed liquid propane that is often mixed with about 10% of other gases such as butane, propylene, butylenes, and mercaptan to give the colorless and odorless propane a smell.

**Logic probe**  A type of tester that can detect either power or ground. Most testers can detect voltage but some cannot detect if a ground is present without further testing.

**Low-grade heat**  Cooling system temperatures that are very close to the temperature of the ambient air, resulting in lowered heat transfer efficiency.

**LP-gas**  See *liquefied petroleum gas*.

**M85**  Internal combustion engine fuel containing 85% methanol and 15% gasoline.

**Magnetic pulse generator**  The pulse generator consists of a trigger wheel (reluctor) and a pickup coil. The pickup coil consists of an iron core wrapped with fine wire, in a coil at one end and attached to a permanent magnet at the other end. The center of the coil is called the pole piece.

**Magnetic sensor**  Uses a permanent magnet surrounded by a coil of wire. The notches of the crankshaft (or camshaft) create a variable magnetic field strength around the coil. When a metallic section is close to the sensor, the magnetic field is stronger because metal is a better conductor of magnetic lines of force than air.

**Magnetic-resistive sensor**  A sensor that is similar to a magnetic sensor but, instead of producing an analog voltage signal, the electronics inside the sensor itself generate a digital on/off signal or an output.

**Malfunction indicator lamp**  This amber, dashboard warning light may be labeled check engine or service engine soon.

**MAP sensor**  A sensor used to measure the pressure inside the intake manifold compared to a perfect vacuum.

**Married coil**  Also called a tapped transformer.

**Mass air flow sensor**  Measures the density and amount of air flowing into the engine, which results in accurate engine control.

**Mechanical force**  A force applied to an object.

**Mechanical power**  A force applied to an object which results in movement or motion.

**Mechanical returnless fuel system**  A returnless fuel delivery system design that uses a mechanical pressure regulator located in the fuel tank.

**Mega**  Million. Used when writing larger numbers or measuring a large amount of resistance.

**Membrane electrode assembly**  The part of the PEM fuel cell that contains the membrane, catalyst coatings, and electrodes.

**Mercury**  A heavy metal.

**Meter accuracy**  The accuracy of a meter measured in percent.

**Meter resolution**  The specification of a meter that indicates how small or fine a measurement the meter can detect and display.

**Methanol**  Typically manufactured from natural gas. Methanol content, including cosolvents, in unleaded gasoline is limited by law to 5%.

**Methanol-to-gasoline**  A refining process in which methanol is converted into liquid gasoline.

**Methyl alcohol**  See *methanol*.

**Methyl tertiary butyl ether**  A fuel oxygenate that is permitted in unleaded gasoline up to a level of 15%.

**Metric bolts**  Bolts manufactured and sized in the metric system of measurement.

**Micro (μ)**  One millionth of a volt or ampere.

**Micron**  Equal to 0.000039 inches.

**Milli (m)**  One thousandth of a volt or ampere.

**Millisecond sweep**  The scope will sweep only that portion of the pattern that can be shown during a 5- or 25-ms setting.

**MSDS**  Material safety data sheet.

**MTHF**  Methyltetrahydrofuron. A component of P-series nonpetroleum-based fuels.

**Multiplexing**  A process of sending multiple signals of information at the same time over a signal wire.

**Mutual induction**  The generation of an electric current due to a changing magnetic field of an adjacent coil.

**Naturally (normally) aspirated**  An engine that uses atmospheric pressure for intake.

**NEDRA**  National Electric Drag Racing Association.

**Negative back pressure**  An EGR valve that reacts to a low pressure area by closing a small internal valve, which allows the EGR valve to be opened by vacuum.

**Negative temperature coefficient**  Usually used in reference to a temperature sensor (coolant or air temperature). As the temperature increases, the resistance of the sensor decreases.

**Nernst cell**  Another name for a conventional oxygen sensor. The Nernst cell is named for Walther Nernst, 1864–1941, a German physicist known for his work in electrochemistry.

**Network**  A communications system used to link multiple computers or modules.

**Neutral position**  A term used to describe the home or the default position of the throttle plate in an electronic throttle control system.

**NGV**  Natural gas vehicle.

**NiMH**  Nickel-metal hydride. A battery design used for the high-voltage batteries in most hybrid electric vehicles.

**Nitrogen oxides (NOx)**  Gases formed during combustion. In the presence of sunlight, $NO_x$ reacts with hydrocarbons in the atmosphere to form ozone ($O_3$) or photochemical smog, an air pollutant.

**Nitrous oxide ($N_2O$)**  type of power adder that uses a gas containing oxygen along with additional fuel to increase engine power.

**Node**  A module and computer that is part of a communications network.

**Noid light**  Designed to electrically replace the injector in the circuit and to flash if the injector circuit is working correctly.

**Nonchecking**  Some General Motors throttle-body units that do not hold pressure.

**Non-methane hydrocarbon**  The standard by which exhaust emission testing for hydrocarbons is evaluated.

**Nonprincipal end**  Opposite the principal end and is generally referred to as the front of the engine, where the accessory belts are used.

**Nonvolatile RAM**  Computer memory capability that is not lost when power is removed. See also *read-only memory (ROM)*.

**OBD**   On-board diagnostic.

**Octane rating**   The measurement of a gasoline's ability to resist engine knock. The higher the octane rating, the less prone the gasoline is to cause engine knock (detonation).

**Oil galleries**   An oil pump, which is driven by the engine, forces the oil through the oil filter and then into passages in the crankshaft and block.

**OL**   Open circuit.

**On-board diagnosis (OBD)**   The computer program within the PCM that performs self-tests of all of the emission-related systems of the vehicle.

**Onboard refueling vapor recovery (ORVR)**   A part of the fuel delivery system designed to reduce evaporative emissions during refueling.

**Opacity**   The percentage of light that is blocked by the exhaust smoke.

**Open-end wrench**   A type of wrench that allows access to the flats of a bolt or nut from the side.

**Open-loop operation**   A phase of computer-controlled engine operation where air–fuel mixture is calculated in the absence of oxygen sensor signals. During open loop, calculations are based primarily on throttle position, engine RPM, and engine coolant temperature.

**Optical sensors**   Use light from a LED and a phototransistor to signal the computer.

**Organic**   A term used to describe anything that was alive at one time.

**ORVR**   Onboard refueling vapor recovery.

**OSC**   Oxygen storage capacity.

**Oscilloscope (scope)**   A visual volt meter.

**OSHA**   Occupational Safety and Health Administration.

**Overhead valve (OHV)**   A type if engine that has the valves located in the cylinder head and the camshaft is located in the engine block. Also called a "cam-in-block"-type engine.

**Oxygen sensor**   A sensor in the exhaust system to measure the oxygen content of the exhaust.

**Oxygen storage capacity (OSC)**   The converter stores, due mostly to the cerium coating in the catalyst, then releases the oxygen during normal reduction and oxidation of the exhaust gases, smoothing out the variations in oxygen being released.

**Oxygenated fuels**   Fuels such as ETBE or MTBE that contain extra oxygen molecules to promote cleaner burning. Oxygenated fuels are used as gasoline additives to reduce CO emissions.

**Ozone**   Oxygen-rich ($O_3$) gas created by sunlight reaction with unburned hydrocarbons (HC) and oxides of nitrogen ($NO_x$); also called smog.

**Palladium**   A catalyst that starts a chemical reaction without becoming a part of, or being consumed in, the process.

**Pancake engine**   See *boxer*.

**Paper test**   Hold a piece of paper or a $3 \times 5$ index card (even a dollar bill works) within 1 inch (2.5 centimeters) of the tailpipe with the engine running at idle. The paper should blow out evenly without "puffing." If the paper is drawn toward the tailpipe at times, the exhaust valves in one or more cylinders could be burned.

**Parameter identification**   The information found in the vehicle data stream as viewed on a scan tool.

**Particulate matter (PM)**   Also called soot and an emission from diesel engines.

**PCM**   The onboard computer that controls both the engine management and transmission functions of the vehicle.

**PCV**   Pressure control valve.

**Peak-and-hold injector**   A type of injector that is typically used for TBI and some port low-resistance injectors. Full battery voltage is applied to the injector, and the ground side is controlled through the computer. The computer provides a high initial current flow (about 4 amperes) to flow through the injector windings to open the injector core. Then the computer reduces the current to a lower level (about 1 ampere).

**Pending code**   Any code(s) that is displayed on a scan tool when the MIL is not on.

**Peripheral pump**   Turbine pump.

**Petrodiesel**   Another term for petroleum diesel, which is ordinary diesel fuel refined from crude oil.

**Petroleum**   Another term for crude oil. The literal meaning of petroleum is "rock oil."

**PFE sensor**   Pressure feedback EGR.

**PHEV**   Plug-in hybrid electric vehicle.

**Pickup coil**   A simple and common ignition electronic switching device in the magnetic pulse generator system.

**Piezoresistivity**   Change in resistance due to strain.

**Pinch weld seam**   A strong section under a vehicle where two body panels are welded together.

**Ping**   Secondary rapid burning of the last 3% to 5% of the air–fuel mixture in the combustion chamber causes a second flame front that collides with the first flame front causing a knock noise. Also called detonation or spark knock.

**Piston stroke**   A one-way piston movement between the top and bottom of the cylinder.

**Pitch**   The pitch of a threaded fastener refers to the number of threads per inch.

**Planar design**   A type of oxygen sensor construction that allow for faster heating of the sensor to help reduce exhaust emissions.

**Plastic optical fiber (POF)**   BUS used in safety critical systems, such as airbags, and uses the time division multiple access (TDMA) protocol, which operates at 10 million bps.

**Platinum spark plug**   A spark plug that has a small amount of the precious metal platinum welded onto the end of the center electrode, as well as on the ground or side electrode. Platinum is a grayish-white metal that does not react with oxygen and therefore, will not erode away as can occur with conventional nickel alloy spark plug electrodes.

**Platinum**   A catalyst that starts a chemical reaction without becoming a part of, or being consumed in, the process.

**Plenum**   A chamber, located between the throttle body and the runners of the intake manifold, used to distribute the intake charge more evenly and efficiently.

**Pliers**   A hand tool with two jaws used to grasp or turn a part.

**Plug-in hybrid electric vehicle (PHEV)**   A type of hybrid electric vehicle that has additional battery storage capacity to allow it to be propelled using battery power alone for many miles.

**Polarity**   The condition of being positive or negative in relation to a magnetic pole.

**Polymer electrolyte fuel cell**   Another term for PEM fuel cell.

**Pop tester**   A device used for checking a diesel injector nozzle for proper spray pattern. The handle is depressed and pop off pressure is displayed on the gauge.

**Port fuel-injection**   Uses a nozzle for each cylinder and the fuel is squirted into the intake manifold about 2 to 3 inches (70 to 100 mm) from the intake valve.

**Positive back pressure**   An EGR valve that is designed with a small valve inside that bleeds off any applied vacuum and prevents the valve from opening.

**Positive crankcase ventilation (PCV)**   A system designed to ventilate the crankcase and recirculate the vapors to the engine's induction system so they can be burned in the cylinders. PCV systems help reduce HC and CO emissions.

**Positive displacement**   All of the air that enters is forced through the roots-type supercharger.

**Potentiometer**   A type of variable resistor.

**Power adder**   A unit such as a turbocharger, supercharger that is used to increase the power output of an engine.

**Power balance test**   Determines if all cylinders are contributing power equally. It determines this by sorting out one cylinder at a time.

**PPO**   Pure plant oil.

**Preconverter**   A small, quick heating oxidation converter.

**Pressure control valve (PCV)**   A valve used to control the fuel system pressure on a demand delivery-type fuel system.

**Pressure differential**   A difference in pressure from one brake circuit to another.

**Pressure relief valve**   A valve located in a power steering pump that uses a check ball, which unseats and allows fluid to return to the reservoir if pressure exceeds a certain volume.

**Pressure transducer**   An electronic device that converts changes in pressures to changes in voltage so that the pressure can be displayed on a meter or scope.

**Pressure vent valve**   A valve located in the fuel tank to prevent overpressure due to the thermal expansion of the fuel.

**Prevailing torque nut**   A special design of nut fastener that is deformed slightly or has other properties that permit the nut to remain attached to the fastener without loosening.

**Primary ignition circuit**   The ignition components that regulate the current in the coil primary winding by turning it on and off.

**Primary winding**   The coil winding that is controlled by the electronic ignition control module or PCM.

**Principal end**   The end of the engine that the flywheel is attached to.

**Programmable controller interface (PCI)**   A type of serial data transmission used by Chrysler.

**PROM**   Programmable read-only memory.

**Propane**   See *liquified petroleum gas*.

**Protocol**   A set of rules or a standard used between computers or electronic control modules.

**Proton exchange membrane**   A low-temperature fuel cell known for fast starts and relatively simple construction.

**Pulse train**   A DC voltage that turns on and off in a series of pulses.

**Pulse width**   A measure of the actual on-time measured in milliseconds.

**Pump cell**   A cell used that contains another zirconia layer with two electrodes used in a dual cell wide-band oxygen sensor.

**Punches**   A type of hand tool used to drive roll pins.

**Pup (mini) converter**   See *preconverter*.

**Purge flow sensor**   Checks for adequate purge flow.

**Purge valve**   The EVAP purge valve is normally closed and is pulsed open by the PCM during purging. The purge valve is connected to intake manifold vacuum using a rubber hose to draw gasoline vapors from the charcoal canister into the engine when the purge valve is commanded open.

**Pushrod engine**   Uses one camshaft for the intake valves and a separate camshaft for the exhaust valves. When the camshaft is located in the block, the valves are operated by lifters, pushrods, and rocker arms.

**PWM**   Pulse-width modulation.

**R & R**   Remove and replace.

**RAM**   Random-access memory.

**Range**   The distance a vehicle can travel on a full charge or full-fuel tank without recharging or refueling. Range is measured in miles or kilometers.

**Raster**   Stacked.

**Ratchet**   A type of hand tool that is used to rotate a socket and is reversible.

**Rationality**   Refers to a PCM comparison of input value to values.

**RCRA**   Resource Conservation and Recovery Act.

**Recall**   A notification to the owner of a vehicle that a safety issue needs to be corrected.

**Reference electrode**   The electrode used in a wide-band oxygen sensor that is exposed to outside (ambient) air and is the signal electrode, also called the ambient air electrode.

**Reference voltage**   A term used to describe the bias voltage applied to an oxygen sensor.

**Reformulated gasoline**   RFG has oxygenated additives and is refined to reduce both the lightest and heaviest hydrocarbon content from gasoline in order to promote cleaner burning.

**Regeneration**   A process of taking the kinetic energy of a moving vehicle and converting it to electrical energy and storing it in a battery.

**Reid vapor pressure**   A method of determining vapor pressure of gasoline and other petroleum products. Widely used in the petroleum industry as an indicator of the volatility of gasoline.

**Reluctor**   A notched metal wheel used with a magnetic sensor to trigger crankshaft or camshaft position.

**Residual check valve**   A valve in the outlet end of the master cylinder to keep the hydraulic system under a light pressure on drum brakes only.

**Residual or rest pressure**   Prevents vapor lock and hot-start problems on these systems.

**Restricted exhaust**   The engine will be low on power, yet smooth.

**Rhodium**   A catalyst that starts a chemical reaction without becoming a part of, or being consumed in, the process.

**Rich indicator**   The higher the CO reading, the richer the air-fuel mixture. High concentrations of CO indicate that not enough oxygen was available for the amount of fuel.

**Right-to-know laws**   Laws that state that employees have a right to know when the materials they use at work are hazardous.

**RMS**   A method of calculating surface roughness using the square root of the average readings squared.

**Roller cell**   Vane pump.

**ROM**   Read-only memory.

**Roots-type**   Named for Philander and Francis Roots, two brothers from Connersville, Indiana, who patented the design in 1860 as a type of water pump to be used in mines. Later it was used to move air and is used today on two-stroke cycle Detroit diesel engines and other supercharged engines. The roots-type supercharger is called a positive displacement design because all of the air that enters is forced through the unit.

**Rotary engine** Operates on the four-stroke cycle but uses a rotor instead of a piston and crankshaft to achieve intake, compression, power, and exhaust stroke.

**Rotary vane pump** The pump consists of a central impeller disk, several rollers or vanes that ride in notches in the impeller, and a pump housing that is offset from the impeller centerline.

**Rotor gap** Measures the voltage required to jump the gap (0.030 to 0.050 inch. or 0.8 to 1.3 mm) between the rotor and the inserts (segments) of the distributor cap.

**Running compression test** A test that can inform a technician of the relative compression of all the cylinders.

**SAE** Society of Automotive Engineers.

**Saturation** The point of maximum magnetic field strength of a coil.

**Schmitt trigger** Converts the analog signal into a digital signal.

**Screwdrivers** A type of hand tool used to install or remove screws.

**Sealed Housing for Evaporative Determination (SHED) test** A test which measures the evaporative emissions from the vehicle after a heat-up period representing a vehicle sitting in the sun. In addition, the vehicle is driven and then tested during the hot soak period.

**Secondary air injection (SAI)** A system that provides the air necessary for the oxidizing process either at the exhaust manifold or inside the catalytic converter.

**Secondary ignition circuit** The components necessary to create and distribute the high voltage produced in the secondary windings of the coil.

**Secondary winding** A winding that has about 100 times the number of turns of the primary winding, referred to as the turns ratio (approximately 100:1).

**Selective catalytic reduction (SCR)** A type of exhaust emission system that uses urea injection to reduce oxides of nitrogen emissions from diesel engines.

**Self-induction** The generation of an electric current in the wires of a coil created when the current is first connected or disconnected.

**Sequential fuel injection** A fuel injection system in which injectors are pulsed individually in sequence with the firing order.

**Serial communications interface (SCI)** A type of serial data transmission used by Chrysler.

**Serial data** Data that are transmitted by a series of rapidly changing voltage signals.

**Service plug** The plug or switch used to disable or de-power the high voltage system in a hybrid electric vehicle (HEV).

**Servomotor** An electric motor that moves an actuator such as the throttle plate in an electronic throttle control system.

**Side-channel pump** Turbine pump.

**Signal electrode** The electrode in a wide-band oxygen sensor that is exposed to outside (ambient) air and also called the reference electrode or ambient air electrode.

**Single-cell** A type of wide-band oxygen sensor that uses a single cell and can be made either in a cup-type of planar design.

**Single overhead camshaft** When one overhead camshaft is used.

**SIP** State Implementation Plan.

**Skewed** An output from a sensor that moves in the correct direction but does not accurately measure condition it is designed to measure.

**Smog** The term used to describe a combination of smoke and fog. Formed by $NO_x$ and HC with sunlight.

**Smog pump** Another name for the secondary air injection (SAI) pump.

**Smoke machine** A machine that generates a stream of smoke that is used to locate leaks

**Snips** A type of hand tool used to cut sheet metal and other thin materials.

**Society of Automotive Engineers (SAE)** Type of diagnostic trouble codes (DTCs) that are the same for all vehicles makes.

**Socket adapter** An adapter that allows the use of one size of driver (ratchet or breaker bar) to rotate another drive size of socket.

**Socket** A tool that fits over the head of a bolt or nut and is rotated by a ratchet or breaker bar.

**Solvent** Usually colorless liquids that are used to remove grease and oil.

**Soot** Another name for particulate matter (PM).

**Spark ignition direct injection** GM's name for GDI system.

**Spark knock** See *Detonation*.

**Spark line** A short horizontal line immediately after the firing line.

**Spark output** The term that Ford used to describe the OBD-II terminology for the output signal from the PCM to the ignition system that controls engine timing.

**Spark plugs** A replaceable part that threads into the cylinder head of an engine and provides a gap where a spark occurs to ignite the air-fuel mixture inside an engine.

**Spark tester** Looks like a spark plug except that it has a recessed center electrode and no side electrode. The tester commonly has an alligator clip attached to the shell so that it can be clamped on a good ground connection on the engine.

**Specific energy** The energy content of a battery relative to the mass of the battery. Specific energy is measured in watt-hours per kilogram (Wh/kg).

**Speed density** The method of calculating the amount of fuel needed by the engine.

**Splice pack** A central point where many serial data lines jam together, often abbreviated SP.

**Spontaneous combustion** A condition that can cause some materials, such as oily rags, to catch fire without a source of ignition.

**SST** Special service tools.

**Standard corporate protocol (SCP)** A network communications protocol used by Ford.

**State Implementation Plan (SIP)** The Clean Air Act Amendments require enhanced Inspection and Maintenance (I/M) programs in areas of the country that have the worst air quality and the Northeast Ozone Transport region. The states must submit to the EPA a State Implementation Plan (SIP) for their programs.

**State of health (SOH)** A signal sent by a module to all of the other modules in the network indicating that it is well and able to transmit.

**Stratified mode** In this mode of operation, the air-fuel mixture is richer around the spark plug than it is in the rest of the cylinder.

**Stoichiometric** The ideal mixture or ratio at which all of the fuel combines with all of the oxygen in the air and burns completely.

**Straight vegetable oil** Vegetable oil, a triglyceride with a glycerin component joining three hydrocarbon chains of 16 to 18 carbon atoms each.

**Stroke** The distance the piston travels in the cylinder of an engine.

**Stud**  A short rod with threads on both ends.

**Supercharger**  Forces the air–fuel mixture into the cylinder for even greater power.

**Superimposed**  A position used to look at differences in patterns between cylinders in all areas except the firing line.

**SWCAN**  An abbreviation for single wire CAN (Controller Area Network).

**Switchgrass**  A feedstock for ethanol production that requires very little energy or fertilizer to cultivate.

**Switching**  Turning on and off of the primary circuit.

**Syncrude**  A product of a process where coal is broken down to create liquid products. First the coal is reacted with hydrogen ($H_2$) at high temperatures and pressure with a catalyst.

**Syn-gas**  Synthesis gas generated by a reaction between coal and steam. Syn-gas is made up of mostly hydrogen and carbon monoxide and is used to make methanol. Syn-gas is also known as town gas.

**Synthetic fuel**  Fuels generated through synthetic processes such as Fischer–Tropsch.

**Tap test**  Involves tapping (not pounding) on the catalytic converter using a rubber mallet.

**Tapped transformer**  See *married coil*.

**Task manager**  A term Chrysler uses to describe the software program that is designed to manage the operation of all OBD-II monitors by controlling the sequence of steps necessary to execute the diagnostic tests and monitors.

**TBI**  Throttle-body injection.

**Technical service bulletin (TSB)**  Information, suggestions, or replacement parts found in a service bulletin. DTCs must be known before searching for service bulletins because bulletins often include information on solving problems that involve a stored DTC.

**Tensile strength**  The maximum stress used under tension (lengthwise force) without causing failure.

**Terminating resistors**  Resistors placed at the end of a high-speed serial data circuit to help reduce electromagnetic interference.

**Tertiary-amyl methyl ether**  An oxygenate added to gasoline that is flammable and can form explosive mixtures with air. It is slightly soluble in water, very soluble in ethers and alcohol, and soluble in most organic solvents including hydrocarbons.

**Test identification (TID)**  The test identification as displayed on a scan tool when observing data using mode $06.

**Test light**  A light used to test for voltage. Contains a light bulb with a ground wire at one end and a pointed tip at the other end.

**Tetraethyl lead**  A liquid added to gasoline in the early 1920s to reduce the tendency to knock.

**Thermactor pump**  Another name for the secondary air injection (SAI) pump.

**Thimble design**  A term used to describe the type of oxygen sensor element that is shaped like a cup or thimble.

**Three-way converter (TWC)**  A catalytic converter where all three exhaust emissions ($NO_x$, HC, and CO) are converted to carbon dioxide ($CO_2$) and water ($H_2O$).

**Throttle body temperature (TBT)**  sensor A temperature sensor that is mounted on the throttle body and measures the temperature of the air entering the engine.

**Throttle-body injection (TBI)**  A type of electronic fuel injection system that delivers fuel from a nozzle(s) into the air above the throttle plate.

**Throttle position sensor**  The sensor that provides feedback concerning the position of the throttle plate.

**TID**  Test identification.

**Time base**  Setting how much time will be displayed in each block.

**Top dead center**  The highest point in the cylinder that the piston can travel. The measurement from bottom dead center (BDC) to TDC determines the stroke length of the crankshaft.

**Track**  A track inside an ignition coil is not a short but a low-resistance path or hole that has been burned through from the secondary wiring to the steel core.

**Transistor**  A semiconductor device that can operate as an amplifier or an electrical switch.

**Transmission fluid temperature (TFT)**  sensor A sensor located inside an automatic transmission/transaxle that measures the temperature of the fluid.

**Trigger**  The action of a sensor that turns on or off the current.

**Trigger level**  The start of the display.

**Trigger slope**  The voltage direction that a waveform must have in order to start the display.

**Trip**  An engine-operating drive cycle that contains the necessary conditions for a particular test to be performed.

**True transformer**  See *divorced coil*.

**TSB**  Technical service bulletin.

**Turbine pump**  Turns at higher speeds and is quieter than the other electric pumps.

**Turbo lag**  The delay between acceleration and turbo boost.

**Turbocharger**  An exhaust-powered supercharger.

**Turns ratio**  The number of times that the secondary windings in an ignition coil exceed that of the primary winding.

**TWC**  Three-way converter.

**Twisted pair**  A pair of wires that are twisted together from 9 to 16 turns per foot of length. Most are twisted once every inch (12 per foot) to help reduce electromagnetic interference from being induced in the wires as one wire would tend to cancel out any interference pickup up by the other wire.

**UART**  Universal Asynchronous Receive/Transmit; a type of serial data transmission.

**UART-based protocol (UBP)**  A type of module communication that uses the UART protocol.

**UBP**  UART-based protocol.

**UCG**  Underground coal gasification.

**Ultracapacitor**  A specialized capacitor technology with increased storage capacity for a given volume.

**Underground coal gasification (UCG)**  process performed underground where coal is turned into a liquid fuel.

**Underground storage tank**  A type of oil tank that is located underground.

**UNF**  Unified national fine.

**Universal joint**  A joint in a steering or drive shaft that allows torque to be transmitted at an angle.

**Up-integrated ignition**  Ignition control where all timing functions are interpreted in the PCM, rather than being split between the ignition control module and the PCM.

**Urea**  A chemical used to reduce oxides of nitrogen emission on diesel engines that use an SCR emission control system.

**Used cooking oil**   A term used when the oil may or may not be pure vegetable oil.

**Used oil**   Any petroleum-based or synthetic oil that has been used.

**Vacuum lock**   A condition that can prevent fuel from being supplied tot eh engine from the fuel tank if the tank is not veined.

**Vacuum test**   Testing the engine for cranking vacuum, idle vacuum, and vacuum at 2500 RPM.

**Vacuum**   Any pressure less than atmospheric pressure (14.7 PSI).

**VAF**   Vane air flow.

**Vane airflow (VAF) sensor**   A sensor used measure airflow entering an engine. The vane is deflected by intake airflow.

**Vapor lock**   A lean condition caused by vaporized fuel in the fuel system.

**Variable fuel sensor**   See *fuel compensation sensor*.

**Variable reluctance sensor**   Magnetic sensor.

**VECI**   Vehicle emission control information. This sticker is located under the hood on all vehicles and includes emission-related information that is important to the service technician.

**Vent valve**   This EVAP valve is a normally open valve and is closed only when commanded by the PCM during testing of the system. The vent valve is closed only during testing by the PCM as part of the mandated OBD-II standards.

**VIN**   Vehicle identification number.

**Virtual-flexible fuel vehicle (V-FFV)**   The virtual-flexible fuel vehicle can operate on pure gasoline, E10, E85, or any combination.

**VOC**   Volatile organic compound.

**Volatile organic compounds (VOC)**   Compounds that are usually hydrocarbons that are considered to be harmful emissions.

**Volatility**   A measurement of the tendency of a liquid to change to vapor. Volatility is measured using RVP, or Reid vapor pressure.

**Volumetric efficiency**   The ratio between the amount of air–fuel mixture that actually enters the cylinder and the amount that could enter under ideal conditions expressed in percent.

**Wankel engine**   Rotary engine.

**Washcoat**   A porous aluminum material that makes the surface rough.

**Washers**   Flat metal discs with a hole in the center used under threaded fasteners to help spread the clamping force over a wider area.

**Waste vegetable oil**   This oil could include animal or fish oils from cooking.

**Wastegate**   A valve similar to a door that can open and close. The wastegate is a bypass valve at the exhaust inlet to the turbine. It allows all of the exhaust into the turbine, or it can route part of the exhaust past the turbine to the exhaust system.

**Waste-spark ignition**   Introduced in the mid-1980s, it uses the onboard computer to fire the ignition coils.

**Water–fuel separator**   Separates water and fuel in a diesel engine.

**Wet compression test**   A test that uses oil to help seal around the piston rings.

**Wet system**   A type of nitrous system that adds nitrous gas and gasoline at the same time.

**Wheel motors**   Electric motors that are attached to the wheel and used to propel the vehicle.

**WHMIS**   Workplace Hazardous Materials Information Systems.

**Wide-band oxygen sensor**   An oxygen sensor design that is capable of detecting actual air–fuel ratios. This is in contrast to a conventional oxygen sensor that only changes voltage when a stoichiometric air–fuel ratio has been achieved.

**Wind farms**   A group of windmills.

**Wood alcohol**   See *methanol*.

**World wide fuel charter**   A fuel quality standard developed by vehicle and engine manufacturers in 2002.

**Wrench**   A hand tool used to grasp and rotate a threaded fastener.

# INDEX